PRINCIPLES
OF
MODERN RADAR

PRINCIPLES
OF
MODERN RADAR

Edited by
Jerry L. Eaves
and
Edward K. Reedy

CHAPMAN & HALL

I T P® International Thomson Publishing

New York • Albany • Bonn • Boston • Cincinnati • Detroit • London • Madrid • Melbourne
Mexico City • Pacific Grove • Paris • San Francisco • Singapore • Tokyo • Toronto • Washington

This edition published by Chapman & Hall, New York

Printed in the United States of America

For more information contact:

Chapman & Hall
115 Fifth Avenue
New York, NY 10003

Chapman & Hall
2-6 Boundary Row
London SE1 8HN
England

Thomas Nelson Australia
102 Dodds Street
South Melbourne, 3205
Victoria, Australia

Chapman & Hall GmbH
Postfach 100 263
D-69442 Weinheim
Germany

International Thomson Editores
Campos Eliseos 385, Piso 7
Col. Polanco
11560 Mexico D.F.
Mexico

International Thomson Publishing - Japan
Hirakawacho-cho Kyowa Building, 3F
1-2-1 Hirakawacho-cho
Chiyoda-ku, 102 Tokyo
Japan

International Thomson Publishing Asia
221 Henderson Road #05-10
Henderson Building
Singapore 0315

7 8 9 XXX 01 00 99 98 97

Library of Congress Cataloging-in-Publication Data

Principles of modern radar.
    Includes index.
    ISBN 0-442-22104-5
    1. Radar. I. Eaves, Jerry L. II. Reedy, Edward K.
    TK6575.P74 1987   86-11144
    621.3848   CIP

Visit Chapman & Hall on the Internet http://www.chaphall.com/chaphall.html

To order this or any other Chapman & Hall book, please contact **International Thomson Publishing, 7625 Empire Drive, Florence, KY 41042.** Phone (606) 525-6600 or 1-800-842-3636. Fax: (606) 525-7778. E-mail: order@chaphall.com.

For a complete listing of Chapman & Hall titles, send your request to **Chapman & Hall, Dept. BC, 115 Fifth Avenue, New York, NY 10003.**

PREFACE

This book, *Principles of Modern Radar*, has as its genesis a Georgia Tech short course of the same title. This short course has been presented annually at Georgia Tech since 1969, and a very comprehensive set of course notes has evolved during that seventeen year period. The 1986 edition of these notes ran to 22 chapters, and all of the authors involved, except Mr. Barrett, were full time members of the Georgia Tech research faculty.

After considerable encouragement from various persons at the university and within the radar community, we undertook the task of editing the course notes for formal publication. The contents of the book that ensued tend to be practical in nature, since each contributing author is a practicing engineer or scientist and each was selected to write on a topic embraced by his area(s) of expertise. Prime examples are Chaps. 2, 5, and 10, which were authored by E. F. Knott, G. W. Ewell, and N. C. Currie, respectively. Each of these three researchers is recognized in the radar community as an expert in the technical area that his chapter addresses, and each had already authored and published a major book on his subject. Several other contributing authors, including Dr. Bodnar, Mr. Bruder, Mr. Corriher, Dr. Reedy, Dr. Trebits, and Mr. Scheer, also have major book publications to their credit.

Principles of Modern Radar is organized into an introductory chapter and seven parts. PART 1 addresses the "Factors External to the Radar," including electromagnetic wave reflectivity and propagation processes and the multipath phenomenon and effects. In PART 2, the "Basic Elements of the Radar System" are discussed. The basic radar task and objective of "Detection in a Contaminated Environment" of noise and clutter is the subject of PART 3, which includes chapters on noise, clutter, target models, and threshold detection techniques. PART 4 is titled "Radar Waveforms and Applications," and these four chapters address special techniques that can result in improved radar performance by introducing waveform conditioning and signal processing trades among the time, frequency, and spatial domains. The subjects of tracking in range, angle, and Doppler frequency are presented in PART 5, "Tracking Radar Techniques and Applications." Finally, PART 6 and PART 7—"Target Discrimination and Recognition" and "Radar ECCM", respectively—present important subjects not included in most books on radar. Chapters 20 and 21 address polarimetric techniques for target recognition, a very important radar

v

topic of the 1980s, and Chap. 22 discusses electronic counter countermeasures (ECCM) that should be an integral part of any radar but often are ignored/overlooked by radar designers and authors of radar books.

Many people have made contributions to the publication of this book, and we thank them all. Among these are those who encouraged us to take on the task of editorship, including all of the contributing authors without whom there would be no book. A very special thank-you is extended to Mr. Melvin McGee and to his staff, Mr. Joseph McKee, and Ms. Melanie Luke, for their help in generating and refining the manuscript. Also, Ms. Shirley Washington is due a special thanks for her support of the short course and the book. Finally, we thank Dr. H. Allen Ecker, formally of Georgia Tech and now with Scientific-Atlanta, for he organized and coordinated the first edition of the Principles of Modern Radar short course back in 1969 and started us all on the course that eventually led to the publication of this book.

JERRY L. EAVES
EDWARD K. REEDY

CONTENTS

1
INTRODUCTION TO RADAR

Jerry L. Eaves

1.1 INTRODUCTION

The word *radar* was a code name used by the U.S. Navy in 1940, early in World War II, and is an acronym derived from the phrase *ra*dio *d*etection *a*nd *r*anging. Radar has many uses, and a particular application can involve extremely complex and sophisticated engineering techniques and designs; however, radar is very simple in its most basic form, as shown in Figure 1-1. Its basic objectives are to detect targets of interest and to derive information such as range, angular coordinates, velocity, and reflectivity signature from the detection.

Electromagnetic (EM) energy generated within the transmitter unit is routed to the antenna via the duplexer, a device that permits both transmission and reception of EM waves with a single antenna. The antenna serves as a transducer to couple the EM energy into the atmosphere, where it propagates as an EM wave at the speed of light (approximately 3×10^8 m/s). Generally, the radar antenna will form a beam of EM energy that concentrates the propagating EM wave in a given direction. Thus, the beam can be directed to desired angular coordinates by effectively pointing the antenna in that direction through a combination of mechanical and electrical means.

An object or target located within the antenna beam will intercept a portion of the propagating energy. The intercepted energy will then be scattered in various directions from the target and, in general, some of it will be backscattered in the direction of the radar. The time delay between transmission by the radar and reception of the signal reflected by a target located at a range R is found from the relationship

$$t_d = \frac{\text{distance}}{\text{velocity}} = \frac{2R}{c} \tag{1-1}$$

where c is the velocity of light. This retroreflected energy is called *backscatter*, as opposed to *bistatic scatter* or *EM scatter* in other directions.

A portion of the backscattered wave is intercepted by the radar antenna, and the collected energy is transduced from the atmosphere or propagation medium into the radar receiver via the transmission lines and the duplexer.

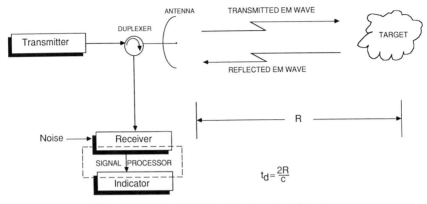

Figure 1-1. Radar principle and basic system elements.

The receiver amplifies the weak received signals and translates the information contained at the radio frequency (RF) to video and/or baseband frequencies. Signals processed by the receiver are then routed to the radar indicator or display, where the data (range, velocity, amplitude, direction, etc.) that were derived within the receiver (signal processor) are presented to the radar operator.

In general, the radar derives target information by correlating the received signal with the transmitted signal. Target information that can be obtained by radar is given in Table 1-1, along with the correlation process from which it is derived.

1.2 BASIC ELEMENTS OF THE RADAR SYSTEM

There are four basic elements in any functional radar: a transmitter, an antenna, a receiver, and an indicator. The basic configuration is illustrated in Figure

Table 1-1. The Derivation of Target Data.

Target Information	Derived by Correlating:		
	Received Signal		Transmitted Signal
Size (radar cross section)	Strength	with	Power
Range	Time delay	with	Time reference
Angular coordinates	Antenna beam position	with	Antenna beam reference
Radial velocity (Doppler)	Radio frequency	with	Frequency reference
Scattering signature	Polarization scattering matrix (PSM)	with	EM wave reference
Identification	Measured PSM with stored signature	for	EM wave reference

1-1. The duplexer appears as an element in the diagram; however, it is more correctly a part of the RF transmission line system than a basic element of the radar. Figure 1-1 suggests that the signal processor may constitute a fifth basic element of a radar system. In many cases, signal processing can be accounted for by some combination of the receiver and indicator operations.

Transmitter. The function of the transmitter is to generate a desired RF waveform at some required power level. The required RF power may be derived directly from a power oscillator such as a magnetron or an extended interaction oscillator (EIO), or it may be derived via an RF amplifier or amplifier chain [traveling-wave tube (TWT) amplifier, crossed-field amplifier, extended interaction amplifier (EIA) solid-state amplifier, etc.]. The waveform is determined by the particular requirements of the system and can range from an unmodulated continuous wave (CW) for a simple moving target indicator (MTI) radar to a complex frequency, phase, and time code modulated wave for some advanced military radars. Radar transmitters are discussed in more detail in Chapter 5.

Antenna. The basic function of the radar antenna is to couple RF energy from radar transmission line into the propagation medium and vice versa. In addition, the antenna provides beam directivity and gain for both transmission and reception of the EM energy. Various radar antenna concepts and techniques are presented and discussed in Chapter 6.

Receiver. The primary functions of the radar receiver are to accept weak target signals, amplify them to a usable level, and translate the information contained therein from RF to baseband. Various receiver configurations are employed, including crystal detector, RF amplifier, homodyne, and superheterodyne; of these, the superheterodyne receiver is by far the most commonly used configuration in radar receivers. Each of these is discussed in Chapter 7. When the receiver's frequency spectral characteristics are optimized to the transmitter waveform, the so-called ideal or matched receiver is produced, which provides a maximum output signal-to-noise ratio. In most cases, the radar designer will choose to trade off a little signal-to-noise performance for some other consideration, such as design simplicity or range measurement accuracy. Thus, the ideal matched receiver is rarely achieved, but it does serve as a commonly used reference for performance comparisons.

Indicator. The primary function and purpose of the radar indicator is to convey target information to the user. The indicator configuration and information format are dependent upon the particular radar application and the needs of the user. Two common indicators are (1) the plan position indicator (PPI), where target range and angle data are displayed on a cathode ray tube for surveillance radar applications and (2) an audio speaker or earphones, where the presence of a moving object is signaled by a Doppler frequency, as in a perimeter alarm radar. Various types of radar indicators and their applications are presented in Chapter 8.

Table 1-2. Pre–World War II Highlights and Milestones in Radar Development.

1886	Hertz demonstrated that radio waves are reflected from both metallic and dielectric objects.
1903	Hulsmeyer (German engineer) detected radio waves reflected from ships.
1922	Marconi made a speech to the Institute of Radio Engineers (now the Institute of Electrical and Electronics Engineers) urging the use of short radio waves for detection of objects.
1922	Tylor and Young at the Naval Research Laboratory (NRL) detected wooden ships with CW radar having a 5-m wavelength.
1930	Hyland (NRL) detected aircraft with CW radar.
1934–1936	Sir Watson-Watt (Britain) and Page (NRL) demonstrated pulsed radar.
Late 1930s	Radar received more attention at the NRL and the Signal Corps in the United States, and at various laboratories in Britain, due to military buildups prior to World War II.
1940	British scientists visited the United States and demonstrated magnetron. They suggested that the United States develop microwave aircraft-intercept and antiaircraft fire-control radars.
November 1940	The Radiation Laboratory was established at the Massachusetts Institute of Technology. It was staffed primarily by physicists, as suggested by the British, and the initial staff of 40 expanded to about 4000 by mid-1945. The research and development activities of this laboratory were documented in many reports, and after World War II a 28-volume set of books (Radiation Laboratory Series) was published to make available to all engineers and scientists the great body of information and new techniques in radar and related fields.

1.3 HISTORICAL DEVELOPMENTS

Although radar did not come into its own until its widespread development and application in World War II, the principle was known and advocated by many now famous scientists of the early 1900s. Table 1-2 provides a chronological list of some highlights and milestones that led to the giant step achieved during World War II.

Since World War II, the use of radar has expanded phenomenally. It has been applied not only to numerous military problems but also to many private and commercial uses. Some of the major postwar developments that stimulated the widespread application and use of radar are as follows:

- High-power klystrons
- Low-noise traveling-wave tube (TWT)
- Parametric amplifiers and masers
- EIO and EIA
- Monopulse
- Over-the-horizon (OTH) radar
- Short pulse techniques
- Synthetic aperture radar
- Phased arrays
- Solid-state devices
- Integrated circuit technology
- Digital computers
- Signal processing
    Pulse compression
    MTI and pulsed Doppler
    Frequency agility
    Polarization techniques
    Digital techniques
    Very-high-speed integrated circuits (VHSIC)

1.4 THE RADAR EQUATION

The radar range equation, or simply the radar equation, is the single most descriptive and useful mathematical relationship available to radar designers and researchers. In its most complete form, the radar equation accounts for not only the effects of each major parameter of the radar system but also those of the target, target background, and the propagation path and medium. Thus, one can use the radar equation to conduct performance and cost studies based on various parameter and scenario trade-offs or conditions for the radar, target, and environment.

A thorough understanding of the radar equation by a practicing radar engineer is essential. The relationship is developed in most radar texts of note and is also included here due to its central importance in radar.

Assume that a transmitter develops P_t watts of power and delivers it to an antenna that radiates the EM energy isotropically (omnidirectionally), as shown in Figure 1-2(a). Since the EM energy radiates in all directions with equal strength, the power density is constant over the surface of an imaginary sphere of radius R with its center located at the radar. Furthermore, the total power over the entire surface of the sphere must be exactly P_t watts (a lossless propagation medium is assumed) due to the conservation of energy principle. Therefore, the power density per unit area at a distance R from the radar is found by dividing the total power on the surface, P_t, by the total surface area, $4\pi R^2$;

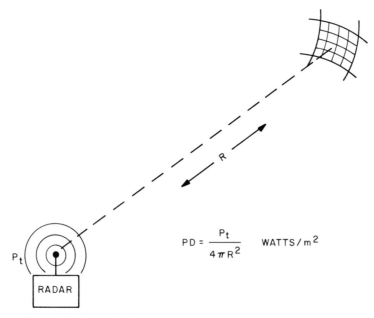

$$PD = \frac{P_t}{4\pi R^2} \quad WATTS/m^2$$

Figure 1-2a. Power density (PD) at range R for P_t watts radiated isotropically.

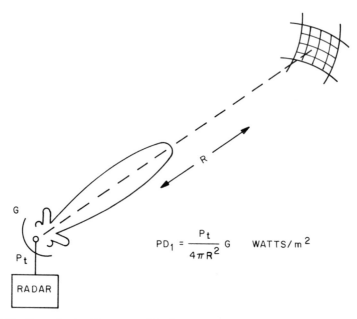

$$PD_1 = \frac{P_t}{4\pi R^2} G \quad WATTS/m^2$$

Figure 1-2b. Power density PD_1 at range R for P_t watts radiated through a directional antenna with gain G.

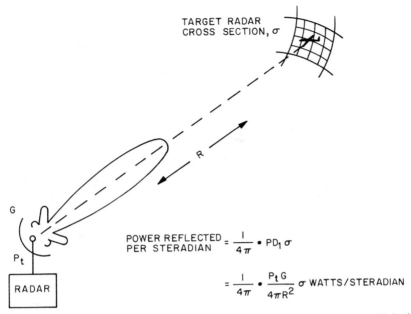

POWER REFLECTED PER STERADIAN $= \dfrac{1}{4\pi} \cdot PD_1\,\sigma$

$\qquad\qquad\qquad = \dfrac{1}{4\pi} \cdot \dfrac{P_t\,G}{4\pi R^2}\,\sigma$ WATTS/STERADIAN

Figure 1-2c. Power per steradian reflected toward radar by a target located at range R with backscatter area σ.

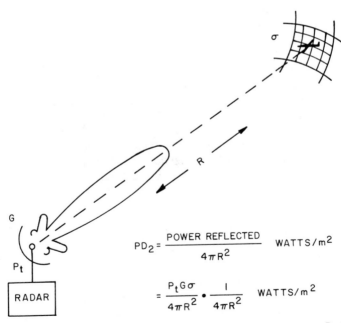

$PD_2 = \dfrac{\text{POWER REFLECTED}}{4\pi R^2}$ WATTS/m^2

$\qquad = \dfrac{P_t\,G\,\sigma}{4\pi R^2} \cdot \dfrac{1}{4\pi R^2}$ WATTS/m^2

Figure 1-2d. Power density, PD_2, at radar due to backscatter from target at range R with RCS σ.

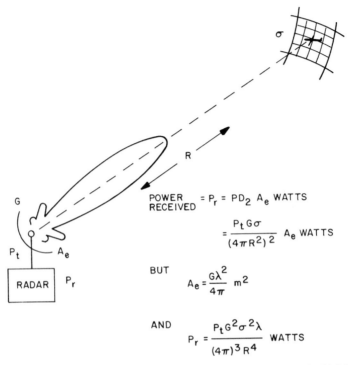

$$\text{POWER RECEIVED} = P_r = PD_2\, A_e \text{ WATTS}$$

$$= \frac{P_t\, G\sigma}{(4\pi R^2)^2}\, A_e \text{ WATTS}$$

$$\text{BUT} \quad A_e = \frac{G\lambda^2}{4\pi}\; m^2$$

$$\text{AND} \quad P_r = \frac{P_t\, G^2 \sigma^2 \lambda}{(4\pi)^3 R^4} \text{ WATTS}$$

Figure 1-2e. Power received by radar due to target located at range R and with RCS σ.

thus, at range R from the radar,

$$\text{Power density} = \frac{P_t}{4\pi R^2} \quad \text{W/m}^2. \tag{1-2}$$

Now if the unity gain omnidirectional antenna is replaced with a directional antenna that has a power gain G_t, a directional beam of energy is generated. As shown in Figure 1-2(b), the power density within the beam at range R is now given by

$$\begin{array}{c}\text{Power density}\\ \text{(at } R \text{ with } G_t)\end{array} = \frac{P_t}{4\pi R^2}\, G_t \quad \text{W/m}^2. \tag{1-3}$$

Next, assume that a target is located within the radiated beam at a range R from the radar, as shown in Figure 1-2c. The propagating EM wave will strike the target and, as a result, the incident energy will be scattered in various directions. Some of the energy will be reflected (backscattered) in the direction of the radar.

The amount of energy reflected toward the radar is determined by the power density at the target and σ, the target's backscatter radar cross section (RCS). σ is a measure of a target's ability to reflect EM waves, and its value is expressed as an area. A target's backscatter RCS is defined as 4π times the ratio of the power per unit solid angle reflected by the target in the direction of the illuminating source (radar) to the power per unit area of the incident wave at the target. Thus,

$$\sigma = 4\pi \frac{P_b/4\pi}{P_t/4\pi R^2} = 4\pi R^2 \frac{P_b}{P_t} \quad \text{m}^2 \tag{1-4}$$

or

$$\frac{1}{4\pi} P_b = \frac{1}{4\pi} \frac{P_t}{4\pi R^2} \sigma \quad \text{W/steradian} \tag{1-5}$$

where $1/4\pi\, P_b$ is the power per unit solid angle reflected in the direction of the illuminating radar.

The difference between a target's RCS and its geometric cross section is often a point of confusion for beginners, since both are defined and expressed as an area. Although there is always a definite relationship between an object's geometric area and its RCS area, it is generally very difficult and not practical to state this relationship explicity. Some insight into the relationship is gained by expressing a target's RCS in terms of the target's equivalent flat plate area.

Suppose that a target's geometric area normal to the radar line of sight is A_n; then the power intercepted and collected by that area is

$$P_c = \frac{P_t}{4\pi R^2} A_n \quad \text{W.} \tag{1-6}$$

That amount of power, P_c, will be reradiated by the target, and since we have defined P_c and A_n for the case of backscatter, the energy reradiated or reflected by A_n will be directed toward the illuminating radar. Thus, the backscattering process provides power gain relative to an isotropic reradiation process. Accordingly,

$$P_b = P_c G_b \quad \text{W}$$

$$= \frac{P_t}{4\pi R^2} A_n G_b \quad \text{W} \tag{1-7}$$

where G_b is the power gain of the equivalent flat plate area, A_n, over an isotropic source. As is shown in Chapter 6, antenna power gain and effective receiving aperture (area) are related as follows:

$$G = \frac{4\pi A}{\lambda^2}.$$ (1-8)

Substituting for G_b, we have

$$P_b = \frac{P_t}{4\pi R^2} A_n \frac{4\pi A_n}{\lambda^2} = \frac{P_t}{4\pi R^2} \frac{4\pi A_n^2}{\lambda^2} \quad \text{W.}$$ (1-9)

Thus, the backscatter RCS of a target with an equivalent flat plate area, A_n, normal to the beam is

$$\sigma = \frac{4\pi A_n^2}{\lambda^2} \quad \text{m}^2.$$ (1-10)

A_n is the geometric area of a conducting flat plate that, when substituted for the target and oriented normal to the radar radiated beam, produces the same backscatter power density as did the target. In general, the equivalent flat plate area, A_n, of a typical target is somewhat less than the actual target area that is normal to the radar beam. This is because typical real targets have range extent; thus, A_n would equal the total normal area of a target only if the EM waves backscattered from all of the range-distributed reflection surfaces were exactly in phase, if none of the backscattered energy resulted from multiple reflections, and if there were no depolarization effects. Examination of the relationship between RCS and the equivalent flat plate area shows that a target's RCS area is proportional to the square of an equivalent geometric area and also that RCS is inversely proportional to the wavelength of the illuminating wave. The relationships between RCS and geometric area for several simple geometric bodies are discussed in Chapter 2.

Referring again to Figures 1-2(b) and 1-2(c), we see that the power density at the target is

$$\frac{P_t G_t}{4\pi R^2} \quad \text{W/m}^2$$ (1-11)

and the power per unit solid angle reradiated in the direction of the radar is

$$\frac{1}{4\pi} \frac{P_t G_t}{4\pi R^2} \sigma \quad \text{W/steradian.}$$ (1-12)

It follows that the power density of the backscattered wave arriving at the radar, as shown in Figure 1-2(d), is

$$\frac{P_t G_t}{4\pi R^2} \cdot \frac{\sigma}{4\pi R^2} \quad \text{W/m}^2. \tag{1-13}$$

The power received by the radar, P_r, is determined by the effective capture area, A_e, of the receiving antenna, as shown in Figure 1-2e, and is given by

$$P_r = P_t G_t \quad \cdot \quad \frac{1}{4\pi R^2} \quad \cdot \quad \sigma \quad \cdot \quad \frac{1}{4\pi R^2} \quad \cdot \quad A_e \quad \quad \text{W.}$$

Power radiated
toward target ⏌
Power density at target ⏌
Equivalent power reradiated toward radar ⏌
Power density of reflected wave at radar ⏌
Power received by radar ⏌ $\tag{1-14}$

As discussed in Chapter 6, the effective capture area and gain of an antenna are related by

$$G = \frac{4\pi A_e}{\lambda^2} \tag{1-15}$$

where λ is the wavelength of the EM wave. Substituting for A_e, we have

$$P_r = \frac{P_t G_t G_r \lambda^2 \sigma}{(4\pi)^3 R^4} \quad \text{W.} \tag{1-16}$$

Most radars employ the same antenna for both transmission and reception; for those cases, the radar equation becomes

$$P_r = \frac{P_t G^2 \lambda^2 \sigma}{(4\pi)^3 R^4} \quad \text{W.} \tag{1-17}$$

This simple form of the radar equation is instructive and useful for making reference and first-order (rough) calculations. For more accurate and realistic performance predictions, one must account for the effects of:

1. The propagation medium and path
2. Atmospheric noise
3. System losses (nonideal components)
4. Thermal noise introduced within the radar

5. Signal processing losses (nonideal)
6. Other losses associated with particular configurations and applications

Many excellent texts address radar performance and the effect of these factors;[1-7] the details of those presentations are not repeated here.

Figure 1-3 depicts a more realistic operational scenario than that shown in Figure 1-1. For the sake of simplicity and briefness, assume that the loss factor, L, accounts for all system, medium, and propagation losses. Also assume that, for a system temperature of 290°K, the system noise factor, F_n, is defined by

$$F_n = \frac{N_o/N_i}{S_o/S_i} = \frac{\dfrac{S}{N}\Big|_{\text{input}}}{\dfrac{S}{N}\Big|_{\text{output}}} = \frac{(\text{SNR})_i}{(\text{SNR})_o} \qquad (1\text{-}18)$$

where N and S indicate noise and signal power levels and i and o indicate input and output, respectively. Thus, the signal-to-noise (power) ratio at the radar receiver output terminals is given by the signal-to-noise ratio at the radar receiver input terminals divided by the system noise factor, F_n:

$$(\text{SNR})_o = \frac{(\text{SNR})_i}{F_n}. \qquad (1\text{-}19)$$

The equivalent thermal noise at the receiver input terminals is generally given by the equation

$$N_i = kTB \qquad (1\text{-}20)$$

where k is Boltzmann's constant, T is the temperature, and B is the receiver bandwidth. Noise factors are defined in terms of a reference temperature, i.e., $T_0 = 290°K$, for which the product kT_o has a value of 4×10^{-21} W/Hz. When the system noise factor is expressed in decibels (logarithmic form), it is generally called the *system noise figure*.

Let L_E and L_I represent all signal losses (relative to the ideal) external and internal to the radar, respectively, and let A represent the radar receiver signal power gain. It follows that the effective input signal power is

$$S_i = \frac{P_r}{L_E} \quad \text{W} \qquad (1\text{-}21)$$

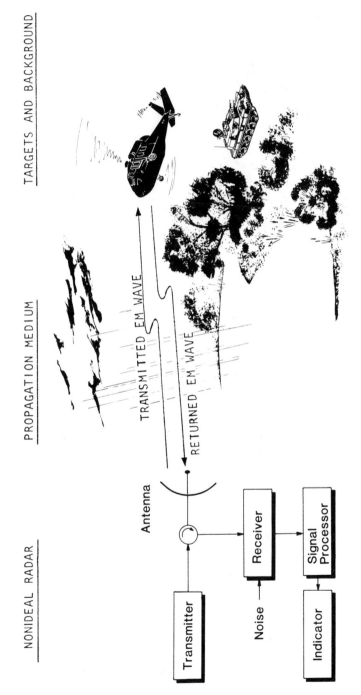

Figure 1-3. Example of a practical operational scenario.

NONIDEAL RADAR

PROPAGATION MEDIUM

TARGETS AND BACKGROUND

Transmitter

Noise

Receiver

Signal Processor

Indicator

Antenna

TRANSMITTED EM WAVE

RETURNED EM WAVE

13

and that the output signal power is

$$S_o = A \frac{S_i}{L_I} = \frac{AP_r}{L_E L_I} = \frac{AP_r}{L} \quad \text{W.} \tag{1-22}$$

Similarly, the output noise power is

$$N_o = AF_n N_i = AF_n kTB \quad \text{W.} \tag{1-23}$$

As illustrated in Figure 1-4, we can now express the radar receiver output signal-to-noise ratio in terms of the radar equation, including the loss and noise factors. Accordingly,

$$(\text{SNR})_o = \frac{S_o}{N_o} = \frac{AP_r/L}{AF_n kTB} \tag{1-24}$$

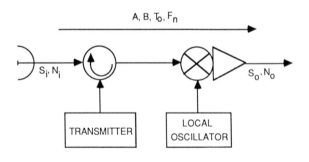

SIGNAL-TO-NOISE RATIO FOR NON-IDEAL CASE

$S_i = \dfrac{P_r}{L_E}$, WHERE L_E IS LOSSES EXTERNAL TO RADAR

$S_o = A \dfrac{S_i}{L_i}$, WHERE L_i IS LOSSES INTERNAL TO RADAR

$\quad = A \dfrac{P_r}{L}$, WHERE $L = L_E L_i$

$N_o = AKT_o B F_n$

$\therefore \left(\dfrac{S}{N}\right)_o = (\text{SNR})_o = \dfrac{A \dfrac{P_r}{L}}{AKT_o B F_n} = \dfrac{P_t G^2 \lambda^2 \sigma}{(4\pi)^3 R^4 K T_o B F_n L}$ WATTS/WATT

Figure 1-4. Radar equation and output signal-to-noise ratio for a realistic scenario.

or

$$(SNR)_o = \frac{P_t G^2 \lambda^2 \sigma}{(4\pi)^3 R^4 kTBF_n L} \qquad W/W. \qquad (1-25)$$

This expression for the output signal-to-noise ratio is a most useful form of the radar equation. The importance of $(SNR)_o$ is readily apparent when one realizes that it is the differential between signal and noise (interference) power levels, rather than the absolute signal power level, that determines the detection performance. For example, one does not improve the detection performance by increasing the output signal power if, at the same time, the noise power is also increased by an equal factor. With the foregoing in mind, the $(SNR)_o$ equation clearly illustrates the importance of a low-noise receiver (low F_n) and minimization of losses both inside and outside the radar.

1.4.1 Maximum Detection Range

The maximum detection range of a target with a specified RCS is an important radar performance measure. After rearranging the radar equation as follows

$$R = \left[\frac{P_t G^2 \lambda^2 \sigma}{(4\pi)^3 kTBF_n L \left(\dfrac{S_o}{N_o}\right)} \right]^{1/4} \qquad (1-26)$$

we may define the maximum detection range as

$$R_{max} = \left[\frac{P_t G^2 \lambda^2 \sigma}{(4\pi)^3 kTBF_n L \left(\dfrac{S_o}{N_o}\right)_{min}} \right]^{1/4} \qquad (1-27)$$

where $\left(\dfrac{S_o}{N_o}\right)_{min}$ is the smallest output signal-to-noise ratio for which a target signal can be recognized. It is quite common for radar engineers to assume that the maximum detection range corresponds to unity $(SNR)_o$. Under that assumption,

$$R_{max} = \left[\frac{P_t G^2 \lambda^2 \sigma}{(4\pi)^3 kTBF_n L} \right]^{1/4} \qquad m. \qquad (1-28)$$

This is a useful form of the radar equation from which one can assess radar performance as a function of radar, target, and environment parameters.

1.4.2 (SNR)$_o$ and R_{max} Expressed in Decibel Form

Radar engineers often prefer to convert the radar equation to the logarithmic form, wherein the various components are expressed in decibels (dB), and then to compute the result using the simple arithmetic operations of addition and subtraction.

If the radar equation components are quantified as power in watts, wavelength in centimeters, radar cross section in square meters, range in meters, and bandwidth in megahertz, then the logarithmic forms are

$$(\text{SNR})_{o\text{dB}} = (P_t)_{\text{dBW}} + 2(G)_{\text{dB}} + 2(\lambda)_{\text{dBcm}} + (\sigma)_{\text{dBm}^2} \qquad (1\text{-}29)$$
$$- 4(R)_{\text{dBm}} - (B)_{\text{dBMHz}} - (F_n)_{\text{dB}} - (L)_{\text{dB}} + 71\text{dB}$$

and

$$(R_{max})_{\text{dBm}} = \frac{1}{4}\left[(P_t)_{\text{dBW}} + 2(G)_{\text{dB}} + 2(\lambda)_{\text{dBcm}} + (\sigma)_{\text{dBm}^2} - (B)_{\text{dBMHz}}\right.$$
$$\left. - (F_n)_{\text{dB}} - (L)_{\text{dB}} - \left[\left(\frac{S_o}{N_o}\right)_{min}\right]_{\text{dB}} + 71\text{dB}\right]. \qquad (1\text{-}30)$$

Some prefer to work with mixed meter-kilogram-second (MKS) and English units, in which the components are expressed as power in watts, wavelength in centimeters, radar cross section in square meters, range in nautical miles, and bandwidth in hertz. For those mixed units

$$(\text{SNR})_{o\text{dB}} = (P_t)_{\text{dBW}} + 2(G)_{\text{dB}} + 2(\lambda)_{\text{dBcm}} + (\sigma)_{\text{dBm}^2} \qquad (1\text{-}31)$$
$$- 4(R)_{\text{dBnmi}} - (B)_{\text{dBHz}} - (F_n)_{\text{dB}} - (L)_{\text{dB}}$$

and

$$(R_{max})_{\text{dBnmi}} = \frac{1}{4}\left[(P_t)_{\text{dBW}} + 2(G)_{\text{dB}} + 2(\lambda)_{\text{dBcm}} + (\sigma)_{\text{dBm}^2}\right.$$
$$\left. - (B)_{\text{dBHz}} - (F_n)_{\text{dB}} - (L)_{\text{dB}} - \left[\left(\frac{S_o}{N_o}\right)_{min}\right]\text{dB}\right]. \qquad (1\text{-}32)$$

1.4.3 Radar Detection

The foregoing discussion has been presented as though the various forms of the radar equation are explicit functions for describing radar performance. In general, this is not true because the radar detection process is statistical due to the following facts:

1. The target signal is usually embedded in and corrupted by random extraneous signals such as atmospheric and thermal noise, electromagnetic interference, background clutter, propagation medium anomalies, electronic countermeasures, and so on.
2. A target's RCS can rarely be characterized as a constant.

Thus, the radar engineer must use statistical techniques to assess detection performance. To illustrate this, consider the simple experiment described in Figure 1-5. Assume that a radar transmits EM pulses at some pulse repetition frequency (prf) and that initially no backscatter energy is received due to an absence of targets; thus, the voltage at the detector output terminals is due solely

OUTPUT FOR (S/N) = 0, i.e., $S_i = 0$

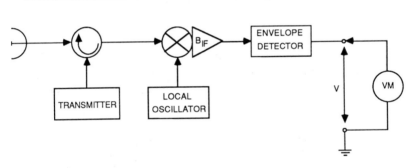

FOR ILLUSTRATION PURPOSES ONLY
(FABRICATED DATA, NOT TO SCALE)

MAKE 100 MEASUREMENTS OF V,
ONCE AFTER EACH TRANSMISSION

FIND RELATIVE FREQUENCY, $\eta_{\Delta v}$, FOR
VOLTAGE INCREMENTS OF $\Delta v = 0.1$ VOLT

V	$\eta_{\Delta v}$	
0 - 0.1	ll	2
0.1 - 0.2	₦l ₦l	10
0.2 - 0.3	₦l ₦l ₦l I	16
0.3 - 0.4	₦l ₦l ₦l ₦l ll	22
0.4 - 0.5	₦l ₦l ₦l llll	19
0.5 - 0.6	₦l ₦l lll	13
0.6 - 0.7	₦l lll	8
0.7 - 0.8	₦l	5
0.8 - 0.9	lll	3
0.9 - 1.0	I	1
> 1.0		0

Figure 1-5. Measurement of envelop detector output voltage for $(SNR)_o = 0$, that is, $S_i = 0$.

to radar system noise. The first step in the experiment is performed by making 100 separate measurements of the detector output voltage at a fixed time delay after each radar transmission. Next, the number of measurements that were within the intervals of 0 to 0.1 V, 0.1 to 0.2 V, 0.2 to 0.3 V, and so on are determined, and a probability histogram is constructed as shown in Figure 1-6. The histogram is a discrete function due to the finite voltage interval (0.1 V) and the limited number of measurements. The discrete probability histogram would approach a continuous probability density function if the voltage increment and the number of measurements were made to approach zero and infinity, respectively, as shown in Figure 1-7. Thus, the curve shown in Figure 1-7 is the probability density function of the detector output voltage for $(SNR)_o = 0$, that is, when no target signal is present.

Next in the experiment, assume that the voltmeter is replaced by a threshold detector and indicator, as shown in Figure 1-8. The operating characteristics of the threshold detector are defined as follows:

$$V_o = V \text{ when } V \geq V_T \tag{1-33}$$

$$= 0 \text{ when } V < V_T. \tag{1-34}$$

Given the probability density function defined and shown in Figures 1-7 and 1-8, one can calculate the probability that the envelope detector output voltage, V, will equal or exceed the threshold voltage, V_T. By definition, that probability

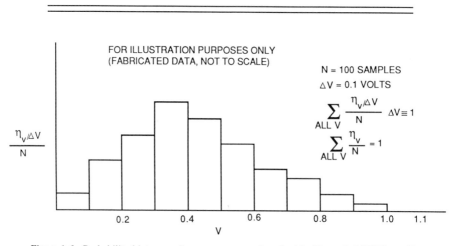

Figure 1-6. Probability histogram for measurements described in Figure 1-5 [$(S/N)_o = 0$].

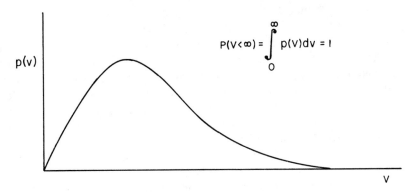

$$\text{P. HISTOGRAM} = \frac{n_V/\Delta V}{N} \implies \text{P. DENSITY} = p(v)$$

$$N \to \infty$$
$$\Delta V \to 0$$

$p(v)$

$$P(V < \infty) = \int_0^\infty p(v)dv = 1$$

V

FOR ILLUSTRATION PURPOSES ONLY
(FABRICATED DATA, NOT TO SCALE)

Figure 1-7. Definition of a probability density function $[(S/N)_o = 0]$ in terms of a probability histogram.

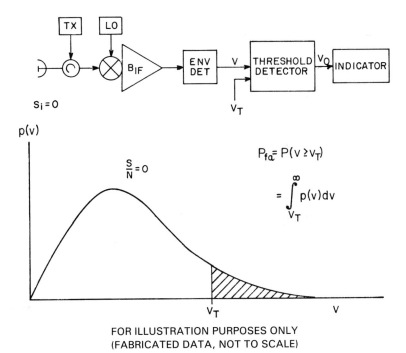

$p(v)$

$$\frac{S}{N} = 0$$

$$P_{fa} = P(v \geq v_T)$$

$$= \int_{V_T}^\infty p(v)dv$$

$S_i = 0$

V_T

V_T

V

FOR ILLUSTRATION PURPOSES ONLY
(FABRICATED DATA, NOT TO SCALE)

Figure 1-8. Single-pulse probability of a false alarm.

is the *single pulse probability of false alarm*, P_{fa}; no target signal is present, that is, $(SNR)_o = 0$; and no integration is employed.

It should now be clear that the value of P_{fa} is dependent upon V_T and that one could set V_T to produce a desired (allowable or tolerable) P_{fa}. Assume for this experiment that V_T was set so that the resulting P_{fa} is 0.1.

Next, assume that the experiment is repeated, but now with a target present at a range corresponding to the exact same time delay after transmission as used before, and that $(SNR)_o = 1$, as calculated from the radar equation given previously. Also assume that the experiment is repeated for $(SNR)_o = 2$ and $(SNR)_o = 4$ and that the resulting probability density functions are as shown in Figure 1-9. For a detection threshold voltage V_{T1} and the resulting $P_{fa} = 0.1$, the probability of detection can be determined for $(SNR)_o = 1, 2,$ and 4. Furthermore, the process could be repeated for other values of P_{fa} by resetting V_T as required. Thus, families of curves can be generated to show the relationship between the probability of detection, P_d, and the signal-to-noise ratio, $(SNR)_o$, versus the allowed probability of false alarm, P_{fa}. Figure 1-10 illustrates the results of the simple experiment just described.

It is important to understand that this example predicts the probability of detecting a steady (nonfluctuating) target for a single pulse (single sample, ob-

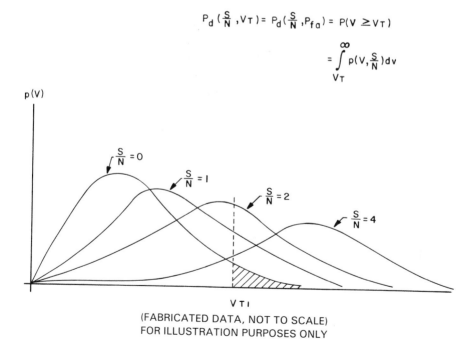

$$P_d\left(\frac{S}{N}, V_T\right) = P_d\left(\frac{S}{N}, P_{fa}\right) = P(V \geq V_T)$$

$$= \int_{V_T}^{\infty} p\left(V, \frac{S}{N}\right) dv$$

(FABRICATED DATA, NOT TO SCALE)
FOR ILLUSTRATION PURPOSES ONLY

Figure 1-9. Single-pulse probability density functions for $(SNR)_o = 0, 1, 2,$ and 4.

$(\frac{S}{N})_o$	$(\frac{S}{N})_{odB}$	V_T	P_{fa}	P_d
0	$-\infty$	V_{T1}	10^{-1}	-
1	0	V_{T1}	-	0.2
2	3	V_{T1}	-	0.4
4	6	V_{T1}	-	0.8
0	$-\infty$	V_{T2}	10^{-2}	-
1	0	V_{T2}	-	0.1
2	3	V_{T2}	-	0.3
4	6	V_{T2}	-	0.6

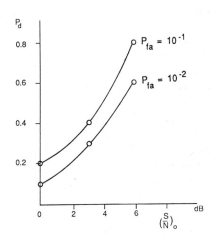

FOR ILLUSTRATION PURPOSES ONLY
(FABRICATED DATA, NOT TO SCALE)

Figure 1-10. Single-pulse probability of detection versus signal-to-noise ratio for allowed false alarm probability of 10^{-1} and 10^{-2}.

servation, look, etc.) and that several other factors which affect the probability of detection were not considered. Some of these factors are listed in Table 1-3. Thus, probability of detection is a function not only of the detection threshold and the allowed P_{fa} but also of the target reflectivity characteristics, the propagation medium and path, the receiver frequency response relative to a matched filter, the number of pulses included in the decision process, and whether coherent or noncoherent signal processing is employed. These factors are discussed in greater detail in Parts 1, 2, and 3 of this book. The remainder of the book, which includes Parts 4 through 6 and Chapter 22, introduces and discusses other important topics related to radar, such as waveforms, signal processing, tracking, target recognition, and electronic counter countermeasures.

Table 1-3. Other Considerations in the Determination of P_d.

- Effects of unmatched receiver
- Effects of target signal (steady or fluctuating)
- Effects of antenna pattern (scan modulation, multipath)
- Interference other than noise (clutter, electromagnetic interference, electronic countermeasures
- Number of samples included in decision
    Coherent or noncoherent?
    Independent or dependent?

1.5 SOME MAJOR FUNCTIONS PERFORMED BY RADAR*

Radars generally are called upon to perform many functions. If the functions are not simultaneous, they are called *modes*. For example, an antiaircraft fire control radar usually has at least two modes: (1) an acquisition mode to enable the radar to locate the target with sufficient accuracy to commence tracking and (2) a tracking mode to furnish position coordinates for aiming guns. Some major functions of radar with brief definitions are as follows:

Acquisition. the action by which a radar locates a target for the purpose of tracking.

Detection. the act or action of discovering, and sometimes locating, some thing or event. For example, special-purpose detection radars might be used to detect an intrusion (unidentified object entering a guarded area or airspace), a nuclear blast, or a rocket launching.

Height finding. the action of determining the height of an airborne object. A height-finder radar usually scans a horizontal fan beam in elevation to determine the elevation angle of the target.

Homing. the action of self-direction toward a given spot or target. A homing radar provides the relative location coordinates of the target.

Mapping. the action of systematically collecting data which allow for the display of a representation of a portion of the earth's surface. Mapping radars are concerned primarily with terrain features and major cultural targets.

Navigation. the action of determining the location of distinctive terrain features, navigational aids (buoys, beacons, etc), and other objects of interest (nearby ships, aircraft, etc.). The acquired information is useful when steering, maneuvering, or controlling a vehicle.

Ranging. the act of determining the distance to an object of interest.

Reconnaissance. the examination or observation of an area or airspace to secure information regarding the terrain, the location of objects of interest, or any other desired information regarding the situation. Reconnaissance usually implies the observation of unfamiliar territory or territory not accessible to continual observation (surveillance).

Search. the action of looking for an object or objects of interest. A search radar ordinarily determines the range and azimuth of objects within its area of detection.

Speed measuring. the action of determining the radial velocity of an object of interest, usually by means of the Doppler effect. For example, police radars usually operate solely in this mode.

Surveillance. the continual observation of an area or airspace. Surveillance usually implies the observation of familiar territory.

Terrain avoidance. the action of controlling an aircraft's altitude and course

---

*Extracted from reference material provided by R. C. Johnson.

so as to fly a path which closely follows the terrain along a generally predetermined course. A terrain-avoidance radar normally scans a solid forward sector in order to sample the upcoming three-dimensional profile. For example, with a terrain-avoidance capability, an aircraft might change course to fly between two mountains rather than fly directly over one of them.

Terrain following. the action of controlling an aircraft's altitude so as to fly a path which closely follows the terrain profile along a predetermined course. A terrain-following radar normally scans ahead in elevation to determine the upcoming terrain profile.

Tracking. the action of continually observing and following the movements of a target. A tracking radar usually "locks onto" the return signal and automatically tracks in angular coordinates and in range.

Track-while-scan. the action of observing and following the movements of a target while continuing to scan in a search or acquisition mode. New position coordinates are obtained with each scan.

1.6 RADAR NOMENCLATURE

Military electronic equipment and systems are designated by the AN nomenclature system as prescribed in Military Specification MIL-N-18307C (ASG) and its amendments. Military radars (and other military electronic systems) are designated by the letters AN followed by a slash, three letters, a dash, and a numeral. Table 1-4 provides a summary of the AN nomenclature designation systems. The system indicator, AN, does not mean that the Army, Navy, and Air Force use the equipment; it simply means that the radar is assigned in the military AN designation systems. The three letters following the dash indicate installation (*F*ixed, *M*obile, *A*ircraft, etc.), type (radar, radio, sonar, etc.), and purpose (detection, communications, identification, etc.), respectively, of a particular electronic system. The numeral following the dash indicates the position of a particular system in the chronological sequence. Subsequent models of a given AN system are designated by a letter following the numeral. For example, AN/TPS-1D designates the fourth model of the first ground transportable search radar that was described under the AN system. Some common AN (and other) designations for various radar applications are given in Table 1-5.[8]

1.7 RADAR AND ELECTRONIC COUNTERMEASURES (ECM) FREQUENCY BANDS*

Originally, during World War II, radar frequency bands were given letter designations for security purposes. However, letter bands continue to be conve-

*Extracted from reference material provided by H. A. Corriher.

Table 1-4. The AN Nomenclature Designation System.

First Letter	Second Letter	Third Letter
Installation	Type of equipment	Purpose
A—Airborne (installed and operated in aircraft)	A—Infrared	A—Auxiliary assemblies (not complete operating sets used with or part of two or more sets or sets series)
B—Underwater mobile, submarine	B—Pigeon	
	C—Carrier (wire)	
C—Air transportable (inactivated, do not use)	D—Radiac	
	E—Nupac	
D—Pilotless carrier	F—Photographic	B—Bombing
F—Fixed	G—Telegraph or teletype	C—Communications (receiving and transmitting)
G—Ground, general ground use (includes two or more ground-type installations)	I—Interphone and public address	
	J—Electromechanical (not otherwise covered)	D—Direction finder and/or reconnaissance
K—Amphibious	K—Telemetering	E—Ejection and/or release
M—Ground, mobile (installed as operating unit in a vehicle which has no function other than transporting the equipment)	L—Countermeasures	G—Fire control or searchlight directing
	M—Meteorological	H—Recording and/or reproducing (graphic meteorological and sound)
	N—Sound in air	
	P—Radar	
	Q—Sonar and underwater sound	L—Searchlight control (inactivated, use G)
P—Pack or portable (animal or man)	R—Radio	
S—Water surface craft	S—Special types, magnetics, etc., or combinations of types	M—Maintenance and test assemblies (including tools)
T—Ground, transportable		
U—General utility (includes two or more general installation classes, airborne, shipboard, and ground)	T—Telephone (wire)	N—Navigational aids (including altimeters, beacons, compasses, racons, depth sounding, approach, and landing)
	V—Visual and visible light	
	W—Armament (peculiar to armament, not otherwise covered)	
V—Ground, vehicular (installed in vehicle designed for functions other than carrying electronic equipment, etc., such as tanks)	X—Facsimile or television	P—Reproducing (inactivated, do not use)
	Y—Data processing	Q—Special, or combination of purposes
W—Water surface and underwater		R—Receiving, passive detecting
		S—Detecting and/or range and bearing
		T—Transmitting
		W—Control
		X—Identification and recognition

Table 1.5. Some Common Radar Types and Nomenclatures.

Surveillance
    Air defense—land based (FPS, MPS), shipboard (SPS, SPY)
    Airborne early warning (APS, AWAC)
    Air traffic control—en route (ARSR), terminal (ASR)
    Mortar and artillery location (TPS)
    Speed measurement—traffic radars
    Intrusion—jungle, burglar alarm (PPS)
    Harbor monitoring—collision avoidance
    Height finders (AHSR)

Tracking, guidance, and navigation—precision approach (TPM)
    Land and shipboard trackers (FPS, FPQ, MPS)
    Airborne interceptor (AI)
    Marine radar
    Missile and gunfire control (SPG, APG)
    Space vehicle
    Missile guidance (SPW, AWG)

Altimetry—aircraft, lunar landing radars, satellite

Mapping—Reconnaisance
    Remote sensing, crops, ice fields

Meteorological—Hydrologic (WRP, CPS)
    Wind field (Doppler radars)
    Clear air turbulence
    Hydrology

nient for generally describing the frequency limits of operation without being specific (which may still be prohibited for security reasons). Furthermore, the band designations are such that the characteristics of radar power sources, propagation, target reflectivity, and so on are generally similar within a band, but these characteristics may differ from band to band. Consequently, letter bands are still useful in describing the operating frequencies of radar. Although the original band designations were unofficial, standardized usage has been defined by organizations such as the IEEE. The designations from IEEE Standard 521-1976 are shown in Table 1-6, along with some notes on other fairly common usages. Additional information on common usage and applications of the radar frequency bands are given in Table 1-7.[8]

The Department of Defense recently issued a directive on designating frequencies used for performing ECM operations. This directive has been implemented in AFR55-44 (identical to AR 105-86, OPNAVIST 3480.9B, and MCO 3480.1), which governs ECM operations in the United States and Canada. The 13 ECM bands are listed in alphabetical order. In practice, each band is divided

Table 1-6. Radar and ECM Frequency Band Designations.

Standard Radar Bands[1]		ECM Bands[2]	
Band Designation[3]	Frequency Range (MHz)	Band Designation	Frequency Range (MHz)
HF	3–30	Alpha	0–250
VHF[4]	30–300	Bravo	250–500
UHF[4]	300–1,000	Charlie	500–1,000
L	1,000–2,000	Delta	1,000–2,000
S	2,000–4,000	Echo	2,000–3,000
C	4,000–8,000	Foxtrot	3,000–4,000
X	8,000–12,000	Golf	4,000–6,000
K_u	12,000–18,000	Hotel	6,000–8,000
K	18,000–27,000	India	8,000–10,000
K_a	27,000–40,000	Juliett	10,000–20,000
millimeter[5]	40,000–300,000	Kilo	20,000–40,000
		Lima	40,000–60,000
		Mike	60,000–100,000

[1]From IEEE Standard 521-1976, November 30, 1976.
[2]From AFR 55-44 (AR105-86, OPNAVINST 3430.9B, MCO 3430.1), October 27, 1964.
[3]British usage in the past has corresponded generally but not exactly to the letter-designated bands.
[4]The following *approximate* lower frequency ranges are sometimes given letter designations: P-band (225–390 MHz), G-band (150–225 MHz), and I-band (100–150 MHz).
[5]The following *approximate* higher frequency ranges are sometimes given letter designations: Q-band (36–46 GHz), V-band (46–56 GHz), and W-band (56–100 GHz).

Table 1-7. Radar Frequency Bands and General Usages.

Band Designation	Frequency Range	Usage
HF	3–30 MHz	OTH surveillance
VHF	30–300 MHz	Very-long-range surveillance
UHF	300–1000 MHz	Very-long-range surveillance
L	1–2 GHz	Long-range surveillance En route traffic control
S	2–4 GHz	Moderate-range surveillance Terminal traffic control Long-range weather
C	4–8 GHz	Long-range tracking Airborne weather detection
X	9–12 GHz	Short-range tracking Missile guidance Mapping, marine radar Airborne intercept
K_u	12–18 GHz	High-resolution mapping Satellite altimetry
K	18–27 GHz	Little use (water vapor)
K_a	27–40 GHz	Very-high-resolution mapping Airport surveillance
Millimeter wave	40–100+ GHz	Experimental

into 10 channels (e.g., A-7) whose width increases with the band frequency width; that is, each channel for Band A is 25 MHz wide, while each channel for Band M is 4000 MHz wide. A key feature of AFR 55-44 is contained in paragraph 1 of Attachment 1 thereto: "The phonetic alphabet will be used to identify the frequency of ECM operations . . . to identify an exact frequency, the frequency will be specified as Band-Channel base (lowest frequency in any channel) plus frequency in megacycles [sic] above the base frequency. Example: 1315 mc/s would be 'Delta 4 plus 15'."

Two widespread misuses of the ECM letter-band designations have developed. First, they have sometimes been used for radars. As stated in IEEE Standard 521-1976: "The letter designations for Electronic Countermeasure operations as described in Air Force Regulation No. 55-44, Army Regulation No. 105-86, and Navy OPNAV Instruction 4530.9B are not consistent with radar practice and shall not be used to describe radar frequency bands." Second, even when used to describe ECM equipment, the band is commonly designated by its alphabetic rather than its phonetic equivalent (e.g., an I-band rather than India-band jammer). Thus, speaking of "a J-band radar" is incorrect on two counts.

Leaders of the radar community are making a vigorous effort to stop the use of the ECM letter designations for radar bands. However, this misuse is becoming somewhat pervasive and may be impossible to correct.

1.8 REFERENCES

1. L. N. Ridenour, *Radar System Engineering,* MIT Radiation Laboratory Series, Vol. 1, McGraw-Hill Book Co., New York, 1947.
2. D. J. Povejsil, R. S. Raven, and P. Waterman, *Airborne Radar,* D. Van Nostrand Company, Princeton, N.J., 1961.
3. M. I. Skolnik, *Introduction to Radar Systems,* 2nd ed., McGraw-Hill Book Co., New York, 1980.
4. D. K. Barton, *Radar Systems Analysis,* Artech House, Dedham Mass., 1976.
5. R. S. Berkowitz (Ed.), *Modern Radar,* John Wiley & Sons, Inc., New York, 1965.
6. F. Nathanson, *Radar Design Principles,* McGraw-Hill Book Co., New York, 1969.
7. M. I. Skolnik (Ed.), *Radar Handbook,* McGraw-Hill Book Co., New York, 1970.
8. F. E. Nathanson, *Radar Range Calculation,* 1979 Short Course Notes, Technology Service Corporation, Silver Springs, Md., 1979.

PART 1
FACTORS EXTERNAL TO THE RADAR

- EM Waves and the Reflectivity Process—Chapter 2
- The Propagation Process—Chapter 3
- The Multipath Phenomena and Effects—Chapter 4

Although a radar itself is an entity, its operational performance is affected by phenomena that are external to it. Thus, the radar designer or analyst needs an understanding not only of the radar proper but also of the principles and effects of those external factors that affect radar performance. Part 1 of *Principles of Modern Radar* is devoted to acquainting the reader with those factors.

Obviously, a radar's performance is dependent on the reflectivity properties of the objects it illuminates. After first introducing and describing the principles of EM waves, Chapter 2 addresses the reflectivity process, including discussions of the polarization scattering matrix and radar cross section. Methods of calculating radar cross section are illustrated by two examples: one using geometric optics techniques and one using physical optics techniques.

The EM waves radiated by the radar must propagate to (and from) the target through a medium that may range from nearly lossless to very lossy. Furthermore, anomalies or gradients in the propagation medium can cause the EM wave to bend (refract) from a straight-line path or sometimes to be trapped in a propagation duct. Particles such as falling hydrometeors, fog, dust, and smoke in the atmospheric propagation medium also affect radar performance by attenuating and backscattering the propagating EM wave. These topics and others are addressed in Chapter 3.

In Chapter 4, the multipath phenomenon and its effects are described and discussed. It is shown that the multipath phenomenon is generally a degrading factor in radar performance due to resulting target signal fading, target range ambiguities, and target angular position ambiguities; however, the multipath phenomenon can also be used to advantage in some applications, such as ground plane reflecting ranges and target height determination.

2
EM WAVES AND THE REFLECTIVITY PROCESS

Eugene F. Knott

2.1 THE EM WAVE PHENOMENON

Everyone is familiar with waves in one form or another. A very common example is the way waves propagate over the quiet surface of a pond when a stone is thrown into the water. Concentric rings expand away from the center of the disturbance, becoming weaker the farther they travel. The wave on the surface of the water is a *transverse* wave, signifying that the actual motion of the water particles (up and down in this case) is at right angles to the direction of propagation. EM waves are also transverse, although no particle motion may be involved. Instead, it is the intensities of the electric and magnetic field strengths that vary in planes transverse to the direction of propagation.[1]

The EM waves emitted by radars are harmonic in time. This phenomenon could be simulated in the still pond if a vibrating plunger replaced the single stone. A continuous stream of expanding wavefronts would flow away from the plunger as long as the plunger was activated. If the wavelength from crest to crest or trough to trough was measured, it would be found to vary inversely with the frequency of the vibrating plunger. The product of the wavelength and frequency would be found to be a constant, and this constant is the velocity of propagation. For EM waves propagating through free space, the velocity of propagation is the familiar speed of light, or 0.2997925 m/ns, very nearly 1 ft/ns. (A nanosecond is a billionth of a second.) The electromagnetic spectrum includes X-rays as well as very low frequencies, as shown in Figure 2-1.

The electric and magnetic field strengths of an EM wave vary sinusoidally with time (t) and distance; if the frequency of the source is denoted f, they have the forms

$$E_x = E_0 \cos (\omega t - kR)$$
$$H_y = H_0 \cos (\omega t - kR)$$

(2-1)

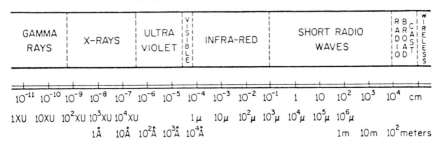

Figure 2-1. The EM spectrum.

where

E_x = electric field in volts per meter
H_y = magnetic field in amperes per meter
E_0 = maximum amplitude of the electric field
H_0 = maximum amplitude of the magnetic field
$\omega = 2\pi f$ = the radian frequency of the wave
$k = 2\pi/\lambda$ = the wave number of the wave
R = distance measured from some origin

Thus, the dependence on time and space is explicit. Although there is nothing imaginary about an EM wave, it is mathematically convenient to use a complex representation, such as

$$E_x = E_0 e^{j(\omega t - kR)}$$
$$H_y = H_0 e^{j(\omega t - kR)}$$
(2-2)

where $j = \sqrt{-1}$. The representation in Eq. (2-1) is the real part of Eq. (2-2). The negative sign of the second term in the exponent of Eq. (2-2) is interpreted as a positive phase angle in engineering; the more negative the number, the more positive the phase angle. The opposite interpretation is used in physics.

The electric and magnetic fields of an EM wave are at right angles to each other and to the direction of propagation, as shown in Figure 2-2. The fields are vector quantities having direction as well as intensity; hence, vector calculus is required in many EM problems. Generally, electric field strength $\overline{E}$ is described in volts per meter and magnetic field strength $\overline{H}$ is described in amperes per meter. The ratio of transverse electric to transverse magnetic field strength is an impedance that is characteristic of the medium in which the wave is propagating. For free space, this ratio is about 377 ohms; for Teflon, it is about 545 ohms.

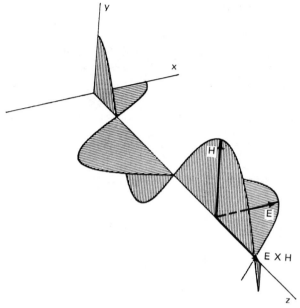

Figure 2-2. A "snapshot" of the electric and magnetic field intensities at a particular moment in time.

The Poynting vector $\overline{S}$ is the vector cross product

$$\overline{S} = \overline{E} \times \overline{H} \qquad (2\text{-}3)$$

and is the power density, say in watts per square meter, of the wave in a plane perpendicular to the direction of propagation. Since field strengths vary sinusoidally with time, so does power density; however, the Poynting vector can be averaged over time to give the complex Poynting vector $\overline{S}*$

$$\overline{S}* = \tfrac{1}{2} \text{ Real } (\overline{E} \times \tilde{H}) \qquad (2\text{-}4)$$

where $\tilde{H}$ is the complex conjugate of $\overline{H}$. Thus, $\overline{S}*$ is a measure of the actual power transfer across a surface, whether that surface is real or imaginary.

The *plane wave* is the simplest to imagine and describe. Surfaces of constant phase are families of parallel planes. The plane wave is nonexistent in the real world but can be closely approximated by a spherical wave at very large distances from a source. Unless the medium is lossy, the plane wave does not decay in strength with increasing distance. Nevertheless, the concept of a plane wave is extraordinarily useful because the phase fronts are flat. The field structure in some uniform transmission lines is essentially that of a plane wave.

The *spherical wave* is probably the most common one found in nature, and as the name implies, surfaces of constant phase are concentric spheres centered on the source. At great distances from the source, the spherical phase fronts can be approximated locally as planar phase fronts. The electric and magnetic field strengths of a spherical wave decay with increasing distance from the source; hence, the power density decays with the *square* of the distance. As will be shown in a later section, radar echoes are received by means of a double traverse of the same path; hence, the received power as given by the radar range equation decays with the fourth power of distance.

The *cylindrical wave* is another type of simple wave structure, and surfaces of constant phase are concentric cylinders centered on a *line source*. Cylindrical waves, like plane waves, are relatively rare in nature, but they are very useful in the analysis of some near-field problems. The field intensities decay inversely with the square root of the distance; hence, the power density decays with increasing distance. In this respect, the cylindrical wave seems to be a hybrid combination of a plane wave and a spherical wave.

A simple plane wave is *linearly polarized* because the electric and magnetic field strengths remain in their respective planes independently of time and distance. For a given direction of propagation, this is also true of spherical wave fronts. Linear polarization is commonly described in terms of the orientation of the electric field vector with respect to the local horizontal or vertical planes and is directly associated with the antenna used to launch or receive a wave. Most radars are vertically or horizontally polarized, but some have the capability to transmit and receive both polarizations.

A pair of plane waves with the same frequency and direction of propagation can be added together to create an *elliptically polarized* wave.[2] This occurs if the two plane waves are of different polarizations and if there is a phase shift between them. A simple example is the addition of a plane wave of amplitude E_1 polarized in the x-direction and one of amplitude E_2 polarized in the y-direction, but advanced in phase by an angle ψ:

$$E_x = E_1 \cos (\omega t - kR)$$

$$E_y = E_2 \cos (\omega t - kR + \psi). \tag{2-5}$$

These are the parametric equations of an ellipse in the xy plane. The vector sum of these two fields is a vector that traces out the ellipse for fixed distance and varying time or for fixed time and varying distance. The net electric field actually rotates in time and space. Thus, linear polarization can be regarded as a special case of elliptical polarization. Another special case is *circular polarization* for which $E_1 = E_2$ and $\psi = \pm \pi/2$ radians. Depending on whether the plus or minus phase shift is used, the wave will have *right* or *left* circular polarization, and the field will rotate clockwise or counterclockwise. Circularly

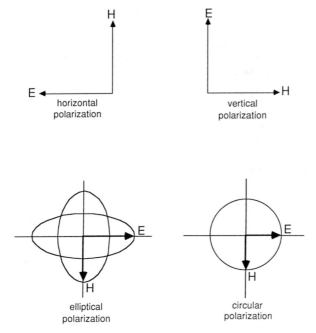

Figure 2-3. Examples of polarization. In each case, the direction of propagation is into the plane of the paper.

polarized waves can be generated with spiral or helical antennas, crossed dipoles, or dual polarized antenna feeds. Figure 2-3 illustrates some of the common polarizations.

2.2 THE REFLECTIVITY PROCESS

So long as a wave propagates in an unbounded medium, it will continue to do so until it meets an obstruction. The discontinuity disturbs the local relationship between the electric and magnetic fields, thereby generating other waves that propagate generally in all directions. Those that travel back to the original source of the wave are of primary interest in radar, since they betray the presence of an obstacle. The obstacle can be one or more real targets, or it can be distributed in space as terrain or vegetation. One of the problems in the design of modern radars is the separation of desired from undesired returns.

A discontinuity can be characterized as a surface separating two media of differing electrical characteristics. The relationship between the fields just inside the boundary and those just outside the boundary can be derived from the integral form of Maxwell's equations. The *boundary conditions* can be stated quite simply: the total tangential electric fields must be continuous across the boundary, while the total tangential magnetic fields from one side of the boundary to the other must differ by precisely the induced surface current density.[3]

The total tangential fields can be represented as the sum of an incident field distribution and a reflected or scattered field distribution. These distributions can be expressed in terms of induced charges and currents, and since the charges and currents are induced by oscillating fields, the charges and currents are also oscillatory. Since oscillating currents and charges radiate energy, the scattered or reflected wave can be regarded as a re-radiated wave.

In describing the reflection process, however, it is often convenient to ignore the concept of induced fields and to establish a direct relationship between the reflected wave and the incident wave that generated the reflection. The theory of *geometric optics* does this.[4] In geometric optics theory, wave motion per se is replaced by *ray tracing* techniques in which the energy flow through narrow beams is considered. The rays represent the direction of propagation, and they are normal to the wavefronts of constant phase.

At a dielectric interface, part of the beam is reflected and part is transmitted across the interface. If the surface is metallic, the beam is completely reflected and no energy is transmitted across the surface boundary. If the surface is curved, the reflected beam may converge to a point or it may diverge, depending on the nature of the curvature. This can occur whether the surface is metallic or not. If it is nonmetallic, the rays transmitted across the boundary are bent (refracted) in a different direction, as shown in Figure 2-4. Geometric optics is the basic tool used in the design of lenses and reflectors (mirrors).

Geometric optics can also be applied to the scattering of radar waves by simple objects. A straightforward application of the theory yields the familiar radar cross section formula for a metallic sphere, $\sigma = \pi a^2$, where a is the radius of the sphere. It should be recognized, however, that the theory is rigidly applicable only in the case of vanishing wavelength. Since the most meaningful yardstick in EM theory is the wavelength, geometric optics is evidently a high-frequency theory in which the typical obstacle dimensions should be much larger than the wavelength. Nevertheless, the theory has been used to predict the return from a soap bubble not much more than a wavelength in diameter with quite acceptable accuracy.

Accounting for the wave nature of EM radiation is more faithful in the theory of *physical optics*. As applied to the scattering of radar waves, physical optics attempts to estimate the induced fields on the surface of an obstacle and integrates the estimated distribution over the surface in the form of a radiation integral. Oddly enough, the induced surface field is estimated by the use of geometric optics and the tangent plane approximation. That is, the induced fields at a point on the obstacle surface are forced to be exactly the geometric optics fields that would have prevailed had the surface been flat and infinite in extent. (This is the tangent plane.)

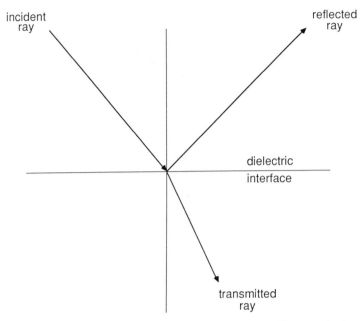

Figure 2-4. Refraction occurs when rays cross dielectric interfaces. The transmitted ray is bent toward the surface normal in going from a less dense to a more dense medium and away from the surface normal in going from a more dense to a less dense medium.

The resulting integral can be evaluated analytically for only a handful of simple cases, such as flat plates and circular cylinders. Most of the analytical effort in the theory is devoted to ways of evaluating the integral rather than characterizing the induced field distributions. Nevertheless, physcial optics is responsible for the familiar formulas for scattering by plates and cylinders, and it predicts the RCS patterns of objects like these quite well if the reflecting surface does not stray too far (say, 20 or 30°) from being perpendicular to the incident wavefronts. Like geometric optics, physical optics is a high-frequency prediction tool and loses accuracy for objects much smaller than three wavelengths or so.

For body orientations such that no surface is closer than 30 degrees or so to being perpendicular to the incident phase fronts, edges become significant sources of diffraction (scattering). The diffraction by an edge can be characterized by geometric optics if the beam "reflected" by a point on the edge is allowed to be distributed along the generators of a forward cone, as shown in Figure 2-5, instead of in a single direction, as from a surface. This theory is called the *geometric theory of diffraction* (*GTD*) and was developed by J. B. Keller in the mid-1950s.[5] Keller's theory has some limitations (it predicts infi-

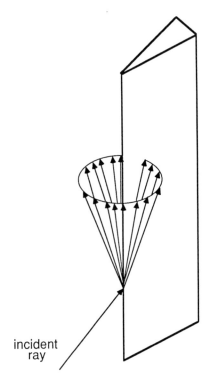

Figure 2-5. Keller's GTD theory introduced the concept of a cone of diffracted rays spawned by a single incident ray hitting an edge.

nite fields along shadow boundaries and reflection boundaries), and others have devised ways of improving it. Nevertheless, to Keller's credit, the theory is widely used because of its simplicity and its superiority to physical optics and geometric optics. Unlike those analytical prescriptions, GTD does have the added feature of accounting for the polarization dependence of the scattering by edges.

2.3   RADAR CROSS SECTION

In estimating the signal strength reaching the input of a radar receiver, one must account for the decay in signal strength as the radiated wave spreads away from the transmitting sources, the excitation of a wave reflection (echo) by a remote obstacle, and the subsequent spreading out of the reflected wave by the obstacle. These effects are all accounted for explicitly in the radar range equation, and the obstacle reflection is characterized by a single function σ, the RCS. The formal definition of RCS is

$$\sigma = \lim_{R \to \infty} 4\pi R^2 \left| \frac{E_s}{E_0} \right|^2 \tag{2-6}$$

where

E_s = scattered field strength
E_0 = incident field strength (usually taken to be a plane wave)
R = distance from the scatterer where E_s is measured.

Since the scattered field varies inversely with R, the RCS is a range-independent function. The incident field is commonly, but not always, taken to be a plane wave.[6,7]

As far as the theory of geometric optics is concerned, one need only invoke the conservation of energy along a flux tube to find the ratio of the scattered power density to the incident power density. Without presenting the details of the derivation, this ratio is

$$\frac{|E_s|^2}{|E_0|^2} = \frac{\rho_1 \, \rho_2}{(r + \rho_1)(r + \rho_2)} \tag{2-7}$$

where ρ_1 and ρ_2 are the principal radii of curvature of the scattered wavefront some distance r from the specular point on the body. Inserting this result into Eq. (2.6),

$$\sigma = 4\pi\rho_1\rho_2. \tag{2-8}$$

For a large, smooth, convex body, the curvature of the wavefront is related to the curvature of the body by

$$\rho_1 = \frac{a}{2} \qquad \rho_2 = \frac{b}{2} \tag{2-9}$$

where a and b are the principal radii of curvature of the body at the specular point. Therefore, Eq. (2.8) becomes

$$\sigma = \pi ab. \tag{2-10}$$

This very simple formula is the geometric optics RCS of the body. For example, the principal radii of curvature of a sphere are both a; hence the RCS of a sphere is simply πa^2. In order for the formula to hold, four general conditions must usually be met. First, the specular point must be on the body; for backscattering, this implies that some surface of the body must be oriented perpendicular to the line of sight. Second, the local radii of curvature at the specular point should be large compared to the wavelength, although "large" may be a bit misleading. The geometric optics of the RCS of a soap bubble, for example, is within 1 dB of the exact value for a bubble as small as two wavelengths in diameter, which is not really large. Third, the specular point cannot be too close to an edge or another discontinuity; this is closely related to the second requirement, because the local radii of curvature are not large or well defined near an edge. And fourth, the radii of curvature at the specular point must be finite.

Let us try, for example, to use Eq. (2.10) to compute the RCS of a flat plate. In this instance, both radii of curvature are infinite, and the formula predicts a doubly infinite RCS. The formula is equally unsuccessful for a singly curved surface, such as a right circular cylinder or a cone. Most radar targets have some singly and doubly curved features, and this simple geometric optics prescription is of no value in such cases. Fortunately, we have another tool: physical optics.

The theory of physical optics assumes that surface currents induced on the body surface by the incident wave are precisely twice the tangential component of the incident magnetic field strength over the lit surfaces and precisely zero over those surfaces shaded from the incident field by other portions of the body. The surface currents are assigned the values they would have had the body been perfectly flat and smooth at the points where incident rays strike the surface. Then, in order to calculate the far scattered field E_s, the induced currents are summed in a surface integral and the results substituted in Eq. (2.6). The result for a rectangular plate which is always viewed in a plane perpendicular to a pair of opposite edges is

$$\sigma = 4\pi \left(\frac{A \cos \theta}{\lambda}\right)^2 \left(\frac{\sin (ka \sin \theta)}{ka \sin \theta}\right)^2 \tag{2-11}$$

where

$A = ab = $ area of the plate
$k = 2\pi/\lambda = $ free-space wave number
$\theta = $ angle of incidence as measured between the surface normal of the plate and the line of sight
$a = $ distance between the pair of edges that is normal to the incident wave
$b = $ distance between the edges inclined to the incident wave.

Note that the broadside return is proportional to the square of the plate area and inversely proportional to the square of the radar wavelength. The surface integral can be evaluated exactly for only a small number of simple shapes, of which the flat plate is one, but in many cases an approximate solution can be obtained. Note that even in those cases in which the integral can be evaluated exactly, the result itself is an approximation by virtue of the assumptions made about the induced surface currents.

The RCS is a measure of the magnitude of the reflection and is expressed as an effective area. A very common RCS unit is square meters, but RCS is sometimes expressed in square wavelengths. Expressing the reflection as an area is useful, because it is necessary only to multiply an incident power density (watts per square meter) at the target by the effective target area (RCS) to obtain the power scattered back to the radar. This converts the job of estimating the received power to a communications problem, wherein the signal strength at the receiver can be calculated from the transmitted signal strength.

The RCS of an obstacle bears no relationship to the physical cross section and, depending on the nature of the target, the RCS can be orders of magnitude larger or smaller than the physical cross section. The RCS depends on the size, shape, composition and orientation of the obstacle, and on the frequency and polarization of the incident wave. Thus, in general, the scattering is completely described only by a *polarization scattering matrix* that accounts for the orthogonally polarized components of the scattered fields for an arbitrary incident polarization.

Formally, the scattering matrix is a description of the coefficients in the following system of equations for the received field components expressed in terms of the incident field components:[8,9]

$$E_{1R} = S_{11}E_{1I} + S_{12}E_{2I}$$
$$E_{2R} = S_{21}E_{1I} + S_{22}E_{2I}. \qquad (2\text{-}12)$$

Here E_{1R} and E_{2R} are the received field components and are orthogonal to each other, and E_{1I} and E_{2I} are the incident field components, also orthogonal to each other. The matrix elements are directly proportional to the square root of the RCS and are, in general, complex numbers. The linear polarizations in Eq. (2-12) are orthogonal, but otherwise general; it is common practice to choose horizontal and vertical polarizations for actual measurement of the field components of polarization scattering matrix.

Since each of the four elements is a complex number, a complete description of the scattering matrix requires no fewer than eight numbers (amplitude and phase of four quantities). Reciprocity arguments can be invoked for the off-diagonal elements

$$S_{12} = S_{21} \qquad (2\text{-}13)$$

to reduce this to six numbers; and if one of the phase angles is arbitrarily subtracted from all matrix elements, we can get by with a minimum of five numbers in the matrix (three amplitudes and two phase angles).

The scattering matrix for a single target orientation is rarely of interest; the *scattering pattern* as a function of the aspect angle is more descriptive. Hence, the complete scattering matrix does not consist of five (or eight) numbers, but five (or eight) *functions*. These can be measured at a suitably instrumented RCS facility, or they can be calculated (estimated) using some of the analytical techniques mentioned above. The polarization scattering matrix is discussed further in Chapter 20, along with polarimetric signal-processing techniques for target discrimination.

In a very few instances, the wave equation can be solved exactly.[2] The wave equation is a set of second-order differential equations that results when one of the electric and magnetic field variables is eliminated by substitution. The wave equation for source free regions, for example, is

$$\nabla^2 \, \overline{E} + k^2 \, \overline{E} = 0 \tag{2-14}$$

where $k = 2\pi/\lambda$ is the propagation constant of the medium in which the wave propagates. This second-order vector differential equation has three spatial components in general, but simplifications can often be made.

The wave equation can be solved exactly for some simple bodies whose surfaces coincide with a coordinate of one of the eight elementary coordinate systems. Examples are spheroids (circular, oblate, prolate), infinite cylinders (circular, elliptic, hyperbolic, parabolic), and the half-plane. Despite the availability of exact formal solutions, the numerical evaluation of many of the functions involved is a formidable task at best.[6]

Certain simplifications can be made if the object is small compared to the wavelength, and in some cases the scattering can be expressed in terms of induced electric and magnetic multipoles.[10] A relatively new technique is the method of moments, in which the integral equations for the induced surface fields are reduced to a system of homogeneous linear equations.[11] A numerical matrix inversion leads directly to the solution for the induced fields, and these are integrated (summed) to compute the far scattered field. Unfortunately, the memory limitations of most modern computers restrict the method to general bodies not much larger than 1 wavelength and to certain symmetrical bodies of roll symmetry not much larger than 10 wavelengths. Nevertheless, there are some frequencies and target sizes such that the method of moments is useful.

An examination of the RCS characteristics of a few simple targets is highly instructive. A hierarchy of scattering shapes exists, as summarized in Table 2-1, ranging from highly retrodirective scatterers, such as corner reflectors, to diffuse scatterers, such as apexes. Corner reflectors formed by two or three orthogonal faces are highly reflective because the incident wave is directed back toward the source for a wide range of viewing angles. Flat plates have large

Table 2-1. Hierarchy of Scattered Features.

Geometry	Type	Freq. Dep.	Size Dep.	Formula	Remarks
	Square trihedral corner retro-reflector	F^2	L^4	Maximum $\sigma = \dfrac{12\pi\, l^2}{\lambda^2}$	Strongest return; high RCS due to triple reflection
	Right dihedral corner reflector	F^2	L^4	Maximum $\sigma = \dfrac{8\pi a^2 b^2}{\lambda^2}$	Second strongest; high RCS due to double reflection, tapers off gradually from the maximum with changing θ and sharply with changing ϕ.

Table 2-1. (*Continued*)

Geometry	Type	Freq. Dep.	Size Dep.	Formula	Remarks
	Flat plate	F^2	L^4	Maximum $$\sigma = \frac{4\pi a^2 b^2}{\lambda^2}$$ Normal Incidence	Third strongest; High RCS due to direct reflection, drops off sharply as incidence changes from normal.
	Cylinder	F^1	L^3	Maximum $$\sigma = \frac{2\pi ab^2}{\lambda}$$ Normal Incidence	Prevalent cause of strong, broad RCS over varying aspect (θ), drops off sharply as azimuth (ϕ) changes from normal. Can combine with flat plate to form dihedral corner reflector.
	Sphere	F^0	L^2	Maximum $$\sigma = \pi a^2$$ Normal Incidence	Prevalent cause of strong, broad RCS peaks other than those due to large openings in target body. Energy defocused in two directions.

Table 2-1. *(Continued)*

Geometry	Type	Freq. Dep.	Size Dep.	Formula	Remarks
	Straight edge normal incidence	F^0	L^2	$f(\theta,\phi)L^2$ θ – aspect θ_{int} – interior dihedral angle between faces meeting at edge	Limiting case of 2-dimensional curved plate mechanism as radius shrinks to 0. Prevalent cause of strong, narrow RCS peaks from supersonic aircraft.
	Curved edge normal incidence	F^{-1}	L^1	$f(\theta,\theta_{int})\, a\lambda/2$ $a \geq \lambda$	Limiting case of 3-dimensional curved plate mechanism as principal radius shrinks to 0. The function f is the same as in mechanism 3.
	Apex	F^{-2}	L^0	$\lambda^2 g(a,P,\theta,\phi)$ a,β – interior angles of tip θ,ϕ – aspect angles	Limiting case of previous mechanism as a shrinks to 0. For $a=\beta$, the tip is that of a cone. For $a=0$, the tip is the corner of a thin sheet, or fin.

Table 2-1. (Continued)

Geometry	Type	Freq. Dep.	Size Dep.	Formula	Remarks
	Discontinuity of curvature along a straight line, normal incidence	F^{-2}	L^0	$\dfrac{\lambda^2}{64\pi^3}\left(\dfrac{1}{a}\right)^2\{1+(\tfrac{dy}{dx})^2\}^{-3/2}$ $a \geq \lambda$ $(1/a)$ – jump in reciprocal of dy/dx – slope of surface w.r.t. incident ray	Strongest of an infinite sequence of discontinuities. Very weak mechanism which together with 6 shares dominance of nose-on RCS of cone sphere.
	Discontinuity of curvature of a curved edge	F^{-3}	L^{-1}	$f(\theta,\phi)\dfrac{\lambda^3 b}{a^2}\{1+(\tfrac{dy}{dx})^2\}^{-3/2}$ $f(\theta,\phi)$ – function of aspect b – radius of edge $>\lambda$	Important mechanism for traveling wave back-scatter where RCS of discontinuity is augmented by gain of traveling wave structure. Dependences are based on dimensional considerations.
	Discontinuity of curvature along an edge	F^{-4}	L^{-2}	$g(\theta,\phi)\lambda\{\tfrac{1}{a}\}^2\{1+(\tfrac{dy}{dx})^2\}^{-3/2}$ $g(\theta,\phi)$ – function of aspect	Important mechanism for traveing wave backscatter where RCS of discontinuity is augmented by gain of traveling wave structure. Dependences are based on dimensional considerations.

returns when incidence is perpendicular to the surface, but the return decays rapidly as the orientation is swung away from that direction. The RCS of a circular cylinder is constant in a plane perpendicular to the cylinder axis but displays a plate-like pattern in a plane containing the axis. The return from a sphere is smaller than that from a plate or cylinder of comparable size because the scattered wave diverges in both orthogonal planes. The scattering from the tip of a cone or the corner of a plate is quite small, and the base of the cone is, in fact, the dominant scatterer.

The RCS of a complicated body is obviously more complex. A complex target such as an aircraft contains several dozen significant "scattering centers" and myriad less significant scatterers such as seams, ports, and rivet heads. Because of this multiplicity of scatterers, the net RCS pattern typically exhibits a rapid scintillation with aspect angle due to the mutual interference as the various contributions go in and out of phase with each other. Figure 2-6, a measured pattern of a scale model of a commercial jetliner,[12] illustrates these characteristics. The larger the target in terms of wavelengths, the more rapid these scintillations become.

In the nose-on region, the pattern is dominated by reflections from jet intake ducts. Ducts, like corner reflectors, are reentrant structures and have large RCSs.

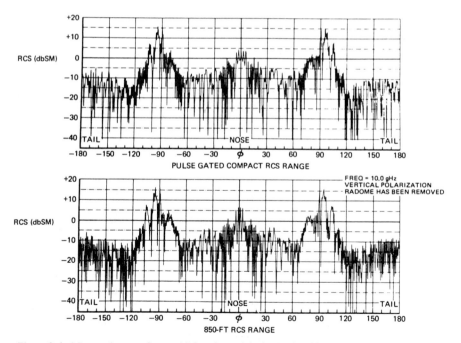

Figure 2-6. Measured return from a 1/15 scale model of a Boeing 737 commercial jetliner at 10 GHz, vertical polarization. (By permission, from Howell, ref 12; © 1970 IEEE)

If there is a radar antenna in the nose of the craft, this too will contribute to the nose-on return because antennas are efficient reflectors along their boresight directions. A few degrees away from nose-on incidence, the leading edges of the wings become major scatterers. If the radius of curvature of the leading edge is large compared to the wavelength, the return is large for any polarization of the incident wave. If not, the leading edge return is stronger when the incident electric polarization is parallel to the edge than when it is perpendicular to it. If the wing is seen from slightly below or slightly above, the trailing edge can be a strong contributor, but unlike the leading edge return, the trailing edge return can be stronger when the electric polarization is perpendicular to the edge than when it is parallel to the edge.

As the aspect angle swings near the broadside view, the sides of the fuselage and the engine pods become dominant sources. At broadside incidence, the vertical fin stands out. If it is viewed from considerably above the plane of the wings, the scattering from the dihedral corner formed by the upper wing surface and the fuselage is strong. Similarly, a dihedral corner may be formed by the vertical fin and the horizontal stabilizer. Somewhere between the broadside and tail-on aspects, the trailing edges of the wings light up, especially for the electric polarization parallel to the trailing edges. At the tail-on aspect, the returns from the engine exhaust ducts will be large. At any intermediate aspect angle where a relatively long surface is seen near grazing incidence, large returns may be observed due to the *traveling wave* phenomenon.[13] They occur primarily when the polarization is perpendicular to the surface, and they are closely related to the surface waves launched by traveling wave antennas.

Thus, the RCS characteristics of a large, complex object like an aircraft will vary considerably from one aspect to another. The interference due to several reflections changing phase with respect to one another can swing the net return by one or two orders of magnitude over a fraction of a degree change in viewing angle. The total dynamic range of the RCS pattern can be as much as 80 dB.

Turning to a different kind of target, a ship is a large collection of echo sources. The RCS of some ships can exceed a *square mile* (64.1 dB above 1 m^2), due in part to the multipath environment provided by the sea surface. The designers of ships have traditionally had very little interest in controlling or reducing ship echoes, as revealed by their fondness for bringing decks and bulkheads together at right angles. In addition, many topside surfaces are vertical, forming efficient dihedral corner reflectors with the mean sea surface.

Figure 2-7 shows the RCS patterns of a large ship at two different frequencies.[14] These patterns were measured as the ship steamed in a circle while in the beam of a multifrequency shore-based radar. The three traces in each diagram represent the 20, 50, and 80 percentile levels, and the measured data are of the average returns over a 2° aspect angle "window." Note that the patterns are not very dependent on frequency and that the mean return attains local maxima at the broadsides, bow-on, and stern-on aspect angles.

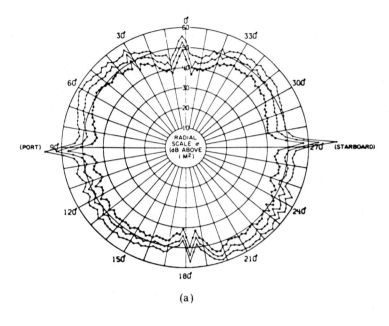

(a)

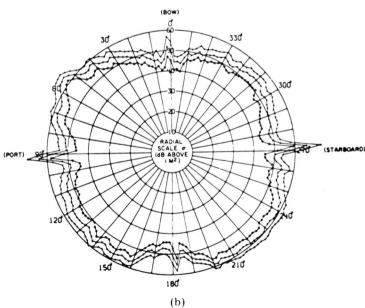

(b)

Figure 2-7. The upper diagram is the RCS pattern of a large naval auxiliary ship measured at 2.8 GHz, and the lower diagram is for the same ship at 9.225 GHz, both for horizontal incident polarization. (By permission, from Skolnik, ref 14; © 1980 McGraw-Hill Book Company.)

The significant scatterers on a ship depend on the range between the radar and the ship. For all except broadside incidence, the *hull* is not the dominant scatterer even when it is not occluded by the horizon. The superstructure and masts are the dominant scatterers, and every ship has a large assortment of fixtures and equipment located topside. Examples are vents, cleats, hoists, railings, capstans, lockers, pipes and conduits, stiffening plates, ladders, catwalks, and hatches. A ship is a very complicated target, and even as its masts disappear over the horizon, its radar antennas—which have large RCSs—are the last major contributors to disappear.

2.4 REFERENCES

1. Dale R. Carson and Paul Lorrain, *Introduction to Electromagnetic Fields and Waves*, W. H. Freeman and Co., San Francisco, 1962.
2. Simon Ramo and John R. Whinnery, *Fields and Waves in Modern Ratio*, John Wiley and Sons, Inc., New York, 1960.
3. Julius A. Stratton, *Electromagnetic Theory*, McGraw-Hill Book Co., New York, 1941, chap. IX.
4. Francis A. Jenkins and Harvey E. White, *Fundamentals of Optics*, McGraw-Hill Book Co., New York, 1957.
5. R. C. Hansen (Ed.), *Geometric Theory of Diffraction*, IEEE Press, New York, 1981.
6. J. J. Bowman, T. B. A. Senior, and P. L. E. Uslenghi (Eds.), *Electromagnetic and Acoustic Scattering by Simple Shapes*, North-Holland Publishing Co., Amsterdam, 1969.
7. George T. Ruck, Donald E. Barrick, William D. Stuart, and Clarence K. Krichbaum, *Radar Cross Section Handbook*, 2 vols., Plenum Press, New York, 1970.
8. O. Lowenschuss, "Scattering Matrix Representation," *Proceedings of the IEEE*, vol. 53, August 1965, pp. 988–992.
9. J. W. Crispin, R. E. Hiatt, F. B. Sleator, and K. M. Siegel, "The Measurement and Use of Scattering Matrices," Report No. 2500-3-T, Radiation Laboratory, The University of Michigan, Ann Arbor, Michigan, February 1961.
10. Ralph E. Kleinman, "The Rayleigh Region," *Proceedings of the IEEE*, vol. 53, August 1965, pp. 848–856.
11. Roger F. Harrington, "Matrix Methods for Field Problems," *Proceedings of the IEEE*, vol. 55, February 1967, pp. 136–144.
12. N. Alleyne Howell, "Design of Pulse Gated Compact Radar Cross Section Range," paper given at the 1970 G-AP International Symposium at Ohio State University, September 14–16, 1970; figures were published in the Program and Digest, IEEE Publication 70c 36-AP, pp. 187–195.
13. Leon Peters, Jr., "End-Fire Echo Area of Long, Thin Bodies," *IRE Transactions on Antennas and Propagation*, vol. AP-6, January 1958, pp. 133–139.
14. Merrill I. Skolnik, *Introduction to Radar Systems*, 2nd ed., McGraw-Hill Book Co., New York, 1980, p. 45.

3
THE PROPAGATION PROCESS

Donald G. Bodnar

3.1 INTRODUCTION

The usual concept of EM* wave propagation is one in which rays of various intensities are launched in radial directions by a transmitting antenna. The shape of the EM beam (pattern) is independent of the distance from the transmitting antenna (along any radial) but in general varies with the angle from any radial. This picture of EM energy traveling in straight lines with a constant pattern shape is valid only when the transmitting antenna exists in an otherwise empty universe. The presence of other objects in the universe modifies this "free-space" behavior. The modifying objects that are considered here are the earth and its atmosphere. Some of the mechanisms by which the earth and its atmosphere influence EM waves are these:[1,2]

1. Reflection from the earth's surface
2. Diffraction due to the earth's curvature
3. Ionospheric reflection
4. Atmospheric refraction

In addition, the propagating EM waves can be subjected to molecular absorption, scattering by precipitation, diffraction by hills and mountains, polarization rotation, and pulse dispersion.

As shown in Figure 3-1, the atmosphere can be roughly divided into three layers for propagation studies: the troposphere, the stratosphere, and the ionosphere. The troposphere has electrical properties that vary with altitude and weather conditions and is influential in bending EM waves. The stratosphere has a negligible effect on EM waves. The ionosphere can reflect signals which are low in frequency. In addition to the atmosphere, the earth itself can reflect and bend EM waves.

---

*EM waves include microwaves as well as the lower frequencies.

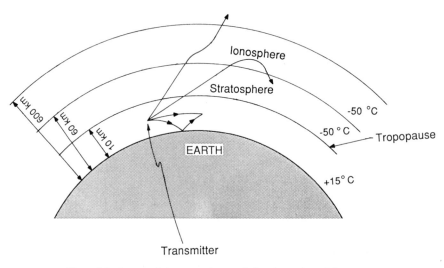

Figure 3-1. Layers of the atmosphere and directions of possible ray travel.

3.2 DIFFRACTION AND INTERFERENCE

The effect of the earth and its atmosphere on antenna pattern shape can be roughly separated into two regions of influence: the interference region and the diffraction region (see Figure 3-2). The straight line from the transmitter that separates these two regions is tangent to the earth. The interference region is that region of space illuminated by both a direct ray from the transmitter and a ray reflected from the earth's surface. The diffraction region is that region of space below the horizon in which the existence of a received field cannot be explained in terms of energy traveling in straight lines from the transmitter. The radio waves must somehow be bent to arrive in the diffraction region of space.

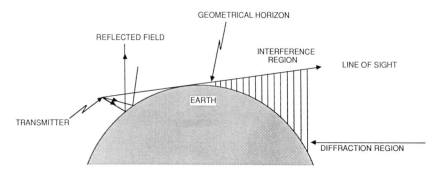

Figure 3-2. The interference and diffraction regions of space for a curved, smooth earth without an atmosphere.

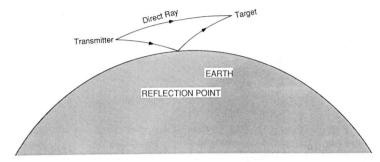

Figure 3-3. Reflection produced by the earth's surface in the presence of the atmosphere.

In the interference region, energy reaches the receiver or target from both a direct ray and a ground reflected ray (see Figure 3-3). These two fields interfere with each other constructively or destructively, depending on the phase relationship between them. This multipath interference causes nulls and peaks in the radiation pattern of the antenna in addition to those predicted for the free space behavior of the antenna. The multipath effect is discussed further in Chapter 4.

The mechanism by which radio waves curve around edges and penetrate the shadow region behind an opaque obstacle is called *diffraction*. This effect can be explained in terms of Huyghens' principle, which states that every elementary area of a wavefront (see P in Figure 3-4) is a center which radiates in all directions on the front side of the wavefront.[2] The intensity of this radiation varies as $(1 + \cos \theta)$. The fields from all of these Huyghens' sources add to produce a field along line AA', as indicated by the curve in Figure 3-4; note that diffraction not only permits energy to enter the shadow region but also

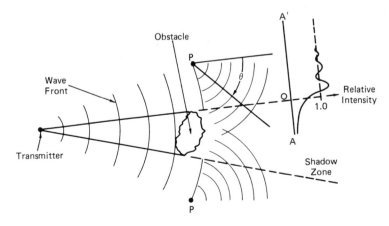

Figure 3-4. Diffraction around an obstacle. (From ref. 2, p. 332)

influences the field outside that region. As the frequency of the incident field increases, there is less penetration into the shadow region. Hence, low frequency radars are more effective than high frequency ones in "seeing" targets behind mountains. The earth itself is an obstacle for radio waves. Diffraction of energy around the earth produces a field in the diffraction region shown in Figure 3-2.

3.3 REFRACTION OF EM WAVES

Energy may also enter the shadow region due to tropospheric refraction and ionospheric reflection. *Refraction* is the change in the direction of travel of radio waves due to a spatial change in the index of refraction. The index of refraction *n* is equal to

$$n = \frac{c}{v_p} \tag{3-1}$$

where

c = speed of light in a vacuum
v_p = wave phase velocity in the medium

3.3.1 Atmospheric Refraction

Averaged over many locations and over long periods of time, it has been found that the index of refraction of the troposphere decreases with increasing altitude, as shown in Figure 3-5. The refractivity N is defined as

$$N = (n - 1) \times 10^6. \tag{3-2}$$

Eq. (3-1) shows that v_p increases as n decreases. Because of this relationship, the part of a transmitted wave that is at a higher altitude travels faster than the part that is closer to the ground. Thus, a wave tends to be bent downward under normal atmospheric conditions.

An approximation to N which is accurate up to about 20 to 30 GHz is

$$N = \frac{77.6}{T} \left(p + \frac{4810e}{T} \right) \tag{3-3}$$

where

T = air temperature in degrees Kelvin
p = total atmospheric pressure in millibars
e = partial pressure of water vapor in millibars.

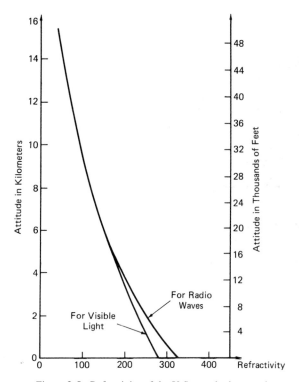

Figure 3-5. Refractivity of the U.S. standard atmosphere.

Notice that meteorological measurements using radiosondes, for example, can be used to determine N.

The path of a ray traveling through the atmosphere can be determined from Snell's law. For a first approximation, the atmosphere is modeled as a series of planar slabs, each having a constant index of refraction (see Figure 3-6), and the earth is approximated by a plane. From Snell's law, one obtains

$$n_0 \cos \alpha_0 = n_1 \cos \alpha_1 = n_2 \cos \alpha_2 \cdots = n_i \cos \alpha_i \qquad (3\text{-}4)$$

where n_i is the index of refraction of the ith slab. Thus if n_i ($i = 0, 1, 2, \ldots$) and the angle α_0 at which the wave is transmitted are known, then angles α_i ($i = 1, 2, \ldots$) can be found from Eq. (3-4). In this manner, the path of the ray is determined. The inhomogeneity of the atmosphere and the path of the ray can be better approximated by making the thickness of each slab smaller.

A still better solution to the refraction problem is obtained by assuming that the earth is spherical, instead of planar, and that the electrical properties of the atmosphere are constant between concentric spheres (see Figure 3-7). For such

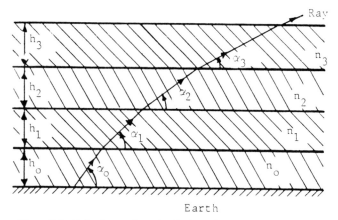

Figure 3-6. Path of a ray through a horizontally stratified atmosphere.

a radially stratified medium, Snell's law assumes the form

$$n_0 r_0 \cos \alpha_0 = n_1 r_1 \cos \alpha_1 = n_2 r_2 \cos \alpha_2 = \cdots = n_i r_i \cos \alpha_i. \quad (3\text{-}5)$$

If α_0 and the index of refraction are known as a function of height, then the remaining α values can be found from Eq. (3-5). The approximation improves as the thickness of each shell is made smaller. The curved path of any ray leaving the transmitter can be determined using the above techniques. As shown in Figure 3-8, the usual effect of refraction is to yield measured ranges and

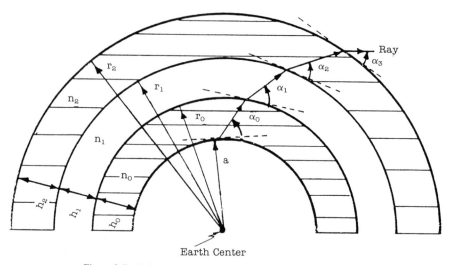

Figure 3-7. Path of a ray through a radially stratified atmosphere.

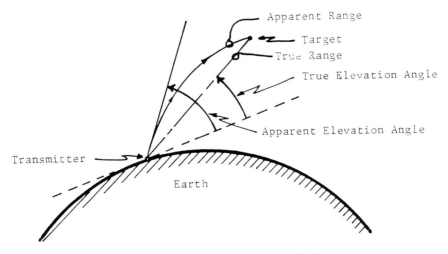

Figure 3-8. Refraction effects on target location.

elevation angles that are larger than the true values. The range measurement is too large because the time of travel is longer along the curved path. The resulting elevation angle and range errors for a spherically stratified model and a standard atmosphere are shown in Figures 3-9 through 3-12. In most cases, the refraction effects are negligible for angles of transmission greater than 5 to 10° with respect to the horizon.

In reality, the atmosphere is not radially stratified. The index of refraction versus height curve varies for different points on the earth's surface. Bean et al.[3] have compiled information about the index of refraction as a function of height for various locations on the earth's surface. This information can be combined with a generalization of the preceding ray tracing techniques to evaluate refraction effects more accurately, but such an effort is usually required only for very-high-precision radar systems.[4] For most applications, a radially stratified atmosphere is an adequate approximation.

Effective Earth Model

Another approach to ray tracing is to linearize the refractive index versus height curve and then obtain an effective earth model. For a radially stratified atmosphere and nearly horizontal transmission, the curvature of a ray path toward the earth's surface is (see ref. 1, pp. 50–52)

$$\rho = -\frac{dn}{dh} \tag{3-6}$$

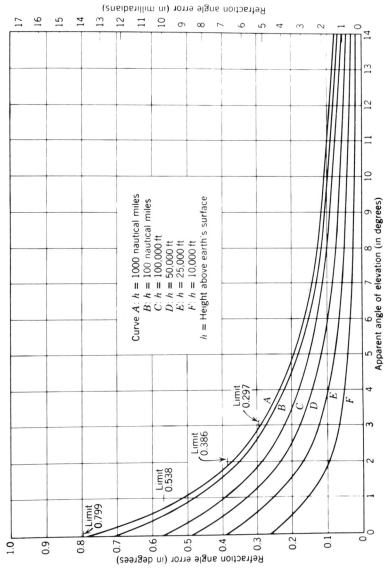

Figure 3-9. Tropospheric refraction errors for a standard atmosphere with 100% relative humidity. (By permission, from Berkowitz, ref 5; © 1965 John Wiley & Sons, Inc.)

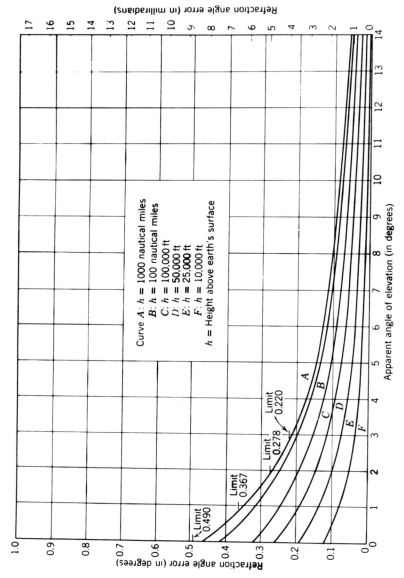

Figure 3-10. Tropospheric refraction errors for a standard atmosphere with 0% relative humidity. (By permission, from Berkowitz, ref 5; © 1965 John Wiley & Sons, Inc.)

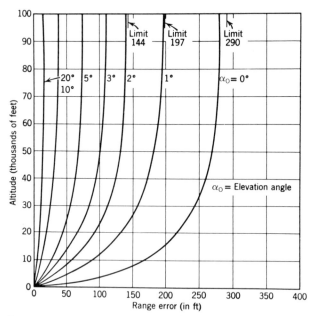

Figure 3-11. Tropospheric range errors for a standard atmosphere with 100% relative humidity, one-way transmission path. (By permission, from Berkowitz, ref 5; © 1965 John Wiley & Sons, Inc.)

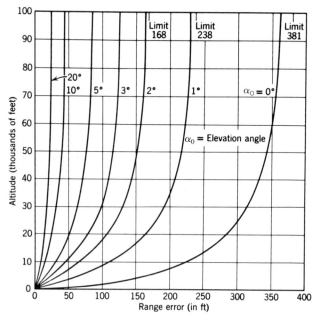

Figure 3-12. Tropospheric range errors for a standard atmosphere with 0% relative humidity, one-way transmission path. (By permission, from Berkowitz, ref 5; © 1965 John Wiley & Sons, Inc.)

where h is the distance above the earth's surface. Since the curvature of the earth's surface is $1/a$, where a is the earth's radius, the curvature of the ray relative to the surface is

$$\rho' = \frac{1}{a} + \frac{dn}{dh} . \qquad (3\text{-}7)$$

For most applications, it would be convenient to be able to plot a ray as a straight line, rather than a curve as shown in Figure 3-13. This can be done approximately by replacing the earth with an effective earth having an effective radius a_e given by

$$\frac{1}{a_e} = \frac{1}{a} + \frac{dn}{dh} . \qquad (3\text{-}8)$$

For this new earth, the ray path is a straight line. The ratio k of the effective radius to the actual earth radius is

$$k = \frac{a_e}{a} \qquad (3\text{-}9)$$

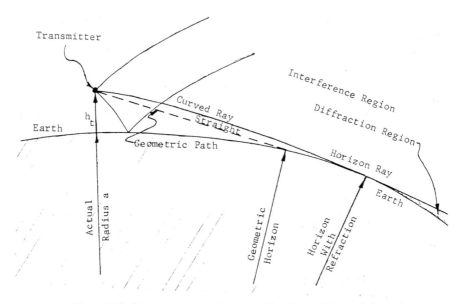

Figure 3-13. Ray curvature over the earth with radius a. (From ref. 2)

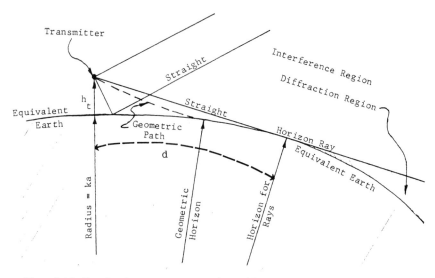

Figure 3-14. Rays in a homogeneous atmosphere with equivalent radius ka. (From ref. 2)

or

$$\frac{1}{k} = 1 + a\frac{dn}{dh}. \tag{3-10}$$

Since $a = 6370$ km and a typical value of $dn/dh = 3.9 \times 10^{-8}$/m at microwave frequencies, a typical value of k is thus 4/3.

The advantage of using an effective earth is that the inhomogeneous atmosphere of the actual earth is replaced by a homogeneous atmosphere over a slightly larger earth. Hence, ray tracing for the effective earth can be done with straight lines, as shown in Figure 3-14, as opposed to curves for the actual earth, as shown in Figure 3-13. The distance d to the radio horizon is, from Figure 3-14,

$$d = (2kah_t)^{1/2} \tag{3-11}$$

while the distance to the geometric horizon is $(2ah_t)^{1/2}$. Our definition of interference region and diffraction region can now be modified to account for the presence of the atmosphere. The radio line of sight in Figures 3-13 and 3-14 now separates the two regions.

Anomalous Propagation

Standard refraction implies that n decreases linearly with height in such a way that $k = 4/3$. (It is important to remember that the standard atmosphere does

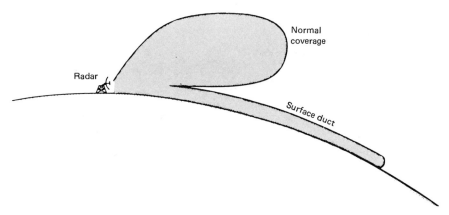

Figure 3-15. Increase in surface coverage by superrefraction. (By permission, from Skolnik, ref 6; © 1962 McGraw-Hill Book Company)

not give standard refraction.) Even though the four-thirds earth model gives only an approximation of the effects of refraction, propagation which deviates from this approximation is called *anomalous propagation*. Three forms of deviation from standard refraction that commonly occur in many parts of the world are subrefraction, superrefraction, and ducting.

When $dn/dh \geq 0$, waves bend away from, instead of toward, the earth. This condition is called *subrefraction*. Its occurrence decreases the ground coverage of a radar.

When dn/dh is more negative than usual, waves are bent more strongly toward the earth. This condition is called *superrefraction*. It increases the ground coverage of the radar (for angles ≤ 1.0 to $1.5°$) but has little effect on high-angle coverage.

If $dn/dh < -16 \times 10^{-8}$/m, the radius of curvature of a transmitted ray will be less than or equal to the radius of curvature of the earth. Such an effect is called *ducting* or *trapping*. A duct traps energy near the earth, as shown in Figure 3-15. A duct can be formed when the temperature increases (temperature inversion) and/or the humidity decreases (moisture lapse) with height. Surface ducts as well as elevated ducts can occur. They can dramatically increase or decrease the radar coverage, depending on whether both the target and the radar are in the duct as opposed to one or both being out of the duct. (See Figure 3-16.)

3.3.2 IONOSPHERIC REFRACTION

The ionosphere is a region of the upper atmosphere that contains a sufficient number of ionized particles to affect radio propagation. The effect of these ionized particles is to try to bend or reflect back toward the earth all radio waves with frequencies less than 30 to 50 MHz. The ionization of the atmospheric

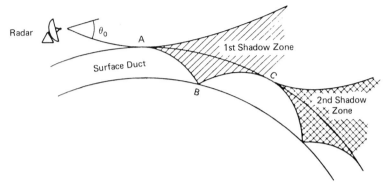

Figure 3-16. Shadow zones produced when the transmitter is above a surface duct. (By permission, from Bean et al., ref 7; © 1968 Dover Publications, Inc.)

gases in the ionosphere is produced primarily by the ultraviolet light from the sun. The ionization increases, but not linearly, with height and tends to have maximum values at particular heights. There are four main regions of relatively high ionization. These four layers are shown in Figure 3-17 and are discussed below.[8,9]

1. The D layer—exists only during daylight hours, at which time it bends and absorbs low-frequency ($f < 3$ to 7 MHz) waves.
2. The E layer—has characteristics similar to those of the D layer, but exists at a higher altitude.
3. The F_2 layer—is the most important layer for long-distance transmission. It bends waves at frequencies below 30 to 50 MHz. Its influence is strongest during the daytime and decreases somewhat at night.
4. The F_1 layer—during the day, the F_2 layer sometimes splits into two layers. The lower layer is called the F_1 layer and the upper one is called the F_2 layer. The F_1 layer is weaker and less influential than the F_2 layer.

The effect of one ionized layer on several rays is shown in Figure 3-18. A ray penetrates farther into the layer as the angle the ray makes with the horizon is increased. Rays transmitted at angles greater than the critical angle do not return to earth. The ionosphere is most influential for low angles of transmission and for low frequencies. The maximum distance than can be covered by a single reflection or hop is 1250 miles for reflection from the E layer and 2500 miles for reflection from the F_2 layer. OTH radars use HF frequencies and one or more hops to obtain radar coverage of regions that are thousands of miles away and that are impossible to cover from such distances with microwave or higher frequency radars.

As discussed in Section 3.3, the bending effect of a region on EM waves can be described by the refractive index n of the region. For the ionosphere, the

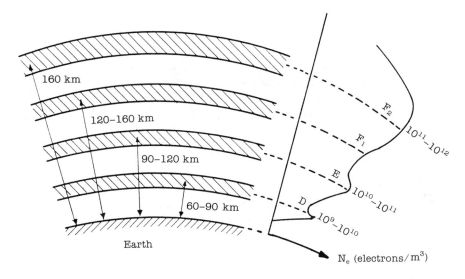

Figure 3-17. The layers of the ionosphere. Electron densities are typical of noontime values.

refractive index neglecting the earth's magnetic field is

$$n = \left[1 - \left(\frac{f_p}{f} \right)^2 \right]^{1/2} \tag{3-12}$$

where f_p is the critical or plasma frequency

$$f_p = \left(\frac{N_e e^2}{4\pi^2 m\epsilon_0} \right)^{1/2} \simeq 8.98 \sqrt{N_e} \tag{3-13}$$

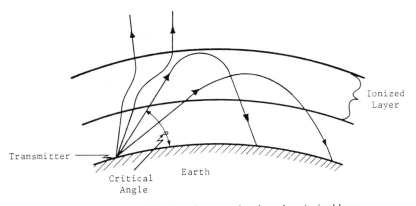

Figure 3-18. Typical behavior of rays passing through an ionized layer.

N_e is the electron density per cubic meter, f is the transmitter frequency in hertz, ϵ_0 is the permittivity of free space, and e and m are the charge and mass, respectively, of an electron. A plot of N_e versus height is shown in Figure 3-17. The peak in the electron number density in a given layer increases in value and decreases in height from morning until local noon. The afternoon behavior of N_e retraces the morning values relative to local noon. Note that a signal will be totally reflected by the ionosphere if its frequency is less than the plasma frequency.

3.4 ATTENUATION

The troposphere and ionosphere attenuate EM signals passing through them in addition to changing their direction of travel as discussed in the preceding section. The magnitude of these effects and their influence on radar range performance is discussed next.

3.4.1 Atmospheric Attenuation

The attenuation of radio waves in the lower atmosphere is due primarily to the presence of both free molecules and suspended particles such as dust grains and water drops which condense in fog and rain. When no condensation or dust particles are present, the attenuation is due to oxygen and water vapor.

The attenuation of the atmosphere can be calculated from the attenuation coefficient, α, in nepers per kilometer, and the effective path length, L, in kilometers. The attenuation coefficient and the effective path length are related to the wave amplitude by

$$A = A_0 \, 10^{-0.05\alpha L} \tag{3-14}$$

where A_0 is the amplitude for $L = 0$. The effect of atmospheric attenuation is to absorb or scatter power in the EM wave, thus reducing both the intensity of the signal that strikes the target and the intensity of the signal returned to the radar. Figure 3-19 shows typical values of atmospheric attenuation versus frequency.[10] Atmospheric attenuation is relatively small below about 3 GHz and becomes much more pronounced at millimeter wave frequencies. A relative peak occurs in the absorption around 22 GHz due to water vapor absorption. An oxygen absorption line at 60 GHz produces a great deal of attenuation for operation at or very near this frequency. A relative minimum in the absorption occurs at 35 GHz and between 70 and 110 GHz, making these desirable frequency regions in which to operate.

Figure 3-20 shows a curve of attenuation in rain for different rainfall rates, and Figure 3-21 shows attenuation in fog for different fog intensities.[8,10] Figure 3-20 shows that as the rainfall rate increases, attenuation increases very rapidly.

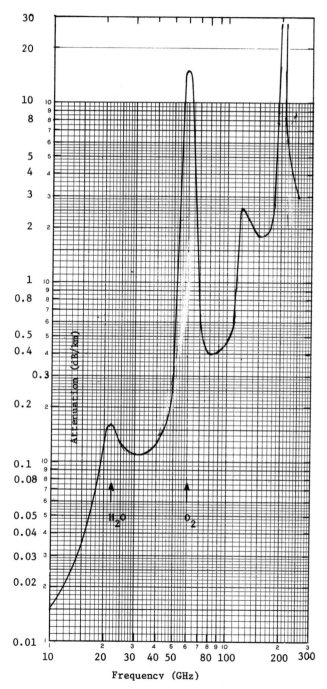

Figure 3-19. Absorption coefficient due to atmospheric gases and water vapor at sea level. (By permission, from Straiton et al., ref 10, © 1960 IEEE)

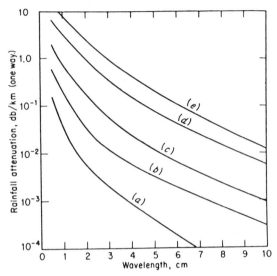

Figure 3-20. One-way attenuation of EM waves in rain at a temperature of 18°C. (a) Drizzle—0.25 mm/hr; (b) light rain—1 mm/hr; (c) moderate rain—4 mm/hr; (d) heavy rain—16 mm/hr; (e) excessive rain—40 mm/hr. (By permission, from Skolnik, ref 6; © 1962 McGraw-Hill Book Company.)

Fortunately, very high rainfall rates do not persist for long periods in most areas of the world.

Figure 3-22 shows the reduction in free space coverage resulting from any combination of atmospheric attenuation, rainfall, or fog. To use Figure 3-22, first determine the amount of attenuation in decibels per kilometer due to atmospheric gases and due to either rain or fog. Then find the free space coverage of the radar and go to the attenuation curve that is the sum of these two attenuations. Then read the range performance when that amount of attenuation is present.

3.4.2   Ionospheric Attenuation

An EM wave passing through an ionized medium imposes a periodic force on the electrons, causing them to vibrate. Signal energy is absorbed when these vibrating electrons collide with other particles and give up their energy. The attenuation in decibels is given by (ref. 8, Section 5-7)

$$A = 20 \log \left[\exp \left(- \int_{s_1}^{s_2} k \, ds \right) \right] \qquad (3\text{-}15)$$

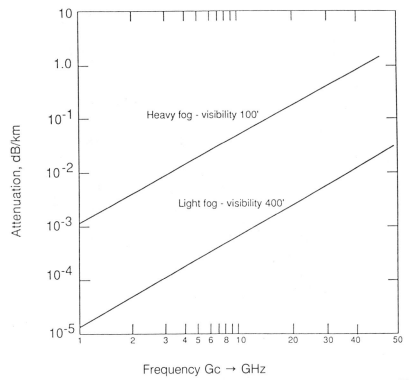

Figure 3-21. One way attenuation of EM waves in fog at a temperature of 18°C. Heavy fog: $\overline{M}$—2.3 g/m³, visibility k—100 ft; light fog: $\overline{M}$—0.032 g/m³, visibility—400 ft. (By permission, from Skolnik, ref 6; © 1962 McGraw-Hill Book Company)

or

$$A = -8.68 \int_{s_1}^{s_2} k \, ds \qquad (3\text{-}16)$$

where k is the absorption coefficient of the medium and s_1 and s_2 are the limits of the path.

There are two types of absorption in the ionosphere: nondeviating and deviating.

Nondeviating absorption is absorption in a medium where the index of refraction is approximately unity. When the index of refraction is much less than 1, resulting in considerable deviation of the wave from its original direction of propagation, the absorption is called *deviating*.

When the earth's magnetic field is neglected and the transmitted frequency

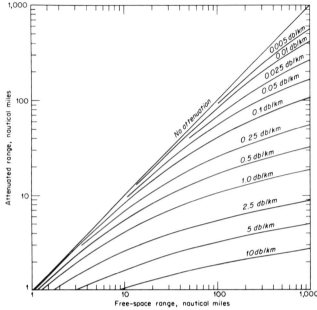

Figure 3-22. Reduction of the radar range due to attenuation along the propagation path between transmitter and target. Parameters on curves are one-way attenuation. (By permission, from Skolnik, ref. 6; © 1962 McGraw-Hill Book Company)

is greater than the electron collision frequency, k is given by

$$k = \frac{N_e e^2 v}{8\pi^2 cnm\epsilon_0 f^2} \tag{3-17}$$

where N_e is the electron density, c is the speed of light, e is the electron charge, m is the electron mass, v is the angular collision frequency, f is the transmission frequency, ϵ_0 is the permittivity of free space, and n is the index of refraction.

3.5 REFERENCES

1. H. R. Reed and C. M. Russell, *Ultra High Frequency Propagation*, Chapman and Hall, Ltd., London, 1966.
2. C. R. Burrows and S. S. Attwood, *Radio Wave Propagation*, Academic Press, Inc., New York, 1949.
3. B. R. Bean, B. A. Cahoon, C. A. Samson, and G. D. Thayer, *A World Atlas of Atmospheric Radio Refractivity*, U.S. Government Printing Office, Washington, D.C., 1966.
4. D. E. Kerr, *Propagation of Short Radio Waves*, Boston Technical Publishers, Lexington, Mass., 1964, pp. 41–46.
5. R. S. Berkowitz, *Modern Radar Analysis, Evaluation and System Design*, John Wiley & Sons, Inc., New York, 1965, Part V.

6. M. I. Skolnik, *Introduction to Radar Systems*, McGraw-Hill Book Co., New York, 1962, chap. 11.

7. B. R. Bean and E. J. Dutton, *Radio Meteorology*, Dover Publications, Inc., New York, 1968, Section 4.4.

8. J. M. Kelso, *Radio Ray Propagation in the Ionosphere*, McGraw-Hill Book Co., New York, 1964.

9. Y. J. Lubkin, "Problems in Extraterrestial Transmission," *Microwaves*, September 1962, pp. 10–17.

10. A. W. Straiton and C. W. Tolbert, "Anomalies in the Absorption of Radio Waves by Atmospheric Gases," Proceedings of the IRE, May 1960, pp. 898–903.

4
MULTIPATH PHENOMENA AND EFFECTS

H. A. Corriher, Jr.

4.1 INTRODUCTION

Although a great deal of radio and radar theory is concerned with EM wave behavior in *free space*, the actual operating environment is usually not so simple. In the previous chapter, the effects of a real atmosphere were discussed. Here we consider the consequences of having both the radar and the target either on or elevated above the reflecting surface of the earth. The approach is primarily phenomenological, and some approximate theories which have been used for point and vertically extensive radar targets are summarized.

The fundamentals of multipath propagation were developed before and during World War II. Since fairly detailed treatments of the basic points are contained in several readily available references, much of the development is merely sketched here. The reader is referred to Kerr,[1] Durlach,[2] Katzin,[3] and Long[4] for additional information. All four references are useful since their treatments emphasize different points.

4.2 DEFINITION AND EXAMPLES

The IEEE radar definition of *multipath* is: "The propagation of a wave from one point to another by more than one path. When multipath occurs in radar, it usually consists of a direct path and one or more indirect paths by reflection from the surface of the earth or sea or from large man-made structures. At frequencies below approximately 40 MHz it may also include more than one path through the ionosphere."[5]

While many readers may not be intimately familiar with radar multipath, everyone has observed multipath effects on radio and television. For HF radio, the ground wave and sky wave (via the ionosphere) may both be received, with resulting reduction or enhancement of the signal, depending on the relative phase and amplitude characteristics of the two RF fields at the receiving antenna. Television reception of a weak station can be affected by an aircraft flying nearby so that the picture and/or sound fade in and out. Also, a "ghost" image may appear to the right of the main picture. FM reception by a car radio sometimes exhibits signal losses which cannot be compensated for by the receiver's auto-matic gain control (AGC) when a large building reflects energy into the antenna to compete with that received by direct propagation.

The above examples contain the elements of two types of multipath effects: the simultaneous and the near-simultaneous reception of EM waves which have reached the receiving antenna by both *direct* and *reflected* (longer) paths. De-pending on the relative phases and amplitudes of the several (two or more) simultaneously received components, the result can be a composite EM field which can be near zero or as much as twice that received by the direct path alone. Near-simultaneous reception of "pulse-type" information can result in delayed, but separate, pulses. We have mentioned that reflection from the ionosphere or buildings can produce multipath, but the remainder of this chapter is concerned only with reflections from the surface of the earth. This is a par-ticularly important source of radar multipath, although other sources cannot always be ignored.

4.3   GEOMETRY

4.3.1   Spherical Earth

A complete treatment of radar multipath involves consideration of the radar and target above a spherical earth with the necessary geometric relations as shown in Figure 4-1. Although the geometry is somewhat complex, the calculations are straightforward and may be easily performed by digital computers. How-ever, a simpler approach is taken here, and the reader who needs more exact results should consult references 1–4 and 6.

A greatly simplified geometry for the general case of specular (i.e., mirror-like) reflection from a spherical earth is illustrated in Figure 4-2. Assume that a radar antenna at height h_1 is illuminating a point on a target at height h_2. The total EM field* at this target point due to the direct and indirect rays can be represented as the phasor

$$\overline{E} = E_0[f(\theta_1)\exp(-jkR_1) + \overline{\Gamma}f(\theta_2)\exp(-jkR_2)]$$

$$= E_0[f(\theta_1)\exp(-jkR_1) + \mathfrak{D}\mathfrak{R}\rho f(\theta_2)\exp[-j(kR_2 + \phi)]] \qquad (4\text{-}1)$$

where

E_0 = magnitude of the free-space EM field at the target if the antenna axis is pointed directly toward the target [$f(\theta) = 1$]

$f(\theta_1)$ = relative value of the antenna field pattern for the direct ray

---

*This is actually the field for one polarization and assumes no depolarization of the indirect ray upon reflection at the surfce. For a more exact treatment, see references 2 and 6.

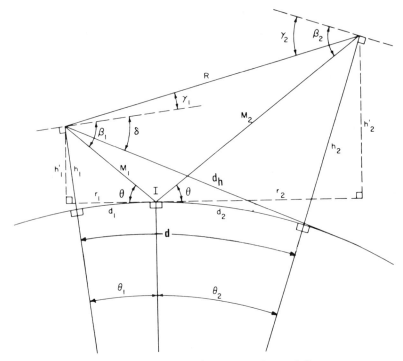

Figure 4-1. Spherical earth parameters. (From ref. 2).

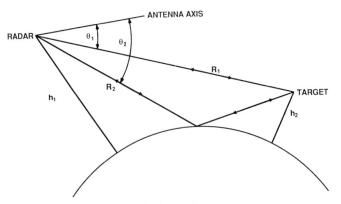

Figure 4-2. Spherical earth geometry.

$f(\theta_2)$ = relative value of the antenna field pattern for the indirect ray
R_1, R_2 = direct and indirect path lengths, respectively
$\overline{\Gamma}$ = $\mathfrak{D}\mathfrak{R}\, \rho \exp(-j\phi)$ = overall reflection coefficient of the earth (see Eq. (4-5) below)
$k = 2\pi/\lambda$

The symbols with overbars are phasors (i.e., complex numbers), and the symbols without overbars are magnitudes (i.e., absolute values).

Because of reciprocity, energy scattered by the target back toward the radar will traverse the same two paths, and will be subject to the same reflection effects and antenna gains as was the illuminating energy. Thus, the received EM field at the radar will be proportional to $\overline{E}^2$ as defined by Eq. (4.1), and the power received will be proportional to E^4.

4.3.2 Flat Earth

It is also worthwhile to simplify further the geometry to the case of a flat earth, as shown in Figure 4-3. In this figure, it is assumed that the radar is sufficiently far from both the target and the reflection point on the surface that the direct and indirect rays are parallel. Note that the target, T, has an image, T', beneath the earth's surface at a distance equal to the height of the target itself above the surface. This geometry allows the path difference, ΔR, to be easily approximated as

$$\Delta R \approx 2h_2 \sin\theta. \tag{4-2}$$

For the flat earth and small grazing angles, R_1 and R_2 are each approximately equal to the radar range, R, so that $\sin\theta$ may be expressed in terms of the range

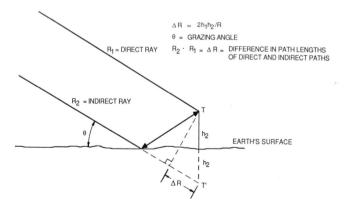

Figure 4-3. Flat earth geometry. (By permission, from Long, ref 4; © 1983 Artech House, Inc.)

and radar height, h_1, as

$$\sin \theta \approx h_1/R. \tag{4-3}$$

Substituting Eq. (4-3) into Eq. (4-2) gives *for the flat earth at low grazing angles* the useful approximation

$$\Delta R \approx 2h_1h_2/R. \tag{4-4}$$

4.4. NONGEOMETRIC PARAMETERS

4.4.1 Antenna Pattern

Equation (4-1) uses the antenna gain functions $f(\theta_1)$ and $f(\theta_2)$ corresponding to the relative gains in the directions of the direct and indirect paths, respectively. If the antenna has a relatively large main lobe, these two values can be essentially equal for low grazing angles. In contrast, if the antenna is highly directive and neither its narrow main lobe nor a major sidelobe illuminates the surface, then $f(\theta_2) \approx 0$ and consequently there are no multipath effects.

4.4.2 Overall Reflection Coefficient

A fundamental consideration in treating the reflectivity of radar targets in the presence of the earth is that both direct illumination and indirect illumination via reflection off the surface are involved. Although the direct path may be considered as propagation in free space (actually, as modified by the presence of the real atmosphere), the indirect path has the additional effect of surface reflection to be taken into account. It is convenient to treat the modifications to amplitude and phase that occur upon reflection by using an overall complex reflection coefficient, $\overline{\Gamma}$, which is itself the product of three factors, as follows:

$$\overline{\Gamma} = \overline{\Gamma}_0 \, \mathfrak{D} \mathfrak{R} \tag{4-5}$$

where

$\overline{\Gamma}_0$ = Fresnel reflection coefficient
$\mathfrak{D}$ = divergence factor
$\mathfrak{R}$ = roughness factor

Fresnel Reflection Coefficient

The complex Fresnel reflection coefficient, $\overline{\Gamma}_0$, of a smooth, plane surface can be expressed as

$$\overline{\Gamma}_0 = \rho \exp(-j\phi) \tag{4-6}$$

where

ρ = magnitude

ϕ = phase shift on reflection

The coefficient $\overline{\Gamma}_0$ depends on many parameters, including wavelength, grazing angle, and polarization. Usually the polarization dependence is treated by defining two coefficients, $\overline{\Gamma}_{0H}$ and $\overline{\Gamma}_{0V}$, for horizontal and vertical polarizations, respectively. These may be computed using the Fresnel equations and the electrical properties of the surface.[1-4]

For a smooth, flat surface and horizontal polarization, $\rho \approx 1$ and $\phi \approx \pi$ ($\overline{\Gamma}_0 \approx -1$) for all grazing angles and all commonly used radar wavelengths. For vertical polarization, $\rho \approx 1$ and $\phi \approx \pi$ only for very low grazing angles. (These approximations are particularly good for the sea surface.) For vertical polarization, as the grazing angle is increased, ϕ decreases and becomes essentially zero at vertical incidence. Concurrently, ρ decreases until $\phi = \pi/2$ and increases thereafter. The angle for which $\phi = \pi/2$ and ρ is at a minimum for vertical polarization is frequently called the *pseudo-Brewster angle*.

Divergence Factor

The Fresnel reflection coefficient as defined above for a smooth, plane surface is not completely suitable for treating reflection from the surface of the earth. Where long path distances are involved, the curvature of the earth can have a significant effect on the actual reflection coefficient. Rays reflecting from a convex surface diverge after reflection, causing an apparent lowering of the magnitude of the reflection coefficient. This effect is commonly accounted for by multiplying $\overline{\Gamma}_0$ by a scalar divergence factor, $\mathcal{D}$, which takes on values between 0 and 1. (For a detailed discussion of the divergence factor, see reference 1, pp. 113–115.) The cross-sectional area of the radar beam is increased from the value Q that it would have for reflection by a plane surface to a larger value Q' upon reflection by the spherical surface. These effects are shown in Figure 4-4. Letting Ω be the solid angle subtended by Q, $\mathcal{D}$ is defined as

$$\mathcal{D} \equiv \lim_{\Omega \to 0} (Q/Q')^{1/2}. \qquad (4-7)$$

Roughness Factor

Roughness of the surface also lowers the effective magnitude of the reflection coefficient by scattering radiation in nonspecular directions. The effect of roughness on the magnitude of the reflection coefficient has been accounted for by use of a scalar roughness factor, $\mathcal{R}$, having values between 0 and 1.[1]

The Rayleigh roughness criterion from optical theory is commonly used to

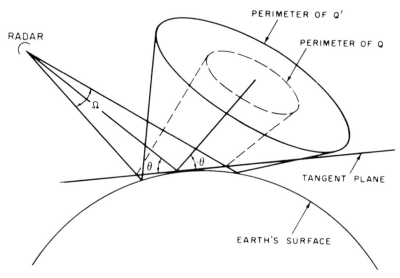

Figure 4-4. Divergence factor. (From ref. 2).

estimate the maximum surface irregularity that will not significantly lower the reflection coefficient (reference 1, p. 411). This criterion states that if the surface irregularities introduce path-length variations substantially less than one wavelength ($\lambda/4$ is frequently assumed), then the surface will reflect essentially as a smooth surface. For a ray incident at a grazing angle of θ on a surface with maximum peak-to-trough variations of h_s, this requires that

$$h_s \sin \theta \leq \lambda/8. \tag{4-8}$$

Any time that Eq. (4-8) is satisfied, the surface can be assumed to be essentially smooth.

Note that even a rough surface, such as the sea with reasonably large waves or swells, will still appear smooth for sufficiently small grazing angles or long enough wavelengths. For a given h_s and λ, there is a grazing angle below which all of the reflected energy can be assumed to be concentrated in a specularly reflected wave. If the grazing angle is many times larger than this angle, the specularly reflected wave may disappear entirely, with the reflected energy being scattered diffusely into the hemisphere above the surface (see, e.g., reference 7, p. 90).

4.5 RAY PATHS IN TWO-WAY (RADAR) PROPAGATION

When surface reflections are involved, as shown in Figure 4-2, there are four possible radar–target ray paths in two-way propagation. Two paths involve either

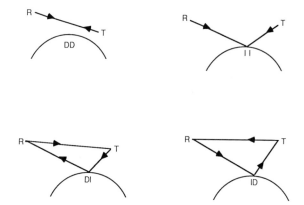

Figure 4-5. Four possible radar–target ray paths when surface reflections are involved. (After ref. 2).

direct or indirect rays only, and two involve both direct and indirect rays. If the direct path is indicated by D and the indirect path by I, the four possibilities for two-way transmission are DD, DI, ID, and II, as illustrated in Figure 4-5. Since the paths are of different lengths, it follows that the four return pulses from a single target point will not all coincide in time at the receiver. Figure 4-5 shows that the four paths have only three different path lengths: DD is the shortest, II is the longest, and the two paths DI and ID have the same length, intermediate between the other two. Due to these differences in path length, the radar may receive pulses at three different times.

If the difference in path length between direct and indirect paths is denoted by ΔR and the velocity of propagation by c, then the pulses traveling the DI or ID paths will lag behind the pulse traveling the DD path by $\Delta R/c$, and the pulse traveling the II path will lag by $2\Delta R/c$. In many practical cases, the pulse length, τ, will be very long compared to $\Delta R/c$ and the three return pulses will overlap considerably, as shown in Figure 4-6(a). In this situation, coherent interference can occur between energy transmitted along the different paths and received simultaneously, so that the concept of lobe-type illumination is valid (see below).

The pulse length τ also can be less than the time difference $\Delta R/c$, and the radar then receives three pulses (the DI and ID pulses are received together), separated in time as shown in Figure 4-6(b). This effect is usually important only for short pulse lengths when the grazing angle is not too small and the target height is at least moderately large. For example, a grazing angle of 10° and a target height of 50 ft will give a time difference $\Delta R/c$ of about 35 ns. It should be noted that while the DD ray is not affected by the surface reflectivity, the magnitudes of the DI and ID rays will be multiplied by Γ and that of the II ray by Γ^2. Target reflectivity may also differ at the two different ray angles.

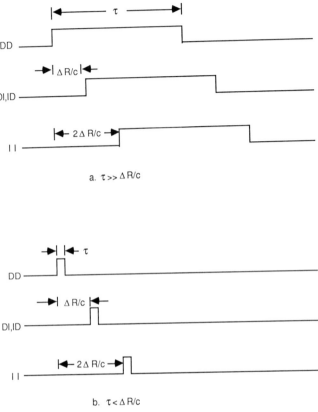

Figure 4-6. Time relationships of multipath received pulses for long and short pulse lengths. (After ref. 2).

It is frequently assumed that the reradiation pattern of the target has a large enough lobe width so that both direct and indirect paths are equally illuminated by the reflected energy. For some geometries, this requries a very large lobe width, which is not compatible with an efficient scatterer. If a scatterer has a diffraction minimum in the direction of one path or the other, the effect of that path is diminished. A target such as a dihedral or trihedral corner reflector with corner angles of exactly 90° is a highly efficient retrodirectional scatterer, but very little energy is reflected in other directions. Thus, energy may be propagated along the DD and II paths, but not along the DI or ID path.

In the case of a high-resolution radar which is imaging a complex target, the presence of multipath can cause individual scattering elements to be multiply imaged. This distortion can cause confusion in image interpretation.

4.6 EXAMPLES OF MULTIPATH AND COHERENT INTERFERENCE

Figure 4-7 shows the effect of multipath on a 1 μs pulse being received with *one-way* propagation. The two paths are direct and indirect, and the delay in reception of the pulse propagated along the longer indirect path increases from zero in the upper left to 1.5 μs in the lower right. The effect is that the received pulse is first stretched and then eventually the delayed pulse is received separately as ΔR becomes large.

The effect of coherent interference between two identical unresolved targets (radio towers) is shown in Figure 4-8. While this is not a strict example of multipath, the analogy is that two overlapping pulses of EM energy are received with the relative phase varying to produce both constructive and destructive phasor addition (and, of course, the cases in between). Considering only the right-hand pulses on each trace, the upper one shows constructive interference in the high-amplitude central section of the return. Here, the return pulses overlap and interface constructively so that the *power* of the received backscatter

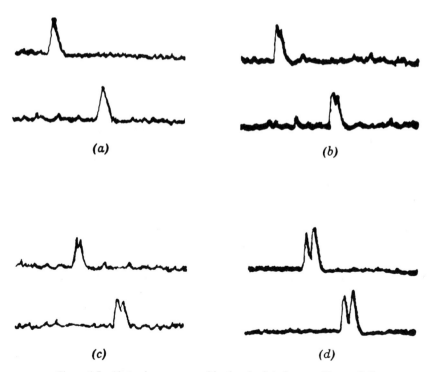

Figure 4-7. Air-to-air one-way multipath pulse interference. (From ref. 8).

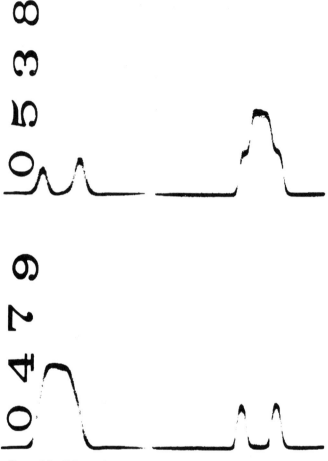

Figure 4-8. Coherent interference from two radio towers. (From ref. 9).

should be four times that received from a single tower.* (The A-scope deflection is nonlinear in power, so the amplitude shown in misleading quantitatively.) The lower trace shows a central section reduced to zero when the overlap of the two return pulses interfere destructively. As is pointed out in reference 9 (p. 74), the towers swaying in the wind had to move only 1 in. with respect to each other, that is, $\lambda/2$ (or a phase change of π radians), to cause a change from constructive to destructive interference.

---

*The two towers were considered to be *point* targets of equal RCS so that their combined RCS would vary between 0 and 4 times the RCS of one as the returns interfered destructively and constructively, respectively.

4.7   INTERFERENCE LOBES

The possibility of coherent EM wave interference (phasor addition) is inherent in Eq. (4-1). As illustrated previously, if the phases of the two phasor terms are identical (relative phase difference = 0), then they will add arithmetically in *constructive* interference. If the phases differ by π, then the arithmetic difference between the terms will show *destructive* interference. Of course, the magnitudes of the terms will also affect the magnitude of the resultant, but, in the two extreme cases, the sum or difference of two EM *fields* of equal magnitude would be 2 or 0, respectively.

The addition of phasors for *one-way* multipath propagation is illustrated in Figure 4-9. For each polarization, the electric field at the target is composed of two components: $\overline{E}_D$, the electric field propagated along the direct path, and

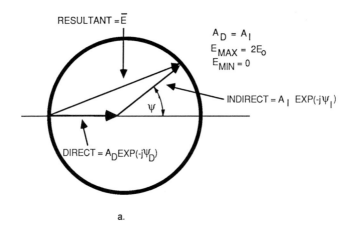

a.

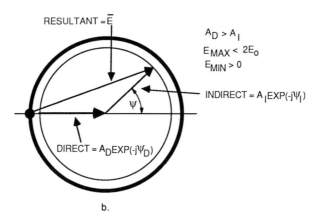

b.

Figure 4-9. Phasor addition of EM fields for one-way multipath.

$\overline{E}_I$ the electric field propagated along the indirect path. These components are taken to be received simultaneously and may be identified in Eq. (4-1) as

$$\overline{E}_D = E_0 f(\theta_1)\exp(-jkR_1)$$

and (4-9)

$$\overline{E}_I = E_0 \mathfrak{D}\mathfrak{R}\rho \, f(\theta_2)\exp(-jkR_2 + \phi).$$

Eqs. (4-9) may be rewritten in terms of amplitude A and relative phase ψ for each component as

$$\overline{E}_D = A_D\exp(-j\psi_D)$$

and (4-10)

$$\overline{E}_I = A_I\exp(-j\psi_I).$$

For the usual case of the antenna gain higher in the direction of the direct path, and $\mathfrak{D}$, $\mathfrak{R}$, and ρ less than or equal to unity, $A_D \geq A_I$. The absolute value of the relative phase between the components is

$$\psi = |\psi_D - \psi_I|.$$

Figure 4-9(a) illustrates the situation for $A_D = A_I$ so that the *magnitude* of the resultant electric field, $\overline{E}$, varies from 0 (destructive interference) to two times the free-space value (constructive interference). As shown in the figure, all intermediate values are taken as the absolute value of the relative phase, ψ, varies between 0 and π. Figure 4-9(b) illustrates the situation for $A_D > A_I$, and neither perfect cancellation nor reinforcement occurs. That is, the total field varies between values greater than 0 and less than twice the free-space field.

A phasor diagram could be drawn for *two-way* radar propagation, but all four of the paths shown in Figure 4-5 would have to be used. Coherent interference will occur if two or more EM waves propagated along the various paths are present simultaneously at the receiver.

For a particular target range and multipath conditions, a plot of electric field strength as a function of height would exhibit a minimum at the surface, would increase smoothly to a maximum value, and would then decrease to a minimum.* This alternating behavior would be repeated if height were increased further, and the interference pattern (or vertical interference lobes) would be a function of wavelength, polarization, grazing angle (i.e., range and radar

---

*This interference pattern is similar to the Lloyd's mirror effect of classical optics.

height), electrical properties of the surface, divergence factor, and surface roughness. Examples for flat ($\mathcal{D} = 1$), smooth ($\mathcal{R} = 1$), seawater at 35 GHz ($\epsilon = 22.4 - j33.0$) are shown in Figure 4-10 for two grazing angles (4.0° and 0.7°) and two polarizations (horizontal and vertical).[10] Note that the height scale is in inches, so that even the broader structure at $\theta = 0.7°$ has gone through a full lobe within the first 15 in. from the surface.

The existence of vertical interference lobes and their calculation has long been important in establishing the height of communication antennas (e.g., see reference 8 or 11), and plots of these lobes are sometimes called *height-gain curves*. Computer plotting of vertical-plane coverage diagrams has been described in an NRL report.[12] We note here that Figure 4-10 is plotted in one of

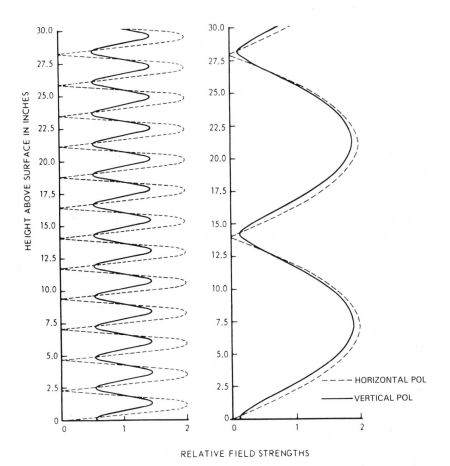

RELATIVE FIELD STRENGTHS

Figure 4-10. Vertical lobe patterns over seawater for grazing angles of 4.0° (left) and 0.7° (right) at 35 GHz. (From ref. 10)

the three common forms used to present the effects of multipath interference in the vertical plane—namely, field strength (or power) as a function of height.

4.8 PATTERN PROPAGATION FACTOR AND EFFECTIVE RCS

When multipath produces a lobed variation in the vertical illumination of the target rather than multiple images, it is convenient to introduce the pattern propagation factor, $\bar{F}$. Although this factor is usually defined (for example, in reference 1, p. 35) to be "the ratio of the magnitude of the actual field at the target to the magnitude of the field which would exist if the target were in free space and the antenna pointed directly at it," it is more useful to consider it as a phasor (complex) quantity which is the ratio of the two phasors representing the actual field and the assumed free-space field at a point, so that

$$\bar{F} = \bar{E}/\bar{E}_0. \tag{4-11}$$

The power received from a point target over the earth can be obtained by multiplying the usual free-space radar equation by F^4:

$$P_r = F^4[P_t G^2 \lambda^2 \sigma'/(4\pi)^3 R^4] \tag{4-12}$$

where the symbols have their usual meanings, except that σ' is primed to emphasize that it is the *free space* RCS and F is the magnitude of the phasor pattern propagation factor. Use of F^4 then permits the concept of RCS as usually defined for free-space situations to be retained.*

Any attempt to measure the cross section of a target on or near the surface actually results in measuring an effective cross section:

$$\sigma_{\text{eff}} = F^4 \sigma'. \tag{4-13}$$

An estimate of σ' could be obtained from such a measurement, provided $\bar{F}$ were known with reasonable accuracy. This concept is quite valid if the target is a point target and the geometry, propagation conditions, and so on are accurately known. Considerable difficulty has been encountered, however, in trying to apply the concept to vertically extensive targets (see Section 4.9.2).

---

*This discussion is appropriate only for transmitting and receiving the same linear polarization in a monostatic situation where the surface reflection does not involve depolarization. Bistatic situations, for example, could have $(F_a)^2$ to the target and $(F_b)^2$ back to the radar. Elliptical polarization may be considered to involve simultaneous horizontal and vertical components with 90° of time phase between them. These situations involve more complicated expressions than F^4 to account for propagation effects (e.g., see reference 2 or 6).

4.9   POINT AND VERTICALLY EXTENSIVE RADAR TARGETS

4.9.1   General Assumptions

Several approaches have been taken in solving the problem of a target extending upward from on or near the surface and (usually) having a uniform free–space cross section as a function of height. Much of this work has been carried out using several simplifications, among them being the previously discussed assumptions of a flat earth ($\mathfrak{D} = 1$), a smooth surface ($\mathfrak{R} = 1$), perfect reflection ($\rho = 1$ and $\phi = \pi$ so that $\overline{\Gamma}_0 = -1$), and maximum antenna gain on both paths ($f(\theta_1) = f(\theta_2) = 1$). These assumptions are particularly applicable for low grazing angles and horizontal polarization. With these assumptions, Eq. (4-1) can be rewritten as

$$\overline{E} = E_0[\exp(-jkR_1) - \exp(-jkR_2)] \tag{4-14}$$
$$= E_0 \exp(-jkR_1)[1 - \exp(-jk\Delta R)] = \overline{E}_0[1 - \exp(-jk\Delta R)]$$

where $\Delta R = R_2 - R_1$.

It is convenient in the work that follows to approximate ΔR, as is done in Section 2-2 of reference 1 for the case where h_1 and h_2 are small compared to R.* The results are

$$\Delta R \approx 2h_1 h_2 / R, \tag{4-15}$$

which is the same as the approximate result obtained in Eq. (4-4).

Using Eq. (4-11), we may identify $\overline{F}$ as the bracketed factor in Eq. (4-14). While the use of F^4 as a scalar in Eq. (4-12) is entirely satisfactory when both σ' and F^4 refer to the overall target, we note that $\overline{F}$ must be treated as a phasor in arriving at an average F^4 for a vertically extensive target which itself requires phase relationships to be taken into account (i.e., a coherent target).

4.9.2   Models of Point and Vertically Extensive Targets

Expressions have been summarized elsewhere[13] for the values of F^4 for a point target above the surface and a uniform target extending vertically from the surface, assuming perfect reflectivity and a smooth, flat earth so that $\overline{\Gamma} = -1$. Three models of the vertically extensive target have been considered: (1) a collection of points having uniform distribution as a function of height with all points scattering noncoherently; (2) a coherent scatterer with the resultant field

---

*R is the radar (slant) range, which for purposes of use in the radar range equation is approximately equal to either R_1 or R_2.

having uniform phase but varying amplitude with height; and (3) a coherent scatterer with the direct ray assumed to have uniform phase, but with phase variation with height of the reflected ray retained. The results given below are all expressed in terms of the dimensionless parameter

$$\Delta = k\Delta R \approx 4\pi h_1 h_2 / \lambda R. \tag{4-16}$$

For a target extending upward from the surface, the height, h_2, is to be interpreted as the maximum height of the target; that is, the target extends from 0 (the surface) to h_2. Results are given below for a point target and three extended target models, along with references to original sources.

Point target[1-3]*

$$F_0^4 = 16 \sin^4(\Delta/2). \tag{4-17}$$

Noncoherent extended target[14-16]

$$F_1^4 = 6\left[1 - \frac{\sin \Delta(4 - \cos\Delta)}{3\Delta}\right]. \tag{4-18}$$

Coherent extended target (uniform phase of resultant illumination)[1,17,18]

$$F_2^4 = 4[1 - (\sin \Delta)/\Delta]^2. \tag{4-19}$$

Coherent extended target (uniform phase of direct illumination)[13]

$$F_3^4 = 1 - \frac{2 \sin \Delta (2 - \cos \Delta)}{\Delta} + \frac{(3 \cos \Delta - 5)(\cos \Delta - 1)}{\Delta^2}. \tag{4-20}$$

The assumptions and approximations used in the analyses give results which are particularly easy to compare, since each F^4 is a function only of $\Delta \approx 4\pi h_1 h_2 / \lambda R$. Plots of Eqs. (4-17) to (4-20) are shown in Figure 4-11; note that each form of F^4 increases rapidly as Δ increases from 0. (Note that an increase in Δ may be interpreted as an *increase* in either radar or target height or as a *decrease* in either wavelength or range. In any case, an increase in Δ implies an increase in the number of interference lobes of the illuminating field which are subtended by the target.)

---

*The point-target model using a spherical earth and actual surface reflectivity usually gives good predictions for the vertical coverage of an air-search radar.

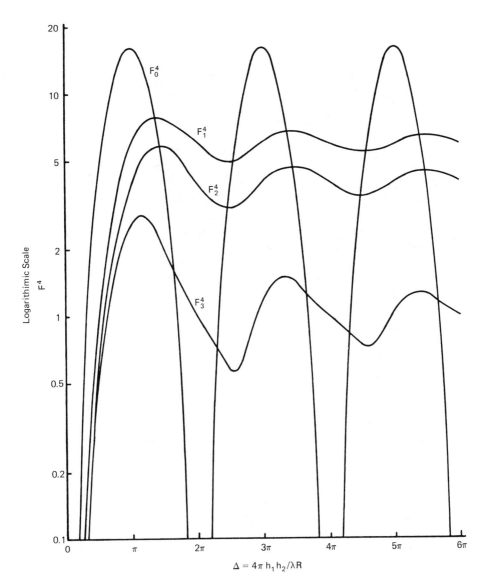

Figure 4-11. Variation of F^4 versus Δ for a point target and noncoherent and two coherent models of a vertically extensive target. (From ref. 13)

Although F_0^4 for the point target oscillates between values of 0 and 16, each of the other target models exhibits oscillations of diminishing amplitude about an asymptotic value of F^4 as

$$\Delta \to \infty: F_1^4 \to 6, F_2^4 \to 4, \text{ and } F_3^4 \to 1.$$

(The asymptotes of the two latter forms are suspect, as discussed below.) Considering the behavior as $\Delta \to 0$, we note that the curve of F_0^4 remains above the other curves, F_1^4 remains above F_2^4 and F_3^4, and the latter two merge and finally follow one common curve.

To a good approximation, as $\Delta \to 0$ all the curves of F^4 can be represented as being of the form $K\Delta^4$, where K is a constant. Because of the R^{-1} factor contained in Δ, this approximation gives an additional factor of R^{-4} in the modified radar equation, Eq. (4-12). This additional factor then leads to the prediction of a region of R^{-8} dependence between received power and range, as discussed by many authors. This region is at a sufficiently low grazing angle (long range or low target) that the target lies below the maximum of the lowest interference lobe.

4.9.3 Transition Range

It has become customary to associate the term *transition range*, designated R_t, with the range at which the value of F^4 using the small-Δ approximation becomes equal to its asymptotic value. (Note, however, that it is in this very region that the approximation departs most from the actual curve.) This, then, is taken as the point where the variation in return power versus range changes from an R^{-4} behavior to R^{-8}. R_t may be expressed in terms of the wavelength and the heights of the radar and target as

$$R_t = K_n h_1 h_2 / \lambda. \tag{4-21}$$

The value of R_t is strongly affected by the choice of model. For the point model, the transition range may be taken to be that range at which the target height equals the height of the maximum of the lowest interference lobe; this gives the lowest value for R_t of the various models considered, $K_0 = 4$. (An alternative is to use the intersection of the small-Δ approximation with the peak-value line of $F^4 = 16$ so that $K_0' = 2\pi = 6.28$.) Also, the distribution of cross section with height affects R_t, and the values of K_n in Eq. (4-21) assume uniform cross section per unit height. For this assumption, other values of K_n corresponding to the curves in Figure 4-11 are $K_1 = 5.37$, $K_2 = 5.13$, and $K_3 = 7.26$.[13]

Since gross approximations are necessary in defining target scattering properties, the point-target estimate is perhaps as useful as any. Effective target height (which directly affects transition range) is particularly difficult to establish for vertically extensive targets. Also, earth curvature and propagation conditions strongly affect R_t. We note further that the existence of an R^{-8} region (and a consequent R_t) is the direct result of having an interference lobe in the illuminating field, and such a lobe is expected at low grazing angles for virtually any radar. However, for the region under this lowest lobe maximum, the specific value of the large exponent (-8) is the result of the approximation to F^4 and should not be expected to give a result exactly matching experimental data which also include propagation effects.

4.10 INTERACTION OF NONUNIFORM ILLUMINATION AND NONUNIFORM TARGET REFLECTIVITY

From various studies, we are led to a generalized representation of a complex target as having a nonuniform distribution of cross section across the projected area being illuminated. Obviously, a uniform, coherent scatterer (flat plate) is not a very exact model for a ship on water, a tank on land, or any complex target on a reasonably reflective surface since, in the vertical direction, this nonuniform cross-section distribution is also illuminated by a nonuniform field if multipath conditions exist. Figure 4-12 depicts magnitude variations in the vertical plane of both target reflectivity and illuminating fields that might represent some possible situations. Although only magnitudes are shown, the illuminating fields are labeled as $|\overline{F}\overline{E}_0|$ to emphasize that both $\overline{F}$ and $\overline{E}_0$ are phasors. The target characteristics are labeled in terms of scattering length, $\overline{l}$, which is also a phasor (whose magnitude equals $(\sigma/4\pi)^{1/2}$; see reference 1, 3, or 7).

This approach is closely analogous to that taken in predicting the cross sec-

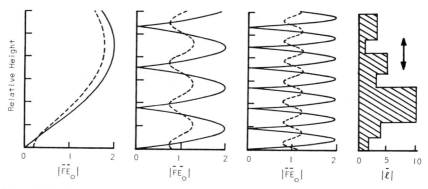

Figure 4-12. Schematic representation of variation in the magnitude of illuminating field and target scattering length with height above the surface. (Solid lines represent horizontal polarization and dashed lines vertical polarization.) (From ref. 13)

tion of aircraft, except that the illuminating field here is not uniform; it varies in both amplitude and phase across the target or target elements. As stated previously, unless the lobes are very small, the aircraft may be assumed to be a uniformly illuminated "point" target, albeit not with free-space illumination values.

Figure 4-12 also illustrates how the form of a nonuniform field incident on a nonuniform target might vary as the illuminating lobes vary in different circumstances. The entire target might be illuminated by part of a single interference lobe, or several interference lobes might illuminate only one portion of the target. While not intended to depict precisely a particular circumstance, the difference in illumination between horizontal polarization (solid lines) and vertical polarization (dashed lines) is also shown. Relative motion might occur between field and target and cause variations in effective cross section, as suggested by the vertical arrows. This combination of variation in target reflectivity *and* nonuniform illumination would cause large variations in return power as a function of time.

4.11 RETURN POWER VERSUS RANGE

4.11.1 Vertically Extensive Target

We have stated that it does not appear possible to determine a specific value for the transition range R_t at which the power returned by a vertically extensive target on or near the surface changes from a slow decrease with increasing range of about R^{-4} (as expected for free space) to a more rapid decrease of about R^{-8}. However, the general behavior of a change in the slope of the curve of return power versus range at some point within the radar horizon has been confirmed experimentally, although severe departures from the approximate theory are also common.

Figure 4-13 illustrates some World War II experimental data obtained by the British using a high-sited radar with low resolution.[15,19] Each data point is the maximum return in a 20-second interval and is thus likely to be on the high–return tail of the statistical distribution of returns. Dashed lines on the figure represent theoretical range variations using the noncoherent theory of Eq. (4-18), assuming that the phase of the resultant of the direct and indirect rays illuminating the target is linearly proportional to height. In computing the phase for different heights as a function of range, spherical earth methods were used, and the curves were adjusted for variation in effective earth radius (i.e., different propagation conditions) to give the best fit.

The target for the data of Figure 4-13 was an ex-American four-stack destroyer,* the *HMS Lancaster*, and the radar was the S-band CA No. 1 Mk II

---

*Presumably one traded to the British during World War II for U.S. bases in Bermuda.

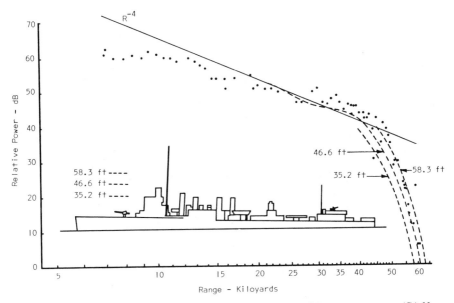

Figure 4-13. Range variation of return power for the destroyer *HMS Lancaster*, stern-on (CA No. 1 Mk II (Star) radar, antenna height 470 ft). (From refs. 13, 15, and 19)

(Star) with an antenna height of 470 ft. The destroyer was tracked stern-on from about 6000 to 60,000 yards, and the theoretical curve chosen as best fitting the data is for a target height of 46.6 ft and an effective earth radius of 5000 mi (approximately standard). The weather on this day was windy, and propagation was generally observed to be normal. Experimental points tend to lie above the theoretical curve at ranges near its knee, then show oscillatory behavior, and finally tend to fall below the curve at shorter ranges.

Power versus range plots from many other studies show the same general characteristics as those of Figure 4-13. Data points frequently are scattered, and considerable leeway is offered in fitting theoretical curves. While data for lower frequencies do not clearly exhibit it, S– and X–band data fairly often have the oscillatory pattern of return power versus range illustrated here. This pattern could be interpreted as being caused by a major contribution from a sensibly flat scattering center elevated above the surface, but the data are too sparse to make this much more than a conjecture. Also seen in this illustration is a drooping of return power at long ranges such that the data points fall below even an R^{-8} relationship.

The tendency for short-range data to fall below an R^{-4} line fitted to points closer to the transition range has been explained as being due to a number of possible causes. One of the most plausible reasons is that the target (or the sea at the point of reflection of the indirect ray) is not illuminated at the maximum

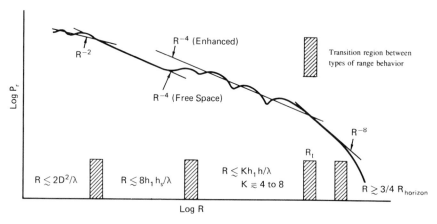

Figure 4-14. Conjectural range variation of peak received power for a vertically extensive coherent scatterer. (From ref. 13)

of the antenna vertical beam pattern. A second possible reason is that the reflection coefficient of the sea departs from "perfect" (i.e., $\overline{\Gamma} \neq -1$) as the angle of incidence departs from grazing. This may be caused by roughness, by polarization (Brewster angle) effects, or by the divergence of rays reflected from the spherical surface. Still another effect which may occur when peak returns are being considered is that, as the range becomes short enough, sufficient phase variation occurs across a dominant coherent scatterer that the effective return is reduced,[3,20] and that return power can vary about as R^{-2} if the illumination of the scatterer is not complete and varies with the angular pattern of the antenna.

Consideration of theoretical and experimental aspects of the problem has led us to make some general conjectures about the variation of return power with range for a vertically extensive target near the surface: The basic scheme is illustrated in Figure 4-14. The return in the R^{-4} region is "enhanced" at intermediate ranges over the "free space" value by approximately a factor of 6 (7.8 dB), the asymptotic value of F_1^4 in Section 4.9.2, since several lobes are averaged over a noncoherent target. The reader is strongly cautioned against drawing more than qualitative conclusions about the situation from this figure, since some of the variations it illustrates may not be correct for all experimental conditions and others are based purely on speculation. In particular, the various regions shown may overlap or be entirely absent. This conjectured variation is discussed in more detail in reference 13, along with the possible changes in statistical distributions of return power for a target at various aspects.

4.11.2 Point Target

Returning to consideration of a point target over a reflecting surface, we may examine the modified radar range equation shown in Eq. (4-12). Note that return power varies as R^{-4} in the free-space range equation but that the F^4 factor also varies with range, as shown (for a perfectly conducting flat earth) in Eqs. (4-17) through (4-20) and in Figure 4-11. For a point target over a non-perfectly-conducting flat earth, the expression[4]

$$F^4 = [1 + \rho^2 + 2\rho \cos (k\Delta R + \phi)]^2 \tag{4-22}$$

should be used instead of Eq. (4-17). When this form of F^4 is combined with the R^{-4} range fall–off of return power, a plot such as that shown in Figure 4-15 results. (The calculation and plotting of such curves are discussed in references 12, 21, and 22.)

Note that in Figure 4-15 return power is in dB and range is plotted on a linear scale so that the R^{-4} line (dashed) is a curve. If range is plotted as a logarithmic quantity, then the R^{-4} characteristic becomes a straight line, as shown in Figure 4-16. Figure 4-16 also shows the effect on the interference lobes of a major change in frequency; the "fat" lobes are for L-band, and the "thin" lobes are for X-band. Note that, for all other parameters fixed, the lowest interference lobe moves closer to the surface as the wavelength becomes shorter. Figures 4-15 and 4-16 illustrate a second common form in which the effect of multipath illumination may be shown—namely, power (or field) as a function of linear or logarithmic range.

Figures 4-15 and 4-16 show calculated received power under multipath conditions, while Figure 4-17 shows experimental results which do not exhibit such regular behavior. For Figure 4-17, a Cessna Skyhawk was flown with a large corner reflector in the cockpit pointed at the radar.[24] The corner reflector return

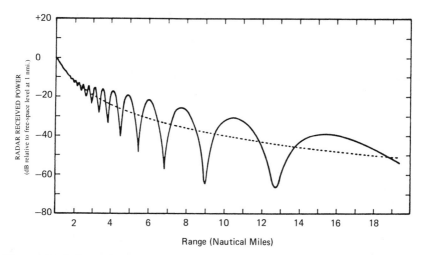

Figure 4-15. Radar received power versus linear range at X-band: radar at 80 feet; target at 100 feet. (From ref. 23)

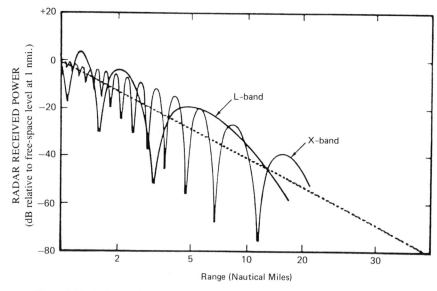

Figure 4-16. Radar received power versus log range for L- and X-bands. (From ref. 23)

should have swamped that of the overall aircraft, but this may not have happened as planned and the effective cross section may have varied due to target behavior in addition to the interference lobes. At any rate, that the lowest lobe was formed is obvious, and there are numerous other experimental results (see, for example, reference 1) which show the general validity of multipath theory.

4.12 MAXIMUM DETECTION RANGE UNDER MULTIPATH CONDITIONS

Figure 4-18 shows multipath illumination for X-band operation at an antenna height of 80 ft viewing an aircraft target at 100 ft.[23] The horizontal lines show the relative returns required for three radars with different powers and receiver sensitivities to detect a target with the same RCS at a value of SNR = 15 dB. That is, with the performance of Radar 1 as a reference, Radar 2 is worse by some 20 dB, and Radar 3 is worse by 30 dB.

The impact on detection range is immediately obvious. Radar 1 would make its initial detection at 18 nmi and would continue to track throughout the lowest interference lobe until the target reached 13.6 nmi. Radar 2 would not make its initial detection until the target penetrated the third interference lobe (near its peak) at 8.2 nmi and would lose track at 7.4 nmi when the target left this lobe. Radar 3 would not make its initial detection until the target penetrated the fifth interference lobe at about 5.1 nmi and would lose track almost immediately at 4.9 nmi. Of course, if the target were flying at a high velocity, the time spent

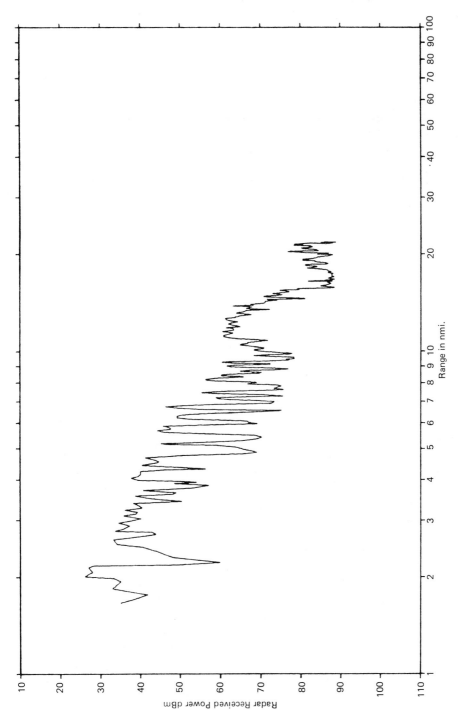

Figure 4-17. Received power versus log range for a Cessna Skyhawk with corner reflector: radar at 80 ft, target at 100 ft. (From ref. 24)

97

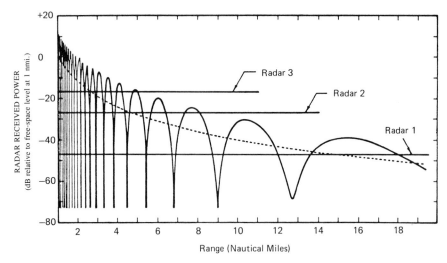

Figure 4-18. Radar received power versus range at X-band: antenna at 80 ft, target at 100 ft. Radar lines are for SNR = 15 dB with a 1 m² target at 1 nmi. (After ref. 23)

in the lower lobes would be very short, and reliable track could not be maintained until the target had reached essentially continuous fine lobes (say, the seventh and higher) at about 3.5 nmi. Also, a scanning radar would have to have one or more looks in the target direction during the time that the target was in a lobe for detection to occur.

The fact that a change of a few dB in radar transmitter or receiver performance, target cross section, or cross section reduction or enhancement can cause (or prevent) detection for a significant portion of the lowest interference lobe can have a profound effect on *maximum* detection range. This is especially important since at least the lowest lobe is usually formed for radars operating with low antenna and/or target heights over either sea or land.

Figure 4-19 shows a hypothetical result for two radar detections requiring, say, at least 7 dB SNR. On the right, the target is detected in the lowest (first) lobe from 14.5 to 12.0 nmi, then in the second lobe from about 9.0 to 6.5 nmi, and so on. On the left, the target and radar combination gives a detection performance which is only 3 dB worse, but now the lowest lobe is essentially missed and detection is delayed until the second lobe is entered at 8.5 nmi and departed at about 7.0 nmi, a very large change in initial detection range.

Multipath propagation via reflection from the earth's surface leads to vertical interference lobes whose varying target illumination results in peaks and nulls which modify the free-space power versus range relation which is proportional to R^{-4}. Since the location of the interference *nulls* can be changed only by modifying radar or target height, wavelength, or (to a small degree) polariza-

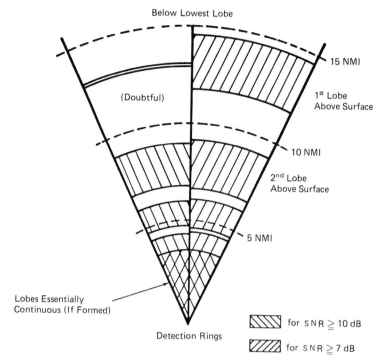

Figure 4-19. Comparison of two multipath detections with a radar/target performance difference of 3 dB. (From ref. 23)

tion, it is only mildly tongue-in-cheek to point out the following fundamental fact:

Signal processing requires a signal to process.

Two standard methods of showing the interference effects of multipath were illustrated previously in Figures 4-10, 4-15, and 4-16. A third common method is illustrated in Figure 4-20, where the interference lobes themselves are particularly obvious. Methods for calculating these lobes are discussed in many references (e.g., references 1, 8, and 12), and what is normally plotted are *contours* of either field strength or SNR for a particular radar and target. Figure 4-20 is not quantitative, but simply shows the lobes of high illumination separated by regions of low illumination. This figure also shows the altitude line for an aircraft at a height of 30,000 ft. The insert at the lower right shows the blip-scan ratio as the target flies through the lowest four lobes: the value is near

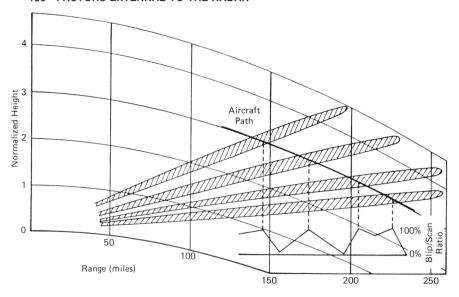

Figure 4-20. Comparison of a theoretical lobing pattern and experimental blip-scan ratio for an aircraft target.

100% when the target is near a lobe maximum and goes much lower (sometimes near zero) when the target is in a lobe minimum.

4.13 OTHER MULTIPATH EFFECTS

The objective of this chapter has been to use primarily phenomenological descriptions to give a unified overview of the physical processes involved in simultaneous and near-simultaneous multipath effects. The majority of the discussion has been concerned with the modification of target illumination which affects detection range. We have only briefly mentioned the fact that a single target scatterer (scattering element) may produce multiple returns separated in range [Figures 4-6(b) and 4-7]. Also, the position of the radar centroid (aim point) of a large target on land or sea is affected by the phase and amplitude distribution of the illumination.

An important consequence of multipath (and one which was observed and investigated in the early days of radar) is that a tracking radar operating at low grazing angles over a reflective surface can see both the target above the surface and its image below the surface (see Figure 4-3). The antenna tracks satisfactorily in azimuth but, if the surface is perfectly reflective, the elevation track is along the surface which is the centroid of the target and its exact image. This is a matter of strong operational interest in air defense, and numerous studies and experiments have been conducted.

Reference 25 contains a good collection of reprinted papers dated through 1974, and a June 1974 paper by Barton[26] ably summarizes the open literature on low-angle radar tracking to that date. Barton's treatment of diffuse scattering is also considered in references 27 to 29. Examples of other studies are given in references 30 to 44 for the reader who may wish to pursue this subject further.*

4.14 CONCLUDING REMARKS

It has been emphasized that the effective RCS of a target on or near the earth is modified from its inherent (free-space) cross section by the presence of the surface in a manner that may or may not be determinable. The use of the pattern propagation factor $\overline{F}$ to account for multipath illumination of the target allows us to conceptualize effective cross section σ as the product of F^4 and the free-space cross section σ'. Ideally, we should be able to determine the F^4 which prevailed during an experimental measurement of σ, remove it to obtain σ', and then apply another F^4 to establish the appropriate effective σ for system analysis. But this cannot be done precisely. The measured cross section data that are available usually have F^4 bound inextricably into the results, and this fact must be appreciated. Although F^4 may be either greater or less than unity, it is always less than unity for a low enough grazing angle so that return power falls off faster than R^{-4} in this region, with R^{-8} being a fair approximation for some cases.

In addition to modification of free-space target cross section, multipath effects may cause pattern distortion in high-resolution radar imagery. The problem of a reflected image target in low-angle tracking is another manifestation of multipath.

In systems design or interpretation of experimental measurements, the possibility of having multipath should be considered. Although multipath may be neglected in many situations, it is also likely that numerous cases of unusual results could be at least partially explained by considering such effects.

4.15 REFERENCES

1. D. E. Kerr (Ed.), *Propagation of Short Radio Waves*, MIT Radiation Laboratory Series, vol. 13, McGraw-Hill Book Co., New York, 1951.
2. N. I. Durlach, "Influence of the Earth's Surface on Radar," Technical Report 373 (ESD-TDR-65-32), Contract AF 19(628)-500, AD-627 635, Lincoln Laboratory, Massachusetts Institute of Technology, January 1965.

---

*The reader who is interested in additional references on multipath will find it useful to search computerized data bases such as INSPEC, COMPENDEX, SCISEARCH, NTIS, and DTIC. Although these data bases cover only the literature since about 1968, they identify hundreds of papers and reports which at least touch on the subject of multipath. A significant number of these references pertain to radar.

3. M. Katzin, "Reflection and Transmission of Radio Waves," in *Airborne Radar*, D. J. Povejsil, R. S. Raven, and P. Waterman (Eds.), D. Van Nostrand Company, New York, 1961, chap. 4.

4. M. W. Long, *Radar Reflectivity of Land and Sea*, 2nd Edition, Artech House, Inc., Dedham, MA, 1983, chap. 4.

5. *IEEE Standard Radar Definitions, IEEE Std 686-1982*, Institute of Electrical and Electronics Engineers, New York, 1982.

6. W. K. Rivers, S. P. Zehner, and F. B. Dyer, "Modeling for Radar Detection," Final Report on Contract N00024-69-C-5430, AD-507 375L, Engineering Experiment Station, Georgia Institute of Technology, December 31, 1969.

7. Petr Beckmann and Andre Spizzichino, *The Scattering of Electromagnetic Waves from Rough Surfaces*, Macmillan Co., New York, 1963.

8. H. R. Reed and C. M. Russell, *Ultra High Frequency Propagation*, 2nd ed., Chapman & Hall, Ltd., London, 1964.

9. L. N. Ridenour (Ed.), *Radar System Engineering*, MIT Radiation Laboratory Series, vol. 1, McGraw-Hill Book Co., New York, 1947.

10. J. G. Boring, E. R. Flynt, M. W. Long, and V. R. Widerquist, "Sea Return Study," Final Report on Contract NObsr-49063, Engineering Experiment Station, Georgia Institute of Technology, August 1957.

11. F. E. Terman, *Radio Engineer's Handbook*, McGraw-Hill Book Co., New York, 1943.

12. L. V. Blake, "Machine Plotting of Radio/Radar Vertical-Plane Coverage Diagrams," Report 7098, AD-703 211, Naval Research Laboratory, June 25, 1970.

13. H. A. Corriher, Jr., B. O. Pyron, R. D. Wetherington, and A. B. Abeling, "Radar Reflectivity of Sea Targets," vol. I, Final Report on Contract Nonr-991(12), AD-829 538L, Engineering Experiment Station, Georgia Institute of Technology, September 30, 1967.

14. R. Beringer, J. G. Carver, and C. W. Hoover, Jr., "Radar Echoes from Air-Laid Mine Splashes," Technical Report 5 on Contract Nonr-609(02), ATI-204 613 or TIP C-8394, Edwards Street Laboratory, Yale University, March 15, 1952.

15. M. V. Wilkes, J. A. Ramsay, and P. B. Blow, "Theory of the Performance of Radar on Ship Targets," Air Defence Research and Development Establishment (Great Britain), Joint Report ADRDE R04/2/CR252 and CAEE 69/C/149, AD-396 529 (DTIC reference only), July 1944.

16. T. S. Kuhn and P. J. Sutro, "Theory of Ship Echoes as Applied to Naval RCM Operations," Radio Research Laboratory, Harvard University, Report 411-93, AD-15 371, July 1944.

17. M. Katzin, "Radar Cross Section of Ship Targets," Report RA-3A 213A, ATI-37 552, Naval Research Laboratory, January 1944.

18. O. Stuetzer, "Research and Measurements on Aircraft Equipment Against Sea Targets," paper in "Electromagnetic Wave Propagation and Absorption," ATI-23 143, Translation F-TS-1973-RE, Air Materiel Command, August 1944.

19. M. V. Wilkes and J. A. Ramsay, "A Theory of the Performance of Radar on Ship Targets," *Proceedings of the Cambridge Philosophical Society*, vol. 43, 1947, pp. 220-231.

20. G. Beck, "Entfernungsabhangigkeit der Radarruckstrahlung in Erbdodennahe (Range Dependence of Radar Return in the Vicinity of the Earth)," Technischen Hochschule Munchen (Germany), summary of a doctoral dissertation, 1965.

21. L. V. Blake, "A Guide to Basic Pulse-Radar Maximum-Range Calculation: Part 1—Equations, Definitions, and Aids to Calculation," Report 6930, AD-701 321, Naval Research Laboratory, December 23, 1969.

22. L. V. Blake, "A Guide to Basic Pulse-Radar Maximum-Range Calculation: Part 2—Derivations of Equations, Bases of Graphs, and Additional Explanations," Report 7010, AD-703 211, Naval Research Laboratory, December 31, 1969.

23. H. A. Corriher, Jr., and F. B. Dyer, "Radar Detection of High-Speed, Low-Flying Aircraft," *Record of the Seventeenth Annual Tri-Service Radar Symposium*, Fort Monmouth, N.J., May 25-27, 1971.

24. G. W. Ewell, N. T. Alexander, and E. L. Tomberlin, "Investigations of the Effects of Polarization Agility on Monopulse Radar Angle Tracking," Final Technical Report, vol. I, Engineering Experiment Station, Georgia Institute of Technology, June 1971.

25. D. K. Barton, *Radars, Volume 4: Radar Resolution & Multipath Effects*, Artech House, Dedham, Mass., 1975.

26. D. K. Barton, "Low-Angle Tracking," *Proceedings of the IEEE*, vol. 62, June 1974, pp. 687–704.

27. R. J. Papa, J. F. Lennon, and R. L. Taylor, "Need for an Expanded Definition of Glistening Surface," Technical Report RADC-TR-82-271, AD-A130 431, October 1984.

28. R. J. Papa, J. F. Lennon, and R. L. Taylor, "Multipath Effects on an Azimuthal Monopulse System," *IEEE Transactions on Aerospace and Electronic Systems*, vol. AES-19, no. 4, July 1983, pp. 585–597.

29. R. V. Ostrovityanov and F. A. Basalov (William Barton (Transl.) and David K. Barton (Ed.)), *Statistical Theory of Extended Radar Targets*, Artech House, Inc., Dedham, Mass., 1985.

30. W. D. White, "Low-Angle Radar Tracking in the Presence of Multipath," *IEEE Transactions on Aerospace and Electronic Systems*, vol. AES-10, November 1974, pp. 835–852.

31. A. M. Peterson, J. F. Vesecky, et al., "Low-Angle Radar Tracking," Technical Report JSR 74-7, Contract DAHC15-73-C-0370, ARPA Order No. 2504, Stanford Research Institute, February 1976.

32. D. K. Barton, "Low-Angle Tracking," *Microwave Journal*, December 1976, pp. 14–24.

33. *Record of DARPA Low-Angle Tracking Symposium, December 1–2, 1976*, vols., I and II, General Research Corp., Washington, D.C., January 1977.

34. D. K. Barton, "Radar Multipath Theory and Experimental Data," *RADAR-77*, October 25–28, 1977, IEE Conference Publication Number 155, pp. 308ff.

35. A. V. Mrstik and P. G. Smith, "Multipath Limitations on Low-Angle Radar Tracking," *IEEE Transactions on Aerospace and Electronic Systems*, vol. AES-14, January 1978, pp. 85–102.

36. C. E. DuFort, "An Adaptive Low-Angle Tracking System," *IEEE Transactions on Antennas and Propagation*, vol. AP-29, September 1981, pp. 766–772.

37. L. W. Pickering and J. K. DeRosa, "Refractive Multipath Model for Line-of-Sight Microwave Relay Links," *IEEE Transactions on Communications*, vol. COM-27, no. 8, August 1979, pp. 1174–1182.

38. M. L. Meeks, *Radar Propagation at Low Altitudes*, Artech House, Inc., Dedham, Mass., 1982. (See also: M. L. Meeks, "Radar Propagation at Low Altitudes: A Review and Bibliography," MIT/Lincoln Laboratory, Project Report CMT-9, 21 July 1980.)

39. B. A. Wyndham and J. M. Shaw, "The Reduction of Multipath-Induced Azimuth Errors in a Monopulse Secondary Surveillance Radar," *IEE Colloquium on Characterization and Mitigation of Multipath Interference Effects*, London, 20 May 1983, IEE Digest No. 52, pp. 8-1 to 8-4.

40. D. Moraitis and S. A. Alland, "Effect of Radar Frequency on the Detection of Shaped (Low RCS) Targets," *Record of the IEEE 1985 International Radar Conference with Supplement*, Arlington, VA, 6–9 May 1983, pp. 159–162.

41. S. Haykin and J. Kesler, "Adaptive Canceller for Elevation Angle Estimation in the Presence of Multipath," *IEE Proceedings*, vol. 130F, no. 4, June 1983, pp. 303–308.

42. Y. Karasawa and T. Shiokawa, "Characteristics of L-Band Multipath Fading Due to Sea Surface Reflection," *IEEE Transactions on Antennas and Propagation*, vol. AP-32, no. 6, June 1984, pp. 618–623.

43. E. Hanle, "Some New Aspects on Low-Elevation Radar Coverage," *Record of the IEEE 1985 International Radar Conference with Supplement*, Arlington, VA, 6–9 May 1985, pp. 163–168.

44. K. D. Ward and C. J. Baker, "Microwave Scattering from the Sea Surface," *IEE Colloquium on Propagation Above and Below the Waves*, London, 25 November 1985, IEE Digest No. 105, pp. 3-1 to 3-5.

PART 2
BASIC ELEMENTS OF THE RADAR SYSTEM

- Radar Transmitters—Chapter 5
- Radar Antennas—Chapter 6
- Radar Receivers—Chapter 7
- Radar Indicators and Displays—Chapter 8

There are four basic elements in any functional radar: transmitter, antenna, receiver, and indicator.

The function of the transmitter is to generate a desired RF waveform at some required power level. The required RF power may be derived directly from a power oscillator such as a magnetron or an extended interaction oscillator (EIO), or it may be derived via an RF amplifier or amplifier chain (traveling wave tube amplifier, crossed-field amplifier, extended interaction amplifier, solid-state amplifier, etc.). The waveform is dictated by the particular system requirements and can range from an unmodulated continuous wave (CW) for a simple moving target indication (MTI) radar to a complex frequency, phase, and time code modulated wave for some advanced military radars. Radar transmitters are discussed in Chapter 5.

The basic function of the radar antenna is to couple RF energy from the radar transmission line into the propagation medium and vice versa. In addition, the antenna provides beam directivity and gain for both transmission and reception of the EM energy. Various radar antenna concepts and techniques are presented and discussed in Chapter 6.

The primary functions of the radar receiver are to accept weak target signals, amplify them to a usable level, and translate the information contained therein from RF to baseband. Various receiver configurations are employed, including the crystal detector, RF amplifier, homodyne, and superheterodyne; the superheterodyne receiver is by far the most commonly used. Each of these configurations is discussed in Chapter 7.

The primary function and purpose of the radar indicator is to convey target information to the user. The indicator configuration and information format are dependent upon the particular radar application and the needs of the user. Two common indicators are (1) the plan position indicator (PPI), where target range and angle data are displayed on a cathode ray tube for surveillance radar applications, and (2) an audio speaker or earphones, where the presence of a moving object is signaled by a Doppler frequency, as in a perimeter alarm radar. Various types of radar indicators and their applications are presented in Chapter 8.

5
RADAR TRANSMITTERS*

George W. Ewell

5.1 INTRODUCTION

The transmitter is one of the basic elements of a radar system. Figure 5-1 illustrates the relationship between the transmitter, duplexer, antenna, and receiver in a simple radar system. As noted in Chapter 1, the primary function of the radar transmitter is to generate the RF signal that is radiated by the antenna and scattered by the target. The RF signal scattered toward the radar (backscatter) is intercepted by the antenna and routed to the receiver for processing to enable target detection.

The basic units of a radar transmitter are an RF power source, a modulator, and a power supply as shown in Figure 5-2. The two basic methods of generating RF power are shown in Figure 5-3. The primary difference between the two methods of generating RF power is the power output of the RF oscillator. In the power oscillator approach, the power oscillator generates a relatively high-power RF signal that is applied directly to the antenna. In the master-oscillator power-amplifier approach, the master oscillator generates a relatively low-power RF signal that is amplified to the appropriate power level before it is applied to the antenna.

The RF signal generated by the transmitter may be CW or pulsed, and its amplitude and frequency are usually designed to fulfill specific requirements of the radar system. Radar transmitter power ratings range from milliwatts to terawatts, and the power source may be vacuum tube or solid state. Transmission pulse lengths range from nanoseconds to milliseconds.

The characteristics of readily available or buildable transmitters influence the selection of operating frequency, transmission waveform, duty cycle, and other important parameters of a given radar application. The transmitter is often a large, heavy, expensive component of the radar system; it may consume considerable power and require periodic maintenance and replacement. Thus, the proper choice of transmitter can strongly influence the size, operating cost, maintainability, reliability, and performance of the radar system. Therefore, a thorough knowledge of transmitter types and characteristics is essential to the

---

*Much of the material in this section is a condensed version of reference 1, and the reader is referred to that source for additional detail.

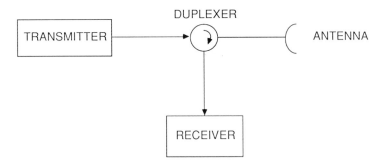

Figure 5-1. Simple radar system.

practicing radar system engineer, and a thorough, professional transmitter design effort is necessary for a successful radar system.

Some of the factors to consider in radar transmitter selection and design include:

- Peak power
- Average power
- Pulse length
- Pulse repetition frequency
- Bandwidth
- Spurious outputs
- Cost
- Life
- Tunability
- Arcing
- Stability (amplitude and phase)
- Distortion
- Efficiency
- Size and weight
- Gain
- Required dynamic range
- Reliability and maintainability

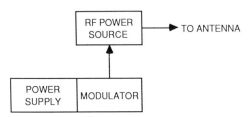

Figure 5-2. Basic transmitter units.

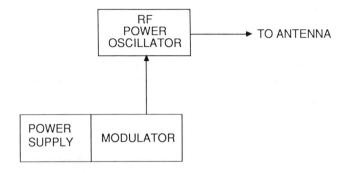

a. Power oscillator transmitter configuration.

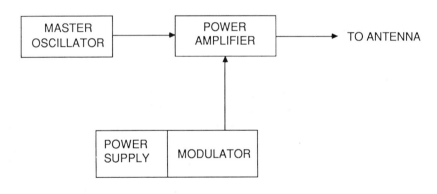

b. Master-oscillator power-amplifier transmitter configuration.

Figure 5-3. Basic transmitter configurations. (a) Power oscillator transmitter configuration. (b) Master-oscillator power-amplifier transmitter configuration.

There is an intimate relationship between the proper operation of the microwave tube and the device used to provide the appropriate voltages and currents for tube operation. Required pulse rise, fall, flatness, ability to withstand arcs, and changes in terminal characteristics with changes in frequency are all important characteristics of the tube that affect the design of the modulator or pulser. The following sections provide an introduction to the principal types of radar transmitters and emphasize the characteristics that are important in both system design and modulator selection and specification.

5.2   TRANSMITTER POWER OSCILLATORS AND AMPLIFIERS

5.2.1   Overview

Transmitter power sources are usually one of two types: oscillators or amplifiers. Self-excited oscillators such as the magnetron and various semiconductor bulk-effect devices—Gunn effect devices, impact avalanche and transit time (IMPATT) devices, and limited space charge accumulation (LSA) devices—are suitable as power sources for many radar applications. Amplifiers such as the crossed-field amplifier (CFA), the klystron, the traveling-wave tube (TWT), and transistor amplifiers are useful as final power amplifiers in applications such as a coherent MTI system, a pulse-compression radar, or a pulse-Doppler system where the transmitter source is usually a stable, low-power device.

Many pulsed radar applications require transmitters capable of providing considerable peak and average power levels. Thermionic vacuum tubes are commonly used in applications requiring high output power. Another approach involves combining (summing) the outputs of a number of low-power devices. Thus, both tube and solid-state sources have application in high-power radars.

5.2.2   Magnetrons[1–4]

General Magnetron Characteristics

The magnetron oscillator was the first practical high-power source for pulsed microwave radars. It is still widely used today in noncoherent radar applications. The magnetron oscillator is characterized by small size, light weight, reasonable operating voltages, good efficiency, rugged construction, and long life.

The magnetron converts energy extracted from a constant electric field to an RF field. A cutaway view of one possible magnetron configuration is shown in Figure 5-4. The principal parts of the magnetron are the cathode, the anode block, the interaction space, cavities, and the output coupling. The magnetron is energized by applying a magnetic field in a plane perpendicular to the plane of the figure and an electric field between the anode block and the cathode; therefore, the electric and magnetic fields are crossed in the interaction space.

Electrons traveling through the interaction space between the cathode and the anode block absorb energy from the DC field and deliver it to an alternating RF field generated by exciting the resonant cavities. The RF field energy is then coupled to the output.

Other possible magnetron cavity configurations include "rising sun," coaxial, and inverted coaxial interaction geometries. The coaxial configuration has become popular in recent years due to its improved frequency stability and long life. The cutaway view of a coaxial magnetron presented in Figure 5-5 shows the outer coaxial cavity, which provides the high-Q resonant element, and the

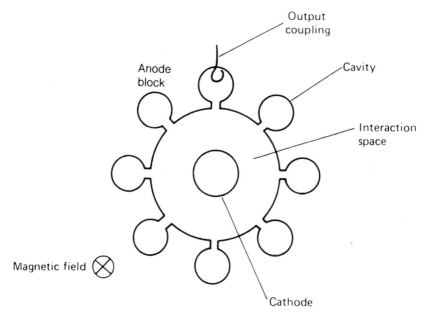

Figure 5-4. A cross-sectional schematic representation of the anode-cathode region of a microwave cavity magnetron. (By permission, from Ewell, ref. 1; © 1981 McGraw-Hill Book Co.)

coupling slots and vanes, which couple energy into the interaction region. As in all magnetrons, the interaction takes place in the region surrounding the cathode, a region of crossed magnetic and electrical fields.

For a given cavity configuration, a specific ratio of magnetic and electric fields must be applied to obtain oscillation at the desired frequency. Figure 5-6 illustrates typical magnetron voltage/current relationships for a constant magnetic field. A conventional magnetron does not draw appreciable current until the cathode-to-anode potential reaches a critical voltage, called the *Hartree voltage*. As shown in Figure 5-6, this critical voltage is normally about 90% of the operating voltage. The magnetron draws some current and produces some RF output power even when the applied voltage is well below the Hartree voltage. Thus, care should be taken to ensure that the modulation voltage is promptly removed after the desired pulse has been generated.

Operation of the magnetron with voltage levels below or above the operating point can result in *moding*—operation in a mode other than the normal mode—which is usually undesirable. When moding occurs, the magnetron's efficiency may deteriorate greatly and produce oscillations at more than one frequency. In a well-designed magnetron, the various modes are well separated in both voltage and frequency, and the magnetron can be made to operate stably in the desired mode if good transmitter design procedures are followed. Continued operation in a wrong mode can permanently damage a magnetron.

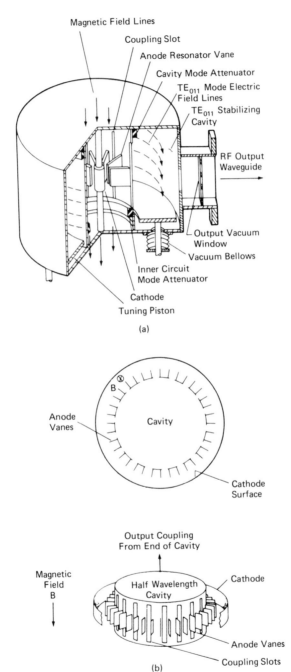

Figure 5-5. Schematic representations of (a) coaxial and (b) inverted coaxial magnetrons.

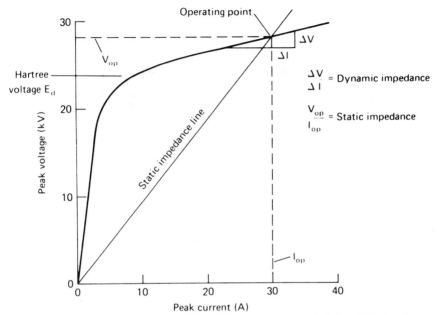

Figure 5-6. Voltage/current relationships for a microwave magnetron, including definition of magnetron static and dynamic impedance. (By permission, from Ewell, ref. 1; © 1981 McGraw-Hill Book Co.)

The change in magnetron output frequency for a given change in magnetron anode current with a fixed RF load is called the *pushing figure* (MHz/A). Another important performance parameter of magnetrons is the *pulling figure*, which is defined as the difference between the maximum and minimum frequencies of oscillation when the phase angle of the RF load having a voltage standing-wave ratio (VSWR) of 1.5 is varied through 360° of phase. These magnetron performance characteristics may be derived from graphic representations such as the Rieke diagram, which presents load VSWR and phase angle in polar coordinates, with lines of constant frequency and power output superimposed. More recently, plots like the one shown in Figure 5-7 are often used to characterize tube performance.

During normal magnetron operation, electrons leave the cathode, transfer energy to the RF field, and are collected by the anode. An appreciable portion of the electrons, however, do not reach the anode but are returned to the cathode and produce additional cathode heating. If the cathode heater power is not reduced or other measures taken, this *back heating* can result in excessive cathode temperatures and damage to the tube. In practice, tube heater power is reduced as average output power is increased. A number of high-power magnetron tubes can be operated at high duty cycles with no heater input power once operation has been started. The reduction of heater power in accordance with increases

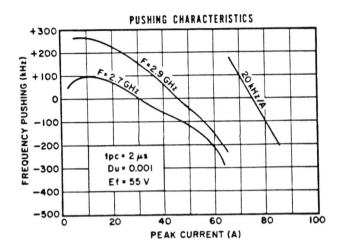

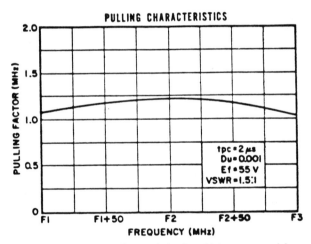

Figure 5-7. Pushing and pulling characteristics for a high-power coaxial magnetron.

in tube output power, the *heater schedule*, is normally part of the tube data provided by the magnetron tube manufacturer.

Frequency-Agile Magnetrons

Frequency agility is desirable for a number of radar applications, and a number of frequency-agile magnetrons have been developed. Frequency agility is achieved by changing the resonant frequency of the magnetron cavity; there are a number of means by which this may be accomplished.

One approach may be loosely called *electronic tuning*. Electronic tuning using an electron beam to produce a variable reactance, tuning by using auxiliary magnetron diodes in each of the magnetron cavities, and tuning of a coaxial magnetron by loading the cavity with PIN diodes have all been attempted but have the drawback of a quite limited tuning range.[5] On the other hand, multipactor-tuned magnetrons have a number of cavities that can be coupled to a given anode, and each cavity can be turned on or off independently, so that if there are n cavities, there are 2^n combinations or frequencies of oscillation that may be selected. While showing some promise, multipactor-tuned magnetrons remain laboratory devices at this time.[5] While it is possible to build a true voltage-tuned magnetron, satisfactory operation has been achieved only at power levels of a few hundred watts. Thus, electronic tuning has limited application in frequency-agile radars.

A second magnetron tuning technique involves placing piezoelectric material in the cavity and applying a voltage across these crystals to change the cavity dimensions. A third magnetron tuning technique may be classed as a reciprocating technique and is usually implemented by incorporating a ring which moves along the axis of the cavity near the end of the resonant structure. Finally, dielectric structures can be inserted in the cavity and rotated to tune the magnetron. Most of the readily available devices that have found widespread system usage are of either the reciprocating or the rotary type.

The reciprocation-tuned magnetron has shortcomings, including limited tuning rates, tuning nonlinearities, and susceptibility to vibration and fatigue. A rotating type of structure was developed to circumvent some of these difficulties. One early technique was based on the rotation of a notched tuning element in the back wall of the magnetron cavity. The passage of slots across the cavity openings resulted in changes in magnetron frequency. This rotary tuning method has a wide frequency excursion and a rapid tuning rate, but it is not readily adaptable to coaxial structures and does not readily yield an accurate measure of magnetron frequency.

CW Magnetrons

CW magnetrons have applications in Doppler and electronic countermeasure systems, and tubes producing from 100 to 500 W at frequencies as high as the X-band have been produced for these purposes. Additionally, CW magnetrons have been produced for nonradar applications such as microwave ovens. These tubes normally operate in the 2450-MHz industrial heating band and produce outputs of 1 to 10 kW at this frequency. The design of these tubes differs somewhat from that of other CW magnetrons,[4] primarily because of the need for economical production, simple power supplies, and the ability to operate into mismatched loads.

Typical Magnetron Characteristics

The conventional magnetron can operate at frequencies through X- or K_u-bands. Above these frequencies, rising-sun or inverted coaxial magnetrons are often used. The voltage–current relationships of all these magnetrons are quite similar, except that the rising-sun magnetron may sometimes exhibit a low-voltage mode, making a rapidly rising voltage pulse quite desirable for such tubes. Table 5-1 is a representative tabulation of some commercially available magnetrons.

5.2.3 Crossed-Field Amplifiers[3, 4, 6, 7]

General CFA Characteristics

The crossed-field amplifier (CFA) is characterized by reasonable operating voltages, good efficiency, wide bandwidth, and long life, but low gain. The CFA is often utilized in Doppler systems, phased-array radars, frequency-agile radars, and pulse-compression systems. It derives its name from the fact that, as in magnetrons, the RF-DC interaction region is a region of crossed electric and magnetic fields.

There are two general types of CFAs: injected-beam and distributed-emission. The latter is of more interest due to its high power capability. The electrons are injected into the interaction region by an electron gun in the injected-beam CFA and are emitted by the cathode or *sole* in the distributed-emission

Table 5-1. Some Commercially Available High-Power Pulsed Magnetrons.

Tube Type	Center Frequency (GHz)	Peak P_o	Maximum Duty Cycle	Peak Voltage (kV)	Peak Current (A)
M 545	1.290	25.0 MW	0.0025	52	260
3M901	2.765	4.7 MW	0.001	75	135
SFD344	5.60	1.0 MW	0.001	37.5	65
VF 11	9.250	1.0 MW	0.0015	30	70
7208 B	16.5	100 kW	0.001	22	20
VF 20	16.5	400 kW	0.0015	26	40
SFD326	24	120 kW	0.0005	14	30
SFD327	34.86	150 kW	0.0005	23	22
BL235	52.5	10 kW	0.0012	14.5	9.5
DX221	69.75	10 kW	0.00055	—	—
M 5613	95.5	2 kW	0.0002	10	9
DX252	120	2.5 kW	0.0002	10	11

(By permission, from Ewell, ref. 1; © 1981 McGraw-Hill Book Co.)

CFA. Due to the limited application of injection beam CFAs, the following discussion addresses only distributed-emission CFAs. CFAs may be operated either CW or pulsed, depending upon the particular design.

The CFA may be constructed in either a circular or linear format, as shown in Figure 5-8. In the circular format, the electron stream from a distributed-emission CFA can also be allowed to reenter, or recirculate, through the interaction region and enhance efficiency. In the reentrant configuration, the feedback electrons may be bunched to increase RF feedback or debunched to decrease RF feedback. RF feedback, like any feedback, opens up the possibility of oscillation as electron bunching occurs; therefore, the electron stream is sometimes debunched to eliminate the possibility of oscillation while retaining

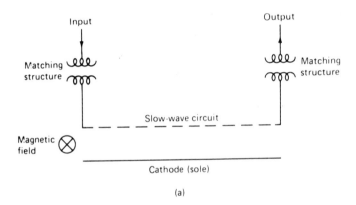

(a)

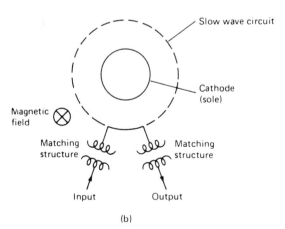

(b)

Figure 5-8. Simplified schematic representation of principal portions of (a) linear-format and (b) circular-format CFAs. (By permission, from Ewell, ref. 1; © 1981 McGraw-Hill Book Co.)

the enhanced efficiency of the reentrant CFA configuration. The linear-format CFA tubes are, of course, of a nonreentrant type.

The interaction in the CFA may be with either the forward or the backward wave. Each of these configurations has specific properties. The backward-wave CFA derives its name from the characteristics of the slow-wave structures used in such tubes. The terminal characteristics of backward-wave and forward-wave CFAs are substantially different.

Recent developments in CFA design involve injection of the input signal into the cathode region to obtain substantially increased gain. Gains on the order of 30 to 40 dB have been demonstrated by using this approach, but devices with this design are still in the development stage at this time.[8]

CFA Operating Characteristics

Because the operating frequency of the CFA is controlled by the input RF drive, beam impedance and hence terminal characteristics change with frequency. The applied voltage and current measured at the tube will then vary with frequency as functions of the modulator and tube impedance. The terminal characteristics of a backward-wave CFA shown in Figure 5-9 indicate the sensitivity to frequency and modulator characteristics. The backward-wave CFA should be operated with a modulator that approximates a constant-current source to minimize

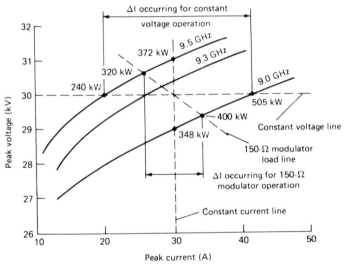

Figure 5-9. Changes in voltage, current, and power output for a backward-wave CFA as frequency is varied. Interaction is plotted for a constant-current modulator, for a constant-voltage modulator, and for a modulator with an internal impedance of 150 Ω. Note the reduction in power variation achieved with the 150-Ω and constant-current modulators. (From ref. 6)

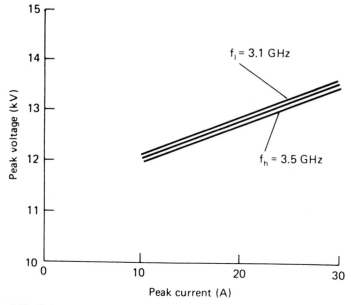

Figure 5-10. Cathode voltage-current characteristics of a forward-wave CFA. (From ref. 6)

changes in output power as the frequency is changed. Figure 5-10 shows the terminal characteristics for a forward-wave CFA; operation from a constant-voltage source is desirable for such a device.

There is considerable similarity between the terminal properties of a CFA and those of a magnetron. In particular, there is a very definite biased-diode characteristic associated with a CFA. Note that such operation is normally achieved only when adequate levels of RF drive are applied at the input to the CFA. The application of voltage with inadequate drive may result in tube operation in undesirable modes and substantial variation in the terminal voltage and current characteristics.

One important property of the CFA is its ability to operate from a DC power supply. In this mode of operation, forward-wave amplifiers having large constant-voltage bandwidths and completely cold secondary emitting cathodes are utilized. A constant DC voltage is applied to the tube, which remains essentially an open circuit until RF energy entering the tube causes the secondary emission process to begin. The amplification process continues until the RF drive pulse is removed. At the termination of the RF drive pulse, the accumulated space charge can support noise-like oscillations in reentrant tubes. An additional control electrode is included to remove this space charge and terminate the RF output.[9] The accumulated space charge gives no trouble to reentrant CFAs; these tubes can be made entirely self-pulsing.

Typical CFA Characteristics

Several types of commercially available CFAs are listed in Table 5-2. Some general comments about CFAs are in order. The reentrant types are more efficient than the nonreentrant types, the gain of the linear tubes can be increased by increasing device length, and CFAs with no RF feedback are less prone to spurious oscillation than those with RF feedback. Several other parameters, such as phase and amplitude stability, are important in comparing CFAs to other tubes for similar applications.[9, 10] The choice of CFA configuration also affects the modulator design; constant-voltage pulsers are desired for forward-wave amplifiers, and constant-current pulsers are desired for backward-wave devices.

The insertion phase of a CFA is a function of drive level and changes in cathode voltage and current. Typical values for CFAs are about 2° of change per decibel change of drive level and from 1 to 5° per percent change in cathode voltage.

CFA output noise power spectral densities as low as -115 dBm/MHz have been measured, with typical values of approximately -106 dBm/MHz. A number of intrapulse noise measurements of CFAs indicate that representative values for both amplitude modulation (AM) and phase modulation (PM) noise range from -60 to -30 dBm/MHz, where values have been referenced to an equivalent CW value to remove the effects of the duty cycle on measured data.

In summary, the CFA is a coherent amplifier characterized by relatively broad bandwidth and high efficiency but low gain. It possesses a high degree of phase stability and may be largely self-pulsing in some applications. Advantages which are particularly important in mobile systems are its relatively small size and weight and its low operating voltages. The tube provides excellent reproducibility, and operating times in excess of 5000 hr have been achieved in numerous systems. Thousands of CFAs have been delivered and operated in the field; most of these were high-power, cathode-pulsed tubes installed in radar systems providing peak power outputs between 100 kW and 5 MW and average power outputs in excess of 1 kW.[7]

Table 5-2. Some Commercially Available High-Power Pulsed CFAs.

Tube Type	Center Frequency (GHz)	Peak P_0 (MW)	Frequency Range (GHz)	Maximum Duty Cycle	Peak Voltage (kV)	Peak Current (A)	Gain (dB)
1AM10	1.288	1.8	1.225–1.350	0.02	46	50	9.2
QKS1452	2.998	3.0	2.994–3.002	0.0015	47	100	
SFD222	5.65	1.0	5.4–5.9	0.001	35	60	18
SFD237	5.65	1.0	5.4–5.96	0.01	35	60	13
QKS506	9.05	1.0	8.7–9.4	—	40	45	7
SFD236	16.5	0.1	16–17	0.001	14	23	17

(By permission, from Ewell, ref. 1; © 1981 McGraw-Hill Book Co.)

5.2.4 Klystrons[11]

General Characteristics

High-power pulsed klystron amplifiers are characterized by high gain, high peak power, and good efficiency, but they have relatively narrow bandwidths and require relatively high voltages. The large physical size of most klystrons and their associated modulators and X-ray shielding have limited the application of klystrons in many radar systems. High-power klystrons are also used in high-energy linear accelerators; the klystrons originally developed for the Stanford Linear Accelerator were the first truly high-power microwave amplifiers and formed the basis for a number of large, ground-based, coherent radar systems.

The schematic representation of a high-power cavity klystron shown in Figure 5-11 illustrates the separation of the principal parts of the device. This separation permits each area to be optimized individually and contributes to the relatively high level of performance that has been achieved with klystron amplifiers.

There is no interaction between the various portions of the RF circuitry within the klystron amplifier, except that which occurs as the electron beam propagates from the cathode to the collector. The injection of electrons into the interaction

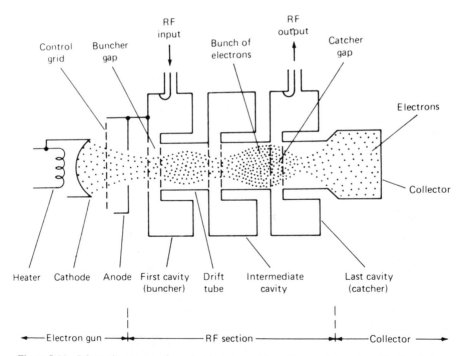

Figure 5-11. Schematic cutaway view of a klystron amplifier. (By permission, from Ewell, ref. 1; © 1981 McGraw-Hill Book Co.)

structure and the focusing of the electron beam as it passes through the cavities are of central importance in klystron operation.

The electron beam is generated by an electron gun of a particularly simple form for a tube that is cathode pulsed. For many applications, however, the inclusion of an appropriate control electrode in the electron gun permits considerable simplification of the modulator. The three principal types of electron-gun control electrodes are the modulating anode, the control-focus electrode, and the shadow grid. A primary figure of merit for these various control electrodes is the cutoff amplification factor μ_c, defined as the negative ratio of control-electrode voltage to anode voltage at the point where the emission is safely cut off. Another figure of merit is the total amplification factor μ of the gun, which is the ratio of the control-electrode voltage swing to the beam voltage. It is desirable that each of these amplification factors be high, yet it is necessary to maintain a well-focused beam to reduce heating from electrons impinging on the tube body. The intercepting and shadow-gridded types have the largest values of μ_c and μ.

The shadow-gridded tubes are unsatisfactory for extremely high-power applications because of electron beam spreading and the resultant heating of the tube body. Modulating-anode structures are normally selected for such applications, in spite of the extremely large voltage swings required for their operation.

Focusing of the electrons leaving the gun area into a narrow beam is particularly critical. Since the beam length-to-diameter ratio is on the order of 100 to 1, small variations in electron spreading can cause appreciable electron interception and tube heating. High-power CW applications often require near-perfect beam transmission, sometimes on the order of 99%. Beam focusing for lower-duty-cycle, pulsed applications is less critical; 85 to 95% beam transmission values are more representative for these applications.

The klystron uses a magnetic field to focus the electron beam through the interaction region. The three basic methods of producing this focusing magnetic field are a solenoid, a permanent magnet, and a periodic permanent magnet. Because solenoid focusing provides the greatest degree of control over the beam, it is used for most high-power tubes. Although the solenoid produces an extremely uniform field, its size and the bulk of the associated power supplies may be objectionable. Size and weight also limit the use of permanent magnet focusing to tubes having short interaction regions. Periodic-permanent magnet (PPM) focusing and electrostatic focusing have been applied to klystrons to reduce size and weight, but they have not yet found wide acceptance.

Klystron Operating Characteristics

The klystron appears to follow the familiar space-charge-limited diode laws; that is, $I = kV^{3/2}$ (where k is the perveance, a constant for a particular tube), and the power is given by $kV^{5/3}$. For cathode-pulsed tubes, the load may often be approximated as a linear resistance paralleled by the tube cathode capacitance. More detailed consideration must involve the tube nonlinearities, but the differences thus produced are normally small.

The amplitude modulation sensitivity due to a change in beam voltage may be readily derived by differentiating the expression for power output of the klystron

$$\frac{dp}{dV} = \frac{5}{2}\frac{P}{V} \qquad (5\text{-}1)$$

and for phase modulation

$$\frac{d\theta}{dV} = \frac{\theta_0}{2V_0} \qquad (5\text{-}2)$$

In Eq. (5-2), θ_0 is the phase length of the klystron and is typically on the order of 6 to 10 radians for each of the drift spaces between the klystron cavities. If AC voltage is used on the heaters of the klystron, the typical result is an AM sideband approximately 50 dB below the carrier; operation with DC heaters alleviates this particular difficulty.

Typical Klystron Characteristics

The characteristics of several high-power klystrons are summarized in Table 5-3. The klystron is typically a narrow-band amplifier, but the bandwidth may be increased at the expense of tube gain by stagger tuning of the cavities.

Klystrons are typically rather low-noise amplifiers. A representative background-noise level for PM noise is 125 dB/kHz below the carrier, and the AM noise is typically 10 dB below that level. Ion oscillation in CW or long-pulse tubes is a potential source of noise that can produce spurious sidebands as great as 50 dB below the carrier; this can be prevented by proper care during fabrication and evacuation of the tube. Other spurious responses can result from power-supply ripple voltages or ringing on the various electrodes of the klystron.

5.2.5   Traveling-Wave Tubes[11–13]

General TWT Characteristics

The high-power TWT is a microwave amplifier characterized by high gain, large bandwidth, relatively low efficiency, and high operating voltages. In addition to the RF input and output connections, most TWTs have a body electrode, a collector, a cathode or electron gun, an interaction region, and some means of focusing the electrons into a linear beam.

Table 5-3. Some Commercially Available High-Power Klystrons.

Tube Type	Center Frequency	Peak P_0 (MW)	Frequency Range	Maximum Duty Cycle	Peak Voltage (kV)	Current (A)	Gain (dB)
VA-1513	0.4–0.45 GHz	20	15 MHz BW	Approx. 0.015	230	280	40
TV2023	1.3 GHz	40	1.2–1.4 GHz	0.001	300	230	53
F2049	2.856 GHz	30	2.856 GHz	0.008	290	295	50
VKS-8262	2.9 MHz	5.5	5 MHz BW	0.001	125	88	50
SAC42	5.65 GHz	3.3	5.4–5.9 GHz	0.002	135	112	23.5
SAX191	9015	1.25	8.83–9.2 GHz	0.0048	85	50	50

(By permission, from Ewell, ref. 1; © 1981 McGraw-Hill Book Co.)

A cutaway schematic representation of the TWT is given in Figure 5-12. The heart of the TWT is the delay-line structure, which is shown as a helix; helices are, in fact, used extensively at lower power levels. Simple helices are not often used at higher power levels due to their limited power-handling capability; instead, either the ring-and-bar or ring-loop configuration, the two-tape contrawound helix, or the coupled-cavity TWT shown in Figure 5-13 may be used.[14] Approximately 90% of all high-power TWTs employ a coupled cavity structure because of its excellent electrical characteristics (its impedance, bandwidth, and mode structure), its mechanical simplicity, its shape (which is well suited for PPM focusing), its ruggedness, and the versatility of its scaling with frequency, power, and bandwidth.

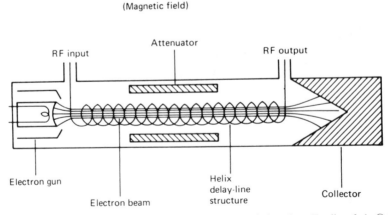

Figure 5-12. Longitudinal-section view of a TWT. (By permission, from Ewell, ref. 1; © 1981 McGraw-Hill Book Co.)

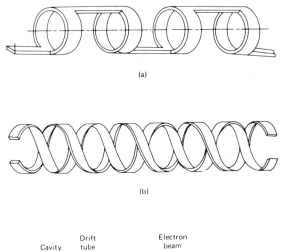

(a)

(b)

Cavity Drift tube Electron beam

Coupling hole θ

(c)

Figure 5-13. TWT interaction circuits: (a) ring bar, (b) contra-wound-helix, and (c) coupled-cavity approaches. (By permission, from Mendel, ref. 13; © 1973 IEEE)

Electron beam focusing is critical to proper TWT performance, as it is with the klystron. The two methods most commonly used in high-power tubes are PPM focusing and solenoid focusing. In solenoid focusing, the coils are typically foil wound as an integral part of the tube to hold mechanically induced variations in the magnetic field to considerably less than 1%. The PPM approach does not provide the degree of focusing that a solenoid-focused TWT achieves, but its smaller size and weight and the absence of additional power supplies and power consumption make it attractive for many applications.

Relatively low efficiency is a principal drawback of the TWT. There are two primary ways that this problem has been attacked. The first approach involves depressing the collector, that is, operating with the collector at a potential below that of the body. Multiple depressed collectors (as many as eight) may be used to improve efficiency at the expense of circuit complexity. The second approach is *velocity resynchronization*, which is achieved by taking the last portion of the periodic structure and increasing its potential by a relatively small amount. One of the significant problems with velocity resynchronization involves insulating that portion of the structure without introducing additional RF reflections and mismatches. Increases in efficiency levels of 50 to 60% have been achieved by depressing the collector and using resynchronization.

TWT Operating Characteristics

The TWT is a space-charge-limited diode, as is the klystron, and shows variations in power output and in phase with variations in beam voltage similar to those shown by the klystron. Sensitivities of a representative TWT to changes in various electrode voltages are summarized in Table 5-4. Note that the insertion phase of a TWT is somewhat more sensitive to variations in supply voltage than the insertion phase of a klystron because of the longer effective electrical length of the TWT. The noise figure and noise power spectral density near the carrier frequency of the TWT are not dissimilar to those of the klystron.

Since TWTs are often operated in saturation, the transfer function of the device is nonlinear, as shown in Figure 5-14. Such nonlinearity may adversely influence the operation of a pulse-compression system or introduce undesired intermodulation products.

Typical TWT Characteristics

A list of some commercially available high-power TWTs is given in Table 5-5. Most of these tubes are representative of the maximum power output readily available in each frequency band, except for the QKW1132A, which is included as a tube typical of one that might be used to drive a higher-power

Table 5-4. Typical Sensitivities for a 10-kW, 5% Bandwidth, 60-dB-Gain TWT.

Voltage Parameter	AM	FM
Cathode to body	(0.5 dB)/1%	30°/1%
Cathode to anode or grid[1]	(0.1 dB)/1%	5°/1%
	(0.15 dB)/1%	7°/1%
Cathode to collector	(0.02 dB)/1%	0.5°/1%
Heater (dynamic)	(0.00005 dB)/1%	0.001°/1%
Solenoid	(0.00001 dB)/1%	0.0005°/1%
Drive power		2.2°/(1 dB)

(By permission, from Ewell, ref. 1; © 1981 McGraw-Hill Book Co.)
[1]This assumes that either an anode or a grid is used to control beam current.

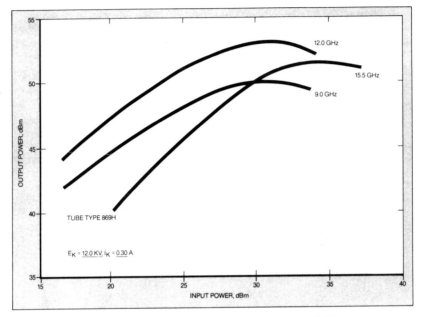

Figure 5-14. TWT transfer characteristic. (From ref. 14)

amplifier. A plot of the range peak and average powers for TWTs in field use is shown in Figure 5-15, but all of these designs are subject to improvement with additional developmental effort.

5.2.6 Twystron Amplifiers

The twystron is a device having a broader bandwidth than a klystron and higher efficiency than a TWT. It is a hybrid device consisting of a klystron driving section and a TWT output section to achieve increased bandwidth with a high level of efficiency. This combination achieves a large bandwidth, as shown schematically in Figure 5-16. Representative characteristics for some twystron amplifiers are given in Table 5-6. Its operation and sensitivities are similar to those for high-power TWTs and klystrons, but in general gain and phase are not as uniform as those that can be achieved for some coupled-cavity TWTs. Phase excursions of $\pm 25°$ from linearity over a 10% bandwidth are representative, but smaller values have been achieved.

5.2.7 Extended-Interaction Devices

Extended-interaction klystrons, which couple between the resonant cavities, have been developed to improve the power-handling capability of the klystron.

Table 5-5. Some Commercially Available High-Power TWTs.

Tube Type	Center Frequency (GHz)	Peak P_0		Frequency Range	Maximum Duty Cycle	Peak		Gain (dB)
						Voltage (kV)	Current (A)	
QKW1490	1.3	700	kW	1250–1350 MHz	0.006	42	14	47
VTS-5754A1	3.0	125	kW	2.9–3.1 GHz	0.007	45	35	20
TPOM4131	2.85	750	kW	2.7–3 GHz	0.02	53	21	50
560H	3.2	250	kW	3.1–3.3 GHz	0.002	130	94	35
VA146	5.65	4.5	MW	5.4–5.9 GHz				
752H	8.9	150	kW	8.4–9.4 GHz	0.01	50	15	16
8716H	9.1	120	kW	9.0–9.2 GHz	0.0025	43	13.5	50
QKW-1132A	9.25	1.5	kW	8.5–10 GHz	0.01	11	1.5	34–40
750H	9.5	25	kW	9.0–10.0 GHz	0.01	24	5.5	47
893H	16.25	100	kW	16.0–16.5 GHz	0.005	62	7.8	50

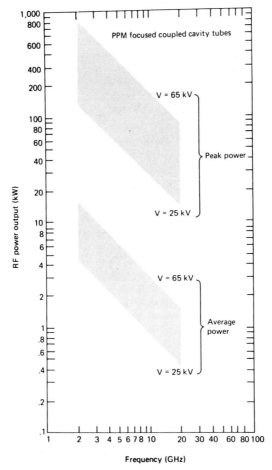

Figure 5-15. Ranges of peak and average powers for pulsed TWTs in field use. (By permission, from Mendel, ref. 13; © 1973 IEEE)

An extended-interaction klystron in one application achieved over 1 MW average power at X-band frequencies.

An increasing application of the klystron extended-interaction oscillator (EIO) is being found at millimeter wavelengths, where the EIO has proven to be one of the more reliable sources of RF energy in the several-kilowatt range at frequencies up to several hundred gigahertz. A cross-sectional schematic view of an EIO is shown in Figure 5-17. The separation of the cathode region from the interaction region permits low cathode-current densities even at millimeter wavelengths, thus providing a highly reliable tube.

Recent development efforts have resulted in an extended-interaction amplifier

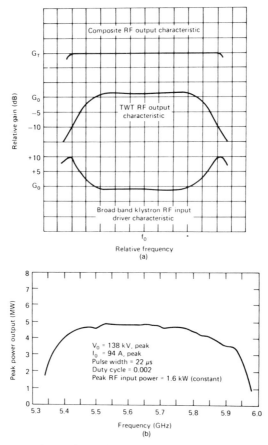

Figure 5-16. Frequency characteristics of a twystron amplifier showing (a) the principle of obtaining a more uniform frequency response by using different input and output circuits and (b) the frequency response actually obtained. (By permission, from Straprans et al., ref. 11; © 1973 IEEE)

(EIA) operating in the 95-GHz region. This EIA is mechanically tunable over a 1-GHz range and has an instantaneous 3-dB bandwidth of 200 MHz. The tube has achieved a 2.3-kW peak RF power with a 33-dB gain. The device uses a samarium-cobalt magnet and occupies a volume of less than 90 in.[3]. The tube is cathode pulsed and, like many cathode-pulsed linear amplifiers, may exhibit a period of self-oscillation during turn-on and turn-off; close attention to modulator rise and fall times may minimize the effects of these undesired oscillations.

Control-gridded millimeter-wave EIOs and EIAs employing so-called aperture grids for control of the beam simplify modulator requirements. Such control electrodes require 1- to 3-kV swings and have relatively low capacitance.

Table 5-6. Some Twystron Characteristics.

Center Frequency (GHz)	Peak Power (MW)	Average Power (kW)	Gain (dB)	1.5-dB Bandwidth (MHz)	Efficiency (%)	Duty Cycle	Pulse Width (μs)	Beam Voltage (kV)	Beam Current (A)
2.7–2.9	2.5	5.0	37	200	35	0.002	6	117	80
2.715–2.915	3.0	8.5	42	200	40	0.0025	10	126	82
2.75–2.95	5.0	12	40	200	40	0.0025	10	125	92
2.9–3.1	2.5	5	37	200	35	0.002	7	117	80
3.015–3.215	3.0	8.5	42	200	40	0.0025	10	126	82
5.4–5.9	2.5	20	36	500	33	0.004	20	135	93
5.4–5.9	3.2	10	36	500	33	0.002	12.5	135	95

Source: Reference 16.

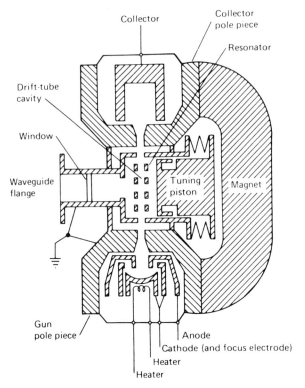

Figure 5-17. Cross-sectional view of a millimeter-wave EIO, showing the removal of the cathode from the interaction region. (Varian Associates)

Bonded grids being developed for these devices should reduce required grid voltage swings but will result in higher grid capacitance. No working tubes employing bonded grids have been delivered.

5.2.8 Gyrotrons[17]

The gyrotron, a new class of microwave tube, has been developed in recent years. This device is based on the interaction between an electron beam and microwave fields due to the cyclotron resonance condition. Figure 5-18 shows power levels achieved for some long-pulse and CW gyrotron sources.

Unlike linear beam tubes, the electron motion in a gyrotron is orbital, and much of the energy is transverse to the longitudinal axis. This permits interaction to take place in a large microwave circuit and explains the high powers achievable.

Gyrotron devices may be made to function as amplifiers, and some typical

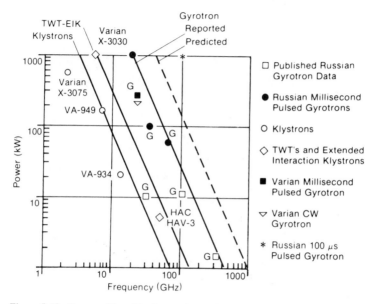

Figure 5-18. Power achieved by long-pulse or CW power sources. (From ref. 17)

configurations are summarized in Table 5-7. The amplifier configuration being developed most extensively is the gyro-TWT.[17] Since the gyro-TWT has a relatively large overall length, its applicability is limited. Further developments include 50-kW tubes operating in the 95-GHz region, with operating voltages in the 50- to 200-kV region.

5.2.9 Solid-State Transmitters

A number of solid-state transmitter sources characterized by high reliability and low cost have recently been developed. Since solid-state devices are capable of only relatively low peak powers, their usefulness is limited to applications where they may be combined or where only low levels of transmitted power are required. Typical of the capabilities of transistor amplifiers are single-package units capable of generating 600 W at a frequency of approximately 1 GHz.[18] The output of a number of such devices can be combined, as for example, in the case of the PAVE PAWS radar system, which uses 3584 individual transmitter modules in a phased-array configuration; each module is capable of generating a peak power of 440 W over a frequency range of 420 to 450 MHz.[19] Another example of combining is a 1250- to 1350-MHz system that uses 40 modules having a peak-power capability of 650 W each to provide a 25-kW peak in a single-output waveguide.[20]

Table 5-7. Reported Gyrotron Operating Conditions and Output Parameters.

Model No.	Mode of Oscillation	Wavelength (mm)	CW or Pulsed	Harmonic Number	β-Field (kG)	Beam Volts (kV)	Beam Amps	Output Power (kW)	Measured Efficiency (%)	Theoretical Efficiency (%)
1	TE_{021}	2.78	CW	1	40.5	27	1.4	12	31	36
2	TE_{031}	1.91	CW	2	28.9	18	1.4	2.4	9.5	15
2	TE_2	1.95	Pulsed	2	28.5	26	1.8	7	15	20
3	TE_{231}	0.92	CW	2	60.6	27	0.9	1.5	6.2	5

(By permission, from Ewell, ref. 1; © 1981 McGraw-Hill Book Co.)

At higher frequencies, gallium arsenide field effect transistor (GASFET) amplifiers are often used to obtain typical capabilities of 5 W peak over a 4- to 8-GHz band,[21] 15 W at 6 GHz,[22] and 10 W at 8 GHz.[23] As with other transistor amplifiers, power combining may be used to increase the total effective power. Successful power-combining techniques have been used at frequencies through K_u-band.[24-26]

For higher-frequency operation, IMPATT diodes may be used as amplifiers or oscillators. The outputs of several amplifiers may be combined to increase the total output power.[27] Care must be employed in using IMPATT diodes, since they do appear to be rather noisy and have a high degree of sensitivity to both current and voltage changes.[28-31] If extreme stability is required, injection locking of the device may considerably improve the spectrum. At the present time, IMPATTs can provide 5 to 10 W at 95 GHz,[32] and laboratory development models have provided 3 W at 140 GHz.[33] IMPATTs at 225 GHz are under development.[34]

5.2.10 Other RF Power Sources

Other types of RF power sources are sometimes encountered in radar systems. Conventional vacuum tube transmitters are sometimes used at lower frequencies, but are rarely used at frequencies above 3 GHz. A number of devices that show promise but have never been fully developed include the Ubitron, beam plasma amplifiers, and relativistic beam devices.

The desire for long-range, high-resolution radar systems has stimulated interest in the direct generation of extremely short, very-high-power pulses. Gigawatt and terawatt peak power pulses become possible for pulse lengths less than approximately 10 ns. Techniques for generating such pulses vary from Hertzian generators (spark gaps) to relativistic beams, and a number of researchers are actively engaged in this field.[35]

5.2.11 Comparison of RF Power Oscillators and Amplifiers

As is now evident, there is a considerable variety of RF power oscillators and amplifiers to choose from when designing a radar transmitter. For a pulsed, high-power, noncoherent source, the magnetron is almost always the suitable choice; also, if a pulsed, noncoherent, frequency-agile source with capabilities within the tuning rates and frequency ranges of magnetrons is desired, either a reciprocation- or a rotary-tuned magnetron can be chosen. For a narrow-bandwidth, coherent system, the klystron might well be the preferred approach. But if a wider-band coherent system is desired, the TWT, the twystron, and the CFA are all suitable candidates. At higher frequencies, an EIO or a gyrotron might also be a possibility.

A comparison of the efficiencies available from a number of different high-

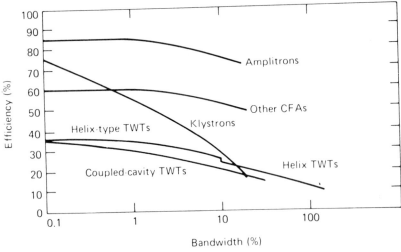

Figure 5-19. Efficiency versus bandwidth for various tubes, illustrating the trade-off of efficiency for bandwidth. (By permission, from Straprans, ref. 36; © 1976 IEEE)

power sources is presented in Figure 5-19. Bandwidth capabilities of microwave tubes in the S- to X-band ranges as a function of power level are shown in Figure 5-20, and it is particularly interesting to note the increase in bandwidth at higher peak powers for klystrons and twystrons. Table 5-8 contrasts the characteristics of a number of available sources.[1]

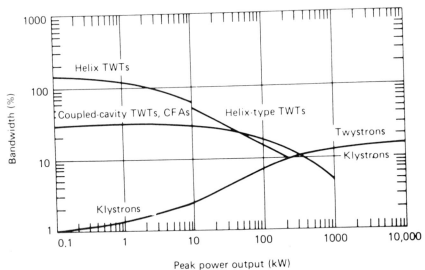

Figure 5-20. Bandwidth versus power output for various tubes. Note particularly the increase in bandwidth for twystrons and klystrons at the higher power levels. (By permission, from Straprans, ref. 36; © 1976 IEEE)

Table 5-8. Radar Transmitter Tube Comparisons.

Tube Type	P_0	Efficiency	Instantaneous BW	Frequency Range	Gain
Magnetron	High	High	N/A	VHF-> 100 GHz	
CFA	High	High	Large	VHF-K_μ band	Low
TWT	Moderate	Low	Large	VHF and up	High
Klystron (magnetic focus)	High	High	Small	VHF-K_μ band	High
Klystron (electrostatic focus)	Moderate	—	Large	L-X band	High
Twystron	High	Moderate	Intermediate	L-C band	
EIO	Moderate	Low	—	mm wave	
Vacuum tubes	Moderate	—	—	VHF	

(By permission, from Ewell, ref. 1; © 1981 McGraw-Hill Book Co.)

Selection of the RF power source is only a part of the transmitter specification process, since suitable voltages and currents must be applied to the device terminals to permit it to operate satisfactorily. The radar transmitter device that provides these required characteristics is called the *pulser* or *modulator*, and is discussed next.

5.3 MODULATORS[1]

5.3.1 General Modulator Characteristics

The modulator, or pulser, provides a voltage or current waveform that permits the selected microwave power source to operate in a proper manner.[37,38] Important considerations in the selection and design of a modulator for a particular transmitter tube include:

- Pulse length
- Pulse repetition frequency
- Operating voltage and current
- Tube protection (from arcs)
- Spurious modes
- Pulse flatness (amplitude and phase)
- Cost
- Size and weight
- Efficiency
- Reliability and maintainability

There are several basic types of modulators and numerous offshoots of each type. All pulse modulators have one characteristic in common: they contain some means for storing energy in either an electric or a magnetic field and a switch to control the discharge of energy into the load. The energy in the energy storage element must be periodically replenished from the power supply, and an isolating element to prevent interaction is often included.

The two principal types of pulse modulators are the line type and the hard tube. In the line-type modulator, all of the stored energy is dissipated in the load during each pulse. In the hard-tube modulator, only a fraction of the stored energy is dissipated during each pulse. The following discussion emphasizes high-power applications, but the same general techniques are also applicable to lower-power, solid-state devices.[39-41]

5.3.2 Hard-Tube Modulators

The hard-tube modulator is essentially a class C amplifier. A simplified diagram of a hard-tube modulator is shown in Figure 5-21. A positive pulse on the grid turns on the tube, and the capacitor couples the resulting step in plate voltage to the load. The hard-tube modulator is usually larger and more complex than other types of modulators, but it is also more versatile because it is not strongly influenced by load characteristics and its output pulse length may be easily changed.

Several variations of the basic hard-tube modulator include the transformer-coupled hard-tube modulator, the capacitor- and transformer-coupled hard-tube modulator, and the series-discharge, parallel-charge, hard-tube modulator with capacitive coupling. The conventional capacitively coupled hard-tube modulator is by far the most common configuration.[42] A number of different loads may be attached to such a modulator, including a resistive load (which is usually a good approximation of a klystron or TWT), a biased-diode load such as a mag-

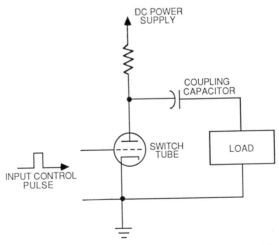

Figure 5-21. Simplified schematic diagram of a hard-tube modulator.

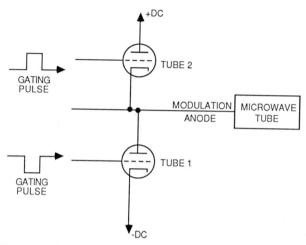

Figure 5-22. Simplified representation of a floating-deck modulator. (By permission, from Ewell, ref. 1; © 1981 McGraw-Hill Book Co.)

netron with resistive charging, or a biased-diode load with an inductive recharging path.

A special form of hard-tube modulator is sometimes used with tubes such as TWTs and klystrons that contain a modulating anode. This is the floating-deck modulator and is illustrated schematically in Figure 5-22. When the microwave tube is off, tube 1 is turned on and tube 2 is nonconducting. The microwave tube is turned on by turning tube 1 off and tube 2 on. At the end of the pulse, tube 2 is turned off and tube 1 is turned on. The gating pulses to each tube may be capacitively coupled, transformer coupled, optically coupled, or coupled by means of RF energy. Floating-deck modulators are typically used with large klystrons, and modulating swings of 1 to 200 kV are not uncommon for this application. Also, floating-deck modulators are sometimes used with cathode-pulsed tubes directly when a very wide range of pulse widths with rapid rise and fall times must be accommodated.

5.3.3 Line-Type Modulators

The line-type modulator derives its name from the similarity of the behavior of its energy storage element to that of an open-circuited transmission line.[37,43] If a length of transmission line having one-way propagation time $\tau/2$ is connected as in Figure 5-23, charged to voltage V, and then discharged through its characteristic impedance Z_0, a pulse of length τ and amplitude $V/2$ will be generated across the load. For practical pulse lengths and voltages, the required bulk of cable often becomes excessive; in practice, a network of lumped inductors and capacitors is often used. Such a network is shown in Figure 5-24 and is called

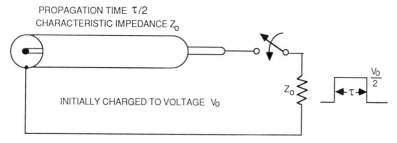

PROPAGATION TIME $\tau/2$
CHARACTERISTIC IMPEDANCE Z_0

INITIALLY CHARGED TO VOLTAGE V_0

Figure 5-23. Generation of a rectangular pulse by discharge of a charged, open-circuited transmission line into its characteristic impedance. (By permission, from Ewell, ref. 1; © 1981 McGraw-Hill Book Co.)

a *pulse-forming network (PFN)*. The PFN resembles the lumped-circuit approximation of the actual transmission line, but there are some significant differences. The PFN shown in Figure 5-24 is sometimes called the *Guillemin E-type network*. It consists of equal-valued capacitors and a continuously wound, tapped coil for which physical dimensions are chosen so as to provide the proper mutual coupling at each mesh. The number of sections is chosen to provide the desired rise time.

The load for a line-type modulator often requires extremely high voltages. The voltages on the PFN are often reduced by inserting a step-up transformer

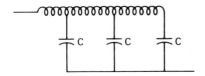

Important relationships are:

$$Z_o = \sqrt{\frac{L_n}{C_n}} \qquad L_n = \frac{\tau Z_o}{2}$$

$$\tau = \sqrt{L_n C_n} \qquad \frac{l}{d} = \frac{4}{3}$$

$$C_n = \frac{\tau}{2 Z_o} \qquad \frac{L_c}{L} = 1.1 \text{ to } 1.2$$

where C_n = total network capacitance
L_n = total network inductance
τ = pulse width at 50% points
Z_o = characteristic impedance
n = number of sections

L = inductance per section L_n/n
L_c = inductance of section on closed end
C = capacitance per section C_n/n
l = length of coil in one section
d = coil diameter

Figure 5-24. Guillemin E-type pulse-forming network circuit arrangement and key design parameters. (By permission, from Ewell, ref. 1; © 1981 McGraw-Hill Book Co.)

between the PFN and the load. Bifilar secondary windings on the step-up transformer often provide a convenient means to provide heater current for the transmitting device.

A device must be used to isolate the switch from the power source. A resistor could be used, but it would limit the maximum efficiency to 50%. An inductor is often used because of the increased efficiency obtained, and it is then possible to charge the PFN to approximately twice the DC supply voltage. A charging diode will prevent discharge of the PFN once it is charged. A more typical line-type pulser includes a shunt diode and resistor to damp out any reflected voltage due to load mismatch, as shown in Figure 5-25.

The choice of pulse width and impedance determines the properties of the PFN. The pulse width is usually set by system requirements, but the impedance of the PFN can be varied over a fairly wide range by the choice of the output–transformer turns ratio. A choice of too high an impedance results in high voltages on the PFN. An extremely low impedance level results in high currents, high switch tube drops, and losses in the pulse transformer primary that make it difficult to achieve efficient operation. When hydrogen thyratron switches are used, PFN impedances of 25 and 50 ohms have proved to be a reasonable compromise between high currents and high voltages. A variety of hydrogen thyratrons have been designed so that their power-handling capability is maximized at impedances of 25 and 50 ohms. If silicon-controlled rectifiers (SCRs) are being used as the switch for the line-type modulator, a lower-impedance PFN may be necessary because of the limited peak forward voltage-handling capability of most SCRs.

5.3.4 Solid-State Modulators

While there are many possible ways in which solid-state devices can be used in modulators, the most common has been to employ the SCR as a closing switch

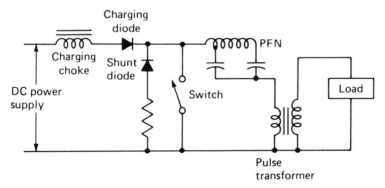

Figure 5-25. Simplified schematic representation of a typical line-type modulator. PFN = pulse-forming network. (By permission, from Ewell, ref. 1; © 1981 McGraw-Hill Book Co.)

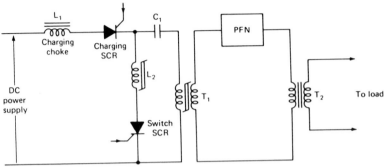

Figure 5-26. Magnetic-SCR line-type modulator using an SCR switch and saturating magnetic devices. Bias windings have been omitted for clarity, but L_2, T_1, and T_2 are all magnetically biased. T_2 need not be biased, but if it is not, a resistor-diode combination must be connected across the primary to provide a low-impedance path for charging the PFN. (By permission, from Ewell, ref. 1; © 1981 McGraw-Hill Book Co.)

to initiate discharge of an energy storage element. Probably the simplest implementation is to replace the thyratron in a line-type modulator with an SCR; unfortunately, the low voltage rating and limited rate of current rise of most SCRs have limited this approach. This difficulty can be circumvented by inserting a saturating magnetic core in series with the SCR to delay full current

Table 5-9. Stability Factors.

Type	FM or PM Sensitivity	Ratio of Dynamic to Static Impedance	Current or Voltage Change to 1% Change in HVPS Voltage	
			Line-Type Modulator	Hard-Tube Modulator
Magnetron	$\dfrac{\Delta F}{F} = 0.001 + 0.003\,\dfrac{\Delta I}{I}$	0.05–0.1	$\Delta I = 2\%$	$\Delta I = 10\text{–}20\%$
Magnetron (stabilized)	$\dfrac{\Delta F}{F} = 0.0002 + 0.0005\,\dfrac{\Delta I}{I}$	0.05–0.1	$\Delta I = 2\%$	$\Delta I = 10\text{–}20\%$
Backward-wave CFA	$\Delta\phi = 0.4\text{–}1°$ for 1% $\Delta I/I$	0.05–0.1	$\Delta I = 2\%$	$\Delta I = 5\text{–}10\%$
Forward-wave CFA	$\Delta\phi = 1\text{–}3°$ for 1% $\Delta I/I$	0.1 –0.2	$\Delta I = 2\%$	$\Delta I = 5\text{–}10\%$
Klystron	$\dfrac{\Delta\phi}{\phi} = \dfrac{1}{2}\dfrac{\Delta E}{E}\quad \phi \approx 5\lambda$	0.67	$\Delta E = 0.8\%$	$\Delta E = 1\%$
TWT	$\dfrac{\Delta\phi}{\phi} \approx \dfrac{1}{3}\dfrac{\Delta E}{E}\quad \phi \approx 15\lambda$ $\Delta\phi = 20°$ for 1% $\dfrac{\Delta E}{E}$	0.67	$\Delta E = 0.8\%$	$\Delta E = 1\%$
Triode or tetrode	$\Delta\phi = 0\text{–}0.5°$ for 1% $\dfrac{\Delta I}{I}$	1.0	$\Delta I = 1\%$	$\Delta I = 1\%$

(By permission, from Ewell, ref. 1; © 1981 McGraw-Hill Book Co.)

Table 5-10. Modulator Comparison.

| Type | Switch | Flexibility | | Pulse Length | | Flatness | Working Voltages | Application | Size and Cost |
		Duty	Pulse Width	Long	Short				
Line type	Thyratron	Limited by charging circuit	No	Large PFN	Good	Ripples	High	Most common	Small size
	SCR						Medium		Smallest size higher cost
Hard tube	Capacitor-coupled	Limited	Yes	Large capacitor	Good	Good	High	Fairly common	Large
	Transformer-coupled	Limited	Yes	Capacitor and transformer gets large	Good	Fair	High	Not often used	Large
	Modulation-anode (floating-deck)	No limit	Yes	Good	OK	Excellent	High	Usually high-power	Quite large
	Grid	No limit	Yes	Good	Good	Excellent	Low	Widely used at low power	Small and inexpensive
SCR-magnetic	SCR and magnetic cores	Limited by charging circuit and magnetic cores	Normally fixed	Large PFN	Losses may be high	Ripples	Low in initial stages	Becoming more common	Small; initial design costs high

(By permission, from Ewell, ref. 1; © 1981 McGraw-Hill Book Co.)

flow until the SCR has fully turned on. This approach is quite limited due to high losses, low voltage operation, and pulse length limitations.

The most satisfactory high-power solid-state modulator for many applications is the so-called SCR-magnetic modulator. This device uses a saturating magnetic core as a switch to provide improved reliability over the hydrogen thyratron switch or to accommodate higher peak currents or more rapid rate of rise of current than can be accommodated by solid-state controlled rectifiers.[44-47]

A simplified schematic diagram of one type of magnetic modulator is shown in Figure 5-26. Operation of the system is critically dependent upon the saturable reactors used. These reactors are typically coils wound as gapless toroids. They are designed to have a high unsaturated inductance and a very low saturated inductance, and to make a rapid transition from the unsaturated to the saturated condition. Such a device can act as a switch and may be useful for two purposes: to decrease the rate of rise of current through the SCR and to reduce the peak current requirements of the switch.

5.3.5   Comparison of Modulator Types

The different modulator and modulator-source combinations often produce systems having different characteristics. Table 5-9 summarizes the frequency and phase stability factors achievable with various combinations of power tube and modulator types. The various modulator types themselves often vary widely in parameters such as size, weight, cost, and pulse fidelity; Table 5-10 compares some characteristics of the three most common types of modulators.

5.4   REFERENCES

A good introductory reference on the subject of transmitters is the book *Radar Transmitters* (McGraw-Hill, 1981) from which much of the material in this chapter was taken. In addition, Skolnik's *Radar Handbook* has a brief section on transmitters which is quite well done. The classic reference on modulators is Volume 5 of the RadLab series, *Pulse Generators*, and more recent information is given in the Modulator Symposia, reference 38. The following references should be consulted for specific additional information.

1. George W. Ewell, *Radar Transmitters*, McGraw-Hill Book Co., New York, 1981.
2. G. B. Collins, *Microwave Magnetrons*, RadLab series, Vol. 6, McGraw-Hill Book Co., New York, 1948.
3. E. C. Okress, *Crossed-Field Devices*, Vols. I and II, Academic Press, New York, 1961.
4. E. C. Okress, *Microwave Power Engineering*, Vols. I and II, Academic Press, New York, 1968.
5. A. H. Pickering, "Electronic Tuning of Magnetrons," *Microwave Journal*, vol. 22, no. 7, July 1979, p. 73.
6. SFD Laboratories, "Introduction to Crossed-Field Amplifiers," Union, N.J., April 1967.

7. J. F. Skowron, "The Continuous-Cathode (Emitting-Sole) Crossed-Field Amplifier," *Proceedings of the IEEE*, vol. 61, no. 3, March 1973, pp. 330–356.

8. G. McMaster and K. Dudley, "Final Report for Cathode-Driven, High Gain, Crossed Field Amplifier," Final Report on Contract no. N00039-72-C-0166, AD B004553, Raytheon Corp., Waltham, Mass., April 15, 1975.

9. W. Smith and A. Wilczek, "CFA Tube Enables New-Generation Coherent Radar," *Microwave Journal*, vol. 16, no. 8, August 1973, p. 39.

10. Litton Electron Tube Division, "Litton Distributed Emission CFA's," vol. 2, no. 516, 5CCA871, 1971.

11. A. Staprans, E. W. McCune, and J. A. Ruetz, "High-Power Linear-Beam Tubes," *Proceedings of the IEEE*, vol. 61, no. 3, March 1973, pp. 299–330.

12. M. Chodorow, E. L. Ginzton, I. R. Neilsen, and S. Sonkin, "Design and Performance of a High-Power Pulsed Klystron," *Proceedings of the IRE*, vol. 41, no. 11, November 1953, p. 1587.

13. J. T. Mendel, "Helix and Coupled-Cavity Traveling-Tube Waves," *Proceedings of the IEEE*, vol. 61, no. 3, March 1973, pp. 280–298.

14. *Hughes TWT and TWTA Handbook*, 1981.

15. A. D. LaRue, "The TWYSTRON Hybrid TWT," Varian Associates, Palo Alto, Calif., December 23, 1963, p. 2.

16. "Twystron Amplifier Catalog," no. 3725, Varian Associates, Palo Alto, Calif., 1979.

17. R. R. Berry, "Introduction to Gyrotrons," *Military Electronics/Countermeasures*, October 1980, pp. 52–60.

18. "Pulsed Power Transistor: 1 hp at 1 GHz," *Microwave Journal*, vol. 22, no. 6, June 1979, p. 34.

19. D. J. Hoft, "Solid State Transmit/Receive Module for the PAVE PAWS Phased Array Radar," *Microwave Journal*, vol. 21, no. 10, October 1979, p. 33.

20. K. J. Lee, "A 25 kW Solid State Transmitter for L-Band Radars," *1979 IEEE MTT-S International Microwave Symposium Digest*, IEEE cat. no. 79CH1439-9MTT, pp. 298–302.

21. K. Ohta et al., "A Five Watt 4–8 GHz GaAs FET Amplifier," *Microwave Journal*, vol. 22, no. 11, November 1979, pp. 66–67.

22. H. Tserng, "Design and Performance of Microwave Power GaAs FET Amplifiers, *Microwave Journal*, vol. 22, no. 2, June 1979, p. 94.

23. K. Honjo, "15-Watt Internally Matched GaAs FET's and 20 Watt Amplifier Operating at 6 GHz, *1979 IEEE MTT-S International Microwave Symposium Digest*, IEEE cat. no. 79CH1439-9MTT, pp. 289–291.

24. M. Cohn, B. Geller, and J. Schellenberg, "A 10 Watt Broadband FET Combiner/Amplifier," *1979 IEEE MTT-S International Microwave Symposium Digest*, IEEE cat. no. 79CH1439-9MTT, pp. 292–297.

25. K. Niclas, "Planar Power Combining for Medium Power GaAs FET Amplifiers in X/Ku Bands," *Microwave Journal*, vol. 22, no. 6, June 1979, p. 79.

26. C. Rucker, "Multichip Power Combining: A Summary with New Results," *1979 IEEE MTT-S International Microwave Symposium Digest*, IEEE cat. no. 79CH1439-9MTT, pp. 303–305.

27. J. Quine, M. McMullen, and D. Khandelwal, "Ku-Band Impatt Amplifiers and Power Combiners," *1978 IEEE International Microwave Symposium Digest*, IEEE cat. no. 78CH1439-9MTT, pp. 303–305.

28. Y. Bellemare and W. Chudobiak, "Thermal and Current Tuning Effects in GaAs High Power IMPATT Diodes, *Proceedings of the IEEE*, vol. 67, no. 12, December 1979, pp. 1667–1669.

29. F. B. Fank, J. D. Crowley, and J. J. Bernz, "InP Material and Device Development for MM-Waves," *Microwave Journal*, vol. 22, no. 6, June 1979, pp. 86–91.

30. H. J. Kuno and T. T. Fong, "Solid-State MM-Wave Sources and Combiners," *Microwave Journal*, vol. 22, no. 6, June 1979, p. 47.

31. D. Masse et al., "High Power GaAs Millimeter Wave IMPATT Diodes," *Microwave Journal*, vol. 22, no. 6, June 1979, pp. 103–105.

32. K. Chang et al., "High Power 94 GHz Pulsed IMPATT Oscillators," *1979 IEEE MTT-S International Microwave Symposium Digest*, IEEE cat. no. 79CH1439-9MTT, pp. 71–72.

33. Y. C. Ngan and E. M. Nakaji, "High Power Pulsed IMPATT Oscillator Near 140 GHz," *1979 IEEE MTT-S International Microwave Symposium Digest*, IEEE cat. no. 79CH1439-9MTT, pp. 73–74.

34. H. J. Kuno and T. T. Fong, "Hughes IMPATT Device Work above 100 GHz," *Millimeter and Submillimeter Wave Propagation and Circuits, AGARD Conference Proceedings*, no. 245, September 1978, pp. 14-1 and 14-2.

35. P. Van Etten, "The Present Technology of Impulse Radars," *Proceedings of the Radar-77 International Conference*, IEEE Conference Publication no. 155, 1977, pp. 535–539.

36. A. Staprans, "High-Power Microwave Tubes," *Technical Digest of International Electron Devices Meeting*, IEEE cat. no. 76CH1151-OED, December 6–8, 1976, pp. 245–248.

37. G. N., Glasoe, and J. V. Lebacqz, *Pulse Generators*, MIT RadLab series, Vol. 5, McGraw-Hill Book Co., New York, 1948. (Also available in Dover and Boston Technical Publishers editions.)

38. Numerous modulator examples and discussions are contained in the proceedings of the modulator symposia sponsored by the U.S. Army Electronics Command and the Pulse Power Conference:

Proceedings of the 5th Symposium on Hydrogen Thyratrons and Modulators, AD 650 899, 1958

Proceedings of the 6th Symposium on Hydrogen Thyratrons and Modulators, AD 254 102, 1960

Proceedings of the 7th Symposium on Hydrogen Thyratrons and Modulators, AD 296 002, 1962

Proceedings of the 8th Symposium on Hydrogen Thyratrons and Modulators, 1964

Proceedings of the 9th Modulator Symposium, 1966

Proceedings of the 10th Modulator Symposium, AD 676 854, 1968

Proceedings of the 11th Modulator Symposium, IEEE 73 CH 0773-2 ED, 1973

Proceedings of the 12th Modulator Symposium, IEEE 76 CH 1045-4 ED, 1976

Proceedings of the 13th Pulse Power Modulator Symposium, IEEE 78 CH 1371-4 ED, 1978

Proceedings of the International Pulsed Power Conference, IEEE 76 CH 1197-8 REG 5, 1980

Digest of Technical Papers, 2d IEEE International Pulsed Power Conference, IEEE 79 CH 1505-7, 1979

Proceedings of the 14th Pulse Power Modulator Symposium, IEEE 80 CH 1573-5 ED, 1981

Proceedings of the 15th Power Modulator Symposium, IEEE 82 CH 1785-5, 1982

Proceedings of the 1983 Pulsed Power Conference, IEEE 83 CH 1908-3, 1983

Proceedings of the 16th Power Modulator Symposium, IEEE 84 CH 2056-0, 1984

The Conference Records of the Power Modulator Symposia from 1950–1982 have been reprinted and are available from:

Dr. A. S. Gilmour, Jr.

State Univ. of N.Y. at Buffalo

Room 215C Bonner Hall

Amherst, N.Y. 14260

39. K. Chang, G. M. Hayashibara, and F. Thrower, "140 GHz Silicon Impatt Power Combiner Development," *Microwave Journal*, vol. 24, no. 6, June 1981, pp. 65–71.

40. A. B. Kaufman, "Speedy Impatt Pulser Protects Against Failure," *Microwaves*, vol. 24, no. 6, June 1981, pp. 107–109.

41. S. Levinson, "Multiple Output Modulator for Microwave Diode Pulse Power Generators," *Proceedings of the 14th Pulse Power Modulator Symposium*, IEEE 80CH1573-5 ED, pp. 110–114.

42. H. D. Doolittle, "Vacuum Power Tubes for Pulse Modulation," Machlett Laboratories, Stamford, Conn.
43. I. Limansky, "How to Design a Line-Type Modulator," *Electronic Design*, vol. 8, no. 4, February 17, 1960, pp. 42–45.
44. K. J. Busch et al., "Magnetic Pulse Modulators," *Bell System Technical Journal*, vol. 34, no. 5, September 1955, pp. 943–993.
45. G. T. Coate and L. T. Swain, Jr., *High-Power Semiconductor-Magnetic Pulse Generators*, MIT Press, Cambridge, Mass., 1966.
46. E. W. Manteuffel and R. E. Cooper, "D-C Charged Magnetic Modulator," *Proceedings of the AIEE*, vol. 78, January 1960, pp. 843–850.
47. W. S. Melville, "The Use of Saturable Reactors as Discharge Devices for Pulse Generators," *Proceedings of the IEEE* (London), vol. 91, pt. III, 1951, pp. 185–207.

6
RADAR ANTENNAS

Donald G. Bodnar

6.1 INTRODUCTION

Energy from the radar transmitter is sent via transmission lines to the antenna. The functions of the antenna on transmit are to concentrate the energy in a predetermined beam shape and to point this beam in a predetermined direction. On receive, the antenna again forms a beam in a particular direction to gather selectively transmitted energy that has been reflected from various targets. Received energy is sent via transmission lines to the receiver. The transmit and receive patterns of the antenna are usually identical; if they are, the antenna is said to be *reciprocal*. Only one pattern, usually the transmit one, is specified and measured for a reciprocal antenna. If the antenna is not reciprocal (e.g., if it contains ferrite devices) or is nonlinear, then the transmit and receive patterns may be different. In this case, both transmit and receive patterns are specified and measured.

The shape of the antenna pattern in general varies with distance R from the antenna as well as with look direction. The pattern shape over a sphere of constant radius is independent of R if R is large enough. The pattern variation with R can be quantified using D, the diameter of the smallest sphere that completely contains the antenna (usually D is the diameter of the antenna). Although the boundary is somewhat arbitrary,[1] the usually accepted criterion is that the pattern is independent of range when $R > 2D^2/\lambda$, where λ is the wavelength of the transmitted wave. Values of R larger than this value are said to be in the *far field* or *Fraunhofer region* of the antenna. Smaller values of R are said to be in the *near field* or *Fresnel region* of the antenna, in which case the pattern shape changes with distance as well as look direction. The pattern varies more rapidly with the distance as R gets smaller. The remainder of this chapter considers only the far field of the antenna.

6.2 BASIC ANTENNA CONCEPTS[2]

Consider an antenna located at the origin of a spherical coordinate system, as illustrated in Figure 6-1. Suppose that we are making observations on a sphere

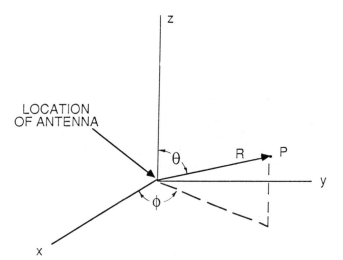

Figure 6-1. Spherical coordinate system.

having a very large radius R, that is, we are in the far field of the antenna. Assume that the antenna is transmitting, and let

P_0 = power accepted by antenna, watts
P_r = power radiated by antenna, watts
η = radiation efficiency, unitless

The above quantities are related as follows:

$$\eta = \frac{P_r}{P_0}. \tag{6-1}$$

Let $\Phi(\theta, \phi)$ = radiation intensity (watts/steradian). The total power radiated from the antenna is

$$P_r = \int_0^{2\pi} \int_0^{\pi} \Phi(\theta, \phi)\sin\theta \; d\theta \; d\phi \tag{6-2}$$

and the average radiation intensity is

$$\Phi_{avg} = \frac{P_r}{4\pi}. \tag{6-3}$$

Let $D(\theta, \phi)$ = directivity (unitless). Directivity is a measure of the ability of an antenna to concentrate radiated power in a particular direction, and is related to the radiation intensity as follows:

$$D(\theta, \phi) = \frac{\Phi(\theta, \phi)}{\Phi_{avg}} = \frac{\Phi(\theta, \phi)}{P_r/4\pi}. \tag{6-4}$$

The directivity of an antenna is the ratio of the achieved radiation intensity in a particular direction to that of an isotropic antenna. In practice, one usually is interested primarily in the peak directivity of the main lobe. Thus, if an antenna has a directivity of 100, it is assumed that 100 is the peak directivity of the main lobe.

Let $G(\theta, \phi)$ = gain (unitless). The gain of an antenna is related to the directivity and power radiation intensity as follows:

$$G(\theta, \phi) = \eta D(\theta, \phi) = \frac{\eta \Phi(\theta, \phi)}{P_r/4\pi} \tag{6-5}$$

and from Eq. (6-1)

$$G(\theta, \phi) = \frac{\Phi(\theta, \phi)}{P_0/4\pi}. \tag{6-6}$$

Thus, the gain is a measure of the ability to concentrate in a particular direction the power accepted by the antenna. Note that directivity and gain are identical for a lossless antenna (i.e., $\eta = 1$).

Let $P(\theta, \phi)$ = power density (watts per square meter). The power density is related to the radiation intensity as follows:

$$P(\theta, \phi) = \frac{\Phi(\theta, \phi)\Delta\theta\Delta\phi}{(R\Delta\theta)(R\Delta\phi)}$$

or

$$P(\theta, \phi) = \frac{\Phi(\theta, \phi)}{R^2}. \tag{6-7}$$

Substituting Eq. (6-6) into Eq. (6-7), we obtain

$$P(\theta, \phi) = G(\theta, \phi)\frac{P_0}{4\pi R^2}. \tag{6-8}$$

The factor $P_0/(4\pi R^2)$ represents the power density that would result if the power accepted by the antenna were radiated by a lossless isotropic antenna.

Let A_e (θ, ϕ) = effective area (square meters). The concept of effective area is easier to visualize when one considers a receiving antenna; it is a measure of the effective absorption area presented by an antenna to an incident plane wave. The effective area is related to gain and wavelength as follows:[3]

$$A_e(\theta, \phi) = \frac{\lambda^2}{4\pi} G(\theta, \phi). \tag{6-9}$$

Many high-gain antennas such as horns, reflectors, and lenses are said to be *aperture-type* antennas. The aperture usually is taken to be that portion of a plane surface near the antenna, perpendicular to the direction of maximum radiation, through which most of the radiation flows. Let η_a = antenna efficiency of an aperture-type antenna (unitless) and let A = physical area of the antenna's aperture (square meters). Then

$$\eta_a = \frac{A_e}{A} \tag{6-10}$$

where A_e is the peak or maximum value of A_e (θ, ϕ). The term η_a is sometimes called *aperture efficiency*.

When dealing with aperture antennas, we see from Eqs. (6-9) and (6-10) that the peak gain of the antenna is

$$G = \eta_a \frac{4\pi A}{\lambda^2}. \tag{6-11}$$

The standard directivity is obtained when the aperture is excited with a uniform amplitude and equiphase distribution. (Such a distribution yields the highest directivity of all equiphase excitations.) For planar apertures in which $A \gg \lambda^2$ and for which radiation is confined to a half space, the standard directivity G_0 is given by

$$G_0 = \frac{4\pi A}{\lambda^2}. \tag{6-12}$$

From Eqs. (6-11) and (6-12), the peak gain of the antenna is

$$G = \eta_a G_0. \tag{6-13}$$

The term η_a is actually the product of several factors, such as

$$\eta_a = \eta\eta_i\eta_1\eta_2\eta_3 \cdots \tag{6-14}$$

The term η is the radiation efficiency as defined in Eq. (6-1) and includes ohmic losses in the antenna. The term η_i is aperture (or antenna) illumination efficiency, which is a measure of how well the aperture collimates the radiated energy; it is the ratio of the directivity that is obtained to the standard directivity. The other factors, $\eta_1 \eta_2 \eta_3 \ldots$ include all other effects that reduce the gain of the antenna. Examples are spillover losses in reflector or lens antennas, phase-error losses due to surface errors on reflectors, random phase errors in phased-array elements, aperture blockage, depolarization losses, and so on.

When the power radiation intensity $\Phi(\theta, \phi)$, the power density $P(\theta, \phi)$, the directivity $D(\theta, \phi)$, and the effective area $A_e(\theta, \phi)$ are plotted after normalizing to the same peak level, they are identical and are referred to as the *antenna radiation pattern*. The main (or major) lobe of the radiation pattern is in the direction of maximum gain; all other lobes are called *sidelobes* or *minor lobes*.

In addition to the peak gain of the main lobe, two very important antenna characteristics are the beamwidth (usually specified at the half-power level, which is approximately 3 dB below the peak of the main lobe) and the maximum sidelobe level. These two characteristics, along with other minor lobes, are indicated in Figure 6-2. Note that the radiation pattern is different for cuts at different angles through the peak of the main lobe.

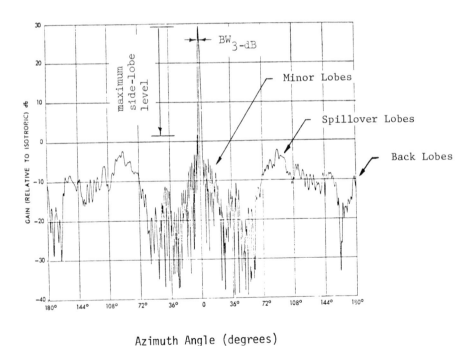

Azimuth Angle (degrees)

Figure 6-2. Typical antenna pattern for a search radar.

6.3 RADIATION PATTERNS AND APERTURE ILLUMINATION FUNCTIONS

Ideally, an antenna radiation pattern would consist of a single main lobe and no minor lobes. In practice, however, it consists of the main lobe and numerous minor lobes, as illustrated in Figure 6-2.

It is important that minor lobes of radar antennas be small compared to the main lobe in order to (1) have an antenna with high directivity, (2) reduce the susceptibility of the antenna to interfering signals, (3) reduce the possibility of detecting a target in a minor lobe, and (4) reduce the probability of interference with other nearby systems. Therefore, the designers of radar systems should be concerned with the minor lobe structure of antennas in addition to the other requirements, such as main lobe gain and beamwidths.

Some important factors which affect an antenna radiation pattern are the aperture illumination function, aperture blockage, and fabrication or random errors. These factors and the importance of minor lobes are discussed in the following sections.

A fictitious surface, called the *antenna aperture*, located on or near an antenna, is often useful in computing the performance of the antenna. Usually the aperture is that portion of a plane surface immediately in front of the antenna through which the major portion of the radiation passes. The distribution of electromagnetic energy from the antenna over the aperture determines the pattern of the antenna. The antenna designer can modify the shape of the pattern by altering the distribution of energy over the aperture.

The three primary performance parameters for an antenna are gain, beamwidth, and sidelobe level. This section presents data on these three parameters for both line source and circular symmetry distributions.[4] The line source is used to represent one plane of a rectangular aperture which has a separable aperture distribution.

The aperture illumination efficiency η_i represents the loss in gain resulting from tapering the aperture distribution to produce sidelobes lower than those achievable from a uniform illumination. For a circular aperture, η_i is given by a single number. For separable rectangular aperture distributions, it equals the product (sum in decibels) of the efficiencies in each of the two aperture directions.

The half-power beamwidth BW of an antenna is related to the beamwidth constant β by[5]

$$BW = 2\,\text{arc}\,\sin\left(\frac{\beta\lambda}{2L}\right) \sim \beta\,\frac{\lambda}{L} \qquad \text{(linear aperture)} \qquad (6\text{-}15)$$

$$BW = 2\,\text{arc}\,\sin\left(\frac{\beta\lambda}{2D}\right) \sim \beta\,\frac{\lambda}{D} \qquad \text{(circular aperture)} \qquad (6\text{-}16)$$

where L is the length of the linear aperture and D is the diameter of the circular aperture. The small-argument approximation for the arc sine is typically used for calculating BW when the aperture is large in terms of wavelengths. Values of β and η as a function of sidelobe level for a number of distributions are given in the following sections.

6.3.1 Continuous Line Source Distributions

The problem of determining the optimum pattern from a line source has received considerable attention. The optimum pattern is defined as the one that produces the narrowest beamwidth measured between the first null on each side of the main beam with no sidelobes higher than the stipulated level. Dolph solved this problem for a linear array of discrete elements using Chebyshev polynomials.[6] If the number of elements becomes infinite while the element spacing approaches zero, the Dolph pattern becomes the optimum continuous line source pattern given by[7]

$$E(u) = \cos(u^2 - A^2)^{1/2} \tag{6-17}$$

where

$$u = \frac{\pi L}{\lambda} \sin\theta$$

λ = wavelength
θ = angle from the normal to the aperture

The beamwidth constant β in degrees and the parameter A are given by

$$\beta = \frac{360}{\pi^2} \left\{ (\text{arc cosh } R)^2 - \left[\text{arc cosh} \left(\frac{R}{\sqrt{2}} \right) \right]^2 \right\}^{1/2} \tag{6-18}$$

where

A = arc cosh R
R = main lobe to sidelobe voltage ratio

This Chebyshev pattern provides a useful basis for comparison even though it is physically unrealizable, since the remote sidelobes do not decay in amplitude. In fact, all sidelobes have the same amplitude in the Chebyshev pattern. The aperture distribution which produces the Chebyshev pattern has an impulse at both ends of the aperture[8] and produces very low aperture efficiency. In addition, the pattern is very sensitive to errors in the levels of these impulses.

Taylor developed a method for avoiding the above problems by approximating arbitrarily closely the Chebyshev pattern with a physically realizable pattern.[7] He approximated the Chebyshev uniform sidelobe pattern close to the main beam but let the wide-angle sidelobe decay in amplitude. Taylor used a closeness parameter $\bar{n}$ in his analysis. As $\bar{n}$ becomes infinite, the Taylor distribution approaches the Chebyshev distribution. By using the largest $\bar{n}$ that still produces a monotonic aperture distribution, one obtains the beamwidth constant and aperture efficiency shown in in Figures 6-3 and 6-4. Notice that the beam-

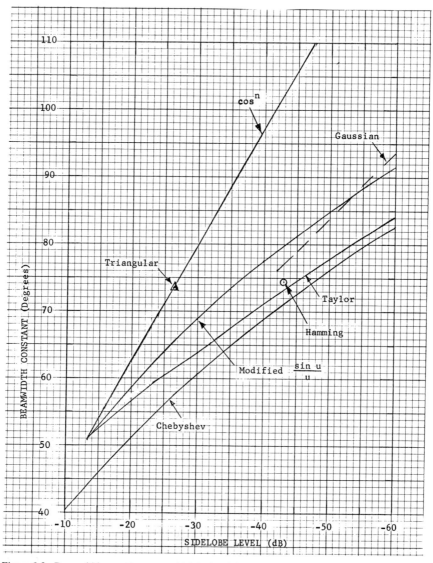

Figure 6-3. Beamwidth constant versus sidelobe level for several line source aperture distributions. (From ref. 4)

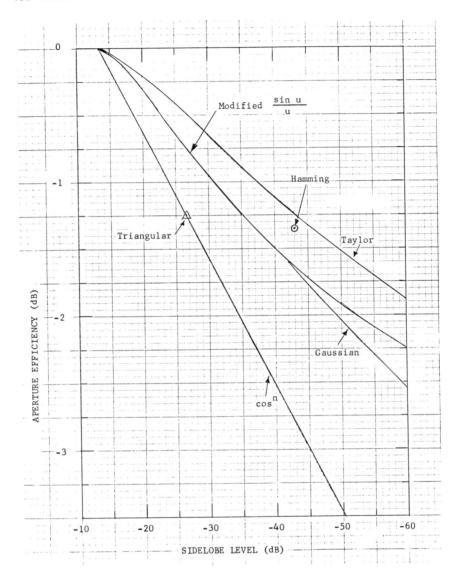

Figure 6-4. Aperture efficiency versus sidelobe level for several line source aperture distributions. (From ref. 4)

width from the Taylor distribution is almost as narrow as that from the Chebyshev distribution while still producing excellent aperture efficiency.

Several other common distributions are also listed in Figures 6-3 and 6-4.[5,9] The advantage of the $\sin[(u^2 - B^2)^{1/2}/(u^2 - B^2)^{1/2}$ pattern and the $\cos^n(\pi x/L)$ distribution is that both the distribution and the pattern for it may be obtained in closed form. This mathematical convenience is obtained at the expense of poorer beamwidth and efficiency performance as compared to the Taylor distribution.

6.3.2   Continuous Circular Aperture Distributions

The Chebyshev pattern described in the preceding section can also be shown to be optimum for the circular aperture. Taylor has generalized his line source

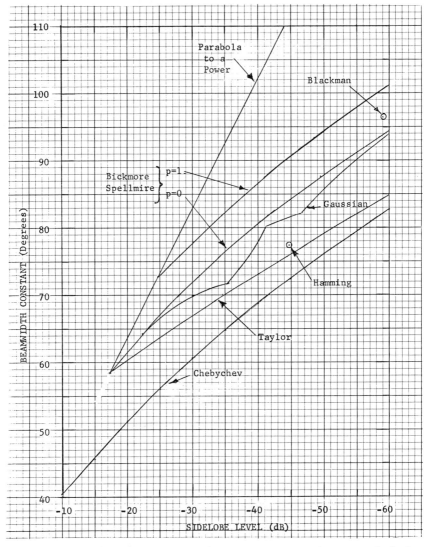

Figure 6-5. Beamwidth constant versus sidelobe level for several circular aperture distributions (From ref. 4)

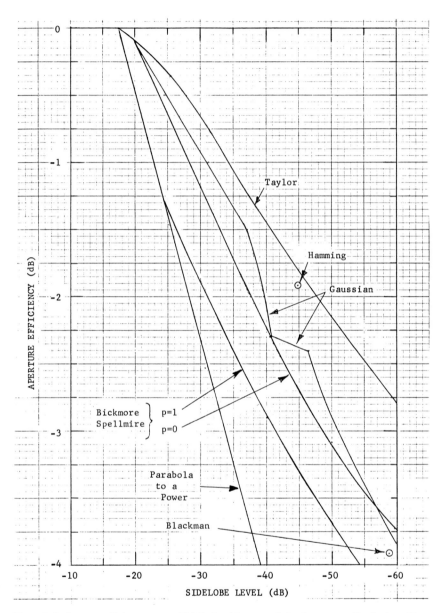

Figure 6-6. Aperture efficiency versus sidelobe level for several circular aperture distributions (From ref. 4)

distribution to the circular case,[10,11] and his pattern approaches the Chebyshev pattern as his closeness parameter $\bar{n}$ for the circular aperture approaches infinity. The beamwidth constant and aperture efficiency shown in Figures 6-5 and 6-6 are obtained using the largest $\bar{n}$ that still produces a monotonic distribution.[12]

The Bickmore-Spellmire distribution[13] is a two-parameter distribution which can be considered a generalization of the parabola to a power distribution.[8] The Bickmore-Spellmire distribution $f(r)$ and pattern $E(u)$ are given by

$$f(r) = p[1 - (2r/D)^2]^{p-1} \Lambda_{p-1} [jA(1 - (2r/D)^2)^{1/2}]$$

$$E(u) = \Lambda_p[(u^2 - A^2)^{1/2}]$$

(6-19)

where

p and A = constants that determine the distribution
Λ = lambda function
$u = (\pi D/\lambda)\sin\theta$.

The Bickmore-Spellmire distribution reduces to the parabola to a power distribution when $A = 0$ and to the Chebyshev pattern when $p = -1/2$.

A Gaussian distribution[12] produces a no sidelobe pattern only as the edge illumination approaches zero. In general, the aperture distribution must be numerically integrated to obtain the far-field pattern. The second sidelobe of this pattern is sometimes higher than the first, which accounts for the erractic behavior of β and η in Figures 6-5 and 6-6.

6.3.3 Blockage

The placement of a feed in front of a reflector results in blockage of part of the aperture energy. In the geometric-optics approximation, no energy exists where the aperture is blocked and the undistributed aperture distribution persists outside the blocked region.[14] A line source of length L with a $\cos^n$ aperture distribution and a centrally located blockage of length L_b produces the gain loss and the resulting sidelobe level shown in Figure 6-7. The corresponding changes are shown in Figure 6-8 for a circular aperture of diameter D and a parabola to a power distribution blocked by a centrally located disk of diameter D_b. Notice that the line source blockage affects the pattern much more rapidly than in the circular case, since the line source blockage affects a larger portion of the aperture. Calculation of strut blockage effects is given by Gray.[15]

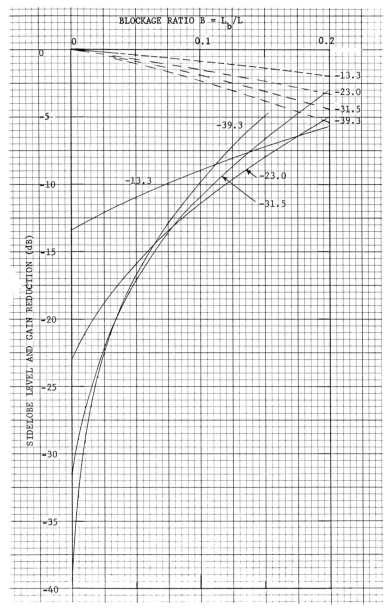

Figure 6-7.   Gain loss (- - - -) and resulting sidelobe level (———) for a centrally blocked line source distribution having a specified unblocked sidelobe level. (From ref. 4)

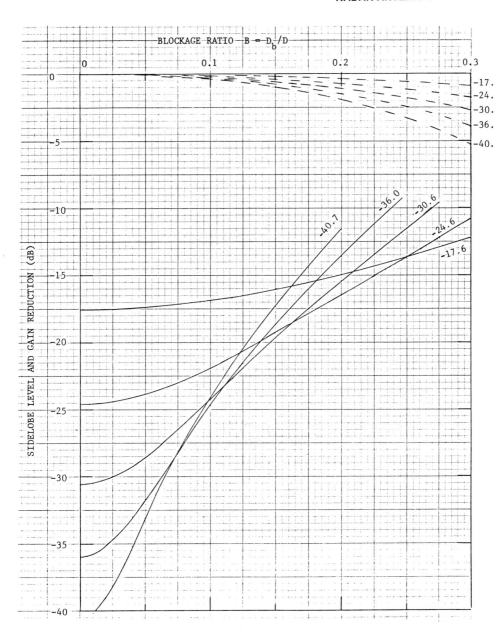

Figure 6-8. Gain loss (−−−−) and resulting sidelobe level (———) for a centrally blocked circular aperture distribution having a specified unblocked sidelobe level. (From ref. 4)

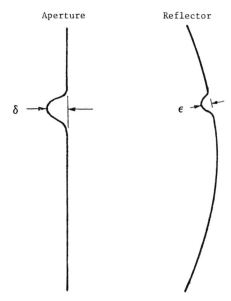

Figure 6-9. Simplified diagram illustrating a reflector error of ϵ producing a phase error of δ in the aperture.

6.3.4   Random Errors

In many applications, very low sidelobes are required. One of the factors which can significantly influence the levels of close-in sidelobes is the effect of random aperture errors. For example, one can envision that the radiation pattern from an antenna consists of a zero-error pattern plus a distributing pattern caused by random errors.

For a shallow reflector, an aperture error is about twice the reflector error due to the decrease in path length traveled to the aperture plane. This fact is illustrated schematically in Figure 6-9, where ϵ and δ are the reflector error and aperture error, respectively. For a continuous aperture, the distance across the aperture where the errors become essentially independent, on the average, is the correlation distance. The average disturbing pattern is dependent on both the root mean square (rms) reflector errors in wavelengths, $(\bar{\epsilon}_\lambda^2)^{1/2}$, and the correlation distance, C.

The effects of random aperture errors on close-in sidelobes can be estimated from disturbing patterns such as those shown in Figure 6-10; these patterns were calculated for three correlation intervals and for various rms reflector errors.[16] For example, suppose an antenna has an rms reflector error of $\lambda/32$, a correlation interval of unity, and a zero error gain of 35 dB. Figure 6-10 shows that

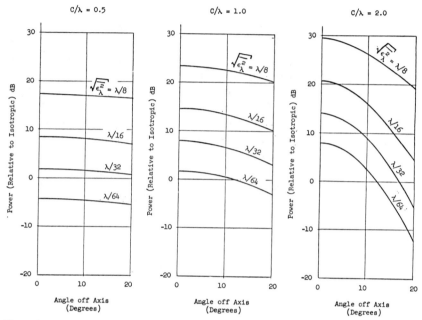

Figure 6-10. Disturbing patterns referenced to an isotropic radiator for the indicated C/λ and for various rms reflector errors in units of wavelengths.

the average disturbing pattern is about 8 dB relative to isotropic at the beam axis. Therefore, the disturbing pattern will be -27 dB relative to the peak of the main lobe.

6.4 RADAR ANTENNA TYPES AND APPLICATIONS

Radar antennas typically are required to have narrow beams (at least in one plane) and relatively low sidelobe levels, and are usually scanned by mechanical movement of the entire antenna structure, by an electromechanical or electronic scanning feed mechanism, or by electronically scanning an array of elements. Although fixed arrays of dipole-type elements were employed in the first radar antennas because of the low frequencies involved, the paraboloidal reflector antenna has been employed in some form in the majority of radar applications. The lens antenna has also seen favor in certain instances, and recently great progress has been made in the development of electronic-scanning phased arrays.

The type of antenna selected for a certain application depends not only on the electrical and mechanical requirements dictated by the radar design specifications but also on cost and risk trade-offs. For example, the antenna requirements for a tracking radar typically call for a pencil-shaped beam with parallel-polarized and cross-polarized sidelobes commensurate with the clutter and jamming environment in which the radar is to be employed. The selection of a paraboloidal reflector antenna versus a scanning phased array, however, depends on cost versus performance and the risk associated with implementing a phased array at the required frequency.

A wide variety of radar antennas have been used in radar systems. A partial list of antennas is given in Table 6-1. Table 6-2 lists some of the parameters that are used to evaluate the suitability of an antenna for a particular application.

Table 6-1. Radar Antenna Types.

Reflector-Type Antennas

Single Reflector	Double Reflector
Paraboloid	Cassegrain
Parabolic cylinder	Gregorian
Parabolic torus	Twist reflector
Shaped reflector	
Pillbox	

Lenses

Dielectric
Metal-plate
Geodesic

Electromechanical Scanners

Organ pipe	Foster
Ring-switch	Lewis
Conical scan	Eagle
Mirror scan	

Array Antennas

Fixed Beam	Electronic-Scanning Beam
Dipoles	Phase Scanning
Waveguide slots	Phase shifters
Broadwall	Time delay
Narrow wall	Frequency scanning
Series feed	Amplitude scanning
Corporate feed	Switched-beam scanning
Center feed	Series-fed array
	Parallel-fed array
	Optical devices

Table 6-2. Important Antenna Parameters.

Electrical

Frequency (and bandwidth)	Impedance (or VSWR)
Gain (and efficiency)	Power-handling capability
Polarization	Scan sector
Beamwidths (and beam shape)	Scan modes and rates
Sidelobes	Electronic counter-countermeasures (ECCM)

Mechanical

Size	Operating Conditions
Weight	Temperature
Reliability	Humidity
Maintainability	Wind loads
Stabilization	Shock loads and vibration
Tolerances	Icing
Manufacturing methods	Dust
	Rain

6.4.1 Parabolic Reflector Antenna

Reflector antennas are extremely important and practical devices for use in radar systems because they offer an economical method of distributing energy over a large aperture area and can produce shaped or pencil beams with high gain. In general, the reflector is used to redirect and reshape energy from one or more point sources located near the focal point into a desired far-field pattern.

The most common reflector shape is the paraboloid formed by rotating a two-dimensional parabola about its focal axis. This shape is particularly useful, since all rays leaving the focal point and striking the reflector are reflected along a path parallel to the focal axis. Additionally, the fields transversing all of these reflected ray paths will be in phase over any plane perpendicular to the axis, since these planes are equidistant from the focal point. Consequently, a spherical wavefront leaving a primary feedhorn will be mapped into a uniform (plane) wavefront by a paraboloid. This constant path length feature indicates that the paraboloid is inherently a broadband device. The paraboloid is reciprocal in that it intercepts an electromagnetic plane wave traveling parallel to its axis and redirects it so that all of the energy passes to the focal point, where it may be collected.

The gain and beamwidth of a paraboloid can be obtained from Eqs. (6-13), (6-14), and (6-16), along with Figures 6-5 and 6-6, once the sidelobe level and antenna efficiency are determined. The sidelobe level can be obtained from a knowledge of the edge taper of the paraboloid. The edge taper is the ratio of the feed power incident on the center of the paraboloid to that incident on its

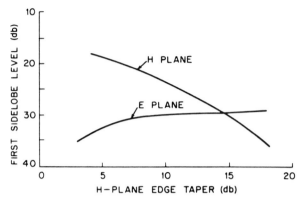

Figure 6-11. First sidelobe level in the two principal planes as a function of amplitude taper across the aperture. (By permission, from Jasik, ref. 17; © 1961 McGraw-Hill Book Co.)

edge. Figure 6-11 shows the sidelobe level produced by edge taper.[17] The E-plane pattern is the pattern in a plane that contains the axis of the reflector and the electric field vector of the feed. The H plane is the orthogonal plane that also passes through the axis of the reflector.

The edge taper is determined by both the beamwidth of the feed and the divergence of energy as it leaves the feed (space attenuation). The taper produced by the feed is found by first determining the angular width of the reflector as seen by the feed (see Figure 6-12). The level of the feed pattern in its on-axis direction relative to the edge level is the feed taper. Empirical feed horn beamwidth formulas that are very useful for reflector antenna design purposes have been developed.[18] For the E plane,

$$BW_{10\text{dB}} = 88 \ \lambda/B \ \text{(degrees)} \qquad B/\lambda < 2.5 \qquad (6\text{-}20)$$

For the H plane,

$$BW_{10\text{dB}} = 31 + 79 \ \lambda/A \qquad A/\lambda < 3 \qquad (6\text{-}21)$$

where

$BW_{10\text{dB}}$ = pattern width in degrees at the -10 dB level
B = E-plane horn aperture
A = H-plane horn aperture

Space taper can be determined from Figure 6-13. The sum of the feed taper and the space taper in decibels represents the edge taper of the antenna. This taper determines the first sidelobe level, as shown in Figure 6-11. The first

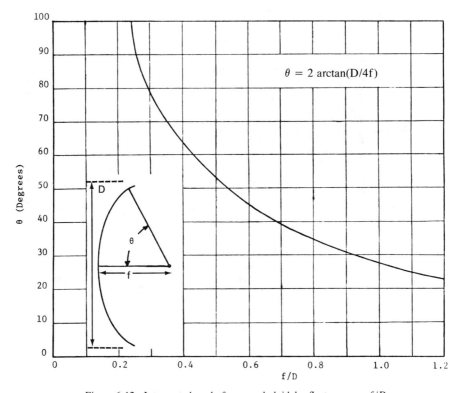

Figure 6-12. Intercepted angle for a paraboloidal reflector versus f/D.

sidelobe level then determines the beamwidth constant and aperture efficiency, as shown in Figures 6-5 and 6-6. An example of overall gain calculation follows. Assume an aperture having a maximum gain of 41.0 dB. The additional losses which result in a practical gain of about 37.2 dB may be budgeted as indicated in Table 6-3.

Many scanning antennas operate on the principle of one or more feeds located at points which are away from the reflector focus. Let θ_f be the angle between the reflector axis and the line connecting the feed and reflector vertex (see Figure 6-14). The beam scan angle θ_s is related to θ_f by the beam deviation factor (BDF):

$$\text{BDF} = \frac{\theta_s}{\theta_f}. \tag{6-22}$$

For a flat mirror reflector, BDF is unity. For a paraboloidal reflector, however, BDF is a function of the f/D ratio, aperture illumination, and scan angle, al-

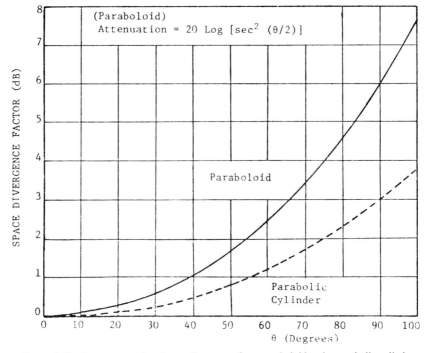

Figure 6-13. Space attenuation due to divergence for a paraboloid and a parabolic cylinder.

though scan angle and illumination effects are less than f/D variations. The results of one analytical investigation[19] are shown in Figure 6-15, where BDF is plotted versus the f/D ratio. The concept of BDF is very important in estimating the feed movement needed to achieve a given far-field scan for a particular reflector antenna. It should be noted that as the beam is scanned off boresight, the antenna gain is reduced and the sidelobes on the side of the beam near the reflector axis are increased substantially (coma lobe), as shown in Figure 6-16. Ruze[20] has developed an empirical formula which gives the number

Table 6-3. Typical Parabolic Antenna Gain Budget.

Maximum aperture gain ($4\pi A/\lambda^2$)		41.0 dB
Losses		3.8 dB
Aperture taper (efficiency)	2.2 dB	
Feed spillover	0.6 dB	
Feed VSWR	0.2 dB	
Blockage	0.5 dB	
Reflector tolerance	0.2 dB	
Polarization	0.1 dB	
	Net gain	37.2 dB

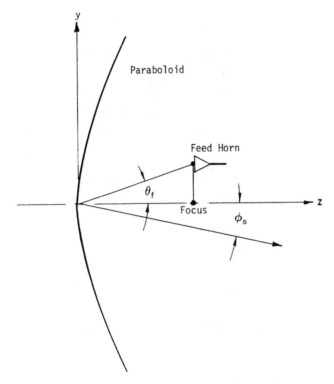

Figure 6-14. Beam scanning with a paraboloidal reflector.

of beamwidths that may be scanned before the coma lobe reaches a level 10.5 dB (corresponding to 1-dB gain loss) below the main beam peak:

$$\text{Number of BWs scanned} = 0.44 + 22(f/D)^2. \qquad (6\text{-}23)$$

6.4.2 Cassegrain Antenna

The Cassegrain antenna is a double-reflector system consisting of a main reflector and a subreflector, as shown in Figure 6-17. In the classical Cassegrain design, the main reflector is a paraboloid and the subreflector is a hyperboloid. One of the foci of the hyperboloid is the real focal point of the two-reflector system and is located at the phase center of the feed; the other focus is a virtual focal point which is located at the focus of the paraboloid. All rays emanating from the real focal point, and reflected from both surfaces, arrive in phase at a plane in front of the antenna.

Another common form of the Cassegrain antenna (and a special case of the

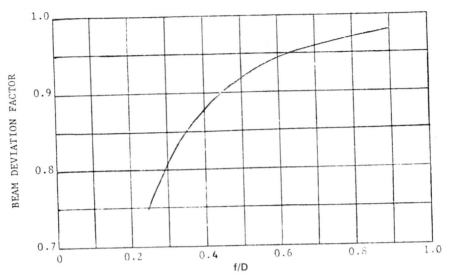

Figure 6-15. BDF versus f/D ratio.

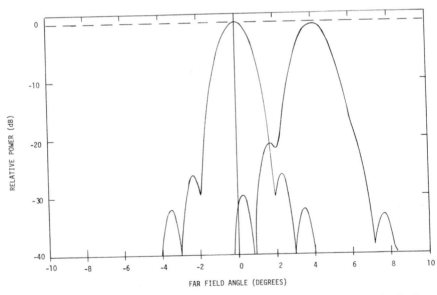

Figure 6-16. Computer-calculated patterns for a paraboloidal section fed by a scanning feed.

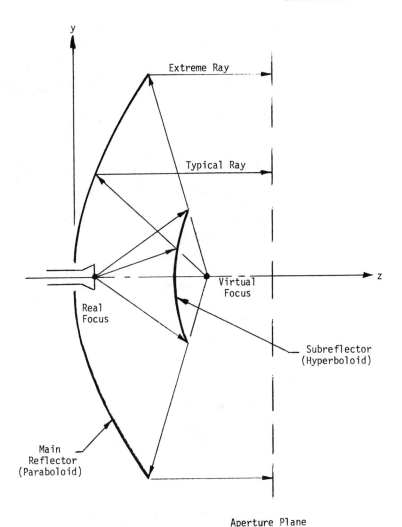

Figure 6-17. Cassegrain antenna geometry.

classical geometry above) is a system which uses a paraboloidal main reflector and a flat subreflector. In the former case, the effective focal length of the two-reflector system is larger than that of the main reflector, while in the latter case it is the same.

The principal advantage of the Cassegrain system for radar applications is that it allows the feed system to be placed behind the main reflector, with an attendant reduction in waveguide plumbing complexity and mechanical moment. The principal disadvantages lie in the more complex geometry (alignment) and in the blockage effects of the subreflector. The blockage effect can be overcome to some extent by the use of polarization twisting schemes in the main reflector and the subreflector.

6.4.3 Phased-Array Antennas

Applications

The phased-array antenna is particularly attractive for radar applications because of its inherent ability to steer a beam without the necessity of moving a large mechanical structure. This ability is an important advantage if the required antenna is large. Other attractive features of the phased-array antenna include the capability to generate more than one beam with the same array and flexibility in the control of the aperture illumination. On the other hand, phased arrays have not found wide acceptance due to their high cost and complexity. The two-dimensional planar array is particularly useful in radar applications. In a rectangular aperture form, it can generate fan beams; in a circular aperture form, it can generate pencil beams. Such an array can also be used to generate multiple beams simultaneously for search and track functions.

Phased-Array Principles

Consider the simple N-element linear array shown in Figure 6-18 with element-to-element spacing of d. For an incoming wavefront from the direction θ, the

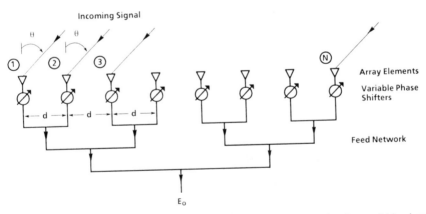

Figure 6-18. N-element linear array. Array elements are indicated by triangles, variable phase shifters by arrow-crossed circles; E_o is the output of the feed summing network.

difference in phase of the signals in adjacent elements is

$$\psi = \frac{2\pi d}{\lambda} \sin\theta. \tag{6-24}$$

Each element lags the element to its right by this amount of phase. Each phase shifter adds to the signal passing through it the negative of the phase received by its array element. In this manner, the phase of the output of each phase shifter is zero (modulo 2π), and the output signal E_0 is the sum of all signals arriving from the direction θ. On transmit, each element phase leads the element to its right by the amount given in Eq. (6-24) in order to keep the beam in the direction θ. Electronic beam scanning is produced by varying ψ using the phase shifters.

Ignoring edge effects, the radiation pattern of a phased array is determined by the product of an array factor and an element pattern. If the element pattern is considered to be isotropic (or the element has constant gain over the region of interest), the array factor determines such radiation parameters as peak gain, beamwidth, and sidelobe levels. For the linear array with uniform phase and amplitude illumination shown in Figure 6-18, the normalized radiation pattern is given by

$$G_a(\theta) = \frac{\sin^2[N\pi(d/\lambda)\sin\theta]}{N^2\sin^2[\pi(d/\lambda)\sin\theta]}. \tag{6-25}$$

If directive elements are used, the resultant array pattern is the product of the element pattern $G_e(\theta)$ and the array factor $G_a(\theta)$, so that

$$G(\theta) = G_e(\theta)G_a(\theta). \tag{6-26}$$

In a two-dimensional rectangular planar array, the radiation pattern may be written as the product of the radiation pattern in the two planes which contain the principal axes of the antenna. The pattern of Eq. (6-25) has maxima whenever $\sin\theta = \pm n\lambda/d$ where $n = 0, 1, 2 \cdots$. The maximum at $\sin\theta = 0$ is the main beam; other maxima are called *grating lobes* and are a result of the periodic nature of the array. Grating lobes appear in real space at $\theta = \pm 90°$ for $d = \lambda$; thus, the element spacing should be less than this value to prevent grating lobes.

If the beam is scanned by adding a uniform phase difference, ϕ, between elements, the array pattern is

$$G(\theta) = \frac{\sin^2[N\pi(d/\lambda)(\sin\theta - \sin\theta_0)]}{N^2\sin^2[\pi(d/\lambda)(\sin\theta - \sin\theta_0)]} \tag{6-27}$$

where $\phi = 2\pi(d/\lambda) \sin\theta_0$ and θ_0 defines the direction of the main lobe. For this case, grating lobes will appear at angle θ_g whenever

$$\pi(d/\lambda)(\sin\theta_g - \sin\theta_0) = \pm n\pi \quad n = 0, 1, 2 \cdot \cdot \cdot \cdot . \tag{6-28}$$

or

$$\sin\theta_g = \pm n(\lambda/d) + \sin\theta_0 \tag{6-29}$$

To scan the beam to $+90°$ with a grating lobe appearing at $-90°$, we must have $d = \lambda/2$. If the scan angle is limited to $\pm 60°$, the elements should be spaced no more than 0.54λ apart.

As the array beam is scanned, the half-power beamwidth increases in the plane of scan. The beamwidth is approximately inversely proportional to $\cos\theta_0$ where θ_0 is the scan angle measured from the normal to the array. This may be expressed as

$$\theta_B(\theta_0) = \frac{\theta_B(0)}{\cos\theta_0} \tag{6-30}$$

where $\theta_B(\theta_0)$ is the beamwidth in the θ direction at scan angle θ_0 and $\theta_B(0)$ is the corresponding beamwidth at broadside. For small scan angles θ_0, the broadside beamwidth $\theta_B(0)$ may be approximated by

$$\theta_B(0) \approx \frac{0.886\lambda}{Nd \cos\theta_0} \quad \text{(radians)}. \tag{6-31}$$

With the broadside beamwidths expressed in degrees, the peak gain for typical illumination functions may be approximated by

$$G(0, 0) \approx \frac{32,000}{\theta_B(0) \, \phi_B(0)} . \tag{6-32}$$

Feed Networks

Feed networks for arrays typically fall into one of two primary categories: space feeds or constrained feeds. The space feed is usually based on an optical illumination device much like a conventional mechanically scanned reflector. As shown in Figure 6-19, this type of feed may use either a transmission or a reflection approach. As may be seen, a planar equiphase front is not used to excite the array. Hence, the array phase shifters may be required to collimate the beam in addition to providing beam steering.

FEED NETWORKS FOR PHASED ARRAYS

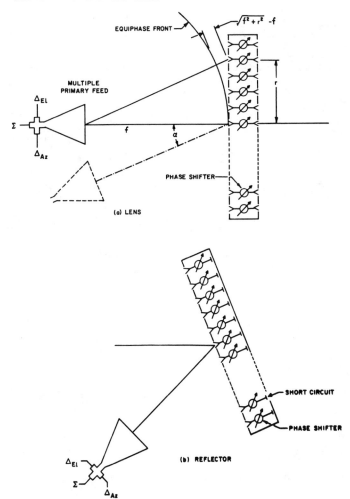

Figure 6-19. Transmission and reflection space feeds. (By permission, from Skolnik, ref. 21; © 1970 McGraw-Hill Book Co.)

There are many possible types of constrained feeds. Some are illustrated in Figure 6-20. Here the constrained feed is further divided into series and parallel feed paths. The parallel constrained feed may be considered to yield equal-length feed paths to all elements, whereas the series feed usually has unequal paths. The phase shift in the feed lines of a constrained feed may be frequency dependent so that the beam scans as frequency changes. Trade-offs involved in selecting a feed network include size, weight, cost, ease of interface with other system components, flexibility in amplitude weighting, loss, and bandwidth.

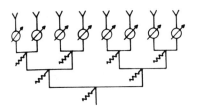

(a) MATCHED CORPORATE FEED

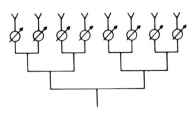

(b) REACTIVE CORPORATE FEED

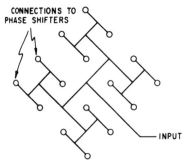

(c) STRIPLINE REACTIVE FEED

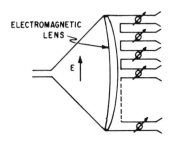

(d) MULTIPLE REACTIVE POWER DIVIDER

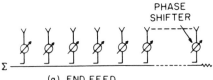

(a) END FEED

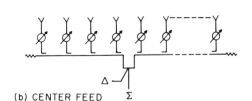

(b) CENTER FEED

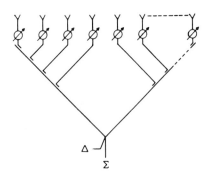

(d) EQUAL PATH LENGTH FEED

(c) CENTER FEED WITH SEPARATELY OPTIMIZED SUM AND DIFFERENCE CHANNELS

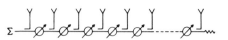

(e) SERIES PHASE SHIFTERS

Figure 6-20. Parallel and series constrained feeds. (By permission, from Skolnik, ref. 21; © 1970 McGraw-Hill Book Co.)

176

Table 6-4. Summary of Phased-Array Bandwidth Criteria.

60° Scanning with Less Than 1 dB One-Way S / N *Ratio Loss*

Feed	CW Bandwidth (%)	Minimum Pulse Length
Equal line	θ_B (degrees)	$2 \times L_a$ (aperture length)
End-fed series	$\dfrac{1}{1 + (\lambda_g/\lambda)} \times \theta_B$	$\dfrac{2}{1 + (\lambda_g/\lambda)} \times L_a$
Center-fed series	$\dfrac{\lambda}{\lambda_g} \times \theta_B$	$2 \times (\lambda/\lambda_g) \times L_a$
Space/optical	θ_B	$2 \times L_a$

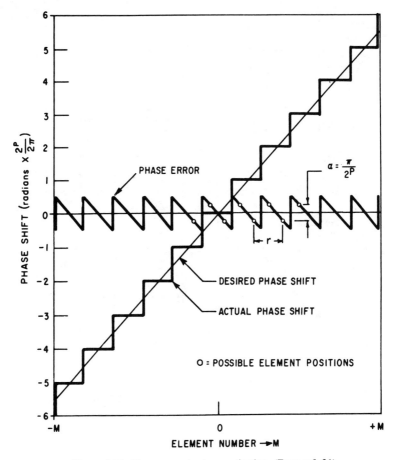

Figure 6-21. Phase error due to quantization. (From ref. 21)

Array Bandwidth

An array that is steered by phase shifters will scan its beam as frequency changes. This beam scanning places a limit on the bandwidth of the array. As an illustration, consider the end-fed series feed shown in Figure 6-20 and assume that it has free-space propagation characteristics. When the frequency is changed, the change in phase across the aperture of length L due to the feed line will be

$$\Delta\Phi = \frac{\Delta f}{f}\frac{2\pi L}{\lambda} \quad \text{(radians)}. \qquad (6\text{-}33)$$

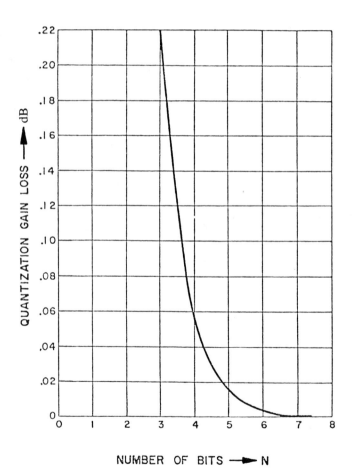

Figure 6-22. Typical phase quantization gain loss.

The corresponding beam scanning will be

$$\Delta\theta_0 = \frac{\Delta f}{f}\frac{1}{\cos\theta_0} \quad \text{(radians)}\tag{6-34}$$

and results in pure beam translation with no distortion. The above result assumes a nondispersive feed line. For a waveguide feed, guide effects must be considered and the beam scanning becomes

$$\Delta\theta_0\ \text{(WG)} = \frac{\lambda_g}{\lambda}\frac{\Delta f}{f}\frac{1}{\cos\theta_0} \quad \text{(radians)}\tag{6-35}$$

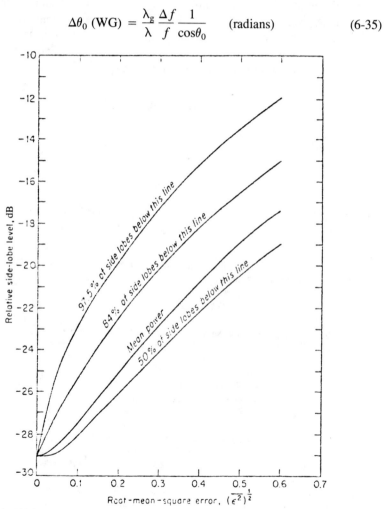

Figure 6-23. Sidelobe distribution due to random errors; 25-element array designed for a 29-dB sidelobe suppression computed at design lobe maxima. (By permission, from Johnson et al., ref. 2; © 1984 McGraw-Hill Book Co.)

where λ_g is guide wavelength. A summary of bandwidth effects in arrays is given in Table 6-4.

Error Effects

It must be noted that any realizable phased array will not achieve perfect phase control at the elements. Phase errors arise due to quantization of the phase shifters into discrete bits of available phase shift and random phase errors in the various components of the array. These errors affect peak gain, beam-pointing accuracy, and peak and rms sidelobes. The concept of phase quantization and the error introduced is illustrated in Figure 6-21. A typical peak gain loss due to quantization is shown in Figure 6-22. The effect of random phase errors on sidelobe distribution is illustrated in Figure 6-23.

6.5 REFERENCES

1. R. C. Hansen (Ed.), *Microwave Scanning Antennas*, Vol. I, Apertures, Academic Press, New York, 1964, pp. 24–46.
2. R. C. Johnson and H. Jasik (Eds.), *Antenna Engineering Handbook*, 2nd ed., McGraw-Hill Book Co., New York, 1984, chap. 1.
3. S. Silver, *Microwave Antenna Theory and Design*, McGraw-Hill Book Co., New York, 1949, section 2.14.
4. D. G. Bodnar, "Materials and Design Data," in *Antenna Engineering Handbook*, 2nd ed., R. C. Johnson and H. Jasik, (Eds.), McGraw-Hill Book Co., New York, 1984, pp. 46-15 to 46-23.
5. Silver, op. cit., chap. 6.
6. C. L. Dolph, "A Current Distribution for Broadside Arrays Which Optimizes the Relationship between Beam Width and Side-Lobe Level," *Proceedings of the IRE*, vol. 34, June 1946, pp. 335–348.
7. T. T. Taylor, "Design of Line-Source Antennas for Narrow Beamwidth and Low Side Lobes," *IRE Transactions on Antennas and Propagation*, vol. AP-3, January 1955, pp. 16–28.
8. J. W. Sherman III, "Aperture-Antenna Analysis," in *Radar Handbook*, M. I. Skolnik (Ed.), McGraw-Hill Book Co., New York, 1970, chap. 9.
9. F. J. Harris, "On the Use of Windows for Harmonic Analysis with the Discrete Fourier Transform," *Proceedings of the IEEE*, vol. 66, January 1978, pp. 51–83.
10. T. T. Taylor, "Design of Circular Apertures for Narrow Beamwidth and Low Sidelobes," *IRE Transactions on Antennas and Propagation*, vol. AP-8, January 1960, pp. 17–22.
11. R. C. Hansen, "Tables of Taylor Distributions for Circular Aperture Antennas," *IRE Transactions on Antennas and Propagation*, vol. AP-8, January 1960, pp. 23–36.
12. A. C. Ludwig, "Low Sidelobe Aperture Distributions for Blocked and Unblocked Circular Apertures, RM 2367, General Research Corp., Santa Barbara, Calif., April 1981.
13. W. D. White, "Circular Aperture Distribution Functions," *IEEE Transactions on Antennas and Propagation*, vol. AP-25, September 1977, pp. 714–716.
14. P. W. Hannan, "Microwave Antennas Derived from the Cassegrain Telescope," *IRE Transactions on Antennas and Propagation*, vol. AP-9, March 1961, pp. 140–153.
15. C. L. Gray, "Estimating the Effect of Feed Support Member Blockage on Antenna Gain and Side-Lobe Level," *Microwave Journal*, March 1964, pp. 88–91.

16. J. Ruze, *Physical Limitations on Antennas*, Technical Report 248, Research Laboratory of Electronics, Massachusetts Institute of Technology, Cambridge, Mass., October 1952.

17. H. Jasik (Ed.), *Antenna Engineering Handbook*, McGraw-Hill Book Co., New York, 1961, chap. 12.

18. Silver, op. cit., p. 365.

19. Silver, op. cit., p. 488.

20. J. Ruze, "Lateral Feed Displacement in a Paraboloid," *IEEE Transactions on Antennas and Propagation*, vol. AP-13, September 1965, pp. 660–665.

21. M. I. Skolnik (Ed.), *Radar Handbook*, McGraw-Hill Book Co., New York, 1970.

7
RADAR RECEIVERS

T. L. Lane

7.1 INTRODUCTION

Radar performance is normally specified in terms of the maximum range that a target of a given radar cross section can be detected. The radar equation, in its basic form, may be written

$$R_{max}^4 = \frac{P_t G_t G_r \lambda^2 \sigma_T}{(4\pi)^3 S_{min}}$$ (7-1)

where

R_{max} = maximum detection range
P_t = transmitter signal power
G_t = transmitter antenna gain
G_r = receiver antenna gain
λ = wavelength of transmitted electromagnetic energy
σ_T = target RCS
S_{min} = receiver minimum detectable signal

If it is assumed that the radar uses the same antenna aperture for transmitting and receiving, then Eq. (7-1) may be rewritten as

$$R_{max}^4 = \frac{P_t GA\,\sigma_T}{(4\pi)^2 S_{min}}$$ (7-2)

where

$$G_t = G_r = G = \frac{4\pi A}{\lambda^2}$$
A = effective antenna aperture

In designing a radar for a particular application, the antenna dimensions are

182

often restricted, leaving the transmitter power (P_t) and the receiver minimum detectable signal (S_{min}) as the key design variables.

Thus, Eq. 7-2 (in its simplest form) may be written

$$R_{max} = \text{constant} \times \left(\frac{P_t}{S_{min}}\right)^{1/4} \tag{7-3}$$

where the value P_t/S_{min} is referred to as the *radar performance figure*. Although Eq. 7-3 does not include such factors as system losses, atmospheric losses, integration gain, and so on, it does indicate the importance of the receiver design in the overall radar target detection process. In addition, since transmitter power is limited and high-power transmitters are costly and usually large, a well-designed receiver can provide the required radar performance with substantial savings in overall radar cost and size.

The major problem facing the receiver designer is how to maximize the radar's ability to detect a target echo in the presence of unwanted signals. The received echoes are often very faint, and the limitation of detection of these faint signals generally depends on the noise contributions of the radar receiver. Referring to Eqs. 7-1 to 7-3, S_{min} is a statistical quantity which involves the probability of detection and the probability of a false alarm, and is expressed as the radar receiver noise times the signal-to-noise ratio required for reliable detection.

In addition to receiver noise, which is the major consideration, other factors such as gain, dynamic range, bandwidth, phase and amplitude stability, and susceptibility to overload and saturation must be considered. The specific design of the radar receiver depends on the operational scenario, the type of transmitted waveform, the characteristics of any interference (unwanted signals), and the type of signal processing to be applied to the received signals.

7.2   BASIC PRINCIPLES

The purpose of the radar receiver is to amplify, detect, and process the desired echoes resulting from the radar transmission. The receiver must provide for separation of the desired echoes from the undesired or unwanted signals. These unwanted signals (interference) may be due to other radars, communication systems, jammers, or noise either received from galactic sources or generated within the radar system itself. Additional interference may occur as a result of echoes due to the interaction of the radar transmission with "clutter," that is, objects other than the desired target.

In many cases, the separation of desired echoes from unwanted signals can be accomplished by proper filtering in the receiver. If this is not done, then special processing must be accomplished prior to the determination of a valid detection (the desired echo). In either case, the received signal power must be of sufficient level for detection to occur in the radar indicator or processor. In the absence of noise, the weakest signal can be detected by providing sufficient gain. The detection process then involves separation of the desired echo from the undesired interference. Noise, however, can never be completely eliminated. For a receiver to be useful, it must be connected to a device for collecting the desired echoes, that is, an antenna. If this antenna is placed inside a black body enclosure at a uniform temperature T, a voltage appears at the terminals of the antenna. This *noise voltage* is due to the random motion of electrons in a finite resistance at nonzero temperatures and is referred to as *white noise*, *thermal noise*, or *Johnson noise*.

If a receiver were perfect and added no internal noise, Johnson noise would be the limiting factor in the detection of faint signals. Receivers, of course, are not perfect, and introduce not only internally generated noise but also spurious signals due to circuit characteristics and instabilities. Regardless of the limiting parameter, reduction of noise will improve the detection capability of any radar system. Therefore, reduction of noise is the key to the design of high-sensitivity, long-range radars.

In the remainder of this chapter, receiver design considerations are discussed, as well as receiver types, general receiver components, and examples of receiver applications. All discussions are confined to receivers for pulsed radars, although the techniques and conclusions usually apply to the design of a general radar receiver.

7.2.1 Receiver Types

Radar receivers can be classified in many ways: by application, construction, function, or configuration. For example, one might describe a radar receiver as being a pulsed-Doppler, wideband, dual-polarized, or log (logarithmic) receiver. Each of these classifications is indicative of one or more of the particular design features, but in general, radar receivers can be grouped into one of four types: superheterodyne, superregenerative, crystal video, and tuned radio frequency (TRF).[1] Of these four, the superheterodyne receiver configuration is used in virtually all radar systems because of its sensitivity, high gain, selectivity, and versatility.

Superregenerative Receiver

The superregenerative receiver is used in applications where simplicity and compactness outweigh the need for low noise reception. Because a single tube may be used for the RF amplifier in the receiver as well as the transmitter source, superregenerative receivers typically are found in radar beacon applications. The superregenerative receiver is an extension of the regenerative amplifier principle, which uses positive feedback for increased gain and selectivity. In the superregenerative receiver, the gain of the circuit is extended beyond the limit of the regenerative amplifier by operating the tube into the region of oscillation for a fraction of time. This results in greatly improved single-stage voltage gains in excess of 1 million. The circuit may be used as either an extremely sensitive tuned detector or as a high-gain RF amplifier.

The advantages of the superregenerative receiver are simplicity, high gain, light weight, low cost, and common transmitter-receiver tube. The disadvantages are gain instability, relatively poor selectivity, reradiation, and high receiver noise level. Usually the first three disadvantages can be overcome by additional circuit techniques and the major disadvantage becomes the high noise contribution, rendering the superregenerative receiver unacceptable for applications requiring high sensitivity.

Crystal Video Receiver

A simple block diagram of the crystal video receiver is shown in Figure 7-1(a). In this receiver, the RF is detected immediately at the output of the antenna, and the detected video is amplified using a high-gain amplifier. This type of receiver derives its name from the general use of crystal detectors in the detection stage. From Figure 7-1(a) the major advantages are obvious: simplicity,

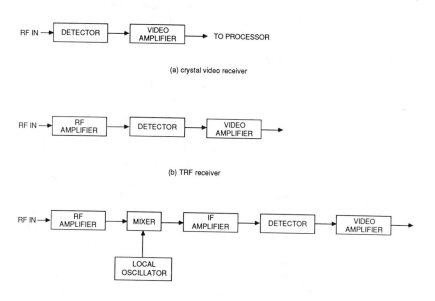

(a) crystal video receiver

(b) TRF receiver

(c) superheterodyne receiver

Figure 7-1. Simple block diagram of (a) a crystal video receiver, (b) a TRF receiver, and (c) a superheterodyne receiver.

small size, and low cost. In addition, since the crystal video receiver contains no RF amplifiers, the RF bandwidth is limited only by the input circuit (i.e., broad RF bandwidth).

The disadvantages of the crystal video receiver are the low sensitivity of the detector and the poor pulse shape of the video amplifier. Sensitivities of crystal video receivers are generally 30 to 40 dB less than those achievable in superheterodyne receivers. This is due to the increase in receiver noise level with decreasing input power to the detector. In addition to the lower sensitivity, all of the gain is at video, with typical gains of 100 to 120 dB required. This high gain requirement results in distortion of the received pulse shape; therefore, this receiver type cannot be used in applications requiring linearity.

TRF Receivers

The TRF receiver is realized by the addition of an RF amplifier prior to the detector in the crystal video receiver [see Figure 7-1(b)]. By adding an RF amplifier, the effect of the noise produced by the detector can be reduced by the gain of the RF amplifier. The overall receiver gain is now distributed between RF and video, reducing the video gain required and thus increasing the receiver's fidelity. Selectivity of the receiver is also improved by the addition of the RF amplifier. TRF receivers are not typically used primarily because of the cost–performance trade-off which can be realized by going one step further in the receiver design: the superheterodyne receiver.

Superheterodyne Receiver

The receiver configuration used in virtually all radar systems is based on the superheterodyne principle. A simple block diagram of the superheterodyne receiver is shown in Figure 7-1(c). In this receiver, the input echo at RF is downconverted to an intermediate frequency (IF) by mixing with a local oscillator (LO) signal; subsequent amplification and detection are performed on the IF signal. In this configuration, the detector prior to the video amplifier is often referred to as the *second detector* and the mixer as the *first detector*, since these are similar devices but operated in different modes. The operation of a detector in the mixing mode results in a much lower conversion loss and is the reason for the excellent sensitivity of the superheterodyne receiver. IF amplification is more cost effective and stable than microwave amplification, and the wider percentage bandwidths occupied by the IF signal simplifies filtering and improves selectability. The superheterodyne receiver is also adaptable, since the local oscillator frequency can be changed to track the transmitter frequency, thus preserving the receiver IF and filtering characteristics.

A more general block diagram of the superheterodyne receiver is shown in Figure 7-2.[2] A radar receiver may not (and in general will not) contain all of

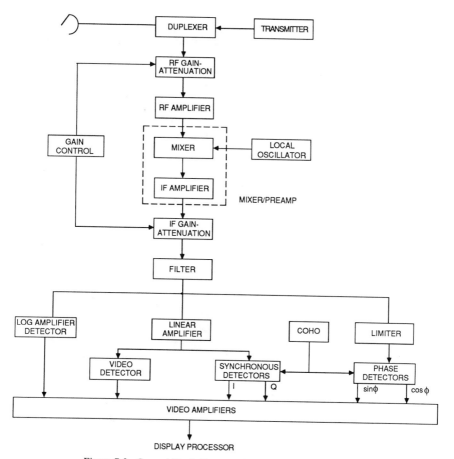

Figure 7-2. General block diagram of a superheterodyne receiver.

the elements shown in this figure, but their general position in the receiver process is indicated. Figure 7-2 is for a pulsed radar system which shares a common transmit/receive antenna with the antenna, transmitter, and duplexer shown in the upper portion of the figure. The duplexer is a device which switches the common antenna between the transmitter (during transmit time) and the receiver (during the listen period). Although shared between the transmitter and the receiver, the duplexer will be considered as part of the receiver in this and later sections and is often referred to as the *transmit-receive (TR) switch*.

The configuration in Figure 7-2 divides the total gain of the receiver into three stages: RF, IF, and video. In many cases, RF amplifiers are not used in modern radar receivers due to the unavailability of low-noise amplifiers and the excellent performance of presently available RF mixers. This is particularly true at the higher frequencies. Typical power gains from the input to the receiver to the output to the processor/display can vary from 100 to 200 dB. In systems where large RF signals may be encountered, RF gain control may be implemented to protect the sensitive mixers from saturation or burnout. Thus, RF gain control (attenuation) may be used in a system that does not require an RF amplifier.

Sensitivity time control (STC) is a technique used in some search radars to reduce the sensitivity of the display or system output due to range. This is accomplished by programming the receiver gain as a function of time (and therefore range) in a pulsed radar system. The gain of the receiver is increased as a function of time after transmission in accordance with the R^{-4} power variation. The adverse effect of the technique is that receiver sensitivity, and therefore the probability of detecting small targets at close-in range, is reduced. This technique may be implemented at either RF or IF stages. Automatic gain control (AGC) is a feedback technique used to adjust the receiver gain in order to maintain proper gain margins in the system tracking loops, which may fluctuate due to variations in the target cross section.

The local oscillator (LO) in a noncoherent radar is a free-running oscillator that may be tuned to the proper frequency via the automatic frequency control (AFC) circuit. This circuit determines the transmitter frequency and adjusts the LO to the transmitter frequency minus (or plus) the IF. This may be done on a continuous long-term basis or on a pulse-to-pulse basis. In a coherent receiver, the stable local oscillator (STALO) is derived from a coherent (usually CW) source used in the generation of the transmitted signal. The incoming RF is mixed with the STALO or LO signal and is down converted to the IF. Immediately following the mixer, the IF signal is amplified, usually by several stages, which may include IF gain control. In many cases, the mixer and first IF amplifier stage are integrated into one unit (mixer/preamp) to optimize mixer-amplifier performance.

The typical IF frequencies range between 30 and 300 MHz but can be as high as several gigahertz, depending on the application. Low IF frequencies are more desirable due to the lower cost of components, as well as gain, dynamic range, fidelity, stability, and selectivity performance. High IF frequencies are used when wide bandwidth is required. Selectivity or filtering is accomplished by proper selection of the IF amplifiers or by filtering after the IF amplifiers to reduce the noise being processed by the receiver and to maximize the receiver signal-to-noise ratio.

Following the selection of IF amplifiers, several types of processing are possible. For noncoherent detection and display, a linear amplifier and detector may be used to provide information to a display or detection circuit. A logarithmic amplifier-detector may be used in cases requiring wide, instantaneous, dynamic range. The logarithmic stage can provide an 80 dB useful dynamic range. In addition, other techniques involving constant false alarm rate (CFAR) detection may be implemented prior to automatic detection to reduce the number of false alarms within the system. The output of the IF amplifier may also be used in a compressive IF filter to perform pulse compression.

For coherent processing, the outputs of the IF amplifier may be processed by a pair of synchronous or phase-sensitive detectors to produce in-phase (I) and quadrature (Q) baseband Doppler signals that can be processed to extract moving targets (MTI) and reduce the effects of stationary clutter. For this detection, a coherent IF oscillator (COHO) is derived from the generation of the coherent transmitted signal. The IF output can also be processed in an IF MTI canceller to suppress the effects of the clutter and enhance the detection of moving targets.

The processing that follows the first down conversion and amplification can obviously vary widely, depending on the application. Regardless of the application, the same basic considerations in the pre-IF detection and processing affect the overall performance of radar detection and processing. These factors are considered in more detail in the following section. The hardware requirements and limitations are examined, and trends in the state of the art are discussed. Finally, several applications and receiver techniques are examined, along with millimeter wave receiver technology.

7.3 RECEIVER PERFORMANCE CONSIDERATIONS

7.3.1 Noise Characteristics

Noise considerations and parameters are usually the first characteristics specified for a radar receiver, although few radars actually employ the lowest-noise receiver available because such a choice may require too great a sacrifice in system performance and cost. The noise contribution of the radar receiver has been reduced sufficiently so that it is seldom a dominant factor in choosing between alternatives. However, the understanding of the receiver noise as the ultimate limitation on radar range performance is important.

The ability to detect received radar echoes is ultimately limited by thermal noise. It has been shown by Johnson and others[3] that the maximum available thermal noise power that can be transferred from a generator, linear network, or general two-port device for matched (conjugate) impedance is given by

$$N_p = kTB \qquad (7\text{-}4)$$

where

N_p = noise power
k = Boltzmann's constant (1.38×10^{-23} W-sec/Kelvin)
T = temperature of input impedance (Kelvin)
B = noise power bandwidth (Hertz)

Eq. 7.4 shows that the available noise power is independent of the impedance and directly proportional to the temperature and bandwidth. If a receiver could be developed that generated no internal noise, the received echoes would have to compete with this thermal noise (N_p). Since the receiver does introduce additional noise, the received echo must be correspondingly stronger.

7.3.2 Noise Figure

The process of detection or down conversion of the received target echoes results in the addition of excess noise due to circuit losses, noise generated within the semiconductors used in the mixers, and noise generated in the IF amplifiers that follow the mixers. A measure of the efficiency and noise contributions of a network is imbedded in the concept of noise factor, F. This may be defined as

$$F = \frac{S_i/N_i}{S_o/N_o} = \frac{S_i}{S_o} \times \frac{N_o}{N_i} \qquad (7\text{-}5)$$

where

S_i = input signal power
S_o = output signal power
N_i = input noise power
N_o = output noise power

Noise figure is the noise factor expressed in decibels:

$$F \text{ (dB)} = 10 \log_{10}\left[\frac{S_i/N_i}{S_o/N_o}\right]. \qquad (7\text{-}6)$$

In this chapter, as in most of the literature, the term *noise figure* will be used to denote both noise factor (decimal units) and noise figure (decibel units). Unless otherwise specified, all equations derived are in decimal units, with the results usually expressed in decibel units.

If a network does not add noise, then it will attenuate or amplify both the input signal and input noise by the same amount, that is, $S_o = S_i G$ and $N_o = N_i G$ when G is the gain of the network. Thus, a noise figure (F) equal to unity (0 dB) indicates that the network introduced no additional noise. If the network does introduce noise (as is always the case), then F is greater than unity (greater than 0 dB). A loss (such as that caused by an attenuator) has an available gain, $G = 1/L$, where L is the loss ratio ($L > 1$). The noise figure of an attenuation at an operating temperature of 290 K is equal to its loss ratio. If its temperature is not 290 K, then its noise figure is expressed as LT/T_0, where T is its temperature. To standardize the value of F, a standard absolute noise temperature of 290 K is used for the noise temperature of the input impedance of the network. The standard symbol T_0 is used to denote 290 K, standard absolute temperature. Thus Eqs. 7-4 and 7-5 can be combined and rewritten as

$$F = \frac{1}{G} \frac{N_o}{N_i} = \frac{1}{G} \frac{kTB}{kT_oB} \qquad (7\text{-}7)$$

where

G = power gain of the network ($G = S_o / S_i$)
T = noise temperature of the output network
T_0 = standard noise temperature of the input network

Since a receiver consists of a number of networks (mixers, amplifiers, attenuators, etc.), it is of interest to examine the case of two or more networks in cascade. The overall noise figure of the combination may be computed as[4]

$$F_T = F_1 + \frac{F_2 - 1}{G_1} \qquad (7\text{-}8)$$

where

F_T = overall noise figure
F_1 = noise figure of the first network
G_1 = power gain of the first network
F_2 = noise figure of the second network

As an example, let us assume that $F_1 = 5$ dB, $F_2 = 15$ dB, and $G_1 = 17$ dB. Converting to decimals and using Eq. 7.8, we find that

$$F_T = 3.16 + \frac{31.62 - 1}{50} = 3.77$$

or

$$F_T = 5.8 \text{ dB}.$$

It is obvious that if the first-stage network has adequate gain, the noise figure of the total network is primarily determined by the first stage. The amount of gain required by the first stage can be determined such that the total noise figure is no more than a certain number of decibels (X dB) above F_1 (i.e., $F_T \leqslant F_1$ + X dB) by solving Eq. 7-8 for gain under these conditions:

$$G_1 \geqslant \frac{F_2 - 1}{F_1} \times \left[\frac{1}{10^{X/10} - 1} \right]. \tag{7.9}$$

Thus, if in the example F_T is required to be no more than 0.5 dB greater than F_1,

$$G_1 \geqslant \frac{31.6 - 1}{3.16} \times \left[\frac{1}{10^{0.5/10} - 1} \right] = 79.36$$

or

$$G_1 \geqslant 19 \text{ dB}$$

and the total two-stage noise figure is $\leqslant 5.5$ dB.

For two or more stages, the noise figure for cascaded networks can be expressed by the general equation for N networks:

$$F_N = F_1 + \frac{F_2 - 1}{G_1} + \frac{F_3 - 1}{G_1 G_2} + \cdots + \frac{F_N - 1}{G_1 G_2 \cdots G_{N-1}}. \tag{7-10}$$

In almost all applications of the superheterodyne receiver, the RF signal is immediately down converted to IF, and therefore the noise level of the receiver is determined by the mixer/IF network. A mixer noise figure can be expressed similarly to that in Eq. 7-7, but the effective noise temperature is usually different from the temperature of its environment since it is not a passive network. The ratio of the effective noise temperature to the standard noise temperature is defined as N_R. Thus, mixers have noise figures given by $F_M = L_C N_R$, where L_C is the conversion loss of the mixer. To obtain the overall noise figure of the mixer and IF amplifier, the cascade noise figure is computed as

$$F_o = F_M + \frac{F_{IF} - 1}{G_M} \tag{7-11}$$

where

F_o = overall mixer noise figure
F_M = noise figure of the mixer
F_{IF} = noise figure of the first IF amplifier
G_M = $1/L_C$

so that the overall noise figure is expressed as

$$F_o = L_C N_R + \frac{F_{IF} - 1}{1/L_C} = L_C(N_R + F_{IF} - 1). \qquad (7\text{-}12)$$

These factors are summarized in Figure 7-3[4,5] for a mixer with RF losses preceding the mixer. The mixer/IF noise figure, then, is an indication of the output noise level of the receiver in receiver designs where the mixer/IF stages occur first in the receiver process.

The minimum echo signal power, S_{min}, that can be detected at the output of the mixer/IF is dependent upon the output signal-to-noise ratio, S_o/N_o, required for detection and the level of noise power out of the receiver, so that

$$S_{min} = kT_o B \, F(S_o/N_o)_{min} \qquad (7\text{-}13)$$

which appears directly in the range performance of the radar as

$$R^4_{max} = \frac{P_t G A_e \sigma_T}{(4\pi)^2 kT_o B \, F(S_o/N_o)_{min}}. \qquad (7\text{-}14)$$

Since S_{min} is a critical radar design parameter, it is often expressed (assuming

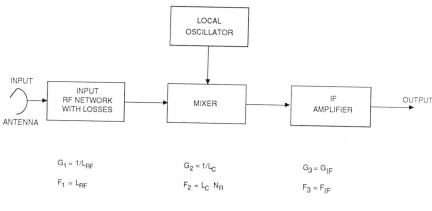

Figure 7-3. Factors affecting the overall receiver noise figure.

a 0-dB S_o/N_o ratio) in decibel watts (dBW) or milliwatts (dBm) as

$$S_{min} \text{ (dBm)} = -114 \text{ dBm} + 10 \log B + F \qquad (7\text{-}15)$$

where

B = noise bandwidth in megahertz
F = noise figure of the mixer/IF in decibels

Thus, for $B = 200$ MHz and $F = 5$ dB, Eq. 7-15 becomes

$$S_{min} \text{ (dBm)} = -114 \text{ dBm} + 10 \log (200) + 5 \text{ dB}$$

$$= -86 \text{ dBm}$$

or

$$S_{min} \text{ (mW)} = 2.5 \times 10^{-9} \text{ mW}.$$

The value of $(S_o/N_o)_{min}$ in Eqs. 7-13 and 7-14 is a statistical quantity and involves knowledge of the target, interference, processor, and application. But in general, a value between 10 and 20 dB is needed for reliable detection. In the example above, S_{min} would then vary between -66 and -76 dBm received echo power necessary for detection.

7.3.3 Radar Receiver Noise Figure

In the superheterodyne receiver, there is an equal response to frequencies both above and below the LO frequency, as shown in Figure 7-4. These two frequencies are displaced from the local oscillator frequency by the IF frequency and are referred to as the *signal channel* and the *image channel* (or *sidebands*). To the mixer, both the signal and the image frequency (which is displaced from the signal frequency by twice the IF frequency) are indistinguishable, and both input frequencies, when mixed with the LO frequency, generate a signal at the IF frequency. In determining the mixer noise figure, a broadband noise source is used which covers both sidebands. Thus, both the image frequency and signal frequency channels respond. The noise figure thus recorded is a broadband noise figure and is referred to as a *double sideband noise figure* (*DSBNF*) or *radio astronomy noise figure* (*radiometer noise figure*). This noise figure is the one most often reported in the mixer specifications and is useful only when the application is for a radiometer-type receiver where the received echo signal occupies both the upper the lower sidebands. In this application, input noise power contains equal contributions in the image and signal bands, and the noise

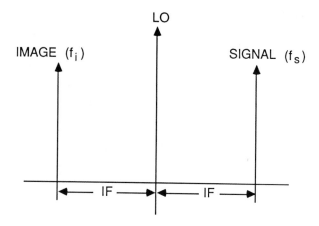

$$f_s = f_{RF}$$

$$f_i = f_{RF} - 2f_{IF}$$

Figure 7-4. Mixer response to both the signal frequency and an image frequency.

figure becomes

$$F_{\text{DSB}} = \frac{1}{G} \frac{N_o}{kT_o(2B)}. \tag{7-16}$$

In a radar, the input signal-to-noise ratio includes only the noise at the signal frequency, and not spurious contributions such as those from an unused image frequency. Thus, the received echo occupies only one of the sidebands, and it is necessary to convert to a radar noise figure referred to as the *single sideband noise figure* (*SSBNF*) and given by

$$F_{\text{SSF}} = \frac{1}{G} \frac{N_o}{kT_o B}. \tag{7-17}$$

The output noise has not changed; therefore, the radar noise figure (SSB) is twice (3 dB higher) that a DSBNF. The noise figure of a mixer depends on its application, leaving one noise figure for radar applications (SSB) and another for radiometer applications (DSB). Vendors usually specify the mixer noise figure in DSBNF; therefore, 3 dB must be added for the radar noise figure.

7.3.4   Dynamic Range

The radar receiver is required to receive and detect signal levels near the receiver noise level for maximum range performance. It must also be able to tolerate echo signals from large-cross-section targets at close range. The receiver front end must be able to transform the large signals faithfully without excessive degradation, and it must be able to survive large power inputs that will cause saturation in the receiver, up to the point where permanent damage (usually diode burnout) is sustained.

Unless some form of RF gain control (attenuation) is used prior to the mixer, the useful dynamic range of the receiver is determined by the mixer noise level and a 1-dB compression point. In general, instantaneous dynamic ranges of 70 to 100 dB are available depending upon the mixer configuration. Care must be exercised to ensure that the various stages following the mixer (i.e., IF amplifiers, detectors, video amplifiers) do not saturate prior to the mixer in order to preserve the full dynamic range. Insertion of RF signal attenuation prior to the mixer during periods of large signal returns will increase the effective dynamic range; however, insertion of any component or network into the receiver front end will result in additional loss prior to the mixer, and therefore will increase the noise level and reduce the receiver sensitivity and dynamic range (see Figure 7-3).

The superheterodyne receiver is basically a linear receiver for signals from the receiver noise level up to a power level of about -10 dBm, depending upon the mixer configuration. Techniques such as STC may be helpful in reducing the effects of close-range targets without degrading the long-range detection performance. The dynamic range is particularly important for processing multiple echoes from each transmitted pulse. The reception of a large signal causing saturation or limiting in the receiver can result in the masking of more distant echoes during the time required for the receiver to recover from the overload.

Bandwidth

The receiver instantaneous bandwidth must be large enough to process the received echo pulses, yet must be as small as possible to reduce the noise and other interferences entering the receiver. The latter is a critical parameter in determining the receiver noise level and sensitivity (S_{min}). The bandwidth term used in calculating S_{min} is the receiver predetection bandwidth and is determined for all practical purposes by the IF bandwidth established prior to the video detection.

Generally, the RF portion of the receiver is sufficiently broadband that the critical response is in the IF section of the receiver. Care must be taken with a broadband RF front end even if out-of-band signals are not passed by the IF. The out-of-band signals could result in saturation of the receiver front end (mixer) or generation of spurious outputs (harmonic products) that appear in the IF passband.

The gain and filtering in the receiver are usually distributed over many stages of varying gains and losses. It is necessary to select IF components for their gain, bandwidth, and output power capabilities. For a maximum overall transfer characteristic, the gain must be distributed so that intermediate stages do not saturate prior to saturation at the RF converter (mixer), thus maintaining the maximum dynamic range. The overall bandwidth of the IF will be less than that of the narrowest-bandwidth component.

7.3.5   IF Selection and Filtering

The selection of the IF to be used is a function of transmitted waveform characteristics, mixer/LO implementation, and availability of IF processing components. The receiver IF chosen must be sufficiently high to provide the necessary bandwidth for the received echoes. The IF is chosen such that the output of the mixer is in a range which produces a satisfactory noise figure. The lower the IF, the higher the mixer- and LO-induced noise become; the higher the IF, the higher the noise contributions from the IF amplifier. In addition, the percentage bandwidth required for filtering the IF becomes larger as the IF is lowered. IFs from 30 MHz to 4 GHz are common in radar applications, but additional IF signal-processing components are available at the lower frequencies. Such items as logarithmic (log) IFs, pulse compression filters, surface acoustic wave devices, and limiters are easier to obtain at 30 to 500 MHz than at 1 to 4 GHz. At the higher millimeter wave frequencies (above 140 GHz), single-ended mixers are the primary down converters, with limited power sources available for LOs. The LO power required is minimized by using waveguide directional filters,[6] rather than couplers, to inject the LO signal into the receiver line. This usually requires a minimum separation of LO and transmitter frequencies of 750 to 1000 MHz, thus requiring a higher IF. Broadband systems also require higher IF frequencies to minimize spurious responses.

A key design parameter in the reduction of receiver noise is in the selection of the IF filter network. If the bandwidth of the receiver is wide compared to the desired echo bandwidth, then excesss noise will enter the receiver. On the other hand, if the bandwidth of the receiver is less than the echo bandwidth, signal energy is lost. In both cases, the signal-to-noise ratio of the receiver is reduced. A network whose frequency response function maximizes the receiver signal-to-noise ratio is called a *matched filter*. The network is designed such that the filter's frequency response function is the complex conjugate of the input pulse. The basic rule of thumb for a pulse radar application is that the receiver bandwidth B should be equal to the reciprocal of the pulse width τ. This is a good approximation for a well-designed superheterodyne receiver in a pulsed radar. In many cases, the bandwidth B is selected to be slightly greater than the pulse width τ to include such things as system instabilities or Doppler effects due to target velocity. If a transmitter produces a 100-ns pulse, the receiver must have at least a 10-MHz instantaneous bandwidth, that is $B = (\tau)^{-1}$ = $(100 \times 10^{-9} \text{ sec})^{-1} = 10^7$ Hz. Other factors determine the final selection of the receiver filter bandwidth. For example, if the transmitted spectrum is greater than 10 MHz, as in a chirped or spread spectrum waveform, or if the frequency stability of either the transmitter or LO is insufficient, then these factors will determine the minimum sufficient receiver bandwidth.

As already stated, narrowband filtering is most easily and conveniently accomplished in the IF section. For superheterodyne receivers with a matched filter incorporated in the IF portion, the video sections will have little effect on the receiver signal-to-noise ratio. Thus, the IF filter bandwidth is typically used in determining the receiver noise bandwidth B and the receiver minimum detectable $S_{\min}$.

7.4 RECEIVER COMPONENTS TECHNOLOGY

7.4.1 Receiver Protection

In most radar applications, one or more RF components are placed between the antenna and the mixer. One commonly used RF component is a consequence of the use of a single antenna for both transmission and reception. This component is called a *duplexer*, and is responsible for protecting the receiver during transmission and for switching the antenna between the transmitter and the receiver. These devices add loss in the receiver channel, and the time for the device to recover after the transmitter pulse limits the minimum range that an echo could be received. A duplexer is simply a microwave double-pole, double-throw switch, but a mechanical switch cannot be used because of the requirements for fast switching (nanoseconds to microseconds).

Many kinds of duplexers are possible, but the most common type used in high-power radars employs power-sensitive gas discharge tubes to direct the transmitted or received energy. Such tubes are referred to as *transmit-receive* (*TR*) and *anti-transmit-receive* (*ATR*) tubes and may be used in a variety of duplexer configurations. A cross-sectional view of a typical gas TR device is shown in Figure 7-5.[7] The high power of the transmitter ionizes the gas, and the tube then behaves as a short circuit to prevent the transmitter power from reaching the receiver. On receive, the tubes are not ionized and will permit the signal to pass, but typically with an insertion loss of about 0.5 to 1.0 dB. One significant problem in duplexers occurs because the normal ionization time of TR tubes is long enough to permit significant "spike" leakage. To prevent this, it is necessary to supply a DC "keep alive" voltage to the tube to shorten the ionization time. The keep alive voltage can shorten the life of the tube and introduce additional noise into the receiver.

Some duplexer configurations are characterized by narrow bandwidths or transmitter leakage due to antenna mismatch. The latter problem can be solved

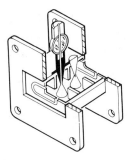

Figure 7-5. Cross-sectional view of a gas TR tube. (From ref. 7)

by placing an additional medium-power TR in the RF line of the receiver. Such a device is also effective in protecting the receiver against nearby sources of radiation which are strong enough to damage the receiver but not strong enough to fire the TR tubes in the duplexers. Figures 7-6 and 7-7 are examples of balanced duplexers using TRs and diode protectors.

Duplexers such as the one shown in Figure 7-6 are constructed using a combination of 3-dB short-slot hybrids and either TR or ATR tubes with optional diode protectors on the input to the receiver. When the transmitter fires, the energy level is sufficient to ionize the gas in the TR and ATR tubes. When this occurs, the TR tube presents a short circuit across the coupler ports; the ATR tube presents a low impedance. The phase characteristics of the 3-dB hybrid are such that essentially all of the power is coupled to the antenna port during tube ionization (transmitter on). The tubes are not ionized during the receive time, and the signal is coupled from the antenna port to the receiver port.

Several types of duplexers which do not employ gas tubes are available; many of them depend on the nonreciprocal properties of ferrite materials for their operation. Ferrite duplexers may be realized using such ferrite devices as Faraday rotators, nonreciprocal phase shifters, or junction circulators. Figure 7-8 is a typical implementation of a ferrite duplexer which uses the phase characteristics of the 3-dB short-slot hybrid, the folded magic tee, and ferrite slabs to control the direction of the TR signals.

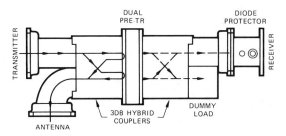

Figure 7-6. Balanced duplexer with dual TR (shown in transmit). (From ref. 7)

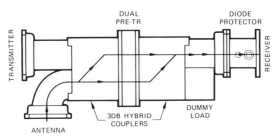

Figure 7-7. Balanced duplexer with dual ATR and diode protector (shown in receive). (From ref. 7)

A receiver diode protector must be used in conjunction with the ferrite duplexer (also referred to as a *ferrite circulator*), since the isolation between the transmitter and receiver ports is only 25 to 30 dB. Under conditions of antenna mismatch, the effective isolation is greatly reduced and additional receiver protection is recommended even for applications using low transmitter power. Ferrite duplexers have the advantage of long life, higher reliability, and shorter recovery time. Active duplexers employing some of the recently developed high-power semiconductor diodes have also been built, but so far their application to radar has been very limited. Passive hybrid junctions may be used as a duplexer, but they cause a combined loss of 6 dB on transmission and reception.

Limiters are components designed to perform the same function as the receiver protector TR discussed above. They are usually passive devices which

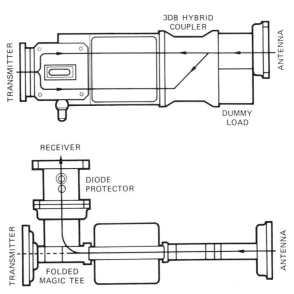

Figure 7-8. Example of a ferrite duplexer (shown in receive). (From ref. 7)

either reflect or absorb essentially all incident RF power above a certain level. Limiters can be built using either ferrite or semiconductor diodes and are generally more reliable than TR tubes. Ferrite limiters depend for their operation on second-order effects which occur in the ferrite material above a threshold power level. Diode limiters operate by making the diode bias level, and consequently the shunt impedance represented by the diode, a function of the incident RF power. Insertion losses of limiters can be made as small as a few tenths of 1 dB.

The use of gas discharge tubes and duplexers is primarily limited to radar frequencies below 35 GHz. In the millimeter-wave (MMW) range, a combination of circulators and Faraday rotational switches is used for receiver protection. In many cases, diode switches have been developed which have extremely fast switching speed and good isolation with relatively low insertion loss. PIN switches (so named because of their construction: P-type, Intrinsic-type, or N-type semiconductor material) have recently become available at frequencies of up to 100 GHz with 5- to 25-ns switching time, 15- to 30-dB isolation, and 2- to 4-dB insertion loss. The major disadvantage of these devices is their low power-handling capabilities—usually less than 1 W CW and 10 W peak. The transmitter power of most MMW systems is low (<5 kW peak) so that these devices used in conjunction with ferrite circulators have been satisfactory. Higher-power gas discharge devices are being developed due to the use of higher pulsed power systems that operate as 95 GHz and higher frequencies. Table 7-1 presents a sample of some receiver protection devices and their characteristics.

7.4.2 Mixers

With the greatly improved noise performance of modern mixer diodes, few superheterodyne receivers use an RF amplifier as a receiver front end. Mixers with double sideband noise figures of 1 dB at 1 GHz to 5 dB at 95 GHz are readily available, and because of reduced complexity, cost, and size, the mixer is by far the perferred receiver front-end component.

In general, mixers are used to convert a low-power signal from one frequency to another by combining it with a higher-power (LO) signal in a nonlinear device.[8-11] The nonlinear mixing process produces many sum and difference frequencies of the signal, LO, and their harmonics. Usually, the difference frequency between the RF and LO signals is the desired output frequency at IF. Devices which are generally used as both mixer and detector microwave devices are the point contact and Schottky barrier diodes. The mixing action is produced by these devices via their nonlinear transfer function, which can be expressed as

$$I = f(V) = a_0 + a_1 V + a_2 V^2 + a_3 V^3 \cdots a_n V^n \qquad (7\text{-}18)$$

Table 7-1. Receiver Protector Characteristics.

Frequency	Device (Single)	Input Power (Peak)	Insertion Loss (dB)	Leakage or Isolation	Recovery Time (μs)
UHF	TR tube	2 MW	2.0	1 W peak	150
	TR tube	10 kW	6.0	2 W peak	50
L-band	TR tube	50 kW	0.4	1 kW peak	10
	Solid state	1 kW	0.4	0.1 avg.	20
	Balanced duplexer	50 MW	0.4	0.35 avg.	300
	Branch duplexer	2 MW	0.7	0.25 avg.	35
S-band	TR tubes	100 kW	0.7	0.3 avg.	5
	TR limiter	100 kW	0.5	0.1 avg.	3
	Ferrite limiter	20 kW	1.4	0.1 avg.	20
X-band	TR tube	100 kW	0.7	0.2 avg.	1.5
	TR limiter	100 kW	0.8	0.1	3
K_a-band	TR tubes	10 kW	0.5	0.15 avg.	1.1
	Ferrite switch	0.5 W avg.	1.1	40 dB	1.0
	TR limiter	100 kW	1.7	0.2 avg.	1.0
	PIN switch	10 W	1.5	40 dB	0.1
W-band	Ferrite switch	0.5 W avg.	1.8	35 dB	1.0
	PIN switch	10 W	1.5	30 dB	0.1

where I and V are the device current and voltage, respectively. For a mixer, the applied voltage is a composite of an RF signal, $V_{RF} \sin \omega_{RF} t$, and an LO signal, $V_{LO} \sin \omega_{LO} t$, in the form

$$V(t) = V_{RF} \sin \omega_{RF} t + V_{LO} \sin \omega_{LO} t \qquad (7.19)$$

where ω_{RF} and ω_{LO} are the RF and LO angular frequencies, respectively. This results in an infinite Taylor power series of mixing products given by

$$I = a_0 + a_1 (V_{RF} \sin \omega_{RF} t + V_{LO} \sin \omega_{LO} T) + a_2 (V_{RF} \sin \omega_{RF} t$$

$$+ V_{LO} \sin \omega_{LO} t)^2 + \cdots + a_n (V_{RF} \sin \omega_{RF} t + V_{LO} \sin \omega_{LO} t)^n. \qquad (7.20)$$

Analysis of the first two components in Eq. 7-20 shows that a DC level, as well as the original signals, is present in the output of the nonlinear device. The primary mixing products, $\omega_{LO} \pm \omega_{RF}$, come from the second-order term and are proportional to a_2 in amplitude. These terms also generate the second harmonics of both signals. The third- and higher-order terms generate products of the form $n\omega_{LO} + m\omega_{RF}$ and higher-order harmonics.

The desired output frequency, the IF (ω_{IF}), can be either the lower or the upper sideband. In most applications, the lower sideband is the sideband of interest and the term *mixer* is used primarily to describe this down conversion process. Note that many other mixing products, called *intermodulation products* (*IMP*), may be present at frequencies within the IF passband. A simple aid often used in system analysis is an IMP chart such as that shown in Figure 7-9,[12] which is a plot of the mixer output frequency spectrum. The vertical axis is the output frequency (F_o) normalized to the LO frequency (F_2); the horizontal axis is the input frquency (F_1) normalized to the LO frequency. The in-band spurs (or spurious responses) may be determined by using the two heavy lines, $F_2 + F_1$ and $F_2 - F_1$, which represent the sum and difference mixing frequencies, respectively. The figure gives all the IMPs up to at least the 10th order ($m + n = 10$). The in-band spurious outputs are defined by the products that cross or are tangential to the sum or difference lines. By using such charts, the re-

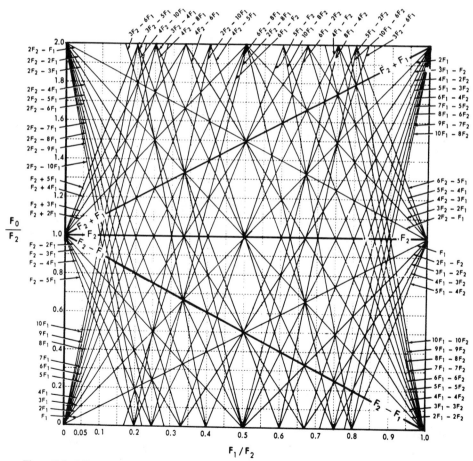

Figure 7-9. Mixer spurious chart. (By permission, from Manessewitsch, ref. 12; © 1976 John Wiley & Sons, Inc.)

ceiver designer can select frequencies with few or no in-band responses. Both in-band and out-of-band spurious levels can be determined for given mixers by using product data sheets such as the one in Table 7-2.[10] This table gives the level of the various IMP for a series of mixer models (WJ-M1, M1D, and M1E) for different input signal levels and three different LO signal levels. The values in the table are in decibels below the fundamental output ($m = n = 1$) case. Using both the IMP chart and the product data sheet, the level of spurious responses and amount of out-of-band suppression needed can be determined. If the in-band IMPs are unacceptable, then new frequencies or mixers must be considered.

Mixer dynamic range is a major parameter and can be ascertained from an input-output characteristic such as the curve in Figure 7-10.[13] It is simply the power range over which the device can be used. The low end is limited by thermal noise and the noise figure of the mixer. The maximum signal level is limited by the saturation of the mixer output. The saturated output is approximately the LO power level minus the mixer conversion loss. The definition more commonly used is the 1-dB compression point of the mixer. This is the input power level at which conversion loss has increased by 1 dB over the low-level conversion loss value. In many applications, however, the useful dynamic range is the spurious-free dynamic range. The upper power level in this case is determined by the input power that generates third-order IMPs that are just equal to the mixer output noise level (see Figure 7-10).

7.4.3 Mixer Configurations

A large variety of mixers has evolved to meet an equally diverse range of applications. Inside the mixer, the means of coupling the RF, LO, and IF signals to the nonlinear device must be provided. Coupling can be achieved in many ways, each with its own characteristics, but in all cases the goal is to provide three independent and isolated ports, one for each signal. In addition, all unwanted responses should be eliminated from the output ports by filtering, phasing, enhancement, or rejection methods. The popular types of mixers that have evolved are summarized in Table 7-3.[8]

The simplest form of diode mixer is a single-ended design with one diode terminating a transmission line (Figure 7-11) with LO power applied through a directional coupler.[8,13] The match between the diode and the input transmission line determines the VSWR for the signal input, as well as the amount of LO power appearing at the signal input port. The amount of LO power at the signal port is commonly measured in terms of LO-to-RF isolation. This is important in many receiver applications, since the LO energy can be radiated by the receiver antenna.

The IF output signal is extracted from the diode by a low-pass filter, eliminating the RF and LO signals. A high-pass filter is needed on the RF side of the diode to prevent loss of IF energy in this direction. This filtering requires

Table 7.2. Example of Typical IMPs in Mixers. (From ref. 12)

Each cell lists three model values (WJ-M1, WJ-M1D, WJ-M1E) for each signal level. Rows are grouped by harmonic of signal (0–7); columns are harmonics of the local oscillator (0–8).

Harmonic of signal	Signal level	0	1	2	3	4	5	6	7	8
7	0 dBm	79 >99 >99	69 >99 >99	80 >99 >99	74 >99 >99	83 >99 >99	63 78 >99	78 >99 >99	60 81 >99	71 99 >99
7	−10 dBm	>90 >90 >90	>90 >90 >90	>90 >90 >90	>90 >90 >90	>90 >90 >90	>90 >90 >90	>90 >90 >90	>90 >90 >90	>90 >90 >90
6	0 dBm	90 >99 >99	86 >99 >99	91 >99 >99	91 >99 97	90 >99 >99	84 >99 >99	93 >99 >99	84 >99 >99	88 >99 98
6	−10 dBm	>90 >90 >90	>90 >90 >90	>90 >90 >90	>90 >90 >90	>90 >90 >90	>90 >90 >90	>90 >90 >90	>90 >90 >90	>90 >90 >90
5	0 dBm	72 93 >99	70 73 96	71 87 >99	52 72 95	77 88 >99	46 66 >99	75 85 >99	45 64 90	73 82 >99
5	−10 dBm	>90 >90 >90	80 >90 >90	>90 >90 >90	71 >90 >90	>90 >90 >90	68 >90 >90	75 >90 >90	65 >90 >90	88 >90 >90
4	0 dBm	80 96 >99	79 80 91	82 96 >99	77 80 92	82 95 90	76 82 95	77 98 87	72 78 94	77 90 87
4	−10 dBm	86 >90 >90	>90 >90 >90	86 >90 >90	88 >90 >90	88 >90 >90	85 >90 >90	86 >90 >90	85 >90 >90	>90 >90 >90
3	0 dBm	51 63 81	49 58 73	53 65 85	51 60 69	55 65 85	48 55 68	54 64 85	53 54 64	58 66 87
3	−10 dBm	67 87 90	68 77 90	69 87 90	50 78 >90	77 >90 >90	47 75 >90	74 85 >90	44 77 89	74 88 >90
2	0 dBm	69 68 64	72 67 71	79 76 62	67 67 70	75 80 63	68 66 70	77 82 61	68 66 62	75 83 64
2	−10 dBm	73 86 73	73 75 83	74 84 75	70 75 79	71 86 80	64 74 80	69 87 77	64 74 82	69 84 79
1	0 dBm	25 25 24	0 0 0	39 39 35	13 11 11	45 50 42	22 16 19	54 59 50	37 19 39	69 59 49
1	−10 dBm	0 0 0	0 0 0	35 39 34	13 11 11	40 46 42	24 14 18	45 62 49	28 19 37	59 53 49
0	0 dBm	24 23 24	36 39 29	45 42 20	52 46 32	63 58 24	45 37 29	60 65 27	71 49 30	64 75 29
0	−10 dBm	26 27 18	28 27 18	35 31 10	39 36 23	50 47 14	41 36 19	53 51 17	49 37 21	51 63 19

Harmonics of signal (row axis)
Harmonics of local oscillator (column axis)

Key (example cell, expanded):

36	39	29
28	27	18

Signal at 0 dBm; local oscillator at +7, +17, and +27 dBm for models WJ-M1, WJ-M1D, and WJ-M1E respectively.

Signal at −10 dBm; local oscillator at +7, +17, and +27 dBm for models WJ-M1, WJ-M1D, and WJ-M1E respectively.

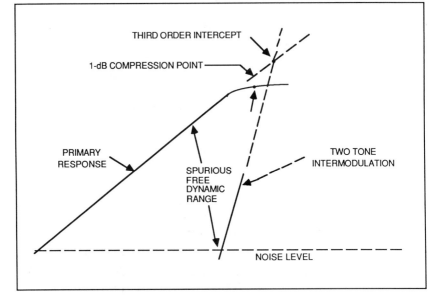

RF INPUT POWER - dBm

Figure 7-10. Mixer input–output relationship.

Table 7-3. Mixer Comparison Guide

Performance Parameter	Single-ended	Balanced (90°)	Balanced (180°)	Double-balanced	Image-reject	Image-Recovery
			Mixer Type			
Conversion Loss	Good 8–10 dB	Good 8–10 dB	Good 8–10 dB	Very Good 6–7 dB	Good 8–10 dB	Excellent 5 dB
VSWR LO, RF	Good, Poor	Good Good	Fair Fair	Poor Poor	Good Good	Good Good
LO/RF Isolation	Fair 12–18 dB	Poor < 12 dB	Very Good > 23 dB	Very Good > 23 dB	Good 18–23 dB	Very Good > 23 dB
LO Power Required (unbiased)	+ 13 dBm	+ 5 dBm	+ 5 dBm	+ 10 dBm	+ 7 dBm	+ 7 dBm
Spurious Rejection	Poor	Fair	Fair Odd: Fair	Good	Fair	Fair
Harmonic Suppression	Poor	Fair	Even: Good	Very Good	Even: Good Odd: Fair	Even: Good Odd: Fair

that the IF always be less than the lowest RF or LO frequency. In addition, if the IF response is extended to DC, the DC component of the voltage will be present at the IF port, as indicated by Eq. 7-20. All harmonics and IMPs will also be present and must be suppressed by filtering, if required.

Many disadvantages of a single-ended mixer can be overcome by combining two circuits in a balanced configuration (Figure 7-12).[10,11,13] A 3-dB hybrid is used to supply RF and LO power to two mixer diodes. Either a 90° or a 180° hybrid can be used. Each has certain advantages and disadvantages, but both balanced designs offer better performance than single-ended mixers. Advantages include reduced spurious responses, cancellation of the DC components at the IF output, and convenient separation of LO and RF inputs.

The first approach, combining two single-ended mixers in parallel and 180° out of phase, suppresses the even harmonics of one of the input signals. The mixer is usually designed to suppress the harmonics of the LO signal. The degree to which the harmonics and the IMPs associated with them are suppressed depends on the degree of balance. This is influenced by the balance of the hybrid and the match of the mixer diodes.

The properties of the 180° hybrid control the characteristics of this class of mixer. If the two output arms of a 180° hybrid are terminated in identical impedances, all reflected power is directed back to the input port. With good diode balance, therefore, the reflected LO power appears at the LO port, but not at the signal port. As a result, LO-to-RF isolation of the 180° balanced mixer is good (typically 20 dB or more). This same property of the hybrid, however, causes the VSWR for the LO port and the signal port to depend on the diode match to the transmission line, as in the case of a single-ended mixer. VSWR is typically 2.0:1 unless care is taken to match the impedance of the diodes carefully to that of the transmission line. As with the single-ended mixer, appropriate filtering is needed around the diodes to separate RF and IF signals for minimum conversion loss.

Balanced mixers designed with 90° hybrids exhibit significantly different RF properties than the components just discussed. The degree of harmonic and IMP suppression in this mixer type is more complicated and depends on the particular IMPs of the RF and the LO signals.

Equal terminations at the 90° output ports of the hybrid result in all reflected power being directed to the fourth, normally isolated, port. Good diode balance, therefore, leads to low VSWR (typically less than 1.5:1) at either the signal or the LO port. The LO-to-RF isolation now depends on the match between diode and transmission line impedances. This specification is normally poor (typically 7 dB) unless careful impedance matching is undertaken. Appropriate filtering is again needed around the diode to separate the RF and IF signals for minimum conversion loss. This type of mixer has been widely used for octave band, single-balanced designs because the 90° hybrid is relatively easy to fabricate in coaxial, stripline, or microstrip transmission media.

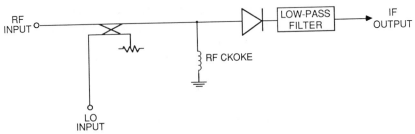

Figure 7-11. Schematic representation of a single-ended mixer.

For a given IF, ω_{IF}, a signal either above or below the LO frequency, ω_{LO} $\pm$ ω_{IF}, produces an IF output. If one of these is considered to be the desired *signal frequency*, then the other is commonly termed the *image frequency*. In many applications, it is desirable to either eliminate or distinguish the image response from the desired signal response. If the IF is high enough and the RF bandwidth narrow enough so that the signal and image frequency spectra bandwidths do not overlap, then the image response can be eliminated by appropriate filtering. This type of design is suitable for narrowband systems where a high degree of image rejection is desirable. For broadband applications, especially for octave bandwidths, filtering cannot be used for image rejection. Here the image frequency is rejected by phasing techniques.

The use of four diodes in a ring (or bridge) network provides cancellation of LO reflections and noise at both the RF and IF ports. The symmetry of the four-diode array provides complete isolation between the RF and LO ports when all diodes are perfectly balanced. This design provides excellent suppression of even harmonics of both the RF and LO signals and reduces the IMPs. A disadvantage of this type of mixer is the increased LO power required to optimize the conversion loss. To achieve optimum performance in this mixer configuration, very broadband transformers are required and the diodes must be well matched and placed close together to avoid inductive parasitics. To achieve the

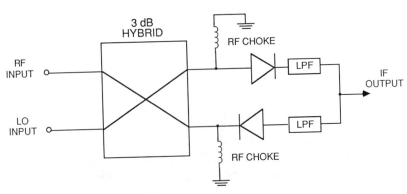

Figure 7-12. Schematic representation of a balanced mixer.

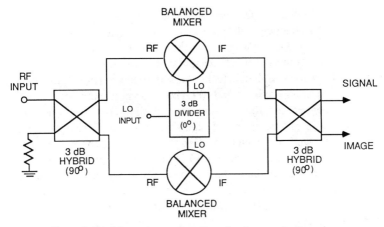

Figure 7-13. Schematic representation of an image rejection mixer.

latter, the diodes are typically constructed in a ring configuration using beam lead technology.

Two single balanced mixers are frequently combined to form an image rejection mixer, as shown in Figure 7-13. The RF signal is fed to the mixers through a 90°, 3-dB hybrid, while the LO signal is applied through an in-phase power divider. The IF outputs of each mixer are then combined through a 90° hybrid. With this arrangement, the signal frequency response appears at one output of the IF hybrid and the image appears at the other. Either the signal or the image (upper or lower sideband) response can then be selected by terminating the appropriate IF output port. The degree of image rejection depends on the amplitude and phase balance between the two mixers. A 20-dB image rejection can typically be achieved for an octave band design. The circuit can be optimized for improved performance for one sideband at the expense of performance of the other sideband. In addition to image rejection, this type of mixer provides good input VSWR and a 3-dB improvement in power handling over other designs. The LO power required is approximately 3 dB higher than that for a balanced mixer, and the minimum conversion loss is higher due to the additional loss of the RF and IF hybrid.

Harmonic and subharmonic mixers are mixers whose LO frequency is a submultiple of the signal frequency. The major advantage of these mixers is in high-frequency applications where local oscillators are relatively expensive, complex, and bulky. Harmonic mixers, because of their high LO power dissipation and high conversion loss, have received little attention, in contrast to the subharmonic mixers, in which excellent noise figures have been demonstrated at up to 230 GHz. Double sideband noise figures of 5 dB at 183 GHz and 6 dB at 230 GHz have been achieved. Most results to the present have been for X2 subharmonic mixers (i.e., $\omega_{LO} = \frac{1}{2} \omega_{RF}$), with promising research beginning with X4 subharmonic mixers at 220 GHz (8.5 dB DSBNF).

Mixer fabrication techniques vary with their frequency of operation. At the lower microwave frequencies (through K_a-band), balanced and double balanced designs in stripline and microstrip construction are common. The trend in lower frequencies is to include the mixer design as part of an overall microwave integrated circuit. Above K_u-band, technology is predominantly waveguide balanced mixers (up to approximately 140 GHz), with continuing development of microstrip, suspended stripline, fin-line, and slotline configurations). The development of an integrated circuit design that is practical, cost effective, and performs equally well at 35 and 95 GHz frequencies is the subject of much research and development.[14] Figure 7-14 is an example of the level of performance of commercially available balanced, beam-lead diode, waveguide mixers. Mixers can be purchased with or without IF amplifiers, with a unit which contains both the mixer and IF amplifier referred to as a *mixer/preamp*. Because of the integrated construction, mixer/preamp performance is generally optimized. The double sideband noise figures shown in Figure 7-14 are for mixer/preamps and include the contribution of the IF amplifier. Typically, the IF amplifier has a noise figure of 1.2 to 1.5 dB. The noise figure performance of

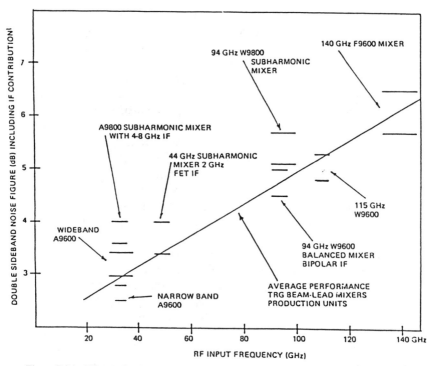

Figure 7-14. Mixer noise figures obtainable in microwave and MMW waveguide mixers.

lower-frequency microwave mixers generally falls in the range from 2 to 5 dB depending on the type of device, bandwidth, and construction.

7.4.4 Detectors

All superheterodyne receivers use at least two stages of down conversion in the detection process. The diodes used for both stages are typically point contact or Schottky diodes, with the only difference being the operating mode. The first down conversion is accomplished by the first detector (mixer), and the resulting information retained at the IF consists of the phase, amplitude, and frequency of the received echo signal. Subsequent detectors retain all or part of the information of the received signal depending on the application. Second-stage detection can be classified as square law detection, synchronous detection, or phase detection.

In the absence of an LO signal, the detector diode will operate in a low-level region in which the output voltage is proportional either to the input RF power or to the square of the RF input voltage. This square law relationship exists for a wide range of input power levels and is called the *square law region*. In this region, the detector conversion loss is a function of the input signal level; therefore, the concept of noise figure does not apply to square law detectors. The output of a square law or video detector is a low-frequency signal at video whose amplitude is a replica of the modulated sideband amplitude of the RF signal.

Tangential signal sensitivity (TSS) is the most common sensitivity rating for detector diodes. It is measured by observing the video voltage out of the detector and increasing the RF input signal until the minimum peak of the video signal plus noise is the same as the maximum peak of the video noise with no RF input. Figure 7-15 is a representation of the TSS level measurement. Diode manufacturers define the TSS signal level to be that RF input power level necessary to raise the output video signal 8 dB above the video noise level. The sensitivity depends primarily on the RF matching structure, rectification effi-

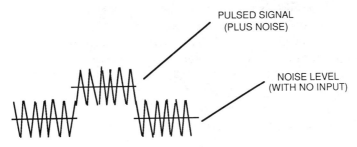

Figure 7-15. Example of an oscilloscope display in determining the TSS.

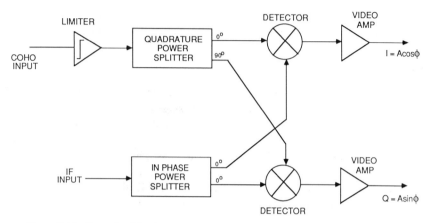

Figure 7-16. Example of in-phase (I) and quadrature (Q) synchronous detection stages.

ciency, output impedance, and noise properties of the diode, as well as the input impedance, bandwidth, and noise properties of the video amplifier. Typically, a video detector can achieve a TSS of approximately -60 dBm for a 1-MHz video bandwidth. A 1-MHz bandwidth superheterodyne receiver, on the other hand, has a maximum sensitivity or minimum discernible signal (MDS) of approximately -109 dBm.

The second detection stage may be operated with an LO input, in which case the detection process is linear. Figure 7-16 depicts this situation, in which the second LO (COHO) is at the same frequency as the IF. In this configuration, the down conversion is called *synchronous detection*, and both phase and amplitude information are retained. Since phase ambiguities may exist when a single synchronous detector is used, detector pairs are usually operated in quadrature. In Figure 7-16, the COHO signal without a phase delay will mix with the incoming IF signal and generate an output voltage proportional to the input amplitude and cosine of the phase angle between the two signals (in phase component). The COHO signal with a 90° phase delay will generate an output voltage proportional to the input amplitude and the sine of the phase angle between the two signals (quadrature component). Processing of the two signals will provide both amplitude and phase angle information.

If the IF input signals are hard limited (i.e., constant amplitude), then only phase information will be retained and the detector is referred to as a *phase detector* rather than a synchronous detector.

7.4.5 LOCAL OSCILLATORS

The superheterodyne receiver utilizes one or more LOs and mixers to convert the received echo to an IF that is convenient for filtering and processing operations. The receiver can be tuned by changing the first LO frequency without disturbing the IF section of the receiver. Subsequent shifts in IF are often accomplished within the receiver by additional LOs, generally of fixed frequency. Most modern radar receivers utilize a solid-state LO, and the Gunn effect oscillator has been by far the most widely used LO for radar frequencies from X-band to 95 GHz. The development of low-noise, wideband Gunn oscillators has led to the replacement of the reflex klystron in many applications. The Gunn has good AM and FM noise characteristics, requires low voltages (5 to 10 Vdc) and current (1 to 3 A), is small, and has a long life. The noise generated by a Gunn oscillator can often be reduced at frequencies within 100 kHz of the carrier by phase locking the Gunn to a low-phase noise oscillator. Figure 7-17 shows the typical power outputs obtainable from CW-Gunn oscillators as a function of frequency. Gunn devices are used as the primary LO source for receivers below 140 GHz. At frequencies above 140 GHz, Gunn devices are used in conjunction with varactor multipliers of up to 300 GHz or as LO power for subharmonic mixers.

The IMPATT is another solid-state device which has been used as a funda-

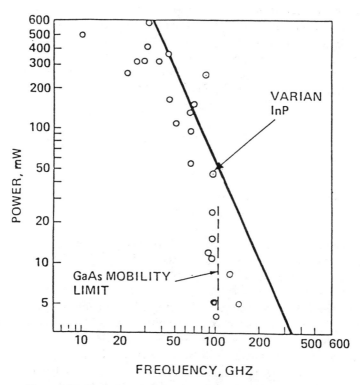

Figure 7-17. Typical output powers for Gunn and InP CW oscillators.

mental LO source at frequencies of up to 220 GHz. These devices are considerably more noisy than the Gunn. Noise generated in the LO is mixed down to the IF and increases the noise figure of the mixer. Care must be taken with the IMPATT to filter and phase lock the LO to minimize the effects of LO noise on the receiver.

In many early radars, the only function of the LOs was to convert the echo frequency to the correct IF. The majority of modern radar systems, however, coherently process a series of echoes from a target. The LOs act essentially as a timing standard by which the echo delay is measured to extract range information that is stable to within a small fraction of the radar wavelength. The processing demands a high degree of phase stability throughout the radar. These processing techniques determine the basic stability requirements of the receiver.

The first LO, generally referred to as a *stable LO (STALO)*, has a greater effect on processing performance than the transmitter. The final LO, generally referred to as a *coherent LO (COHO)*, is often utilized for introducing phase corrections which compensate for radar platform motion or transmitter phase variations.

The stability requirements of the STALO are generally defined in terms of a tolerable phase modulation spectrum. Sources of unwanted modulation are mechanical or acoustic vibration from fans and motors, power supply ripple, and spurious frequencies and noise generated in the STALO. In general, the tolerable phase deviation decreases with increasing modulation frequency, because the Doppler filter is less efficient in suppressing the effects. Limitations on the MTI improvement factor I for the STALO and COHO can be expressed as

$$I = 20 \log_{10} \left[\frac{1}{2\pi\Delta f T} \right] \qquad (7\text{-}21)$$

where

Δf = interpulse frequency change
T = transit time to and from the target

An example of a phase-locked Gunn STALO is shown in Figure 7-18.[15] The stability of the network is determined by the stability of the reference oscillator. In general, temperature-controlled crystal oscillators with stabilities on the order of 10^{-8} to 10^{-12} are used. In addition to frequency stability, phase noise and amplitude noise are important parameters. The reference signal frequency is typically 10 to 100 MHz and is used to phase lock a higher-frequency oscillator, or is multiplied up to a higher frequency which becomes the LO (ω_{LO}) signal to the mixer. A sample of the Gunn oscillator frequency (ω_{RF}) is mixed with the LO signal and down converted to the IF (ω_{IF}). The frequency of the LO is selected to equal the reference oscillator so that the output of the phase

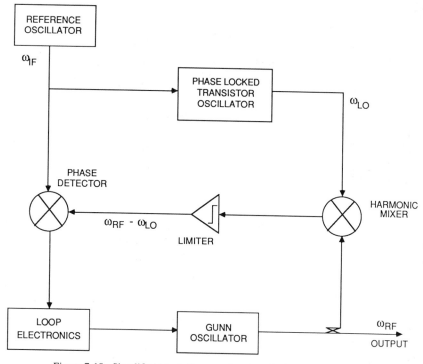

Figure 7-18. Simplified block diagram of a phase-locked Gunn STALO.

detector is a voltage proportional to the drift in ω_{RF}. The output of the phase detector is used to tune the frequency of the Gunn oscillator for zero voltage from the phase detector.

Maintaining a constant IF in noncoherent systems and frequency-agile systems requries that the receiver LO frequency follow the pulse-to-pulse frequency changes of the transmitter with a constant frequency displacement. To achieve this tracking, the LO must be electronically tunable over the full transmitter range, provide means of presetting the LO near the desired frequency prior to each pulse, include an AFC circuit capable of fine tuning the LO with sufficient accuracy during the time of the transmitted pulse, and maintain a stable frequency during the listen time of the radar.

A unit capable of accomplishing all of these tasks is called the *tracking LO* (see Figure 7-19). A detection device attached to the transmitter produces a signal proportional to the instantaneous output signal frequency (frequency analog input). The frequency analog input is processed and used as a frequency preset signal for coarse frequency setting of the varactor-tuned Gunn, voltage-controlled oscillator (VCO). Just prior to the RF pulse, a trigger signal is applied to a synchronizer circuit. The function of this circuit is to break the preset

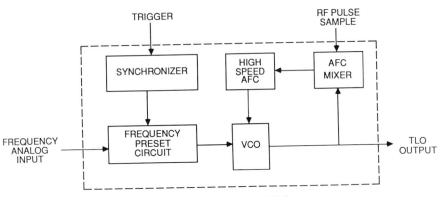

Figure 7-19. Example of a TLO.

signal circuit and hold the VCO frequency constant at the point last indicated by the readout circuitry. During the time of the RF output pulse from the transmitter, a sample of the RF pulse is applied to an AFC mixer together with the VCO output. The AFC mixer output is applied to a high-speed AFC circuit to generate an error signal. The error signal is fed back to the VCO to fine-tune the VCO frequency. The VCO is then held at the fine-tuned frequency during the part of the interpulse period representing the maximum range of the radar. At a point in time after the maximum radar range but prior to the next trigger input, the synchronizer circuit releases control and the VCO frequency is again determined by the transmitter frequency analog preset signal. This process is repeated for each successive RF pulse. In the more sophisticated AFC systems, a correction is applied to the analog preset signal based on the pull-in required on the preceding pulse.

In terms of operating sequence, the preformance of the tracking LO (TLO) closely parallels that of any standard, tunable radar LO/AFC circuit, with one significant exception. In the standard case, the final stabilized LO frequency is achieved by applying incremental AFC correction signals derived from a number of successive RF pulses. The TLO, on the other hand, must correct the VCO frequency to the same final stabilized value using an AFC correction signal derived from a single transmitted RF pulse.

7.4.6 If Amplifiers

Microwave integrated circuit (MIC) techniques have resulted in an excellent selection of low-noise, high-performance linear IF amplifiers. These MICs are designed to be cascadable, and have low noise and stability. Table 7-4 lists some typical MIC units that may be used as building blocks in the receiver IF.[16] The use of these elements allows the IF gain to be distributed throughout the IF section without significantly affecting the overall system noise figure. The

MIC modules are typically available in TO-8 packages designed to be used in 50-ohm microstrips. Other receiver functions obtainable as MIC modules include limiters and PIN attenuators for use in AGC functions at IF.

The IF amplifier must be selected on the basis of the amplifier noise figure, the gain required, and the output power capability. The gain and output power of the IF amplifiers must be selected so that the output of each stage is below saturation under conditions of maximum input signal from the mixer. The noise floor (minimum signal level that can be amplified) can be computed using the equation

$$P_n = -114 \text{ dBm} + 10 \log B + F \qquad (7\text{-}22)$$

where B is the bandwidth in megahertz and F is the amplifier noise figure in decibels. The bandwidth is determined by the filtering in the receiver (prior to the IF amplifier) or the amplifier bandwidth if no filtering is provided.

Figure 7-20 shows the input–output characteristics of a typical MIC amplifier.[17] The fundamental transfer curve has a $1:1$ slope in the linear operating region. Also shown is a plot of the second-order ($2:1$ slope) and third-order ($3:1$ slope) IMPs. Third harmonic spurious responses are the result of $2F_1 - F_2$ or $2F_2 - F_1$ differences and are very likely to appear in the passband of a medium bandwidth or narrowband amplifier. These are spurious outputs due to nonlinear effects of the amplifier when more than one frequency is amplified. Consequently, at high input powers, spurious outputs may rise to undesirable levels. To reduce this effect, one can select an amplifier with a higher saturated output or reduce the input power. In Figure 7-20, the amplifier output intercept point is defined as the point where extension of the first- and second-order responses intersect on the output power scale. In this example, the intercept point is $+30$ dB. Once the intercept point is determined, the spurious free dynamic range of the amplifier can be calculated as[16–18]

$$\text{Spurious free dynamic range (dB)} = \tfrac{2}{3}(P_1 - P_n) \qquad (7\text{-}23)$$

where P_1 is the input intercept point (determined by subtracting the gain from the output intercept point). The spurious free dynamic range is much less than the full dynamic range calculated using the difference between the minimum input signal power at saturation and the minimum detectable signal power (P_n). The minimum input signal power at saturation is determined by the difference between the output signal power at saturation and the amplifier gain, i.e., $P_o - G$. The useful dynamic range of the amplifier will vary depending on the application.

A logarithmic IF amplifier/detector, referred to as a *log amp*, is a device whose output video is proportional to the logarithm of the RF input. Log amps

Table 7-4. Typical Performance Characteristics of MIC Amplifiers.[1] (From ref. 16)

Frequency Response (MHz) Minimum	Gain (dB) Minimum A	Gain (dB) Minimum B	Noise Figure (dB) Maximum A	Noise Figure (dB) Maximum B	Power Output Gain Compression (dBm) Minimum A	Power Output Gain Compression (dBm) Minimum B	Gain Flatness (±dB) Maximum A	Gain Flatness (±dB) Maximum B	Typical Intercept Point for IM Products (dBm)	VSWR (50 ohms) Maximum In	VSWR (50 ohms) Maximum Out	Input Power (±1% Reg.) Volts DC	Input Power (±1% Reg.) Current mA Typ
2 to 500 MHz, low-noise versions (listed in order of increasing noise figure, decreasing gain)													
30-200	15	14.5	2.0	2.5	-3	-4	0.75	1.0	+7	2.0	2.0	+15	8
5-500	22	21	3.0	3.5	+5	+4	1.0	1.0	+15	B = 2.0	2.2	+15	25
5-500	15	15	2.5	3.0	-2	-3	1.0	1.0	+8	A = 2.0 / B = 2.2	2.0 / 2.0	+15	10
10-500	10	9	2.5	3.0	+6	+6	1.0	1.0	+22	2.0	2.0	+15	25
5-500	15	15	3.0	3.5	-2	-3	1.0	1.0	+8	A = 2.0 / B = 2.2	2.0 / 2.0	+15	10
10-500	10	9	3.0	3.5	+12	+12	1.0	1.0	+28	2.0	2.0	+15	35
5-500	15	14	4.0	4.5	+5		1.0	1.0	+10	2.0	2.0	+5	7
5-500	14	13.5	4.0	4.5	-2	-3	1.0	1.0	+11	2.0	2.0	+15	10
5-500	20	19	4.5	5.0	+7	+7	1.0	1.0	+20	2.0	2.0	+15	23
5-500	27	27	5.5	5.5	+6	+5.5	1.0	1.0	+18	2.0	2.0	+15	38
5-500	14	13.5	5.5	6.0	+7	+7	1.0	1.0	+21	2.0	2.0	+15	23

2 to 500 MHz, high-power versions (listed in order of increasing power output, decreasing gain)

5-500	14	13.5	5.5	6.5	+10	+9.5	1.0	1.0	+23	2.0	2.0	+15	35
5-500	23	23	7.0	7.0	+12	+12	1.0	1.0	+25(1) / +36(2)	2.0	2.0	+15	80
5-500	9	8.5	7.0	7.0	+13	+13	1.0	1.0	+27	2.0	2.0	+24	50
2-500	12	11	7.0	7.5	+14	+13	0.7	0.5	+25	2.0	2.0	+15	65
5-500	16	15	5.5	6.0	+14	+13	1.0	0.7	+28	2.0	2.0	+15	44
5-500	16	15.5	6.0	6.0	+14	+14	1.0	1.0	+27(1) / +36(2)	2.0	2.0	+24	50
10-500	10	10	5.0	5.5	+17	+16	0.5	0.5	+34	2.0	2.0	+15	60
5-500	6	6	11.0	11.0	+17	+17	1.0	1.0	+31	2.0	2.0	+24	100
10-500	9	9	8.5	9.0	+18	+18	1.0	1.0	+30	2.0	2.0	+15	95
10-500	14	14	8.5	9.0	+20	+20	1.0	1.0	+35	2.0	2.0	+15	110
10-500	11.5	11	8.5	9.0	+20	+20	1.0	0.7	+35	2.0	2.0	+24	110
10-500	10	10	8.0	8.5	+23	+22	0.5	0.5	+38	2.0	2.0	+15	110
10-500	11	10	9.0	9.5	+26	+25.5	1.0	0.7	+43	2.0	2.0	+15	190

2 to 1000 MHz (listed in order of increasing noise figure, decreasing gain)

10-1000	10	9	4.0	4.5	+6	+6	1.0	1.0	+22	2.0	2.0	+15	25
2-1000	14	13.5	3.5	4.0	-5	-6	0.7	1.0	+10	2.0	2.2	+15	8
10-1000	10	9	4.5	5.0	+12	+12	1.0	1.0	+28	2.0	2.0	+15	35
5-1000	10	9	5.0	5.7	-6	-6	1.0	1.0	+10	2.0	2.0	+5	7
5-1000	14	13.5	5.0	5.5	-2	-3	1.0	1.0	+11	2.0	2.0	+15	10
5-1000	14	13.5	6.5	7.0	+7	+7	1.0	1.0	+21	2.0	2.0	+15	23
10-1000	9	6.5	6.5	7.5	+17	+17	1.0	1.0	+30	2.0	2.0	+15	60
5-1000	9	8.5	8.0	8.5	+13	+13	1.0	1.0	+27	2.0	2.0	+24	50
5-1000	10	9	8.0	8.5	+14	+13	1.0	1.0	+28	2.0	2.0	+15	48
10-1000	6.0	5.5	12.0	12.5	+20	+19	0.7	1.0	+33	2.0	2.0	+15	110

¹Guaranteed specifications 0 to 50°C(A) −54° to +85°C (B).

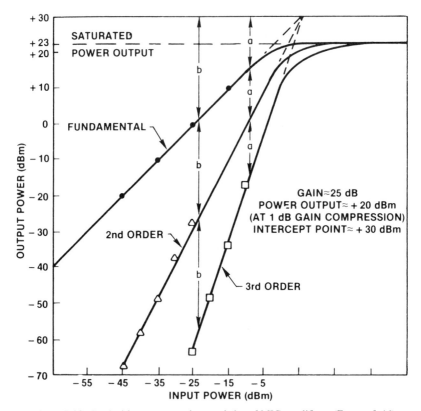

Figure 7-20. Typical input–output characteristics of MIC amplifiers. (From ref. 16)

are found in the IF sections of nearly all radar receivers, primarily due to the extremely wide instantaneous dynamic range (70 to 80 dB) achievable without limiting or overloading. Since log amps require no AGC to achieve the wide dynamic range, they can respond to rapid changes in the received signal levels without loss of signal information. Log amps are used extensively in receivers which must compress signal levels of high dynamic ranges into smaller, more manageable ranges without clipping or loss of information. They are also used in receivers requiring high sensitivity and instantaneous pulse-to-pulse response.

A simple block diagram of a log amp is given in Figure 7-21.[19] Piecewise approximation of a logarithmic function of output versus input is achieved by using cascaded stages of limiting IF amplifier stages and video detectors. Each stage typically has a gain of 10 dB, and the output of each stage is detected and summed in a video delay line. The operation of the log amp can be expressed

LIMITING IF AMPLIFIERS

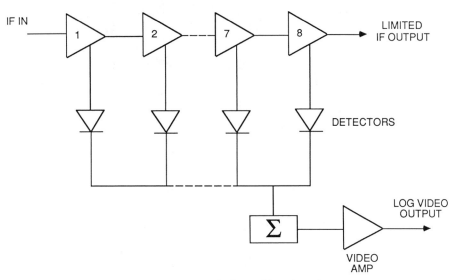

Figure 7-21. Logarithmic amplifier schematic.

as[19]

$$\Delta E_{\text{out}} = K \log_{10} \left(\frac{E_1}{E_2}\right)_{\text{in}} \qquad (7\text{-}24)$$

where

ΔE_{out} = change in output video voltage when input E_1 is replaced by input E_2

K = constant which determines the log slope

$\left[\dfrac{E_1}{E_2}\right]_{\text{in}}$ = ratio of the IF input signals

A plot of IF power input versus video voltage output is shown in Figure 7-22. The transfer characteristic is a straight line when the video output, in volts, is plotted against the input in dBm (linear-log scale). The slope of the curve is a constant on this scale and is expressed in millivolts/decibels. The gain cannot be defined in conventional linear terms, since the change in output voltage is a function not only of the change in input voltage but also of the input power level. That is, changes in low-level signals are amplified; changes in high-level signals are compressed due to the logging action. Because of this logging ac-

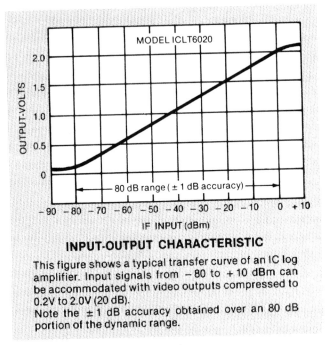

INPUT-OUTPUT CHARACTERISTIC

This figure shows a typical transfer curve of an IC log amplifier. Input signals from −80 to +10 dBm can be accommodated with video outputs compressed to 0.2V to 2.0V (20 dB).
Note the ±1 dB accuracy obtained over an 80 dB portion of the dynamic range.

Figure 7-22. Typical log amp input–output characteristics. (From ref. 19)

tion, care must be exercised in measuring the log amp performance using linear techniques. For example, the 3-dB bandwidth of a log amp is not 0.707 of the output; it is the point where the output is down by $3 \times$ the transfer slope. In addition, the signal-to-noise ratio changes with drive level, and care must be exercised when providing gain ahead of the log amp. As the minimum input level to the log amp increases, the signal-to-noise ratio decreases, as does the effective dynamic range. For these reasons, it is always good practice to maintain the gain in the receiver at a level such that the minimum signal level at the log amp is near the log amp noise level. This may require attenuation to be inserted prior to the log amp.

Log amplifiers are readily available at frequencies of up to 1 GHz. Above 1 GHz, log video amps are typically used. These involve the use of an RF detector in the input, which limits the sensitivity and dynamic range of the amplifiers to typically −45 dBm and 40 dB, respectively. Video bandwidths of both video and IF log amps are selectively narrow, limiting their use in extremely short pulse radars with pulse widths of less than 50 ns.

Regardless of the type of IF amplifier, the overall noise figure of the IF must be considered, especially if attenuation is included in the IF chain. This noise figure should be part of the overall computation of the receiver noise figure using the usual techniques of cascade noise figures.

7.5 APPLICATIONS AND TECHNIQUES

Many modern radar systems incorporate coherent detection processing and/or some spread spectrum transmission. The technique employed depends upon the application, but the detection of moving targets generally requires some coherence to achieve spectral separation of targets and clutter. A wideband signal may be used to enable a low-power, long-pulse transmission to be compressed for improved range resolution and detection. The wideband signal may be synthesized by using a frequency-stepped waveform in conjunction with fast Fourier transform (FFT) processing to achieve the equivalent improved range resolution and detection.

7.5.1 Coherent Detection

The fully coherent receiver can be implemented only when a coherent transmitter has also been used. As shown in Figure 7-23, the RF transmitted signal is formed by the combination of a STALO, usually near the transmitted frequency, and a COHO IF oscillator.[20] The sum of these two signals is formed in an up-converter (or down-converter) and amplified using a pulsed RF amplifier for transmission. On receive, the echo signal is first down-converted by the STALO to an IF signal (60 MHz to 4 GHz) which is the same frequency as the COHO. The IF signal is amplified, filtered, and otherwise processed, and then down-converted to baseband Doppler by mixing with the COHO. A pair of orthogonal mixers is used to produce a set of I and Q signal components. These components are then fed to signal-processing circuitry, which may consist of MTI or pulse-Doppler filtering. This receiver configuration provides the greatest level of coherence and hence processing gain.

In applications where maximum coherence or MTI improvement is not required, a noncoherent transmitter can be employed, reducing radar complexity. Figure 7-24 shows the implementation of two coherent-on-receive systems. There are basically two forms of such systems. The first, shown in Figure 7.24(a), involves phase-locking a COHO to the transmitter phase during the transmitted pulse and using the COHO as the reference to remember the phase of the transmitted signal. In this approach, a sample of the transmitter signal is mixed with a sample of the STALO signal. The difference frequency, $\omega_{RF} - \omega_S$, is selected to be the same as the COHO frequency, ω_c. The difference frequency is then mixed with the COHO signal in the phase detector and the output used to control the phase lock loop (PLL) electronics. Using this scheme, changes in the noncoherent transmitter are detected by the PLL and used to lock

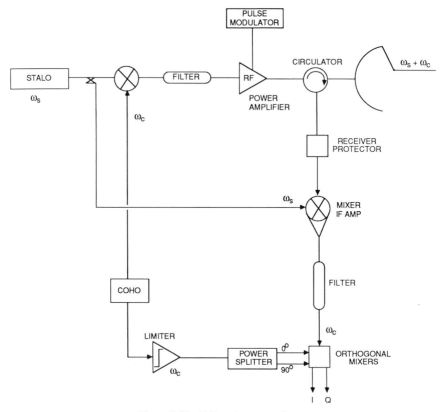

Figure 7-23. Fully coherent receiver.

the phase of the COHO signal to the RF signal. Thus, the radar is coherent-on-receive, since the receive phase tracks the transmitter phase. The performance of this system is limited by the stability which can be maintained in the COHO during the time between transmission and reception and the repeatability of the phase locking of the COHO from pulse to pulse. Improved stability can be realized by using a fixed-frequency COHO, as in the second approach.

The second approach, which can be employed only if the transmitter is tunable, shown in Figure 7.24(b), involves using a highly stable COHO and tuning the transmitter to match it in frequency, then measuring and "remembering" the phase difference between the transmitter and reference oscillator signals for each pulse. This remembered phase factor is used to correct each target echo before applying the signal to the moving target filter. The tuning loop, consisting of the mixer, phase detector, and transmitter tuning electronics, tunes the transmitter to the proper frequency (determined by the STALO and COHO) on each pulse. At the same time within each pulse, the sample and hold circuit

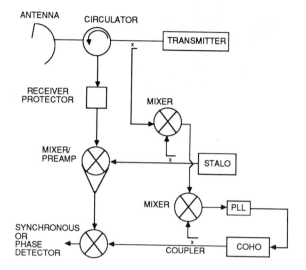

(a) Coherent-on-receive system (tuned COHO)

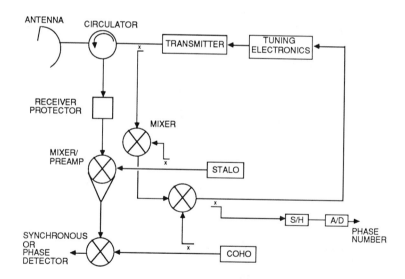

(b) Coherent-on-receive system (tuned transmitter)

Figure 7-24. Examples of coherent-on-receive implementations.

samples the output of the phase detector. This sampled value represents the phase difference between the actual transmitted pulse and the reference oscillator. Since the transmitter is not coherent, this phase difference is not the same for each pulse, but as long as the transmitted phase is known, the fact that it is not constant is irrelevant. The phase video or baseband Doppler which results from mixing the RF return with the STALO and COHO must be modified by the phase number before it is applied to the MTI filter.

7.5.2   Pulse Compression

Pulse compression allows a radar to use a long pulse to achieve high radiated energy and simultaneously to obtain the range resolution of a short pulse. This is accomplished by employing frequency or phase modulation to widen the signal bandwidth. The received signal is processed in a matched filter that compresses the long pulse to a duration that is inversely proportional to the transmitted bandwidth. Figure 7-25[20] shows an example of a linear FM pulse compression radar. The superheterodyne receiver is again employed using a stable but noncoherent LO. The entire RF and IF processing circuitry must be broadband so that phase distortion of the received signal is not introduced. The IF amplifier must have sufficient bandwidth and linear phase characteristics over the band of frequencies being compressed. The compressive filters often used are surface acoustic wave (SAW) devices. These devices are analog processing devices that can be used to obtain a compressed video output equivalent in resolution to a real pulse with the same bandwidth.

7.5.3   Frequency Stepped Coherent Receiver

High-range resolution can be achieved by using a wideband frequency stepped waveform and processing the received echo using FFT techniques. Coherent or noncoherent detection may be used, but noncoherent processing cannot provide the absolute position of the resolved echoes. In addition, coherent processing

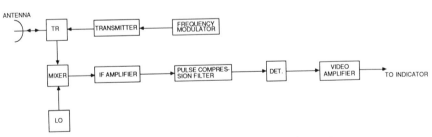

Figure 7-25. Example of an FM pulse compression receiver.

can increase the receiver signal-to-noise ratio. Therefore, the coherent, wideband, frequency stepped system is perferred.

The coherent receiver configuration shown in Figure 7-23 can be modified to provide frequency stepped transmission. This is accomplished by replacing the STALO with a frequency synthesizer whose output frequency is selectable in N discrete steps of Δf step size. The total bandwidth required of the transmitter is now $N \times \Delta f$. This wide bandwidth requirement is also placed on the receiver front end to include the circulator, receiver protector, and mixer (both the RF and LO ports). Eq. 7-25 indicates the relationship between the various frequencies:

$$f_{\text{syn}} = f_0 + i\Delta f; \qquad i = 0,1,2,\cdots N - 1 \qquad (7\text{-}25)$$

and

$$f_{\text{RF}} = f_{\text{syn}} + f_{\text{c}}$$

where

f_{syn} = synthesizer frequency
f_0 = synthesizer minimum frequency
Δf = synthesizer frequency step size
N = number of synthesizer frequency steps
f_{RF} = transmitted frequency
f_{c} = COHO frequency

The transmit waveform is shown in Figure 7-26. Since the synthesizer output is used as the first LO, the output of the receiver mixer is still equal to the frequency of the COHO. This technique permits a series of narrowband signals to be transmitted using a stepped frequency waveform, and thus effectively generates a wideband signal while maintaining a narrowband receiver. The advan-

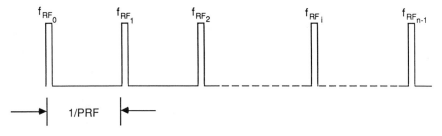

Figure 7-26. Transmitted stepped frequency packet. The time between pulses is the radar interpulse period (PRF)$^{-1}$.

tages are obvious where the resolution theoretically achievable is a function of total signal bandwidth (B_T):

$$\text{Resolution} = \frac{c}{2B_T} \tag{7-26}$$

where

$c = 3 \times 10^{10}$ cm/s
$B_T = $ bandwidth in hertz $= N \times \Delta f$

and the required receiver predetection bandwidth, B, is the inverse of the transmitter pulse width (matched filter approximation).

It can be shown that the signal-to-noise ratio improvement of the stepped frequency technique as compared to a single narrow pulse (wideband) with equivalent resolution is

$$(\text{S/N})_{\text{impr}} = 10 \log_{10}\left(\frac{B_T}{B}\right). \tag{7-27}$$

As an example, for a stepped frequency radar with 128 10-MHz steps and a 100-ns pulse width, the resolution given by Eq. 7-26 is

$$\text{Resolution} = \frac{3 \times 10^{10} \text{ cm/s}}{2 \times 128 \times 10 \text{ MHz}} = 11.72 \text{ cm or } 4.6 \text{ in.}$$

and the signal-to-noise ratio improvement over a single pulse having a width of <1 ns is given by Eq. 7-27 as

$$(\text{S/N})_{\text{impr}} = 10 \log_{10}\left(\frac{1280 \times 10^6}{10^7}\right)$$

$$= 21 \text{ dB.}$$

The disadvantages of using the frequency stepped technique lies in the complexity of the synthesizer and the tight stability (frequency and phase) requirements on all of the components over relatively wide bandwidths. In addition, the synthesizer switching speed may be very slow (on the order of microseconds), and N pulses must be transmitted, received, and processed to achieve the same range resolution achieved in a single high-resolution (wideband) pulse. On the other hand, the high-resolution real pulse (<1 ns pulse width) presents other problems concerned with acquisition and processing of the received echo.

7.5.4 MMW and Sub-MMW Receivers

MMW receiver technology has advanced in recent years to the point where levels of performance for uncooled MMW heterodyne receivers are comparable to those obtained at microwave frequencies. The fabrication of the beam-lead GaAs semiconductors used in these receivers has proven to be well suited for the design of receivers in the 30- to 100-GHz range. This technology will likely lead to the development of hybrid planar microwave integrated circuit techniques in MMW applications in the near future. At present, waveguide devices and techniques have found widespread use up to 100 GHz. The development of MMW integrated circuits has resulted in the use of many different transmission media and technologies. Many designs employ hybrid combinations of suspended stripline, microstrip, fin-line, slotline, and waveguide simultaneously in multifunction modules operating at between 35 and 100 GHz. Table 7-5[21] compares some of the properties of the transmission media used at MMW. Waveguide, suspended stripline, and microstrip have all been used in the successful development of mixers up to 100 GHz. One limiting factor in the performance of MMW receivers has been the degradation in noise figure due to far-off-the-carrier noise of Gunn oscillators used as LOs. A number of MMW radar systems and receivers operating at frequencies of up to 95 GHz have been developed in small prototype quantities. Fully coherent transmitters and receivers are possible at 95 GHz using solid-state transmitters or newly developed EIAs to develop the coherent transmission.

MMW receivers can be configured in much the same way as conventional microwave receivers. In general, they use higher IF frequencies due to limitations in device stability and wider frequency bandwidth in the transmitters. The lower MMW frequency receivers (30 GHz) employ the same technology as microwave receivers in terms of integrated circuit receivers, gaseous duplexers and receiver protection, as well as a wide assortment of RF components such as filters, limiters, circulators, isolators, and PIN switches. At 100 GHz, the technology is less mature. Receiver protectors consist of ferrite circulators and Faraday rotational switches. LO power is minimal and often requires the use of high-power doublers or triplers from microwave frequencies to get the necessary power to drive millimeter wave receivers.

Radar systems are being contemplated and proposed at frequencies above 100 GHz. Again, the transmission medium becomes a severe problem.[22] A number of media have been proposed for these MMW receivers. A brief list and description of these techniques follows:

1. Dominant mode rectangular TE_{10} waveguide: lossy, peak power limited, oversize guide results in mode propagation.
2. Circular TE_{01} waveguide: large dimensions, lower loss, links gyratron output mode; mode filters required.

Table 7-5. Characteristics of MMW Transmission Media.
(From ref. 21)

	Metal Waveguide	Microstrip	Image Line [Planar-Dielectric Waveguides]	Fin Line
Transmission loss	Lowest	Relatively high	Potentially low. Lower than microstrip.	Moderate loss
Size, weight	Large, heavy	Small, light. Often too small for manufacturing.	Intermediate, light	Large, heavy
Dispersion, multimoding	Low dispersion. Normally single-moded.	Dispersive potentially multi-moded.	Dispersive often heavily multi-moded.	Dispersive. Potentially multi-moded.
Solid-state device compatibility and integrability	Unsuitable for integration	Good/fair. Planar devices more suitable.	Good/fair. Still under investigation.	Fair. Suited for beam-lead diodes.
Useful frequency range	All frequency range. Size too small. Above 130 GHz	Up to 60 GHz. Unsuitable beyond.	>60 GHz. Also good below.	30 to 95 GHz
Cost	Very expensive	Low cost	Moderate cost	Low cost
Comments	Design info. available. Very high performance	Design info. available. Shielding may be required.	Lack of adequate theory. Radiation problems.	Circuit interacts with housing. Some design data available.

(By permission, from Deo et al., ref. 21; © 1979 Hayden Publishing Co., Inc.)

3. Microstripline: high loss, upper frequency approximately 200 GHz.
4. Dielectric guides: physical support problems, radiation losses, low loss, components developed limited to 100 GHz.
5. Suspended stripline and fin-line: low loss possible, coupling to waveguide easy, suspended stripline to above 200 GHz, fin-line limited to 100 GHz.
6. H-guides and groove-guides: simple construction, lower loss, low dispersion, inexpensive construction.
7. Quasi-optic techniques: usable to above 300 GHz, physically large, many components developed.

The development of radar receivers for purposes other than laboratory research at near MMW frequencies will require considerable advances in technology and fabrication techniques.

7.6   SUMMARY

In this chapter we have attempted to highlight the key parameters to be considered when designing a radar receiver. Extremely important in the design are overall receiver sensitivity, bandwidth, and dynamic range. In most practical cases, size, weight, and cost are prime drivers in the final design selected for a particular application. All of these factors must enter into the design equation as well as requirements placed on the receiver by the particular transmitter and signal processor selected and the operational environment. As one might imagine, the final design selected involves an iterative process with compromises in most areas.

To minimize the compromises, the receiver designer must constantly interact with the system designers and stay abreast of the rapidly changing receiver component technology. An excellent source for the latter are the various trade magazines and vendor catalogues. Some of these are listed in the references to this chapter and the reader is encouraged to review these as a source of basic information as to the available receiver component technology.

7.7   REFERENCES

1. S. N. VanVoorhid (Ed.), *Microwave Receivers*, Dover Books, New York, 1966.
2. M. I. Skolnik, *Radar Handbook*, McGraw-Hill Book Co., New York, 1970.
3. W. W. Mumford and E. H. Scheibe, *Noise Performance Factors in Communications Systems*, Horizon House, Dedham, Mass., 1967.
4. H. A. Watson, *Microwave Semiconductor Devices and Their Circuit Applications*, McGraw-Hill Book Co., New York, 1969.
5. A. G. Milnes, *Semiconductor Devices and Integrated Electronics*, Van Nostrand Reinhold Co., New York, 1980.
6. G. L. Matthei, L. Young, and E. M. T. Jones, *Microwave Filters, Impedance-Matching Networks, and Coupling Structures*, McGraw-Hill Book Co., New York, 1964.
7. *Receiver Protector Catalog*, Varian Associates, Palo Alto, Calif., 1978.
8. J. F. Reynolds and M. R. Rosenzweig, "Learn The Language of Mixer Specifications," *Microwaves*, vol. 17, no. 5, May 1978, pp. 72–80.
9. R. V. Pound (Ed.), *Microwave Mixers*, MIT Radiation Laboratory Series, vol. 16, McGraw-Hill Book Co., New York, 1948.
10. *Mixers, Switches Hybrids, Transformers Catalog*, Watkins-Johnson Co., no. 1, Palo Alto, Calif.,
11. *RF and Microwave Components Catalog*, Anzac Co., Burlington, Vt., 1982.
12. V. Manessewitch, *Frequency Synthesizers Theory and Design*, John Wiley & Sons, New York, 1976.
13. *Receiving Diode Handbook*, Bulletin No. 4006, Microwave Associates, Burlington, Vt., 1980.

14. A. G. Cardiesmenos, "Practical MIC's Ready for Millimeter Receivers," *Microwave System News*, vol. 10, no. 8, August 1980, pp. 37–51.
15. A. Blanchard, *Phase Locked Loops: Application to Coherent Receiver Design*, John Wiley & Sons, New York, 1976.
16. *Designing with Modular Amplifiers*, Avantek, Inc., Santa Clara, Calif., 1979.
17. *1982 Solid State Microwave Components Sourcebook*, Avantek, Inc., Santa Clara, Calif., 1982.
18. *Solid State Microwave Amplifiers Catalog*, Varian Associates, Palo Alto, Calif., 1982.
19. *MIC Microwave and IF/RF Products Catalog 300*, RHG Electronics Laboratory, Deer Park, NY.
20. M. I. Skolnik, *Introduction to Radar Systems*, McGraw-Hill Book Co., New York, 1980.
21. N. C. Deo, "Milleter Wave IC's Spring from the Lab," *Microwaves*, October 1979, pp. 38–42.
22. D. J. Harris, "Waveguiding Difficult at Near Millimeters," *Microwave Systems News*, vol. 10, no. 12, December 1980, pp. 69–73.

8
RADAR INDICATORS AND DISPLAYS

James A. Scheer

8.1 INTRODUCTION

Radar indicators and displays are used to present the information contained in the radar return signal in a format suitable for operator interpretation. The display may be connected directly to the video output of the radar receiver. Radars that provide information at a greater rate than an operator can assimilate, however, normally use automatic data processors to interpret and condense the radar receiver output. In that event, only the condensed information is displayed.[1]

The radar display is usually a two-dimensional screen that shows the target locations with respect to some reference point. It can, however, assume some other form, such as a light to indicate a status or condition, or a meter to indicate the value of some parameter such as antenna pointing angle or range to a particular target. A numerical readout is another example of a radar display; it may indicate the range, bearing, or speed of a particular target or the status of a radar function.

The information is typically presented on a screen for direct viewing by the operator. In applications where direct viewing is not practical, the operator may view a virtual image, as is the case with lens-aided displays, such as those which are helmet mounted, and head-up displays (HUDs) that project images onto a glass screen in the operator's field of view.

8.2 DISPLAY FORMATS FOR CATHODE RAY TUBE DISPLAYS[2-5]

A variety of display formats are used for presenting radar target information on cathode ray tube (CRT) displays. There are essentially four dimensions available to the display designer: X dimension, Y dimension, brightness, and in some cases color. These dimensions are used in various ways to convey information to the operator. Some common formats are shown in Figure 8-1. Other display formats may be used for special-purpose applications.[5,6]

The *A-scope* is one of the simplest types of radar displays, providing target range and signal strength (RCS) information. The horizontal position on the screen represents the distance to the target, and the vertical deflection relative to the baseline represents target signal strength. The A-scope display is used

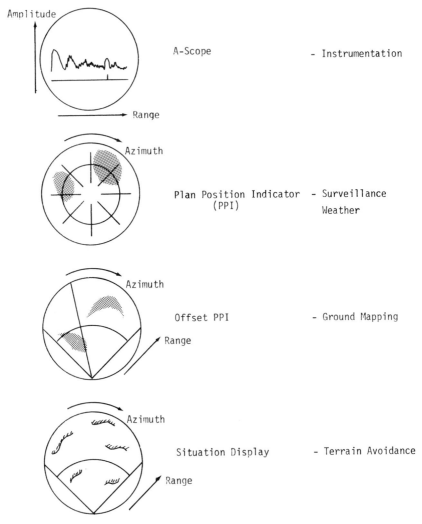

Figure 8-1. Various display types.

with instrumentation and data collection radars in which the antenna is not normally scanning. If the antenna were scanning, the dynamics of the A-scope display would be difficult to interpret.

Most radar sets use a simple azimuth scan and present their data on an intensity modulated range/azimuth (ρ/θ) display in a two-dimensional format. Of the two-dimensional displays, the *plan position indicator (PPI)* is the most commonly used. PPI displays frequently represent the slant range and azimuth coordinates for the various targets in the field of view of the radar by distance

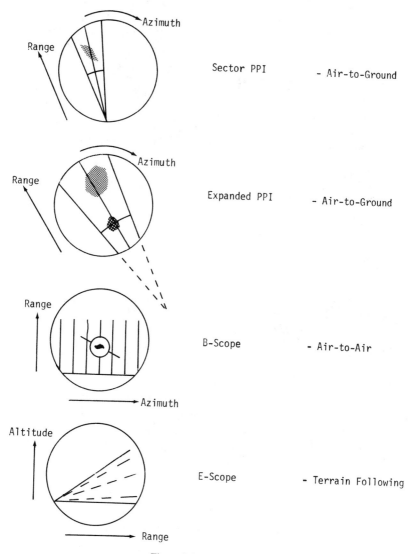

Figure 8-1. (*Continued*)

from the center of the CRT and by the azimuth angle on the tube face, respectively. The result is a map-like presentation. A PPI display is achieved in its simplest form by rotating a range sweep about the range origin in synchronism with the azimuthal scanning of the antenna. The various targets then appear as bright blips on the display as the range sweep rotates in angle. Depending upon the rotational rate of the antenna, the display may have sufficient persistence between successive looks (scan to scan) at given positions to retain most of the original brightness of the intensity modulated return at a particular spot. At low scan speeds, a blip corresponding to a target at a particular range and azimuth may virtually fade away before the antenna beam is again pointed in that direction.

The zero-range origin of the PPI is generally positioned at the center of the CRT to provide an equal field of view in all directions. There are occasions, however, when it is desirable to displace the origin, sometimes far off the face of the tube, to get maximum sweep expansion in a given direction. Such a display is called an *offset PPI*. In surveillance and weather radar applications, the PPI will display a full 360° scan. In forward-looking radar applications, the scan may be limited to a sector and the display is a sector PPI. In some sector PPI displays, the vertex is offset off the screen for an expanded view of a target.

Figure 8-2 is a photograph of a forward-looking attack radar display scope and illustrates the use of an offset PPI display format.

The radar search or surveillance area is often represented in rectangular (X, Y) rather than polar (ρ, θ) coordinates by combining range and angle data to form a *B-scope* display. The B-scope format is obtained by moving a range

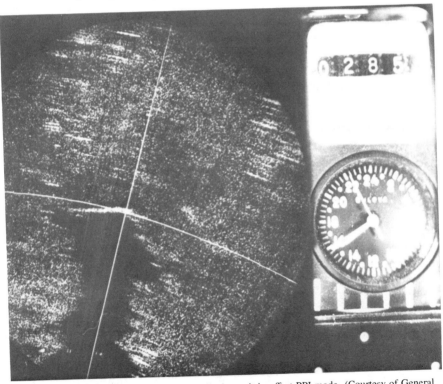

Figure 8-2. Forward-looking attack radar display unit in offset PPI mode. (Courtesy of General Electric Company)

sweep across the CRT display in synchronism with antenna motion so that the origin is stretched into a straight line. The resulting display format is a distorted map where the maximum distortion occurs at the sweep origin.[3]

B-scope displays are often used in air-to-air applications in which the chief considerations are the range and bearing of point targets or groups of targets and where little or no importance is attached to the shapes of extended targets or to the relative locations of widely separated targets. Also, the B-scope display is particularly useful in fire control applications for determining the range and bearing to particular targets. B-scopes are widely used in airborne radars and in short-range ground surveillance systems.

The *E-scope*, another two-dimensional display, presents range on the horizontal axis and elevation on the vertical axis. It is used in terrain-following (TF) systems in which the radar antenna is scanned in the elevation plane in the forward sector to obtain vertical profiles of the terrain ahead of the aircraft. In automatic terrain-following (ATF) systems, the range and elevation data are sent to the ATF computer, which then directs the aircraft around or over terrain obstacles.

8.3 OPERATION OF A CRT

A CRT is an elongated vacuum tube with a cylindrical neck and an enlarged section. It has a flattened front face containing, or in association with, an electron gun, electron beam-forming and -deflecting systems, and a fluorescent screen. The drawings in Figure 8-3 show the physical relationship between these elements and compare, in schematic form, two different systems used for deflecting the electron beam.

The electron gun is placed within the neck of the tube; it consists of a cathode, as a source of electrons, and one or more electrodes which form the beam and create the high electron velocities necessary to excite the phosphor of the display screen. Beam intensity is controlled by a negative control grid and is accelerated by anode grids.

The beam is focused onto a fine spot at the screen by either an electrostatic or a magnetic lens. Deflection of the beam is accomplished either by electric or magnetic fields, as indicated in Figure 8-3.[1]

With the electrostatic deflection technique, a potential is applied to the pair of electrostatic plates mounted inside the tube. As the electron beam passes between the plates, it is deflected upward or downward (or to the right or left) due to the effect of the electric field created by the applied voltage. One set of plates is oriented orthogonal to the other set such that one set of plates controls the vertical deflection and the other set controls the horizontal deflection.

With the magnetic deflection technique, a set of wire coils is arranged around the neck of the tube. When a current is applied to one of the coils, a magnetic field develops inside the tube, causing the electron beam to deflect (change

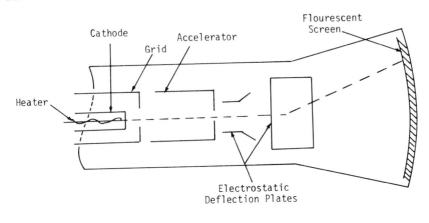

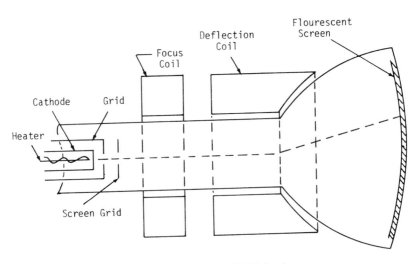

Figure 8-3. Two forms of CRT deflection.

direction). When a current is applied to the other coil, a deflection in the orthogonal direction results. This arrangement of coils is sometimes called a *yoke*, as in a home television set.

A variety of *CRT phosphors* have been used to form the CRT screen. Phosphors differ primarily in their decay times and persistence. The length of image persistence required in the CRT screen depends upon the particular application. For example, a long-persistence screen is appropriate for a PPI presentation where the frame time may be on the order of several seconds, such as with a

Table 8-1. CRT Phosphor Data.

Phosphor	Fluorescence Color	Phosphorescence Color	Persistence	Application
P1	Yellowish-green	Yellowish-green	Medium	Oscilloscope
P2	Bluish-green	Green	Blue-green, medium; long low-level decay	Special oscilloscope and radar
P4	White	White	Medium to medium short	TV
P5	Blue	Blue	Medium short	Photographic
P7	White	Yellowish-green	Blue, medium short; yellow, long	Radar
P11	Blue	Blue	Medium short	Oscilloscope
P12	Orange	Orange	Long	Radar
P13	Reddish-orange	Reddish-orange	Medium	Radar
P14	Purplish-blue	Yellowish-orange	Blue, medium short; greenish-yellow, medium	Radar
P15	Bluish-green	Bluish-green	Very short decay	Flying-spot scanners
P17	Yellow-white to blue-white	Yellow	Blue, short; yellow, long	Oscilloscope and radar
P19	Orange	Orange	Long	Radar
P21	Reddish-orange	Reddish-orange	Medium	Radar
P25	Orange	Orange	Medium	Radar
P26	Orange	Orange	Very long	Radar
P28	Yellow-green	Yellow-green	Long	Radar
P31	Green	Green	Medium, high brightness	Oscilloscope and bright TV

weather radar or other 360° scan surveillance radars. On the other hand, for photographic purposes and TV use when the frame time is less than the response time of the eye (on the order of a tenth of a second or less), phosphors such as a P4 are used. If the phosphor persists longer than the refresh time of the display, smearing can occur. Table 8-1 lists some of the popular CRT phosphors, along with their characteristics and applications.

8.4 DISPLAY SYSTEM

In essence, the technique of illuminating a given point on the screen with a given intensity involves generating the appropriate horizontal and vertical deflection voltage or current and the appropriate grid voltage to give the desired brightness.

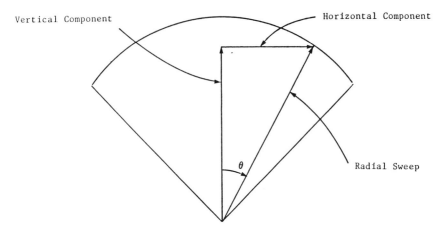

Figure 8-4. Vertical and horizontal components of the sweep.

A vertical sweep and a horizontal sweep must be generated simultaneously to move the spot from the vertex of a PPI display along a radial line representing range at some angle, θ as depicted in Figure 8-4. The amplitude and slope of the vertical sweep are proportional to the cosine of the antenna pointing angle, and the horizontal sweep amplitude and slope are proportional to the sine of the antenna pointing angle. Normally, the information available from the radar includes a synchronization pulse which defines when the sweep must start and a synchronization signal indicating the antenna pointing angle (θ). These signals can be thought of as representing polar coordinate variables (ρ, θ) which have to be transformed to rectangular coordinates (H, V) for the two sweeps. In some systems, the ρ, θ to H, V transformation is done by sophisticated electronic circuits which perform the sine and cosine multiplications. In other systems, a mechanical synchronization system is used to rotate the vertical deflection yoke in synchronism with the antenna scanning characteristics.[6] In this system, only the vertical sweep of constant slope need be generated, since the ρ, θ to H, V transformation is performed in space. That is, the scan line on the CRT follows the yoke which duplicates the antenna scanning. Figure 8-5 shows typical timing relationships among the various signals.

Figure 8-6 shows a simplified block diagram of a typical display system which generates the grid, high-voltage, and deflection signals required to display the radar information on the CRT.

The display system must know at what position on the screen to put a spot and how bright to make that spot. Synchronization signals are sent to the sweep generator from the timing generator such that the sweep begins approximately at transmit time. The appropriate horizontal and vertical sweep components are generated using angle information from the antenna position transducer.

Synchronization Pulses

Radar Video

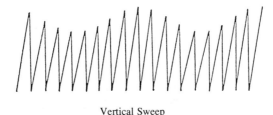

Vertical Sweep

Horizontal Sweep

Figure 8-5. Display timing.

The video signal from the radar receiver (video amplifier) is used to modulate the spot brightness by either grid or cathode voltage control. As the trace sweeps across the CRT face, bright spots appear where detected signals occur; the brightness depends on the strength of the signal.

Figure 8-7 is a photograph of a typical airborne radar display unit. All of the high-voltage power supplies, sweep-generating circuits, and video circuits are included in the unit. Most of the radar display–oriented controls, as well as many other radar function controls, are incorporated on the front panel, making them easily accessible to the operator.

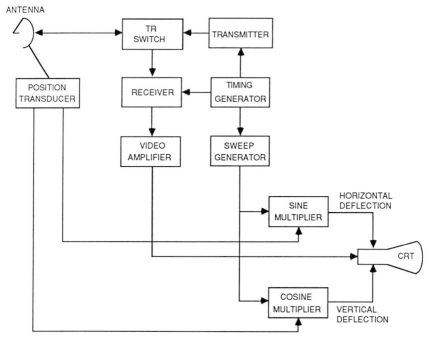

Figure 8-6. Simplified block diagram of a display system.

8.5 DISPLAY CONTROLS

Five radar operator controls typically affect the quality of the display: brightness, video gain, IF gain, sensitivity time control (STC) amplitude, and STC slope. When the display employs a storage scope, an additional persistence control is used to determine how long the radar scene remains on the CRT. The functions of these controls are as follows:

- Brightness: The brightness control determines the brightness level of the darkest spot in the scene. It should be adjusted so that the darkest spot just exceeds the ambient light level.
- Video gain: The video gain control, sometimes called the *contrast control*, determines the brightness of the brightest spot in the scene. It should be adjusted so that the brightest spot nearly saturates the display tube.
- IF gain: Although the IF gain control is a radar receiver control, it affects the display performance and, therefore, must be properly set to optimize the radar display quality. It should be set so that the dynamic range of the radar scene matches that of the display tube.

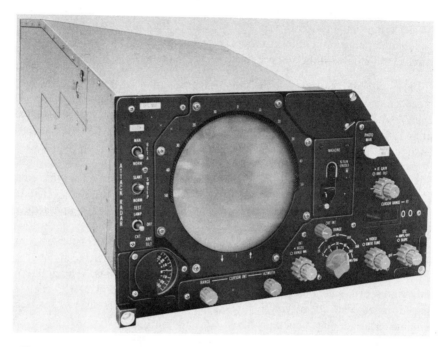

Figure 8-7. A typical airborne radar display unit. (Courtesy of General Electric Company)

- STC controls: The STC controls are also radar receiver controls which change the receiver gain as a function of range, so that the stronger close-in targets can be attenuated to compensate for their high intensity. The STC-amplitude control adjusts the amount of attenuation for the closest or zero range targets. The STC slope control adjusts the rate (in range) at which full receiver gain is achieved.

8.6 COMPUTER-GENERATED DISPLAYS

Various other types of display surfaces exist in addition to the CRT. Two of the most notable are the light-emitting diode (LED) array and a plasma or electroluminescent display. The LED array consists of an array of many LEDs placed very closely together in such a way as to form a square or rectangular matrix. The present state of the art is such that up to 200 rows and 200 columns can be made, although a more common size is 100 × 100. Properly programmed lighting of selected diodes in the matrix forms the image.

A plasma or electroluminescent display involves two grids of fine conductors, one oriented horizontally and one vertically, with a plasma between them. A typical grid configuration is a 512 × 512 array. Photon emission at the intersection of a pair of grid wires is produced by exciting the vertical and horizontal wires associated with the xy coordinate on the screen.

Plasma display systems require complex programmed excitation of the horizontal and vertical conductors. The programming is typically implemented using a microprocessor and a digital memory. Such a display processor has preprogrammed routines for generating dots, lines, alphanumeric characters, and special symbols such as circles and squares.

Some advantages of the plasma display include (1) higher reliability and longer life than the CRT, (2) improved brightness over some CRT types, and (3) ease of programming for a wide variety of displays once the "canned" interface is developed. The plasma display is compatible with modern radar signal-processing technology.

With the advent of microprocessor technology, the use of sophisticated computer-derived display processing is becoming popular. Whereas in the past "live" video was displayed to the operator commonly in a map configuration, many present systems use synthetic video to display moving target tracks, selected targets with identifying symbols, or simply computer-derived steering, aiming, or target engagement symbols.

Computer-derived displays can be generated in a raster-scan format, a script symbol generator format, or a combination of the two. The raster-scan is essentially the same as a television presentation. A large number of closely spaced horizontal lines are scanned in a prescribed sequence, and symbols are generated by varying the intensity at the appropriate instant. Script writing involves deflecting the beam so that the symbols are "drawn" on the screen; the intensity is modulated to create the individual symbols and delete the lines connecting them. A combination of these two approaches involves generating a raster-scan and performing some script writing during vertical retrace time.[6]

Computer-generated displays allow alphanumeric characters to be used as target identifiers, such as with air traffic control systems. They also allow special symbols such as aiming circles and artificial horizon symbols to be used for target engagement and piloting aids.

Many modern radar systems use a microprocessor or minicomputer as a signal processor and target manipulator. It is natural to include the display-processing function as part of the task of the computer. Figure 8-8 shows a block diagram of a surveillance radar system which uses a minicomputer for coordinating the radar and interface functions, calculating filtered moving target tracks, and processing the target information for the plasma display. The resulting display for such a system is shown in Figure 8-9. Not only the target tracks but also other important system parameters such as antenna scan position and target

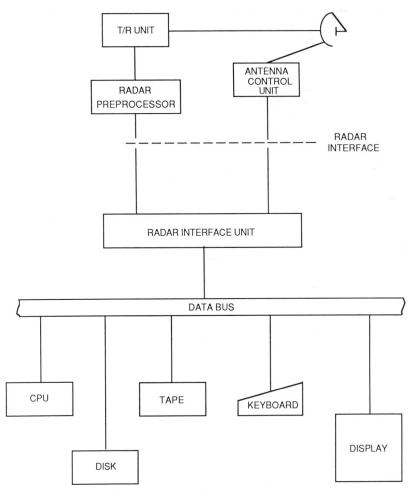

Figure 8-8. Computer-generated block diagram of a display system.

ranges are displayed. In this example, the lower portion of the display is used as a control panel. The control buttons are synthesized on the display face; when the operator "touches" one of the buttons, a set of light-sensitive transistors placed on the edge of the display panel sense where the panel is touched, and the desired control is initiated.

8.7 SCAN-CONVERTED DISPLAYS

Scan-converted displays represent a special case of a computer-generated display. The typical application of the scan converter is to display radar-mapping

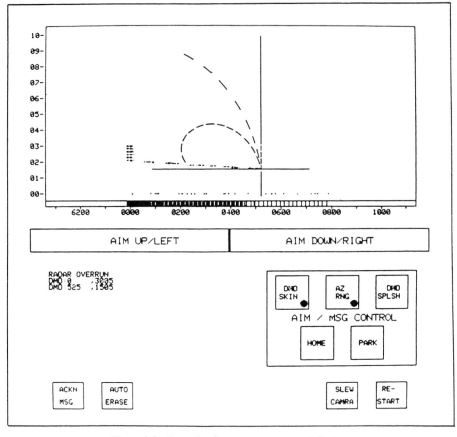

Figure 8-9. Example of a computer-generated display.

information using a conventional TV raster format. Historically, this was accomplished by pointing a TV camera at a conventionally generated radar display and displaying the output on a conventional TV. This was followed by the use of scan converter tubes which had a write gun at one end that was scanned in a fashion appropriate for generating the radar display, and a phosphorus screen and a read gun at the other end that were scanned in a raster-scan format for generating the TV video. These techniques are depicted in Figure 8-10. Microprocessor technology has eliminated the need to generate the radar map by allowing the radar-to-TV transformation to be done by a sequence of polar-to-rectangular calculations and appropriate, sequenced memory write and read cycles. The sweep generation function is replaced by a set of calculations for memory addressing. Figure 8-11 shows a block diagram of a modern digital scan converter. The elimination of the windshield wiper effect in the scan-con-

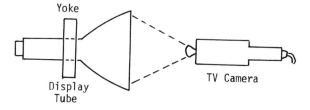

Display Tube/Camera Combination

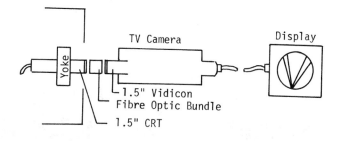

Fibre Optic Coupling

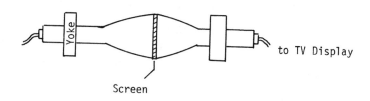

Scan Converter Tube

Figure 8-10. Optical scan converters.

verted display often reduces operator fatigue because of the reduction of target fading and blooming.[7] Figure 8-12 is a photograph of a digital scan converter used in an airborne radar system.

The incorporation of a scan converter allows the use of a single display for various sensors such as a radar, a low light level TV (LLLTV), and a forward-looking infrared (FLIR) system. Without the scan converter, the radar data

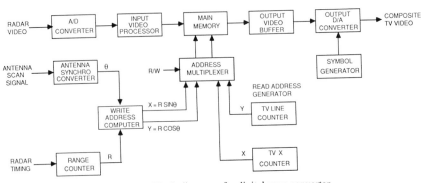

Figure 8-11. Block diagram of a digital scan converter.

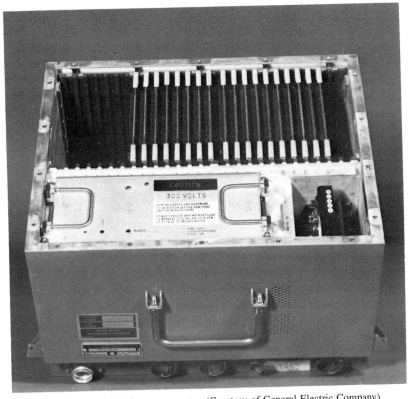

Figure 8-12. A digital scan converter. (Courtesy of General Electric Company)

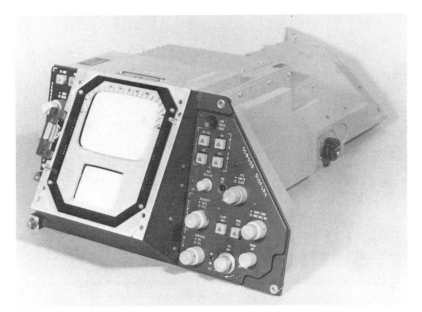

Figure 8-13. Example of a two-tube radar/electro-optical display. (Courtesy of General Electric Company)

would have to be displayed on a separate display system, thus reducing operator efficiency. Figure 8-13 is a photograph of a two-tube radar/electro-optical display.

8.8 SPECIAL DISPLAYS

Color displays add another dimension to the radar screen which can be used to highlight special features of the targets of interest to the operator. In weather radar applications, brighter colors (toward yellow) can be used to signify dense moisture and indicate storm activity. In a computer-generated color display, different colors can be used to identify various types of targets; for example, blue can be used for water, green for land masses, red for moving targets, and yellow for stationary targets. The added dimension of color can be used to reduce the operator fatigue associated with interpretation of the display.[2]

The HUD used with forward-looking radars in tactical aircraft involves projecting the radar data onto a screen which is in the forward field of view of the pilot and between the pilot and the windscreen. It allows the pilot to see the radar data without having to divert his attention completely from the task of flying the aircraft. Applications for this type of display include presentation of ground-mapping and aiming information in cases where the pilot must contin-

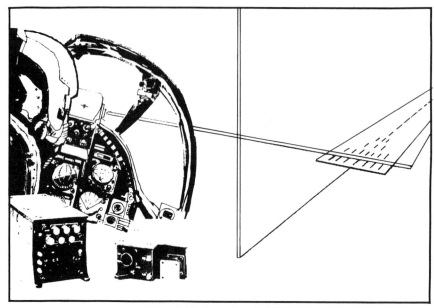

Figure 8-14. Operation of a HUD.

ually look through his windscreen while piloting the craft. Alphanumeric information and aiming cursors are often projected onto the HUD panel for providing additional information to the pilot. Figure 8-14 depicts the operation of a HUD.

8.9 REFERENCES

1. Merrill I. Skolnik, *Introduction to Radar Systems*, 2nd ed., McGraw-Hill Book Co., New York, 1980, chap. 9.
2. H. R. Luxenberg and R. L. Kuehn, *Display Systems Engineering*, McGraw-Hill Book Co., New York, 1968.
3. Louis N. Ridenour, *Radar System Engineering*, MIT Radiation Laboratory Series, Vol. 1, Boston Technical Publishers, Boston, 1964, chap. 6.
4. Theodore Soller, Merle A. Starr, and George E. Valley, Jr., *Cathode Ray Tube Displays*, MIT Radiation Laboratory Series, Vol. 22, Boston Technical Publishers, Boston, 1964.
5. George W. Stimson, *Introduction to Airborne Radar*, Hughes Aircraft Company, El Segundo, Calif., 1983.
6. Merrill I. Skolnik, *Radar Handbook*, McGraw-Hill Book Co., New York, 1970, chap. 6.
7. C. H. Baker, *Man and Radar Displays*, Pergamon Press, New York, 1962.

PART 3
DETECTION IN A
CONTAMINATED ENVIRONMENT

Having considered the basic building blocks of a radar system and how a radar actually functions in the preceding parts of this book, we now turn our attention to some of the practical, real world aspects of radar operation and performance prediction. Unfortunately, radar must operate in a non-ideal environment, one that is corrupted by extraneous or undesirable electrical signals called noise and by radar return signals from "targets" or reflectors which the radar was not intended to detect called clutter.

Chapter 9 presents some basic considerations in the detection of real target returns in the presence of noise starting with a fundamental definition of noise as a random process, followed by discussions of several detection processes and their characteristics. Multiple pulse integration, human operator performance, and an introduction to automatic detection, a subject covered in considerable detail later in Part 3, are also briefly covered in Chapter 9.

The characterization and understanding of radar clutter and its effect on radar performance are absolutely essential if the radar design engineer is to be able to predict accurately the expected performance of a radar and then actually develop hardware to achieve the predicted performance level. Chapter 10 presents one of the most extensive summaries of radar clutter data available in the current radar literature. Standard descriptions used to characterize clutter return properties are also developed along with statistical models of clutter which relate observed differences in radar reflectivity to frequency effects, polarization, depression angles, and other environmental factors.

Radar targets are not perfectly reflecting spheres with radar reflectivity which can be completely described with a single number. Radar targets are normally composed of many individual scattering centers, spatially separated and require relatively complex statistical functions to describe their radar reflectivity properties. Chapter 11 discusses the complex nature of radar targets and develops various statistical models to describe target amplitude fluctuations and then illustrates how these models might be used to predict radar performance.

In the final chapter of Part 3, the various aspects of target detection in a contaminated environment are considered from the standpoint of achieving automatic target detection and, at the same time, rejection of clutter and noise detections or false alarms. The concepts of adaptive-threshold constant false alarm rate (CFAR) processors and distribution free, or non-parametric, CFAR processors are introduced in Chapter 12. The intent of the chapter discussion is to provide insight into some of the problems that can restrict the usefulness of detection processes that are designed without full consideration of key characteristics of the background and interference and target characteristics as discussed in the preceding chapters in Part 3.

9
DETECTION IN NOISE

Jim D. Echard

9.1 OVERVIEW

Detection is the process by which the presence of the sought-after object, or target, is sensed in the presence of competing indications which arise from background echoes (clutter), atmospheric noise, or noise generated in the radar receiver. The characteristics and effects of background clutter are discussed in Chapter 10. Atmospheric and receiver noise voltages result from random electron movement in the atmosphere and in electrically conducting radar components, respectively. These time-varying voltages are called *random processes* and are described by a well-defined mathematical formula. When the target return is added to the noise voltage, the result is also a random process. Random processes are defined by statistical measures such as the probability density and autocorrelation functions. Sometimes the power density spectrum (PDS) is used to define random processes in addition to or instead of the autocorrelation function.

The noise power present at the output of the radar receiver can be minimized by using a filter that passes most of the target return energy while rejecting much of the noise energy. Such a filter, whose frequency response function maximizes the output peak-signal to mean-noise (power) ratio is called a *matched filter*. Matched filter approximations are used extensively in radar systems. The rule of thumb in radar practice is that the matched filter bandwidth should be approximately equal to the reciprocal of the pulse width. This is a reasonable approximation for pulse radars.

Correlation detection is mathematically equivalent to matched filter detection but is implemented differently. In correlation detection, the input signal is multiplied by a delayed replica of the transmitted signal, and the product is passed through a low-pass filter. The matched filter receiver, or an approximation, is generally preferred to the cross-correlation receiver in the vast majority of applications.

The envelope of an IF carrier within a radar receiver contains the information that is used to determine if a target is present or not. The radar detection process is almost always described in terms of threshold detection. If the envelope of the receiver output exceeds a preestablished threshold, a target is said to be present. Of course, with this type of detection, errors are made. Sometimes a target that is present is missed, and sometimes a target is declared to exist when none is present. These two situations are called *missed targets* and *false alarms*, respectively. The threshold detector is sometimes called a *Neyman-Pearson detector*. Other statistical criteria usually discussed when considering detection of targets in noise are the likelihood ratio and inverse probability, but these two types of receivers are seldom implemented in practice.

The portion of the radar receiver that extracts the modulation from the carrier is called the *detector*. One form of detector is the envelope detector, which recognizes the presence of the signal on the basis of the amplitude of the carrier envelope. When this detector is used, all phase information is ignored. It is also possible to design a detector which utilizes only phase information for detecting targets.

The most efficient detector is the coherent detector. With this detector, the reference oscillator signal is assumed to have the same frequency and phase as the input signal and simply provides a translation of the carrier frequency to DC. This detector does not ignore phase information, as does the envelope detector, but passes all of the modulation information to the detector output. Thus, it is not surprising that the coherent detector provides better performance than other detectors. Since the phase of the received signal is usually not known, a variation of the coherent detector called the *synchronous detector*, or *I&Q detector*, is usually utilized. The performance of the I&Q detector is equivalent to that of a coherent detector.

The rate of information production inherent in a typical radar signal is considerably greater than can be handled by a human operator. Thus, the function of the radar display is to aid the operator in efficiently extracting the information that is important to the task. In some radar systems, automatic detection is employed to overcome the limitations of an operator due to fatigue, boredom, or overload. In addition, automatic detection allows the radar output to be transmitted over communication links.

9.2 THE DETECTION PROCESS

Detection is the process by which a target is sensed in the presence of competing indications which arise from background echoes, atmospheric noise, or noise generated in the radar receiver. In radar, the target is sensed by its reflection of RF energy originating in the radar's own transmitter. Noise always accompanies the target signal in each stage of the radar receiver. Most of the noise is generated in various stages of the radar receiver. To some extent, the reflected target energy also competes with atmospheric noise even before entering the radar system.

The properties of target reflection, such as target fluctuations, are important in calculating detection performance, but a simple point source target (nonfluc-tuating) is assumed for this discussion.* Thus, the returning echo is assumed to be an exact reproduction of the transmitted signal, shifted in time by an amount corresponding to the range delay, shifted in frequency by the Doppler shift due to target radial velocity, and modified in amplitude by the geometry of the radar–target situation.

The ability of a radar receiver to detect a weak echo signal is limited by the noise energy that occupies the same portion of the frequency spectrum as the signal energy. The weakest signal the receiver can detect is called the *minimum detectable signal (MDS)*. Detection is based on establishing a threshold level at the output of the receiver. If the receiver output exceeds the threshold, a target is assumed to be present.

9.3   NOISE AS A RANDOM PROCESS

Noise signals are generated by the random movement of free electrons in a conducting medium. In the atmosphere, the electron movements generate an RF radiation which is intercepted by the radar antenna. Inside the radar re-ceiver, noise voltages are caused by random electron movement in electrically conducting components. Since the generation and movement of free electrons is a random occurrence and defies a deterministic characterization, the voltages resulting from this action are described by statistical measures. In particular, since the noise voltage is random and varies with the passage of time, it is called a *random process*. Random processes are part of a well-defined mathematical formula. The discussion which follows uses the commonly accepted terminol-ogy and definitions associated with this mathematical discipline.[1,2]

A typical noise random process is shown in Figure 9-1. The curve shown represents the RF or IF voltage in the radar receiver as a function of time before it is detected. A mathematical description of this random process is given by

$$n(t) = a(t) \cos [\omega_0 t + \theta(t)] \qquad (9\text{-}1)$$

where

$n(t)$ = noise voltage in the radar receiver
$a(t)$ = amplitude of the envelope modulation
ω_0 = carrier frequency in radians/second
t = time
$\theta(t)$ = phase modulation

The only assumption made in Eq. (9-1) is that the bandwidth of the noise

---

*More complicated target models are discussed in Chapter 11.

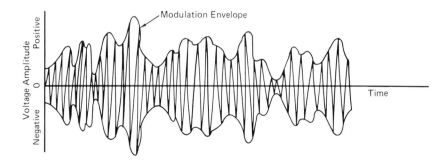

Figure 9-1. Typical noise voltage in the radar receiver as a function of time before detection.

process, B_n, is much smaller than the carrier frequency, ω_0; that is,

$$B_n \ll \omega_0/2\pi \qquad (9\text{-}2)$$

This assumption is almost always satisfied in conventional radars.

Both the amplitude and phase functions, $a(t)$ and $\theta(t)$, of Eq. (9-1) are random processes, but the carrier frequency, ω_0, is not; it is a deterministic value. Of course, t is also deterministic and varies from zero in a linear fashion. For thermally generated noise, the amplitude of the noise voltage at any instant of time, t_1, is statistically described by the Rayleigh probability density function:

$$p(a) = \frac{a}{\sigma_n^2} \exp\left[-a^2/2\sigma_n^2\right] \qquad (a \geq 0) \qquad (9\text{-}3)$$

where a is a shorthand notation for $a(t_1)$ and σ_n^2 is the statistical variance of $a(t_1)$. The phase of the noise voltage at any instant of time is described by the uniform density function:

$$p(\theta) = \frac{1}{2\pi} \qquad (0 \leq \theta \leq 2\pi) \qquad (9\text{-}4)$$

where θ is the shorthand notation for $\theta(t_1)$. These density functions are graphically displayed in Figures 9-2(a) and (b).

Another convenient mathematical representation of the noise voltage is given as

$$n(t) = x_n(t) \cos(\omega_0 t) - y_n(t) \sin(\omega_0 t) \qquad (9\text{-}5)$$

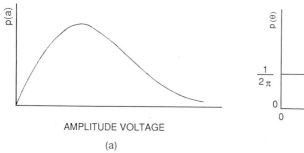
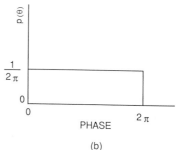

Figure 9-2. Probability density functions of thermally generated noise in (a) amplitude and (b) phase.

where

$n(t)$ = noise voltage in the receiver,
ω_0 = carrier frequency in radians/second,
t = time
$x_n(t)$ and $y_n(t)$ are related to $a(t)$ and $\theta(t)$ as follows:

$$x_n(t) = a(t) \cos [\theta(t)] \tag{9-6}$$

$$y_n(t) = a(t) \sin [\theta(t)]. \tag{9-7}$$

Note that Eq. (9-5) is mathematically equivalent to Eq. (9-1); however, it is sometimes more convenient to use than Eq. (9-1). In Eq. (9-5), $x_n(t)$ and $y_n(t)$ are both random processes. It can be shown that for thermal noise $x_n(t)$ and $y_n(t)$ are statistically independent and are both zero mean, Gaussian (normal) processes with variance σ_n^2, where σ_n^2 is the same as that used in Eq. (9-3). The zero mean Gaussian probability density function is shown in Figure 9-3. This density function is mathematically described by

$$p(x_n) = \frac{1}{\sigma_n(2\pi)^{1/2}} \exp [-x_n^2/(2\sigma_n^2)] \tag{9-8}$$

and

$$p(y_n) = \frac{1}{\sigma_n(2\pi)^{1/2}} \exp [-y_n^2/(2\sigma_n^2)]. \tag{9-9}$$

When a target is present in the radar antenna beam, the reflected signal is added to the noise voltage in the radar receiver and the combined voltages are

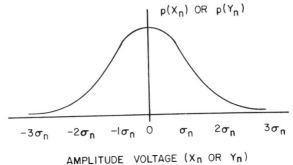

AMPLITUDE VOLTAGE (X_n OR Y_n)

Figure 9-3. Gaussian (normal) probability density function with zero mean and variance σ_n^2.

mathematically represented as

$$s(t) = r(t) + n(t), \tag{9-10}$$

where $r(t)$ is the target return voltage as a function of time and $n(t)$ is the noise voltage as a function of time.

For some types of targets, $r(t)$ may vary randomly with time. For this discussion, however, the target is assumed to provide a constant, nonfluctuating return signal:

$$r(t) = c \cos (\omega_0 t + \phi) \tag{9-11}$$

where both c and ϕ are constants. Thus, Eq. (9-10) may be written as

$$s(t) = c \cos (\omega_0 t + \phi) + a(t) \cos [\omega_0 t + \theta(t)] \tag{9-12}$$

or as

$$s(t) = [c \cos \phi + x_n(t)] \cos \omega_0 t - [c \sin \phi + y_n(t)] \sin \omega_0 t \tag{9-13}$$

whichever is more convenient to use. Note that $s(t)$ is a random process with a deterministic component, that is, $c \cos (\omega_0 t + \phi)$. Thus, $s(t)$ is mathematically described by a probability density function. The magnitude of $s(t)$ is defined by the Ricean (modified Rayleigh) density function:

$$p(\hat{s}) = \frac{\hat{s}}{\sigma_n^2} \exp [-(\hat{s}^2 + c^2)/(2\sigma_n^2)] I_0 (\frac{c\hat{s}}{\sigma_n^2}) \qquad (\hat{s} \geq 0) \tag{9-14}$$

where

$\hat{s}$ = envelope of the target-plus-noise signal $s(t)$
σ_n^2 = variance (power) of the noise voltage
c = peak magnitude of the target return
$I_0(\cdot)$ = modified Bessel function of the order zero

The term $c^2/(2\sigma_n^2)$ is called the *signal-to-noise ratio* and is sometimes denoted as SNR or S/N. It is the ratio of average signal power, $c^2/2$, to noise variance or noise power, σ_n^2. When the signal power is zero, that is, when no target return is present, then Eq. (9-14) is the density function for noise alone and is identical to Eq. (9-3). The probability density function describing the amplitude of the target-plus-noise signal is shown in Figure 9-4 for various signal-to-noise ratios.

In the previous discussion, the random processes of noise alone and target-plus-noise were described by using probability density functions, but the probability density function is not a complete description of a random process. For example, it does not describe how the value of a random process is related at one instant of time to its value at another instant of time. One needs to determine how the random process described by Eq. (9-1) varies from time t_1 to time t_2 and from time t_2 to time t_3, ad infinitum. The ensemble of random variables

$$n(t_1), \; n(t_2), \; n(t_3), \; \ldots \tag{9-15}$$

must be examined to answer this question.

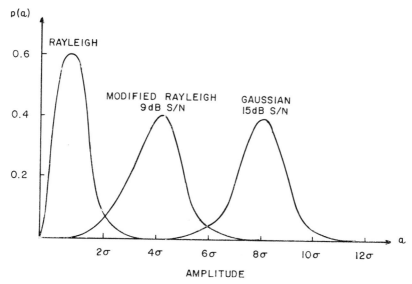

Figure 9-4. Probability density function of target-plus-noise signal envelope for various SNRs.

The mathematical formulation which describes this process is called *auto-correlation*. The autocorrelation function of a random process is mathematically defined as

$$\phi_n(t_1, t_2) = E[n(t_1) \, n(t_2)] \tag{9-16}$$

$$= \int_{-\infty}^{\infty} n(t_1) \, n(t_2) \, p[n(t_1), \, n(t_2)] \, dn(t_1) \, dn(t_2). \tag{9-17}$$

This function is sometimes difficult to evaluate. Another formulation which is easier to use is the time autocorrelation function of a random process, defined as

$$R_n(\tau) = \lim_{T \to \infty} \frac{1}{2T} \int_{-T}^{T} n(t) \, n(t - \tau) \, dt \tag{9-18}$$

where $\tau = t_2 - t_1$. If the random process is ergodic,* then

$$\phi_n(t_1, t_2) = \phi_n(\tau) = R_n(\tau). \tag{9-19}$$

The random variables $x_n(t_1)$ and $x_n(t_1 + \tau)$ formed by fixing the observation times at t_1 and $t_1 + \tau$, respectively, are in general distributed differently. However, in the special case when "stationarity" is assumed, $x_n(t_1)$ and $x_n(t_1 + \tau)$ are identically distributed for all values of τ. More generally, a random process $x_n(t)$ is said to be *stationary* if for any n observation times $t_1, t_2, \ldots, t_n$ and for any value of τ, the distributions of the random vectors $x_n(t_1), x_n(t_2), \ldots,$ $x_n(t_n)$ and $x_n(t_1 + \tau), x_n(t_2 + \tau), \ldots, x_n(t_n + \tau)$ are identically distributed. Therefore, stationarity of a process implies that the statistics of the process are independent of the time origin.

If the ensemble statistics of a process at a fixed time are independent of the time origin and these statistics are equal to the corresponding time statistics obtained from any one of the time functions, the process is said to be *ergodic*. Clearly, a process might be ergodic for certain statistical parameters and not for others. Note that for a process to be ergodic, it must be stationary; however, the converse is not true.

Relating the foregoing discussion to most common radar applications, Eqs. (9-18) and (9-19) apply if the noise and target statistics are stationary. For two parameter statistics such as a Gaussian random process, the first moment (mean) and the second moment (variance) must not vary with time.

The autocorrelation function of $n(t)$, given by Eq. (9-5), can be determined,

---

*For a more thorough discussion of ergodicity, see pages 327 and 471 of reference 1.

using Eq. (9-18), to be

$$\phi_n(\tau) = \phi_m(\tau) \cos(\omega_0 \tau) \qquad (9\text{-}20)$$

where $\phi_m(\tau) = \phi_x(\tau) = \phi_y(\tau)$ is a description of the autocorrelation properties of the modulation impressed on the carrier signal. The autocorrelation function given above is based on the assumption that $n(t)$ is ergodic.

The Fourier transform of $\phi_n(\tau)$ represents the PDS of $n(t)$ and is determined to be

$$\Phi_n(\omega) = \frac{\Phi_x(\omega - \omega_0) + \Phi_x(\omega + \omega_0)}{2} = \frac{\Phi_y(\omega - \omega_0) + \Phi_y(\omega + \omega_0)}{2} \qquad (9\text{-}21)$$

where $\Phi_x(\omega)$ and $\Phi_y(\omega)$ are the PDSs of the carrier modulations $x(t)$ and $y(t)$, respectively. The PDSs of $x(t)$ and $y(t)$ are related to the respective autocorrelation function by the Fourier transform:

$$\Phi_x(\omega) = \int_{-\infty}^{\infty} \phi_x(\tau) \exp^{-j\omega\tau} d\tau \qquad (9\text{-}22)$$

$$\Phi_y(\omega) = \int_{-\infty}^{\infty} \phi_y(\tau) \exp^{-j\omega\tau} d\tau. \qquad (9\text{-}23)$$

The term *white noise* is sometimes used to describe the correlation or spectral properties of noise. If the autocorrelation of the noise is given by

$$\phi_x(\tau) = \phi_y(\tau) = \begin{cases} 1; & \tau = 0 \\ 0; & \tau \neq 0 \end{cases} \qquad (9\text{-}24)$$

then the PDS of this noise is calculated to be

$$\Phi_x(\omega) = \Phi_y(\omega) = 1 \qquad \text{all } \omega \qquad (9\text{-}25)$$

and

$$\Phi_n(\omega) = 1 \qquad \text{all } \omega. \qquad (9\text{-}26)$$

This type of noise is called white since it is very wideband and *all* frequency components of equal amplitude exist in the noise process. This also means that no matter how small τ is assumed to be (except zero), $n(t)$ and $n(t + \tau)$ are uncorrelated. Thus, white noise implies uncorrelated noise.

In practice, noise is never white. However, it may be very wideband in comparison to the bandwidth of the system under consideration and, therefore, may be called white.

The noise which exists at its source of generation (e.g., at the electrical component) is usually very wideband and is generally assumed to be white or uncorrelated. The most significant source of noise in a radar receiver is at its front end near the antenna. This noise is passed through several mixing and amplification processes before reaching the radar detector. The amplifiers used are limited in bandwidth and determine the bandwidth of the noise which reaches the radar detector. Usually the IF amplifiers have the smallest bandwidth in the radar and, thus, determine the bandwidth of the noise. Of course, band-limited noise is correlated.

9.4 MATCHED FILTER[3,4]

A filter whose frequency-response function maximizes the output signal-to-noise ratio is called a *matched filter*. The frequency-response function, denoted by $G(f)$, expresses the relative amplitude and phase of the output of a filter with respect to the input when the input is a pure sinusoid. The magnitude $|G(f)|$ of the frequency-response is the radar-receiver passband characteristic. If the bandwidth of the receiver is wide compared with that occupied by the signal energy, extraneous noise is introduced by the excess bandwidth, which lowers the output signal-to-noise ratio. On the other hand, if the receiver bandwidth is narrower than the bandwidth occupied by the signal, the noise energy is reduced along with part of the signal energy. The result is again a lowered signal-to-noise ratio. Thus, there is a receiver bandwidth at which the signal-to-noise ratio is a maximum. The rule of thumb often used in radar practice is that the receiver bandwidth should be approximately equal to the reciprocal of the pulse width. This is a reasonable approximation for radars with uncoded pulses; however, it is not valid for coded waveforms because their bandwidths are larger than the reciprocal of the pulse width. The specification of the *optimum* receiver characteristic involves the receiver frequency-response function and the shape of the received waveform.

The receiver frequency-response function is given by the response from the antenna terminals to the output of the IF amplifier. The narrow receiver bandwidth most often occurs in the IF amplifier chain. The bandwidths of the RF and mixer stages of the normal receiver are usually large compared with the IF bandwidth. Therefore, the frequency-response function of the portion of the receiver included between the antenna terminals and the output of the IF amplifier is usually taken to be that of the IF amplifier alone. Thus, only the frequency-response function of the IF stage need be used to maximize the SNR. The IF amplifier may be considered as a filter with gain. The response of this filter as a function of frequency is of interest.

For a received waveform $r(t)$, it can be shown that the frequency-response function of a linear, time-invariant filter which maximizes the output peak-signal to mean-noise ratio for a fixed input signal-to-noise ratio is[4]

$$G(f) = R*(f) \exp(-j2\pi f t_1) \qquad (9\text{-}27)$$

where

$$R(f) = \int_{-\infty}^{\infty} r(t) \exp(-j2\pi f t) \, dt$$

$R*(f)$ = complex conjugate of $R(f)$
t_1 = fixed value of time at which the signal is observed to be maximum

The noise that accompanies the signal is assumed to be stationary and to have a uniform spectrum (white noise).* The filter whose frequency-response function is given by Eq. (9-27) is the matched filter.

The frequency-response function of the matched filter is the conjugate of the spectrum of the received waveform, except for the phase shift $\exp(-j2\pi f t_1)$.[4] This phase shift varies uniformly with frequency. Its effect is to cause a constant time delay. A time delay is necessary in the specification of the filter for physical realizability, since there can be no output from the filter until the signal is applied. The frequency spectrum of the received signal may be written as an amplitude spectrum $|R(f)|$ and a phase spectrum $\exp[-j\phi_r(f)]$. The matched filter frequency-response function may similarly be written in terms of its amplitude and phase spectrum $|G(f)|$ and $\exp[-j\phi_g(f)]$. Eq. (9-27) for the matched filter may then be written as

$$|G(f)| = |R(f)| \qquad (9\text{-}28)$$

and

$$\phi_g(f) = -\phi_r(f) + 2\pi f t_1. \qquad (9\text{-}29)$$

Thus, the amplitude spectrum of the matched filter is the same as that of the signal, but the phase spectrum of the matched filter is the negative of the phase spectrum of the signal plus a phase shift that is proportional to the frequency.

The matched filter may also be specified by its impulse response, $g(t)$, which is the inverse Fourier transform of the frequency-response function:

$$g(t) = \int_{-\infty}^{\infty} G(f) \exp(j2\pi f t) \, df. \qquad (9\text{-}30)$$

---

*In practice, the noise only has to be uniform over the frequency passband of the filter for Eq. (9-27) to be valid.

Physically, the impulse response is the output of the filter as a function of time when the input is an impulse (delta function). Substituting Eq. (9-27) into Eq. (9-30) gives

$$g(t) = \int_{-\infty}^{\infty} R^*(f) \exp[-j2\pi f (t_1 - t)] \, df. \qquad (9\text{-}31)$$

Since $R^*(f) = R(-f)$, we have

$$g(t) = \int_{-\infty}^{\infty} R(f) \exp[j2\pi f (t_1 - t)] \, df = r(t_1 - t). \qquad (9\text{-}32)$$

Thus, the impulse response of the matched filter is the image of the received waveform; that is, it is the same as the received signal run backward in time starting from the fixed time t_1.

The impulse response of a practical filter is not defined for $t < 0$ (one cannot have a response before the impulse is applied). Therefore, $t < t_1$. This is equivalent to the condition placed on the transfer function $G(f)$ that there be a phase shift $\exp(-j2\pi f t_1)$. However, for the sake of convenience, the impulse response of the matched filter is sometimes written simply as $r(-t)$.

The output of the matched filter is not a replica of the input signal. However, in detecting signals in noise, preserving the shape of the signal is of no importance. If it is necessary to preserve the shape of the input pulse rather than maximize the output signal-to-noise ratio, some other criterion must be employed.

The output of the matched filter may be shown to be proportional to the input signal cross-correlated with a replica of the transmitted signal, except for the time delay t_1. The cross-correlation function, $R(\tau)$, of two signals, $y(\lambda)$ and $g(\lambda)$, each of finite duration, is defined as

$$R(\tau) = \int_{-\infty}^{\infty} y(\lambda) g(\lambda - \tau) \, d\lambda. \qquad (9\text{-}33)$$

The output, $y_0(t)$, of a filter with impulse response $g(t)$ when the input is $y_{in}(t) = r(t) + n(t)$ is the convolution of $g(t)$ and $y_{in}(t)$:

$$y_0(\tau) = \int_{-\infty}^{\infty} y_{in}(\lambda) g(\tau - \lambda) \, d\lambda. \qquad (9\text{-}34)$$

If the filter is a matched filter, then $g(\lambda) = r(t_1 - \lambda)$ and Eq. (9-34) becomes

$$y_0(\tau) = \int_{-\infty}^{\infty} y_{\text{in}}(\lambda)r(t_1 - \tau + \lambda)d\lambda = R(\tau - t_1). \qquad (9\text{-}35)$$

Thus, the matched filter forms the cross-correlation between the received signal corrupted by noise and a replica of the transmitted signal. The replica of the transmitted signal is "built into" the matched filter via the frequency-response function. If the input signal $y_{\text{in}}(t)$ were the same as the signal $r(t)$ for which the matched filter was designed (that is, the noise is assumed to be negligible), the output would be the autocorrelation function of $r(t)$.

9.5 CORRELATION DETECTION

Eq. (9-35) describes the output of the matched filter as the cross-correlation between the input signal and a delayed replica of the transmitted signal. This implies that the matched filter receiver can be replaced by a cross-correlation receiver that performs the same mathematical operation, as shown in Figure 9-5. The input signal $y_{\text{in}}(t)$ is multiplied by a delayed replica of the transmitted signal $r(t - t_1)$, and the product is passed through a low-pass filter. The cross-correlation receiver tests for the presence of a target at only one time: t_1. Targets at other time delays, or ranges, might be found by varying t_1. However, this requires a longer search time. The search time can be reduced by adding parallel channels, each containing a delay line corresponding to a particular value of t_1, as well as a multiplier and a low-pass filter.

Since the cross-correlation receiver and the matched-filter receiver are equivalent mathematically, the choice of which one to use in a particular radar application is determined by which is more practical to implement. The matched-filter receiver, or an approximation, has been generally preferred in the vast majority of applications.

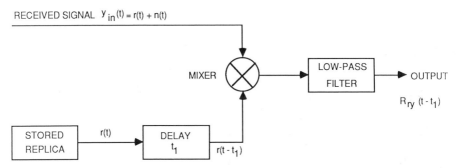

RECEIVED SIGNAL $y_{\text{in}}(t) = r(t) + n(t)$

MIXER

LOW-PASS FILTER

OUTPUT

$R_{ry}(t - t_1)$

STORED REPLICA

$r(t)$

DELAY t_1

$r(t - t_1)$

Figure 9-5. Block diagram of a cross-correlation receiver.

9.6 DETECTION CRITERIA

Detecting signals in the presence of noise is equivalent to deciding whether the receiver output is due to noise alone or to signal-plus-noise. When detection is carried out automatically by electronic means, it cannot be left to chance, but must be specified and built into the decision-making device by the radar designer.

The radar detection process is described in terms of threshold detection. Almost all radar detection decisions are based upon a comparison of the output of a receiver with some threshold level. If the envelope of the receiver output exceeds a preestablished threshold, a target is said to be present. The purpose of the threshold is to divide the output into regions of no detection and detection. In other words, the threshold detector allows a choice between one of two hypotheses. One hypothesis is that the receiver output is due to noise alone; the other is that the output is due to signal-plus-noise. The dividing line between these two regions depends upon the probability of a false alarm, which in turn is related to the average time between false alarms. This situation is shown in Figure 9-6. Here the envelope of the noise is illustrated with its rms value and threshold denoted by horizontal lines. Note that occasionally the threshold is exceeded by the noise voltage—a false alarm occurs.

Two types of errors may be made in the decision process. One kind of error is to mistake noise for a signal when only noise is present. This occurs whenever the noise is large enough to exceed the threshold level. In statistical detection theory, it is called a *type I error*. The radar engineer would call it a false alarm. A *type II error* is one in which a signal is erroneously considered to be noise. This is a missed detection to the radar engineer. The setting of the threshold represents a compromise between these two types of errors. A relatively large threshold will reduce the probability of a false alarm, but more missed detections will occur. The nature of the radar application will influence to a large extent the relative importance of these two errors and, therefore, the setting of the threshold.

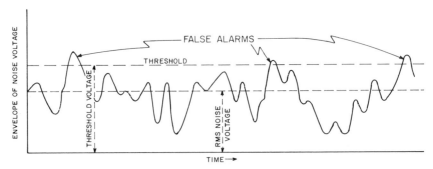

Figure 9-6. Envelope of noise voltage versus time with threshold shown and false alarms noted.

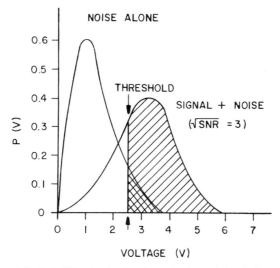

Figure 9-7. Probability density functions of noise and signal-plus-noise.

The threshold level is usually selected so as not to exceed a specified false alarm probability; that is, the probability of detection is maximized for a fixed probability of false alarm. This approach to selecting the threshold level is illustrated in Figure 9-7, where the probability density functions for noise alone and signal-plus-noise are shown. This is equivalent to fixing the probability of a type I error and minimizing the type II error. It is similar to the Neyman-Pearson test used in statistics for determining the validity of a specified statistical hypothesis. Therefore, this type of threshold detector is sometimes called a *Neyman-Pearson detector*. The Neyman-Pearson criterion is well suited for radar applications and is often used in practice.

Two other statistical criteria usually discussed when considering detection of targets in noise are the likelihood ratio and inverse probability, but these types of receivers are seldom implemented in practice. In some cases, the receiver which computes the likelihood ratio is equivalent to one which computes the cross-correlation function or one with a matched-filter characteristic. The inverse probability receiver requires that the a priori probability of a target being present in a particular range cell be known. In practical situations, this information is rarely known. Thus, this type of receiver is difficult to implement and will not be discussed in detail here.

9.7 DETECTOR CHARACTERISTICS

The device which extracts the modulation from the carrier is called the *detector*. As used here, detector implies more than simply a rectifying element. It includes that portion of the radar receiver from the output of the IF amplifier to the input of the indicator or data processor. The effect of the amplifier is not important to our present discussion. The major concern is the effect of the detector on the desired signal and the noise.

One type of detector is the envelope detector, which extracts the signal represented by the amplitude of the carrier envelope. All phase information is ignored. It is also possible to design a detector which utilizes only phase information. An example is one which counts the zero crossings of the received waveform. The zero-crossings detector neglects amplitude information. If the exact phase of the echo carrier were known, it would be possible to design a detector which makes use of both the phase and amplitude information contained in the radar return. It would perform more efficiently than a detector which used either amplitude information or phase information only. The coherent detector uses both phase and amplitude information. Two of these detectors, the envelope and the coherent detector, are considered in this section.

The function of the envelope detector is to extract the modulation and reject the carrier. By eliminating the carrier, all phase information is lost and the detection decision is based solely on the envelope amplitude. The envelope detector consists of a rectifying element and a low-pass filter that passes the modulation frequencies but removes the carrier frequency. The rectifier characteristic relates the output signal to the input signal and is called the *detector law*. Most detector laws approximate either a linear or a square-law characteristic. In the so-called linear detector, the output signal is directly proportional to the input signal envelope. (Actually, the so-called linear detector is a nonlinear device, or else it would not be a detector.) Similarly, in the square-law detector, the output signal is proportional to the square of the input signal envelope. In general, the difference between the two is small, and the detector law in any analysis is usually chosen for mathematical convenience.

For a single pulse, the optimum form* of the detector law can be shown to be

$$y = \ln I_0(bv) \qquad (9\text{-}36)$$

where

$\quad y$ = output voltage of the detector
$\quad I_0(\cdot)$ = modified Bessel function of order zero
$\quad b$ = amplitude of signal voltage divided by rms noise voltage, or signal-to-noise ratio
$\quad v$ = amplitude of IF voltage envelope divided by rms noise voltage

---

*This maximizes the likelihood ratio for a fixed probability of false alarm.

In other words, the voltage amplitude of the pulse must be transformed via the nonlinear curve described by the Bessel function and the natural log. For large signal-to-noise ratios ($b \gg 1$), this detector law is approximately given by

$$y \approx bv. \tag{9-37}$$

Thus, the "linear" detector is a good approximation of the optimum detector law for large signal-to-noise ratios. For smaller signal-to-noise ratios the detector law is approximately

$$y = (bv)^2/4 \tag{9-38}$$

which is the characteristic of a square-law detector.

Hence, it may be concluded that for small signal-to-noise ratios the square-law detector may be a suitable approximation to the optimum detector, while for large signal-to-noise ratios the linear detector is more appropriate. In practice, it makes little difference which of the two detector laws is used. The difference between the square-law and linear detectors is less than 0.2 dB in the required signal-to-noise ratio.

The probability of detection (P_d) versus signal-to-noise ratio for a square-law detector is given by the curves in Figure 9-8, with constant false alarm probabilities (P_n) as a parameter.[12] The target is assumed to be nonfluctuating and the number of pulses summed is 1 (no pulse-to-pulse integration). As indicated before, similar detection curves for a "linear" detector would only be a few tenths of a decibel different.

Suppose one desired to determine the single-pulse signal-to-noise ratio required to provide a probability of detection of 90% with a false alarm probability of 10^{-6}, assuming a nonfluctuating target. The procedure for using the detection curves of Figure 9-8 is as follows. First, select the appropriate false alarm curve. Second, locate a point on this curve corresponding to a detection probability of 90% (horizontal dashed line). Next, follow a vertical line from this point to the abscissa to determine the required signal-to-noise ratio: $+13.1$ dB. This example is applicable only to single-pulse detection but will be extended later to include multiple pulses.

If the output of the receiver is proportional to the logarithm of the input signal envelope, the receiver is called a *logarithmic receiver*. This receiver finds application where large variations in input signals are expected. It might be used to prevent receiver saturation or to reduce the effects of unwanted clutter in certain types of radar receivers.

The coherent detector of Figure 9-9 consists of a reference oscillator feeding a linear multiplier or mixer. The input to the mixer is a signal of known fre-

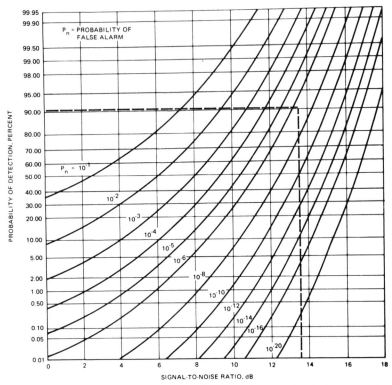

Figure 9-8. Probability of detection (P_d) versus signal-to-noise (power) ratio for a square-law envelope detector and a nonfluctuating target. Single-pulse detection is used. (By permission, from Hovanessian, ref. 12; © 1973 Artech House, Inc.)

quency and known phase plus its accompanying noise. The reference-oscillator signal is assumed to have the same frequency and phase as the input signal to be detected. The output of the mixer is followed by a low-pass filter which allows only the DC and low-frequency modulation components to pass while rejecting the higher frequencies in the vicinity of the carrier. The coherent detector provides a translation of the carrier frequency to DC. It extracts both the modulation envelope and phase and is a truly linear detector, whereas the "linear" envelope detector is not linear in the same sense. Therefore, the coherent detector is more efficient, especially when signal-to-noise ratios are low.

Usually the phase of the received signal is not known, so the detector in Figure 9-9 cannot be used. However, a variation of this type of detector known as a *synchronous detector* is shown in Figure 9-10. It is similar to that of Figure 9-9, except that two mixers and low-pass filters are utilized. The reference-oscillator signal for each mixer is assumed to have about the same frequency as the transmitted signal, but the two reference signals are out of phase with

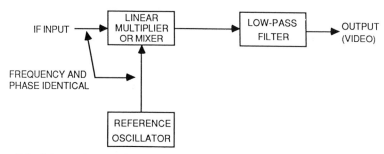

Figure 9-9. Coherent detector. Frequency and phase of reference oscillator matched to that of IF input.

each other by 90°. Usually the reference signal is obtained from the transmitter source so that it is synchronous with or follows the transmitter frequency. With this type of detector, the phase of the received signal may have any value and the modulation information will be fully represented by the in-phase (I) and quadrature (Q) signals. If the IF signal is represented by either of the mathematical forms given by Eqs. (9-1) or (9-5), then the I and Q signals out of the synchronous detector will be given by

$$I = x(t) = a(t)\cos[\theta(t)] \tag{9-39}$$

and

$$Q = y(t) = a(t)\sin[\theta(t)] \tag{9-40}$$

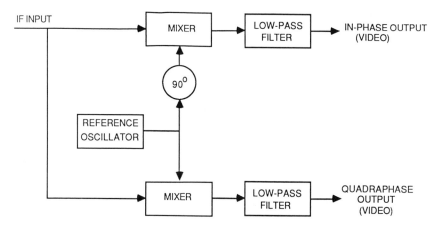

Figure 9-10. Synchronous or I,Q detector. Reference oscillator frequency approximately equals IF input frequency.

where

$a(t)$ = envelope amplitude voltage
$\theta(t)$ = phase voltage

If the IF modulation is represented by complex valued mathematics, the detector output, $o(t)$, can be written as

$$o(t) = I + jQ \qquad (9\text{-}41)$$

or

$$o(t) = x(t) + jy(t) \qquad (9\text{-}42)$$

or

$$o(t) = a(t)\exp j[\theta(t)] \qquad (9\text{-}43)$$

where $a(t)$ and $\theta(t)$ are the amplitude and phase modulations, respectively. Thus, the detector signal, $o(t)$, is proportional to the complex valued modulation envelope at IF.

One of the weaknesses of the I,Q detector just presented is the requirement that each channel be properly balanced in amplitude and phase. If the amplitude gains of the I and Q channels are slightly different or if the reference signals are not quite in quadrature, then errors occur which produce unwanted images in the resulting signal spectrum. Recently, attempts have been made to obtain quadrature signals using other approaches. They have been reported as being quite successful. One approach uses a single analog-to-digital (A/D) converter which samples the IF signal at the video signal bandwidth and then produces I,Q signals by appropriate digital signal processing. For further information, see references 5 to 7.

9.8 MULTIPLE PULSE INTEGRATION

Only in unusual cases will the radar detection decision be based on a single received pulse. Instead, a train of several to several hundred pulses will be received from each target, and the entire train will be processed before it is decided if a target is present. High detection probability can then be obtained even when the single-pulse signal-to-noise ratio is near or below unity. The process by which the pulses are combined is known as *integration*.

In an ideal processing arrangement, the energy from all of the radar returns is added directly in an integrator prior to envelope detection. The result is an effective, or integrated, value of signal-to-noise ratio $(SNR)_i$ equal to the single-pulse signal-to-noise ratio multiplied by n, the total number of pulses in the train. This process is referred to as *coherent integration* because it requires that

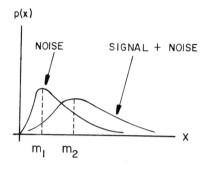
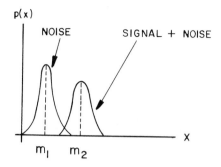

a. BEFORE INTEGRATION

b. AFTER INTEGRATION

Figure 9-11. Probability density functions of noise and signal-plus-noise.

the phase of successive signals be known relative to that of a local reference sinusoid.

Many radars use *noncoherent* integration to improve target detection capabilities without requiring a coherent reference for phase information. The noncoherent integrator can achieve a gain which approaches n in cases where only a few pulses are processed and can be approximated by $n^{0.8}$ over a wide range of conditions. In general, noncoherent integration is not as efficient as coherent integration. However, this disadvantage may be offset by the fact that noncoherent detection and integration are most easily implemented and do not require a coherent reference.

The integration gain is defined for a specific detection and false alarm probability. The gain is calculated by determining the signal-to-noise ratio required before and after integration and taking the ratio (or difference in decibels). The probability density function of noise and target-plus-noise before integration is as shown in Figure 9-11(a). Normalized integration* of radar returns causes the variances of the probability density functions to become smaller according to the amount of statistical independence between radar returns. If the interference is only thermal noise, the variance will decrease by n, the number of samples integrated. The probability density function means remain unchanged. This is illustrated in Figure 9-11(b). The net effect is to increase the separation (in terms of variance) between the two probability density functions. Thus, a smaller signal-to-noise ratio is required to achieve a specific probability of detection and false alarm.

Coherent integration may be implemented in two ways: before or after the signal detector. These two implementations are illustrated in Figure 9-12. If the signal is integrated while it is on an IF carrier, as shown in Figure 9-12(a),

*Normalized integration consists of summing the radar returns and *dividing* by the number summed.

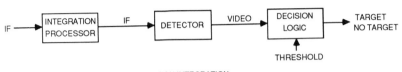

a. PRE-DETECTION INTEGRATION.

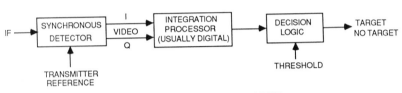

b. COHERENT POST-DETECTION INTEGRATION.

Figure 9-12. Two ways in which coherent integration may be implemented.

coherent integration will occur if the transmitted signal is coherent or phase stable. Coherent integration may be performed with a pulse-to-pulse unstable transmitter source if a synchronous detector with a transmit reference is used, as illustrated in Figure 9-12(b). In this case, integration is performed after the detector utilizing two quadrature signals, I and Q. Since the I and Q signals are at baseband or video, and thus at a lower frequency then the IF signal, they are usually converted to digital signals via an A/D converter and the integrator is implemented digitally.

Noncoherent integration is usually implemented as shown in Figure 9-13. Envelope detection implies that only the magnitude or envelope of the IF carrier modulation signal is removed by the detector. No phase information is recovered by the detector. The envelope may be removed from the IF carrier without distortion by the use of a so-called linear detector. Most often a square-law detector will be used to provide the square of the envelope at its output. Sometimes a logarithmic detector will be used to provide an output which is approximately the log of the modulation envelope.

After the modulation envelope is removed, the energy from all of the pulses is added in the integration processor. This may be performed with an analog or digital integrator. If a digital integrator is used, an A/D converter must be used between the detector and the integrator.

Detection curves similar to those shown in Figure 9-8 can be generated for various numbers of pulses integrated. In Figure 9-8, the probability of detection

Figure 9-13. Implementation of noncoherent integration.

(in percent) is plotted versus the signal-to-noise ratio (in decibels), with the false alarm probability shown as a parameter for single-pulse detection. When the number of pulses is added as a parameter, multiple plots are required. An example is shown in Figure 9-14, where the probability of detection (in percent) is plotted versus the number of pulses integrated with the signal-to-noise ratio (in decibels) as a parameter. These curves apply to only one value of false alarm probability or false alarm number n'. In reference 8, from which these "Meyer plots" are taken, the probability of false alarm (P_{fa}) is related to false alarm number n' as follows: $P_{fa} \approx 0.693/n'$. In Figure 9-14(a) and (b), the false alarm number is given as 6×10^5, yielding a false alarm probability of $P_{fa} \approx 1.155 \times 10^{-6}$, which is very close to the 10^{-6} curve in Figure 9-8. In the curves of Figure 9-14, the target model assumed is nonfluctuating or "Case 0."

Continuing the detection example given earlier, if a probability of detection of 90% is desired with a false alarm probability of about 10^{-6}, then the single-pulse signal-to-noise ratio required is approximately +13.1 dB, as shown in Figure 9-14(a) by the circle. If 10 pulses are transmitted and reflected by a nonfluctuating target and are *noncoherently* integrated, the required single-pulse signal-to-noise ratio (before integration) is found to be approximately +5.3 dB, as indicated by the square in Figure 9-14(b). In other words, radar returns with a single-pulse signal-to-noise ratio of +5.3 dB, when noncoherently integrated, yield the same detection probability as a single radar return with a signal-to-noise ratio of +13.1 dB. Referring to Figure 9-14(b), one notes that 1000 integrated pulses require a single-pulse signal-to-noise ratio of approximately −6.9 dB to give an integrated 90% probability of detection (shown by the triangle).

If *coherent* integration is used, the required single-pulse signal-to-noise ratio in the above example is calculated as follows for 10 and 1000 pulses, respectively:

$$\text{SNR} = +13.1 \text{ dB} - 10 \log_{10}(10) = +3.1 \text{ dB}$$

and

$$\text{SNR} = +13.1 \text{ dB} - 10 \log_{10}(1000) = -16.9 \text{ dB}.$$

Note that coherent integration provides a signal-to-noise ratio "gain" of 2.2 and 10 dB, respectively, as compared to noncoherent integration. That is, smaller single-pulse signal-to-noise ratios are required with coherent integration to achieve a 90% detection probability.

The pulse integration discussed so far is defined as the process of adding or summing amplitudes of radar returns (either video returns from an envelope detector and noncoherent integration, or video returns from an I, Q envelope detector and coherent integration). Another integration scheme sometimes used

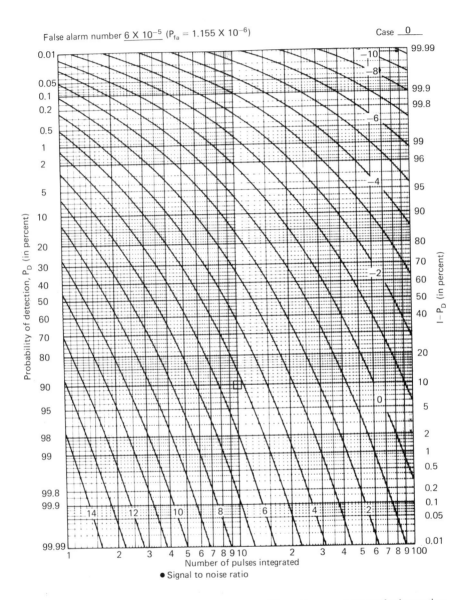

Figure 9-14. Detection curves for a square-law detector followed by noncoherent pulse integration for a nonfluctuating target model. The false alarm probability is $P_{fa} = 1.155 \times 10^6$. (By permission, from Meyer et al., ref. 8; © 1973 Academic Press, Inc.)

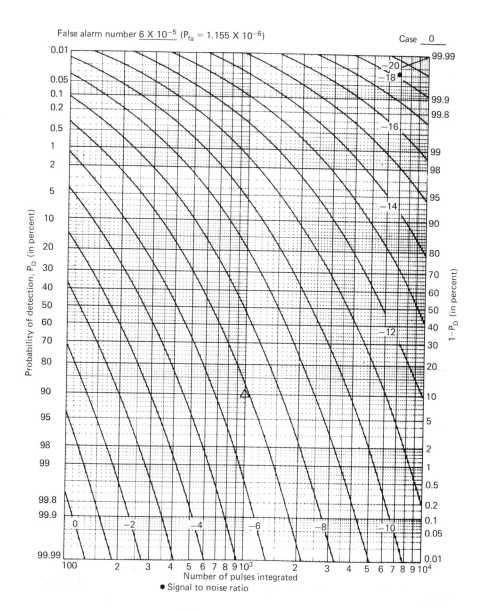

Figure 9-14. (*Continued*)

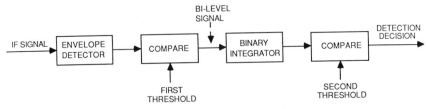

Figure 9-15. Binary pulse integration.

in detection involves quantizing the detected video signal to one bit (two possible voltage levels) and then summing the quantized video. Such a detection process is illustrated in Figure 9-15 and is called *binary* integration.[9-11] If the video is converted from an analog to a digital signal (via an A/D converter), then the integration and second detection decision shown in Figure 9-15 may be implemented digitally. Binary integration lends itself well to automatic detection, which is the subject of Chapter 12.

9.9 HUMAN OPERATOR AND CRT DETECTION

Early in the development of radar, the link between the radar output and the radar operator was the presentation of radar information on a CRT. Several types of radar displays were developed. The A-scope and PPI-scope were some of the most common displays. The A-scope displays radar return amplitudes along the *Y*-axis (vertical) and range or time delay to the target along the *X*-axis (horizontal). Thus, the A-scope displays a rectilinear representation of the range profile of target cross sections. The plan position indicator (PPI) scope displays radar information in a polar format. Range is displayed as distance along the radial, while angle to the target is displayed as a polar angle. Radar displays are discussed more thoroughly in Chapter 8.

An operator's capacity for searching a CRT display and recognizing the presence of target echoes is very limited. The information bandwidth of a human operator is about 20 bits per second, and fatigue and lack of motivation severely reduce this bandwidth.

Based upon empirical and experimental results, operator efficiency is found to be proportional to the square of the single-pulse probability of detection.[4] Thus, the operator's ability to detect a target when the single-pulse detection probability is 0.25 is 1/16th that when the detection probability is near unity. This performance occurs under the best operator conditions, and operator efficiency can be much lower. For these and other reasons, the human operator has been largely replaced by electronic decision circuitry.

9.10   AUTOMATIC DETECTION

The rate of information production inherent in a typical radar signal is considerably greater than can be handled by a human operator. A one-megahertz-bandwidth signal, for example, is capable of conveying information at a rate of two megabits per second, but an operator can accept information at the rate of only 10 to 20 bits per second. Thus, there is a tremendous mismatch between the information content of a radar and the information-handling capability of an operator. When the radar signal detection function is performed by electronic decision circuitry without the intervention of an operator, the process is known as *automatic detection*. One of the chief reasons for employing automatic detection is to overcome the limitations of an operator. In addition, automatic detection allows the radar output to be transmitted more efficiently over communication links, since only detected target information need be transmitted, not the full bandwidth signal (raw video). More information about automatic detection is presented in Chapter 12.

9.11   SUMMARY

The discussion in this chapter should provide the reader with an elementary understanding of how to model the detection process. There are two fundamental descriptions of signal detection in noise. The amplitude variation of the signal-plus-noise must be described as a combination of deterministic and nondeterministic (probabilistic) mathematics. For this purpose, the notions of random processes and probability density functions were presented. In addition, the time variation of signals embedded in random noise must be specified by use of the autocorrelation function or PDS. Once the signal-plus-noise has been adequately described in these terms, the effects of various types of detection and integration processes can be determined. In general, the success with which signals in noise are sensed or detected is measured by two parameters: the probabilities of detection (P_d) and false alarm (P_{fa}). Having determined the P_d and P_{fa} requirements and knowing the noise and signal statistics and autocorrelation functions, one can determine the required signal-to-noise ratio.

In Chapter 1, the relationship between the required signal-to-noise ratio and the maximum detection range is derived and presented. Ultimately, the maximum detection range is the basic description of a surveillance radar system's performance.

9.12   REFERENCES

1. A. Papoulis, *Probability, Random Variables, and Stochastic Processes*, McGraw-Hill Book Co., New York, 1965.

2. W. B. Davenport, Jr., and W. L. Root, *An Introduction to the Theory of Random Signals and Noise,* McGraw-Hill Book Co., New York, 1958.

3. R. S. Berkowitz, *Modern Radar,* John Wiley & Sons, New York, 1965.

4. M. I. Skolnik, *Introduction to Radar Systems,* McGraw-Hill Book Co., New York, 1980.

5. W. M. Waters and B. R. Jarrett, "Bandpass Signal Sampling and Coherent Detection," *IEEE Transactions on Aerospace and Electronics Systems,* vol. 18, no. 6, November 1982, pp. 731–736.

6. D. W. Rice and K. H. Wu, "Quadrature Sampling with High Dynamic Range," *IEEE Transactions on Aerospace and Electronics Systems,* vol. 18, no. 6, November 1982, pp. 736–739.

7. J. L. Brown, Jr., "On Quadrature Sampling of Bandpass Signals," *IEEE Transactions on Aerospace and Electronic Systems,* vol. AES-15, no. 3, May 1979, pp. 366–371.

8. D. P. Meyer and H. A. Mayer, *Radar Target Detection,* Academic Press, New York, 1973.

9. Robert E. Lefferts, "Adaptive False Alarm Regulation in Double Threshold Radar Detection," *IEEE Transactions on Aerospace and Electronic Systems,* vol. AES-17, no. 5, September 1981, pp. 666–675.

10. M. Schwartz, "A Coincidence Procedure for Signal Detection," *IRE Transactions on Information Theory,* vol. IT-2, December 1956, pp. 135–139.

11. B. M. Dillard, "A Moving Window Detector for Binary Integration," *IEEE Transactions on Information Theory,* vol. IT-13, January 1967, pp. 2–6.

12. S. A. Hovanessian, *Radar Detection and Tracking Systems,* Artech House, Inc., Dedham Mass. 1973.

10
CLUTTER CHARACTERISTICS AND EFFECTS*

N. C. Currie

10.1 INTRODUCTION AND DEFINITIONS

Since the earliest days of radar, detection of unwanted objects have interfered with the detection of desired targets. These unwanted returns are known as *clutter*, and there is a substantial amount of literature available on radar clutter dating from World War II to the present. Whether particular returns are desired or not depends on the situation. For example, while one aircraft might be using the pattern of radar returns from terrain by which to navigate, a radar in an AWACS aircraft at higher altitude might be attempting to detect the first aircraft in the presence of this same pattern of ground returns. Therefore, we are led immediately to the following pragmatic definition and consequence:[1]

Definition: *Clutter is a return or group of returns that is unwanted in the radar situation being considered.*

Consequence: *One radar's clutter is another radar's target.*

In most situations, the unwanted clutter returns detected by the radar result from a number of distributed scattering centers which return energy to the radar. As shown in Figure 10-1, an assembly of scattering centers has been illuminated by the radar transmitter, and each such center is reflecting electromagnetic energy toward the receiver.

Considering the reradiated fields (in contrast to the power), each scattering center has associated with it an amplitude, A_i, and a phase, ϕ_i. The polarization properties of the reradiated fields may be represented by components of two

---

*Author's note: Parts of Section 10.1 and Section 10.5 are reprinted from Corriher et al., Chapter XVII, "Elements of Radar Clutter," of *Principles of Modern Radar*, Georgia Institute of Technology, 1972, Reference 1.

basis vectors, say horizontal and vertical (or some other orthogonal pair, such as $+45°$ and $-45°$ or right-hand and left-hand circular). We see therefore that the fields being reradiated by the scattering centers are of a vector-phasor nature, and each vector (polarization) component may be summed in the usual manner for a set of phasors. This is also shown in Figure 10-1, along with the resultant diagram for one of the polarization components.

A physical feeling for the nature of the scattering centers may be obtained by thinking of the specular (mirror-like) reflections seen visually when viewing a wet, leafy tree or the glittering facets on the surface of the sea. Although the eye does not perceive the polarization or the phase characteristics of these reflections of visible light, a radar is sensitive to these additional parameters which describe the reflected fields.

It is also possible to take an approach based on return power by considering the radar cross section (RCS) of each scattering center. By taking the square root of each RCS and assigning a phase term to it, the returns from individual scatterers associated with clutter (or, for that matter, any other complex target)

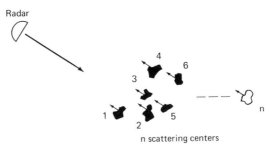

(a) Vector-phasor summation of individual field contributions. (Adapted from ref. 1.)

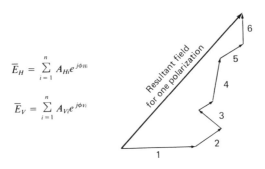

$$\overline{E}_H = \sum_{i=1}^{n} A_{Hi} e^{j\phi_{Hi}}$$

$$\overline{E}_V = \sum_{i=1}^{n} A_{Vi} e^{j\phi_{Vi}}$$

(b) Clutter return from an assembly of scattering centers. (Adapted from ref. 1.)

Figure 10-1.

can be summed to obtain an equivalent overall RCS as follows:

$$\sigma = \left|\left[\sum_{i=1}^{n} (\sigma_i)^{1/2} \, e^{j(\phi_i)}\right]\right|^2 \tag{10-1}$$

In Eq. (10-1), the phase, ϕ_i, is frequently approximated by an expression which takes into account the distance, d_i, of an individual scattering center from a reference plane and the wavelength, λ, as follows:

$$\phi_i = \frac{4\pi d_i}{\lambda} \tag{10-2}$$

For example, if two scatterers with an RCS equal to 10 and 1 m^2 are located 0.0075 m apart, then their equivalent joint RCS at 10 GHz can be calculated as follows. If the 1 m^2 target is at a reference distance of $d = 0$, then from Eq. (10-2)

$$\phi_{10} = \frac{4 \times \pi \times 0.0075}{0.03} = \pi \tag{10-3}$$

$$\phi_1 = \frac{4 \times \pi \times 0.0}{0.03} = 0 \tag{10-4}$$

and from Eq. (10-1)

$$\sigma = [\sqrt{10} \, e^{j\pi} + \sqrt{1} \, e^{j0}]^2 \tag{10-5}$$
$$= [3.16 \times (-1) + 1 \times (1)]^2 = 4.675 \text{ m}^2$$

If the distance d between the 10-m^2 scatterer and the 1-m^2 scatterer is 0.015 m, then $\phi_{10} = 2\pi$ and the equivalent RCS is

$$\sigma = [3.16 \times (+1) + 1 \times (1)]^2 = 17.3 \text{ m}^2 \tag{10-6}$$

Thus, the interaction of these two scatterers within a cell that are a factor of 10 difference in RCS can result in an effective cross section which varies from 5 to 17 m^2, a 5-dB variation.

10.1.1 Definition of Scattering Coefficients

Since a number of scattering centers may be illuminated within the resolution cell of the radar instead of a single point target, it has become standard practice to normalize clutter returns to the area (in the case of surface clutter) or the volume (in the case of precipitation or other volumetric clutter) being illuminated by the radar. Actually, this practice often does not represent the real world situation, since the resolution of many modern radars is sufficiently high that a considerable amount of nonuniform structure is visible in clutter returns. The use of normalized clutter is so pervasive among experimenters, however, that it is important to understand its fundamental characteristics.

For the case of surface scatterers, the backscatter parameter commonly used is the radar reflectivity per unit area illuminated, which was formulated by Goldstein[2]:

$$\sigma^0 = \sigma/A \qquad (10\text{-}7)$$

where

σ^0 = radar reflectivity per unit area (in square meters/square meters)
A = illuminated ground/sea patch (in square meters)
σ = RCS of the illuminated area (in square meters)

Since the units of σ^0 are square meters/square meters, σ^0 is a unitless coefficient and is commonly specified in decibels where $\sigma^0(\text{dB}) = 10 \log (\sigma/A)$ (For example a σ^0 value of 0.1 would be expressed as -10 dB.) Some experimenters use a different definition for RCS known as γ where

$$\gamma = \frac{\sigma^0}{\sin \theta} \qquad (10\text{-}8)$$

(θ is the depression angle.) For large values of θ, $\gamma \approx \sigma^0$.

The backscatter parameter used for meteorological phenomena is the radar reflectivity per unit volume illuminated:

$$\eta = \sigma_v = \frac{\sigma}{V} \qquad (10\text{-}9)$$

where

η = radar reflectivity per unit volume (in square meters/cubic meters)
V = volume of the resolution cell (in cubic meters)
σ = RCS of the resolution cell (in square meters)

The units of η are thus meters^{-1}, and η is often also specified in decibels relative to meters^{-1} (dB/m).

Although σ^0 and η are reflective coefficients independent of the radar, the parameter which is actually measured by a radar is the RCS of the illuminated resolution cell. The reflectivity parameters σ^0 and η are then derived from the measured RCS σ using either Eq. (10-7) or Eq. (10-9) as appropriate. This process requires calculation of either the illuminated area A on the ground/sea surface or the volume V of the hydrometeor resolution cell.

The antenna beam shape is one factor in the determination of either A or V, and both one-way and two-way antenna patterns (3-dB beamwidths, azimuth and elevation) have been used in reported results. Therefore, care must be exercised when comparing reflectivity data from different investigators and different measurement exercises.

In the determination of ground/sea reflectivity, the calculation of the illuminated area A depends on the geometry. Two different geometries can be defined, as determined by the elevation beamwidth ϕ_{el}, the slant range R, and the radar pulse length τ. The pulse length limited case is defined when the radar range resolution length projected onto the ground/sea surface is smaller than the range extent of the area illuminated by the antenna beam, as shown in Figure 10-2. The area of the largest-resolution cell in this case is approximately rectangular and can be calculated by multiplying the projected range resolution length ($c\tau/2$) sec θ (where c is the speed of light and θ is the depression angle) by the width of the illuminated ellipse, that is, $2R \tan (\phi_{az}/2)$, where ϕ_{az} is the azimuth 3-dB beamwidth. In the pulse length limited case, the area A is given by

$$A = 2R(c\tau/2) \tan (\phi_{az}/2) \sec \theta \qquad (10\text{-}10)$$

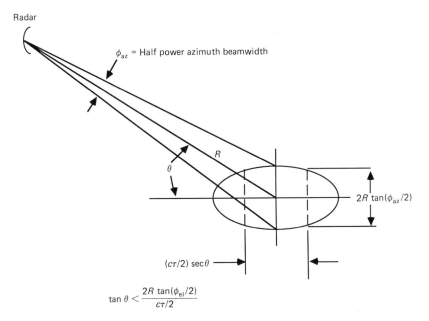

Figure 10-2. Pulse length limited resolution cell. (By permission, from Trebits, ref. 3; © 1984 Artech House, Inc.)

where

$$\tan \theta < \frac{2R \tan (\phi_{el}/2)}{c\tau/2}$$

For beamwidth values less than approximately 10°, the small-angle approximation may be used:

$$A = R(c\tau/2)\phi_{az} \sec \theta \qquad (10\text{-}11)$$

where

$$\tan \theta < \frac{R\phi_{el}}{c\tau/2}$$

The beamwidth limited case occurs when the radar range resolution projected onto the ground/sea surface is larger than the range extent of the area illuminated by the antenna beam, as shown in Figure 10-3. The resolution cell on a flat earth is an ellipse, with a range axis diameter of approximately $2R \tan (\phi_{el}/2) \csc \theta$ and an azimuth axis diameter of approximately $2R \tan (\phi_{az}/2)$. Area A of the resolution cell for the beamwidth limited case is then given by

$$A = R^2 \tan (\phi_{az}/2) \tan (\phi_{el}/2) \csc \theta \qquad (10\text{-}12)$$

where

$$\tan \theta > \frac{2R \tan (\phi_{el}/2)}{c\tau/2}$$

and for small beamwidths

$$A = \frac{\pi R^2}{4} \phi_{az}\phi_{el} \csc \theta \qquad (10\text{-}13)$$

where

$$\tan \theta > \frac{R\phi_{el}}{c\tau/2}$$

The resolution cell for volume clutter is approximately a cylinder with an elliptical cross section, as shown in Figure 10-4. The cell volume V is then calculated by multiplying the elliptical cross-sectional area $\pi/4$ $(R\phi_{el})(R\phi_{az})$ by

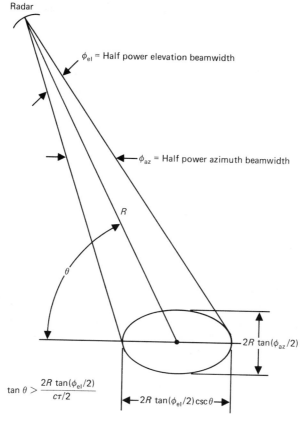

Figure 10-3. Beam limited resolution cell. (By permission, from Trebits, ref. 3; © 1984 Artech House, Inc.)

the range resolution $c\tau/2$:

$$V = \frac{\pi}{4} R^2 \phi_{el} \phi_{az} (c\tau/2) \qquad (10\text{-}14)$$

The value of V calculated from Eq. (10-14) is often reduced by a factor of 2 to account for the two-way illumination of the volume of the resolution cell, calculated from one-way 3-dB beamwidths.

Unlike a stationary object, distributed targets such as land, sea, and precipitation will cause the backscattered radar signal to vary over a wide range of amplitudes, due to the effects of the vector summation of moving scattering centers. Therefore, σ^0 and η will vary in time and space, and statistical parameters are necessary to describe the ''reflectivity.'' In addition, certain statistical

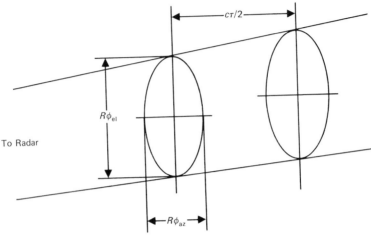

Figure 10-4. Volume resolution cell. (By permission, from Trebits, ref. 3; © 1984 Artech House, Inc.)

parameters such as the mean and median values of σ^0 and η are often confused with each other when data from different sources are compared. Differences between mean and median reflectivity values depend on the measured clutter statistics which in turn depend on the source, geometry, and radar parameters themselves, thus making comparisons all the more complicated.

10.1.2 Clutter Polarization Scattering Matrix

In general, the scattering properties of clutter are dependent on the polarization of the incident radio wave in a manner similar to that of man-made targets. This effect is accounted for by using the polarization scattering matrix to express clutter reradiation polarization effects which are independent of the transmitted polarization, as discussed in Chapter 20. The rectangular scattering matrix can be expressed in terms of the clutter RCS as:

$$S_c = \begin{vmatrix} (\sigma_{HH})^{1/2} \, e^{j\phi_{HH}} & (\sigma_{HV})^{1/2} \, e^{j\phi_{HV}} \\ (\sigma_{VH})^{1/2} \, e^{j\phi_{VH}} & (\sigma_{VV})^{1/2} \, e^{j\phi_{VV}} \end{vmatrix} \tag{10-15}$$

where, for example,

$(\sigma_{HH})^{1/2}$ = magnitude of the scattering element
ϕ_{HH} = the associated phase.

The σ and ϕ subscripts H and V denote horizontal (H) and vertical (V) polarization. The first subscript denotes transmit polarization; the second subscript denotes receive polarization.

Once again, as in the case of point targets, other matrices using circular or elliptical polarizations can also be developed; the circular matrix can be related to the linear one by a simple transformation (see Chapter 20). For the monostatic backscatter case (bistatic angle equal to zero), it can be shown that $(\sigma_{HV})^{1/2} = (\sigma_{VH})^{1/2}$ and $\phi_{HV} = \phi_{VH}$. Thus, the matrix is symmetrical.[4]

Another parameter currently of interest for discriminating between man-made targets and natural clutter is the *polarimetric phase*, ϕ_{H-V}. This parameter, which is $\phi_{HH} - \phi_{VV}$, has been found to be useful in target/clutter discrimination techniques and is usually expressed in terms of its in-phase (I) and quadrature (Q) components, $\sin(\phi_{H-V})$ and $\cos(\phi_{H-V})$. Also, the dot product parameter $[(\sigma_{HH} \, \sigma_{VV})^{1/2} \cos \phi_{H-V}]$ is of interest for certain clutter discrimination schemes.[5]

10.2 GENERAL CHARACTERISTICS OF CLUTTER

A number of known general physical properties of clutter form a framework within which specific sets of data or specific clutter models are expected to agree. These properties are generally divided into those involving the average values and those involving temporal characteristics (amplitude distributions and frequency spectra). These are discussed below.

10.2.1 Average Values

Radar backscatter is caused by the various scattering elements within the resolution cell of a radar. The quantities σ^0 and η are used to provide RCS normalized to the surface area or radar cell volume, as discussed previously. In using the quantity σ^0 or η, it is tacitly assumed that the echo is caused by a large number of scattering elements located within the physical area illuminated by the radar. The previously defined quantities σ^0 and η provide a measure of expected radar return that is normalized to radar cell size.

From a radar scattering point of view, the radar/surface geometry for plane surfaces can be divided into three distinct regions: the near grazing incidence region, the plateau region, and the near vertical incidence region, as discussed by Long[6] and shown in Figure 10-5. Within each of these regions, the dependence of σ^0 on depression angle and wavelength can be described in a general way, but the borders of the three regions change with wavelength, surface characteristics, and polarization.[6]

In the near grazing incidence region, σ^0 increases rapidly with increasing incidence angle. In this region, even rough terrain can appear to be smooth if the Rayleigh roughness criterion is satisfied, that is,

$$\Delta h \sin \theta \leqslant \lambda/8 \qquad (10\text{-}16)$$

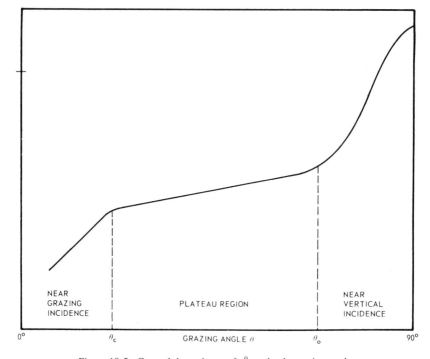

Figure 10-5. General dependence of σ^0 on the depression angle.

where

Δh = rms height of the surface irregularities
θ = angle between the incident ray and the average plane surface
λ = transmitted wavelength

For example, for a frequency of 2 GHz ($\lambda = 0.15$ m) and a depression angle of $1°$, from the Rayleigh roughness criterion, the maximum rms surface roughness for which the surface will appear smooth is given by

$$\Delta h \leq 0.01875/0.0175 \text{ m} \tag{10-17}$$
$$\Delta h \leq 1.07 \text{ m}$$

Thus, an rms surface roughness of greater than 1 m could still appear smooth from a radar scattering point of view.

In the plateau angular region, a clutter surface appears rough, and the scattering is almost completely incoherent. Thus, σ^0 changes very little with depression angle.

Finally, for large incidence angles, specular reflection becomes dominant so that σ^0 increases rapidly with depression angle up to a maximum incidence of $90°$. The magnitude of σ^0 at $90°$ incidence depends in general on the rms surface roughness and the dielectric properties of the clutter.

The angular dependence of σ^0 has been described by Hayes and Dyer[7] as

$$\sigma^0 = K \sin^n \theta \qquad (10\text{-}18)$$

where

$K =$ an arbitrary constant
$n =$ roughly $+1$ (so-called constant γ model)
$\theta =$ incidence angle

The frequency dependence of clutter is generally dependent on the surface roughness in terms of the wavelength. For smooth surfaces relative to the wavelength, the frequency dependence is variable and may be difficult to measure. For surfaces that are rough relative to a wavelength, the frequency dependence is generally considered to vary as λ^{-1} at microwave frequencies. This leads to the possibility of an "ultraviolet catastrophe" at millimeter wavelengths. That is, values of σ^0 at millimeter wavelengths could be extremely large if the λ^{-1} law were followed. Generally, data have shown that the frequency dependence of σ^0 decreases at millimeter wavelengths and may actually change sign, that is, λ^{+n}, $n \leq 1$, above 35 GHz for some types of clutter.[7]

One problem with the use of σ^0 to describe surface clutter is the fact that the concept of a constant reflective coefficient falls apart as the number of scatterers within a resolution cell decreases. This is borne out by clutter measurements which have shown that the RCS of a given clutter cell decreases with decreasing cell size down to a minimmum value; then it becomes independent of the cell size, indicating that discrete scatterers are being resolved.

Polarization effects of clutter are varied, and depend on the clutter type and the depression angle. For example, for low-angle sea return from relatively smooth seas, the average horizontally polarized return is less than the vertically polarized return. As the sea becomes rough, however, this difference decreases. Rain, as another example, reflects horizontal and vertical polarizations approximately the same, but there is a large difference in the return for cross versus parallel circular polarizations due to the fact that spherical raindrops return the opposite-sense circular polarization to the radar from that transmitted (this fact is used to reduce rain returns on many search radars). The return from trees at high angles shows only small differences in returns between either horizontal and vertical or circular polarizations as opposed to sea clutter or rain. These dependencies are discussed in more detail in Section 10-3.

10.2.2   Amplitude Temporal Properties

Since distributed clutter is made up of a large number of discrete scatterers, as discussed in Section 10-1, and since the apparent distances between these scatterers can change either due to the physical movement of the scatterers (such as the wind blowing the branches on a tree) or due to movement of the radar platform or antenna (changing the aspect angle of each scatterer), fluctuations in amplitude with time are frequently observed in clutter returns. The fluctuations can be described in terms of the amplitude statistics and the fluctuation rates or frequency spectra. The amplitude statistics give information on the percentage of time during which the returns have a given range of values, and the frequency spectra yield information on how rapidly the values change.

The amplitude statistics of clutter have been studied for many years. For a large number of roughly equal-sized scatterers with uniformly distributed phases, the amplitude values of clutter are found to be Rayleigh distributed. If one scatterer is dominant, that is, much larger than the others, then the distribution will be Rician. For low depression angles and small-resolution cell sizes, log-normal or Weibull statistics have been observed. Those distributions which have high-value "tails" may result from large scatterers which are shadowed most of the time but occasionally can be observed by the radar. Figure 10-6 compares

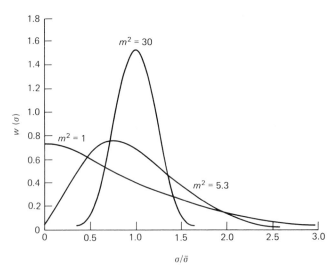

Figure 10-6(a). Plots of Rayleigh and Rician distributions. (Adapted from ref. 8) For Rayleigh:

$$w(\sigma)d\sigma = \frac{1}{\bar{\sigma}} \exp \frac{-\sigma}{\bar{\sigma}}\, d\sigma$$

For Rician:

$$w(\sigma)d\sigma = (1 + m^2)e^{-m^2}e^{-\sigma(1 + m^2)/\bar{\sigma}}J_0(2im(1 + m^2)^{1/2}(\sigma/\bar{\sigma})^{1/2})\frac{d\sigma}{\bar{\sigma}}.$$

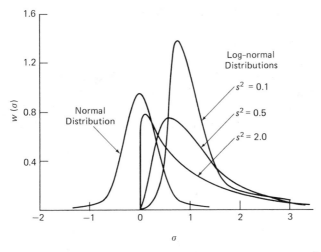

Figure 10-6(b). Comparison of the lognormal distribution with different values of s^2 with a normal distribution. (Adapted from ref. 8)

$$w(\sigma)d\sigma = \frac{1}{\sigma\sqrt{2\pi}s} \exp\left(-\frac{1}{2s^2} \ln\frac{\sigma}{\sigma_m}\right)$$

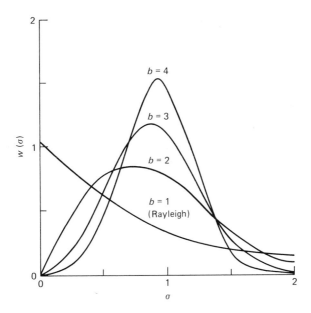

Figure 10-6(c). Weibull distributions for several values of b.

$$w(\sigma)d\sigma = \frac{b\sigma^{(b+1)}}{\sigma_m} \exp\left(-\frac{\sigma b}{\alpha}\right)$$

where $\alpha = \dfrac{\sigma_m^b}{\ln 2}$.

293

the Rayleigh, Rician, lognormal, and Weibull distributions (see Chapter 11 for more information on these distributions).

The frequency spectra of clutter returns have been as intensely studied over the years as amplitude statistics. The frequency spectrum of a fluctuating random signal such as the return from clutter may be expressed as either the power spectral density or the autocorrelation function; the two are Fourier transform pairs. That is:

$$P(f) = \int_{-\infty}^{\infty} R(\tau)e^{j2\pi f \tau}d\tau \tag{10-19}$$

where

$P(f)$ = power spectral density function value at frequency f
$R(\tau)$ = autocorrelation value at time τ

Both the frequency spectrum and the autocorrelation function are used to describe clutter fluctuations, but frequency spectra are generally used more often.

The first attempts to describe clutter spectra by Barlow in 1949 used a Gaussian model of the form[9]

$$W(f) = W_0 e^{-a(f/f_t)^2} \tag{10-20}$$

where W_0 is the mean value of the power density, f_t is the radar operating frequency, and a is a constant which depends on clutter type. Clutter measurements performed in the 1960s and 1970s have indicated that clutter spectral responses can be better described by a power function of the form[10,11,12]

$$\rho(f) = \frac{A}{1 + (f/f_c)^n} \tag{10-21}$$

where

A = mean value of the power density
f_c = clutter spectrum half-power frequency
f = frequency
n = a positive real number

10.3 CLUTTER CHARACTERISTICS

10.3.1 Atmospheric Clutter

Atmospheric clutter is generally due to some form of precipitation, although heavy dust, insects, and birds may sometimes contribute significant scattered energy. In this section, the backscattering properties of precipitation will be discussed, since precipitation is the primary atmospheric scatterer. References 13 and 14 give information on atmospheric scattering from other sources.

Rain Backscatter

Average Values. The backscattering properties of rain depend on the transmitted frequency, the polarization, and the number and size of the raindrops. Since many raindrops are approximately spherical, three scattering regimes are possible, depending on the ratio of the drop diameter to the wavelength: the Rayleigh region, where the ratio of the drop diameter D to the wavelength λ is small; the Mie region, where D/λ is approximately 1; and the optical region, where D/λ is large. In the Rayleigh region, the backscatter increases rapidly with increasing frequency; in the Mie region, the backscatter varies rapidly with small changes in frequency; and in the optical region, the backscatter is independent of frequency and depends only on the size (cross-sectional area) of the drops. Multifrequency measurements performed in 1975 by the U.S. Army Ballistic Research Laboratory (BRL) and the Georgia Institute of Technology indicated that the transition from the Rayleigh scattering region to the Mie region occurs at between 35 GHz and 70 GHz.[11,15] Figure 10-7 gives least mean square fits to the data obtained during that experiment. The figure presents the volumetric backscatter coefficient η as a function of rain rate. Since the average size of a raindrop increases with increasing rain rate, Rayleigh scattering would require that the backscatter vary strongly with rain rate (D/λ increasing). This is the case for 9 and 35 GHz. For 70 and 95 GHz, however, the dependence is weaker, indicating a transition from the Rayleigh scattering region to the Mie or optical scattering regions, where scattering is a function of only the cross-sectional area. Figures 10-8 and 10-9 show the actual data points as compared to the least-square fit curves. The large amount of scatter in the points is attributed to the very dynamic changes which occur in drop size distribution as a function of time. Rain rate and drop size do not have a one-to-one relationship at any given time as illustrated by Figure 10-10. However, on the average, Figure 10-7 shows that the backscatter from rain increases with increasing rain rate and with increasing frequency up to 70 GHz. Above 70 GHz, the dependence of backscatter on frequency appears to reverse, although data above 100 GHz are not currently available to confirm this trend.

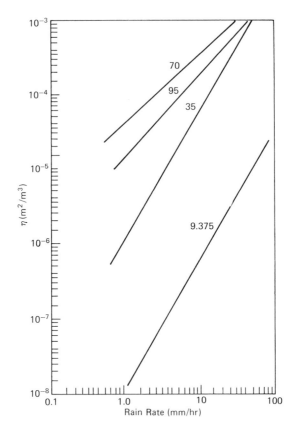

Figure 10-7. Least mean square fit to radar rain backscatter data, VV polarization. (From ref. 11)

Rain Fluctuations. Since raindrops are very small compared to the typical volumetric resolutions of search radars, there are usually large numbers of scatterers within a radar resolution volume, causing rain backscatter to be a very dynamic process. The dynamics of rain can be described statistically in several ways. One can talk about the variations of the return around the mean value; this is usually described by the amplitude distribution or the standard deviation. One can also talk about how rapidly the return fluctuates. This parameter is usually described in terms of the equivalent frequency spectrum or the autocorrelation function.

Rain amplitude distributions were computed by the Georgia Institute of Technology in conjunction with the rain measurement experiment described above (reference 11), and it was determined that the distribution shapes varied with transmitted frequency; they were somewhat wider than a Rayleigh distribution

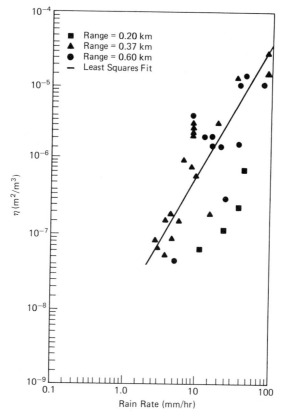

Figure 10-8. Comparison of rain backscatter data with least-square fit (9.375 GHz). (From ref. 11)

at 10 GHz and narrower than a Rayleigh distribution at 95 GHz. Table 10-1 summarizes the standard deviations measured for the returns as a function of frequency, polarization, and rain rate. The rain rate and polarization seemed to have only small effects on the measured standard deviations but the frequency has a very large effect, the standard deviation decreasing with increasing frequency. This is probably because the smaller drops become more significant scatterers as the frequency increases, thus increasing the effective number of scatterers within the radar resolution volume which decreases the dynamics of the return signal.

The frequency spectra for rain were measured during the same experiment in which the amplitude distributions were determined. Since the radar data were not coherent, the spectra measured represented the amplitude fluctuations only. The results from this experiment are shown in Figures 10-11 and 10-12 for 10

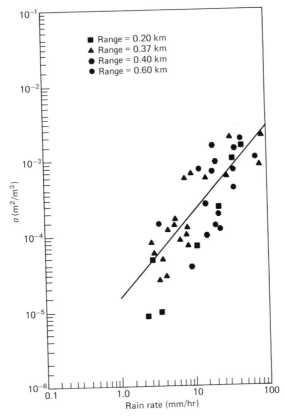

Figure 10-9. Comparison of rain data with least squares fit (95 GHz) (Adapted from ref. 11)

and 95 GHz, respectively. The spectral shapes were found to have a power function rolloff characteristic, as given by Eq. (10-20), rather than the expected Gaussian characteristic. Furthermore, the corner frequency was found to depend on both rain rate and transmitted frequency, and the exponent was found to depend on transmitted frequency only. Table 10-2 summarizes the values of the corner frequency and exponent measured as a function of frequency and rain rate. Note that at 35 GHz and above, the power function becomes a quadratic relationship. The implication is that the use of a Gaussian characteristic to model the spectrum would considerably underestimate the spectral width.

As discussed in Section 10.2.2, correlation functions are another way of describing the temporal variations of clutter returns. The so-called decorrelation time indicates the amount of time required for the composite return from all of the scatterers within the radar resolution volume to change enough to be independent or significantly different. The decorrelation time has been defined by

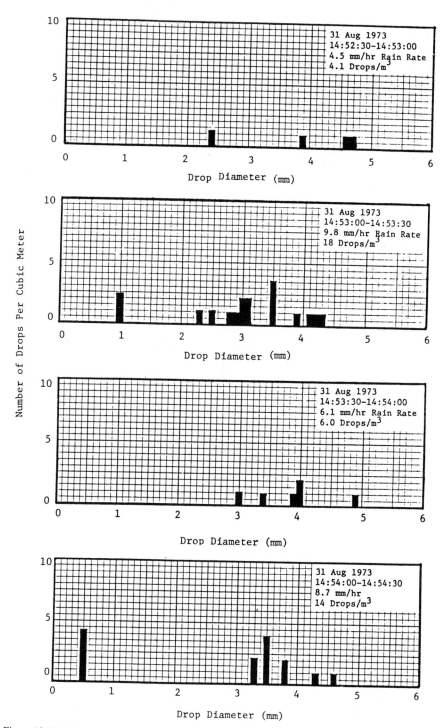

Figure 10-10. Rain drop size distribution for four consecutive 30-sec intervals with a light to moderate rain rate. (From ref. 11)

Table 10-1. Typical Standard Deviation Values for Rain Backscatter.

Frequency (GHz)	Rain Rate (mm/hr)	Standard Deviation (dB)	
		Vertical Polarization	Circular Polarization
10	5	7.5	8.0
	20	8.0	8.5
35	5	3.0	2.7
	20	3.2	2.5
95	5	3.2	3.0
	20	3.0	2.9

Source: Reference 11.

Currie et al.[11] as the time required for the autocorrelation function of the process to decay to either $1/e$ or 36.7% of its maximum value because it was found that almost all clutter autocovariance functions remained below this level once the level was crossed. The autocovariance function is often used instead of the autocorrelation function to describe clutter dynamics. (The autocovariance function is simply the autocorrelation process in which the DC value is normalized before the computation is performed.) Table 10-3 gives values of de-

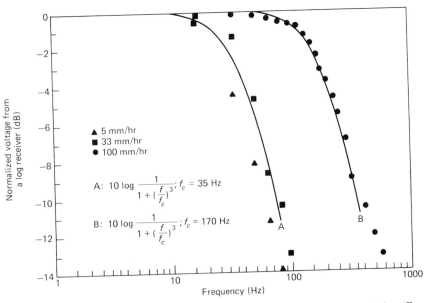

Figure 10-11. Normalized frequency spectrum of rain return at 9.375 GHz, VV polarization. (From ref. 11)

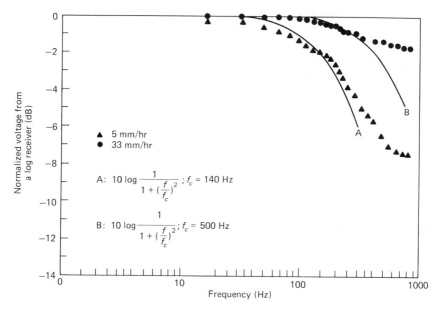

Figure 10-12. Normalized frequency spectrum of rain return at 95 GHz, VV polarization. (From ref. 11)

correlation times that were measured by the Georgia Institute of Technology during the joint rain experiment with BRL. The values were computed by calculating the autocovariance function of rain returns and determining the $1/e$ decay time. As can be seen from Table 10-3, the decorrelation times vary inversely with frequency and rain rate. Figure 10-13 summarizes these values in graphic form and indicates the equivalent maximum sample rate for independent

Table 10-2. Measured Rain Spectral Constants.

Frequency (GHz)	Rain Rate (mm/hr)	f_c (Hz)	n
10	5	35	3
	100	70	3
35	5	80	2
	100	120	2
70	5	175	2
	100	500	2
95	5	140	2
	100	500	2

Source: Reference 11.

Table 10-3. Measured Decorrelation Times for Rain.

Rain Rate (mm/hr)	Decorrelation Time (ms)			
	10 GHz	*35 GHz*	*70 GHz*	*95 GHz*
5	13.8	8.7	4.0	3.5
9	11.6	5.5	3.9	3.1
26	6.7	4.8	3.6	2.5
70	5.8	4.2	2.4	1.8
100	5.3	2.9	2.0	1.3

Source: Reference 11.

samples. The implication from these data is that, for typical radar PRIs of 1 to 10 ms, frequency agility (or some other method of decorrelating the returns) is necessary to ensure that independent samples are obtained from rain backscatter during each PRI period.

Frozen Precipitation

Frozen precipitation, which is normally in the form of snow or hail, also reflects microwave energy. Normally, snow or hail is considered to be a mixture of

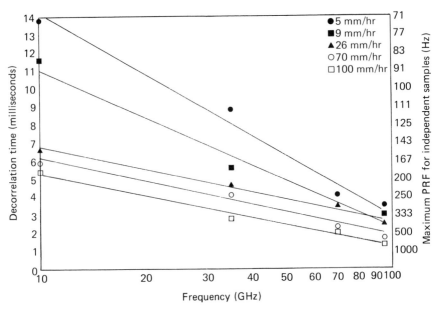

Figure 10-13. Decorrelation time versus rain rate and frequency for rain backscatter. (From ref. 11)

water and ice, and its reflectivity depends on the percentage of water and ice present and the distribution of the water (for example, whether the water is on the surface or distributed uniformly throughout the particles). Very dry snow or hail is generally much less reflective than rain because the dielectric constant of ice is much less than that of water. However, when the particles are mixtures of water and ice, the situation becomes much more complex. The reflectivity of such particles depends on the relative particle sizes, the drop size distributions, and the interactions between the water and ice layers. Thus, hail or snow can have a greater reflectivity than rain in certain situations. One well-known example is the existence of the so-called bright band, in which the radar reflectivity from a rain storm is much greater just below the 0°C temperature altitude than above or below this level. This increase in reflectivity is due to melting of ice particles below the freezing level, which have a higher reflectivity than the unmelted particles above and a higher particle density than the raindrops below. An excellent discussion of the comparison of the reflectivity of rain, snow, and ice particles is given in Battan.[14]

10.3.2 Land Clutter

The characterization of land clutter has been of primary interest for many years to both the military (because many tactical targets are located on land) and the remote sensing community (since the radar properties of land surfaces are used to infer important physical characteristics such as the amount of free water present). The characterization of land clutter is a very difficult problem because there are so many varieties of natural and man-made objects. Many types of scatterers have been grouped into several classes for measurement, including trees, grasses and crops, sand, rocks and desert terrain, and snow-covered ground. These various classes of clutter are discussed below.

Average Values

Average values of clutter have been used for many years to predict radar performance. In the 1950s, some experimenters used γ instead of σ^0, where γ is related to σ^0 by Eq. (10-22):

$$\sigma^0 = \gamma \sin \theta \qquad (10\text{-}22)$$

where θ is the depression angle. (For the beam-filled scenario, γ is equivalent to the ground cross-sectional area intercepted by the beam as mapped onto the plane perpendicular to the line of sight between the radar and the ground. γ is thus approximately independent of the depression angle for small angles.) However, the majority of experimenters have used σ^0 to describe the average radar reflectivity of surface clutter. As might be expected, the average values of clut-

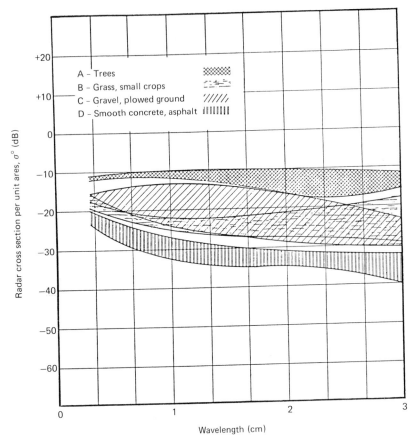

Figure 10-14. Clutter σ^0 between 70 and 20° depression angle as a function of wavelength. (From ref. 7)

ter depend on clutter type, depression angle, frequency, and polarization. Figure 10-14 summarizes the extremes of data for several types of land clutter for depression angles between 20 and 70° (the typical range of the plateau region) as a function of wavelength. This figure indicates an apparently weak dependence on wavelength. This does not mean that there is no wavelength dependence, but rather that other parameters which affect reflectivity are stronger. This is a problem that is often encountered when comparing data sets that were measured at different sites and times. Thus, the only way to determine the dependence of clutter on parameters such as frequency or polarization is to measure the same clutter cell simultaneously at different frequencies and polarizations so that all parameters are fixed except the ones to be compared. In recent years, these types of experiments have been performed by a number of experimenters.

Grass, Crops, and Trees. Extensive data have been obtained on the reflectivity of grass, crops, and trees. Figures 10-15 to 10-18 give summaries of recent data on grass at X-band and K_a-band, and tree data at X-band, K_a-band, and 95 GHz. The data exhibit a strong dependence on depression angles below 10° and show a weaker dependence on depression angles at higher angles, indicating a transition from a smooth to a rough surface at around 10°. The reflectivity of the tall grass and crops in Figures 10-15 and 10-16 is about the same as the tree clutter in Figures 10-17 and 10-18, while the short grass data are about 5 dB lower on the average. Figure 10-18 compares the reflectivity of wet and dry foliage at 35 GHz; a 2- to 3-dB difference is evident. Polarization effects are illustrated by Figures 10-19 and 10-20, which compare σ^0 for linear parallel- and cross-polarizations and odd- and even-bounce polarizations at 35 GHz. The cross-polarized linear returns are about 5 to 7 dB lower than the parallel-polar-

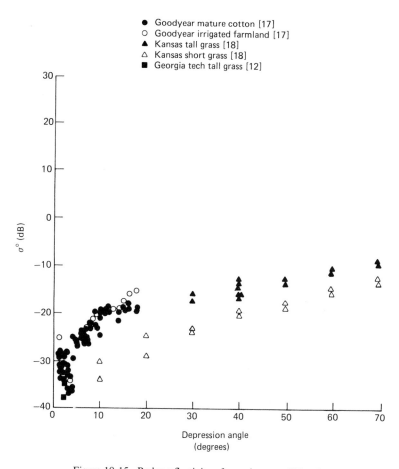

Figure 10-15. Radar reflectivity of grass/crops at X-band.

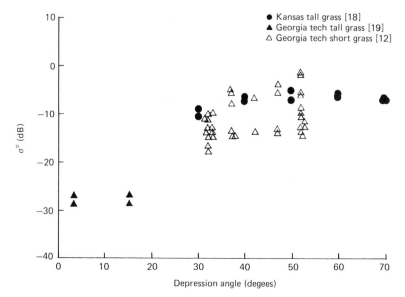

Figure 10-16. Radar reflectivity of grass at 35 GHz.

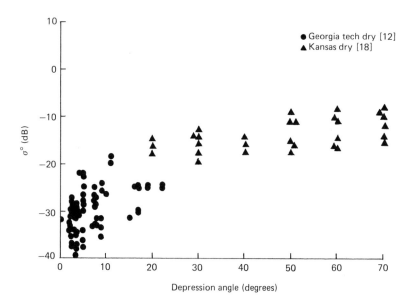

Figure 10-17. Radar reflectivity of trees at X-band. (Adapted from ref. 12)

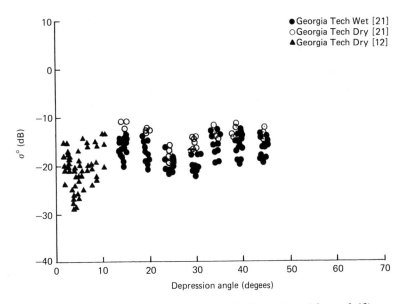

Figure 10-18. Radar reflectivity of trees at 35 GHz. (Adapted from ref. 12)

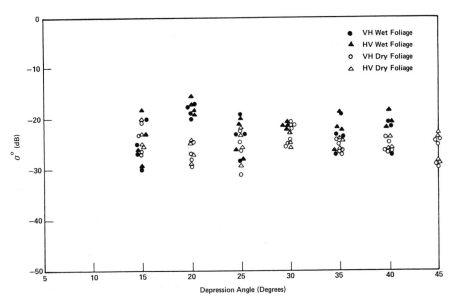

Figure 10-19. Measured radar cross section per unit area of deciduous trees at 35 GHz, VH and HV polarizations, wet and dry conditions. (From ref. 21)

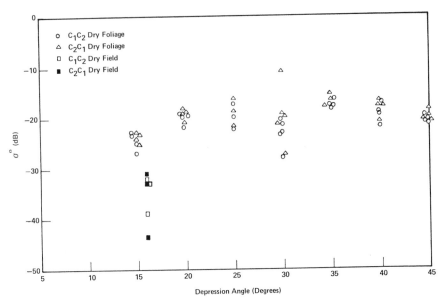

Figure 10-20. Measured radar cross section per unit area of deciduous trees at 35 GHz, (LR) and (RL) polarizations, wet and dry conditions. (From ref. 21)

ized linear returns, and the cross-polarized circular returns are approximately equal to the parallel-polarized returns (indicating that trees are complex scatterers at high depression angles).

Snow-Covered Ground.   For many years, wide variations were observed for the radar reflectivity of snow-covered ground, and these variations were attributed to the effects of objects on the surface under the relatively transparent snow cover. However, in 1976 an experiment was conducted by Currie et al. to relate the reflectivity of snow to physical properties of the snow.[22] This experiment showed that the reflectivity of snow-covered ground could vary rapidly with changing weather conditions, and that the free water content in the surface layer of the snow could be correlated to the reflectivity. Figure 10-21 illustrates this effect; it shows the measured values of σ^0 for the snow at 35 GHz, as well as the measured values of snow free water content as a function of time. During this test period, the sun's rays struck the snow surface at 8:30 A.M. Melting of the snow surface began shortly thereafter, as indicated by the increase in snow free water content. A corresponding decrease in the reflectivity of the snow was simultaneously observed at approximately 10 dB. Cross-polarized measurements were performed during the same experiment, as shown in Figure 10-22,

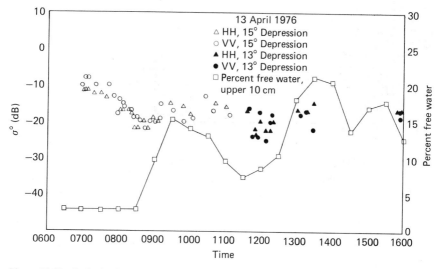

Figure 10-21. Radar backscatter per unit area from two snow-covered fields as a function of time at 35 GHz and percentage of free water present in the top snow layer. (From ref. 22)

and indicated that the cross-polarized return showed a similar change in reflectivity.

Measurements were performed by Stiles et al. at several frequencies at the same site in 1977.[23] Figure 10-23 compares σ^0 as a function of depression angle and free water content at 8, 17, and 35 GHz. The effect of the snow free water content on snow reflectivity is evident at the lower frequencies as well as at 35 GHz, although the effect is not as pronounced at the lower frequencies.

Sometimes very wet snow can have a lower reflectivity than grass alone. Figure 10-24 illustrates this point; it compares the normalized return from short grass at 7.25 GHz to the return from the grass when covered with 12 cm of wet snow.[23] Below 70° depression angle, the snow-covered ground has a lower reflectivity than the grass without snow cover, most likely indicating that the snow is a smoother surface than the grass. This hypothesis is further supported by the fact that the snow-covered grass has a higher reflectivity at nadir (90° depression angle) than the grass alone.

Sand, Rocks, and Desert.   Generally, barren or sparsely vegetated ground is considered to be a relatively benign clutter environment. An exception to this generalization may occur when large rocks or barren hills are present in the environment. Figure 10-25 presents data on soil, sand, and rocks at X-band from three different areas of barren terrain.[24] By comparing these data with

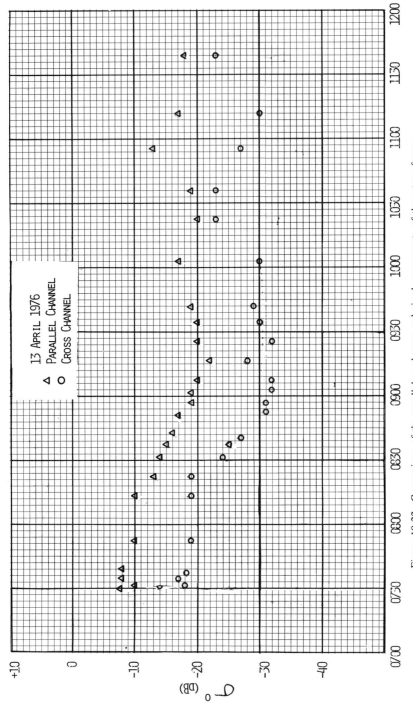

Figure 10-22. Comparison of the parallel- and cross-polarized components of the return from a snow-covered field for vertically polarized transmission at 35 GHz, April 13, 1976. (From ref. 22)

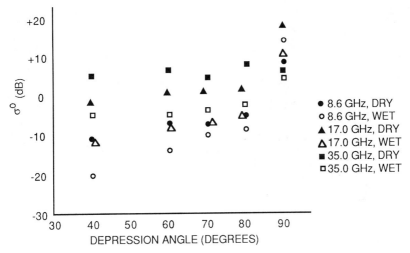

Figure 10-23. Comparison of the snow reflectivity at 8.6, 17, and 35 GHz for low and high snow free water content. (Adapted from ref. 23)

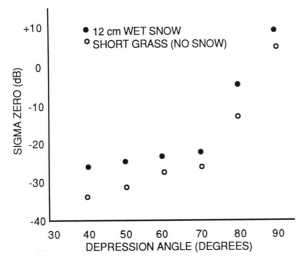

Figure 10-24. Comparison of the backscatter of short grass with short grass covered by 12 cm of wet snow; 7.25 GHz, HH polarization. (Adapted from ref. 23)

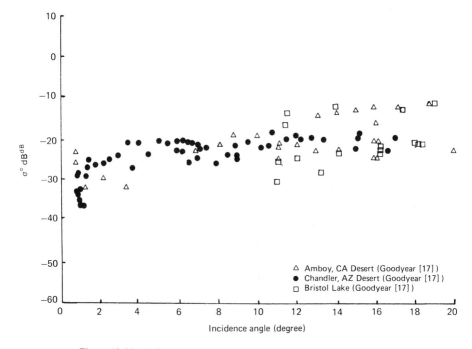

Figure 10-25. Soil, sand, and rock clutter data at X-band. (From ref. 24)

those in the preceding sections, one sees that the reflectivity of sand, rocks, and desert is generally lower than that of trees, grass, and snow.

Spatial Statistics

The concept of σ^0 assumes that clutter is a uniform quantity from a spatial point of view. In reality, however, clutter can vary significantly from point to point as a radar beam scans across the ground. For a scanning radar, the returns from several points on the ground are usually processed before a decision is made as to whether a desired target is present or not. Thus, the spatial statistics are of considerable importance. In general, very few data in the open literature describe the spatial statistics of clutter. However, in 1969 Boothe summarized several such data sets in a report.[25] These data sets included Rocky Mountain terrain at S-band, Alabama woods at L-band, and Swedish forests and cultivated fields at X-band. Boothe determined that these data sets could be described by Weibull distributions of appropriate width. However, the data used by Boothe do not appear to be temporal averages. Thus, they probably represent a composite of temporal and spatial effects, making the data difficult to interpret. Figure 10-26 gives the X-band data for cultivated land for several depression angles, along with the appropriate Weibull slope parameter. Boothe redefined

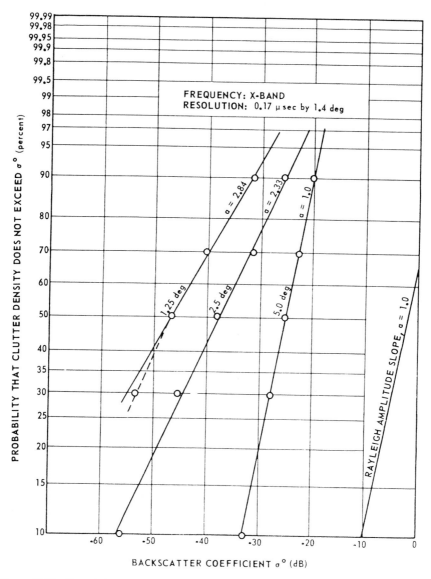

Figure 10-26. Ground clutter spatial distributions for cultivated land in May at several depression angles. (From ref. 25)

Table 10-4. Weibull Parameters for Clutter Distributions.

Figure No.	Terrain type	Frequency band	Depression angle (°)	Resolution ($\mu s \times °$)	Median $\overset{\circ}{\sigma}_m$ (dB)	Slope Parameter b
20	Rocky Mountains	S	—	2.00×1.5	−46.25	3.9
21	Forest (March–May, August)	X	0.7	0.17×1.4	−42.5	3.95
21	Forest (November)	X	0.7	0.17×1.4	−36.4	3.76
22	Cultivated land (April)	X	1.25	0.17×1.4	−47.75	3.3
22	Cultivated land (April)	X	2.5	0.17×1.4	−38.0	1.75
22	Cultivated land (April)	X	5.0	0.17×1.4	−29.8	1.1
23	Cultivated land (May)	X	1.25	0.17×1.4	−46.5	2.84
23	Cultivated land (May)	X	2.5	0.17×1.4	−37.8	2.33
23	Cultivated land (May)	X	5.0	0.17×1.4	−25.3	1.0

Source: Reference 25.

the Weibull distribution as follows:

$$P(\sigma^0) = 1 - \exp[(-(\ln 2)(\sigma^0/\sigma_m^0)^b]\qquad(10\text{-}23)$$

where

σ_m = median value
b = Weibull slope parameter
P = probability that $\sigma^c < \sigma^0$ (σ^0 is some arbitrary clutter value)

The parameter b thus defines a slope on Weibull probability paper. From Figure 10-26 it appears that the Weibull slope parameter increases as the depression angle becomes small for cultivated land. Above $5°$, the distribution becomes a Rayleigh distribution (Weibull slope of 1). An increased slope parameter implies a wider distribution and a much wider dynamic range for the clutter. This increase in dynamic range is most likely due to shadowing effects of a rolling terrain at low depression angles. Table 10-4 summarizes the median values and slope parameters for the other data sets. These data sets appear to be consistent with the cultivated land data.

Snow-covered ground can also exhibit large variations in the returns even at relatively large depression angles. This phenomenon is usually observed as the presence of ''hot spots'' or target-like returns that are present at certain positions in the return from snow-covered ground. Figure 10-27 illustrates this effect. It shows the return from a snow-covered field as a 35-GHz radar antenna

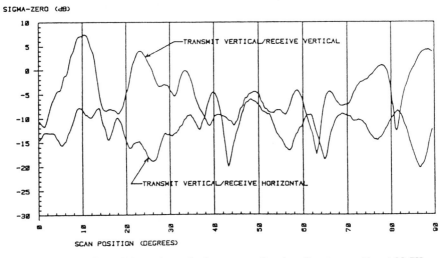

Figure 10-27. Spatial variations of snow backscatter as a function of antenna position at 35 GHz. (From ref. 26)

is scanned in azimuth.[26] Returns as high as 10 to 15 dB above the average return are evident at certain antenna positions. It is thought that these regions of high return are due to constructive reinforcement of the reflections from several layers of snow near the surface. If this is the case, then wideband frequency agility should greatly reduce the effect.

Temporal Properties

The radar returns from many types of land clutter vary with time. However, the types of clutter most likely to vary rapidly with time usually involve some type of foliage such as grass, trees, or crops. Foliage typically contains a large number of small scatterers (leaves or needles) which move in the wind, resulting in temporal variations in the radar returns. As for the case of rain, these variations in the returns can be described in terms of amplitude distributions and spectral responses.

The amplitude distributions of the backscatter from foliage have been found to be approximately Rayleigh in shape for moderate cell sizes and microwave frequencies. Figure 10-28 gives a typical standard deviation measured for deciduous trees at X-band. The distribution is plotted on normal paper so that a lognormal distribution would plot as a straight line. In this case, the distribution bends upward at the higher power levels, which is indicative of a Rayleigh distribution. Also, the standard deviation is 5 dB, which is also indicative of a Rayleigh distribution.

However, as the cell size decreases or the frequency increases above 10 GHz, the distribution becomes wider than Rayleigh. A summary of standard deviations measured for various types of foliage is given in Table 10-5 for data measured for depression angles between 3 and 10°. These data indicate that the distributions are no wider than a Rayleigh distribution up through 35 GHz. However, at 95 GHz, the distributions are wider than Rayleigh (6–7 dB) and their shape appears to be much closer to lognormal, as shown by Figure 10-29. Thus, a fundamental change in the clutter temporal statistics appears to take place at 95 GHz.

Measurements were performed by Fishbein et al. in 1967 to determine the spectral bandwidths of foliage at X-band.[10] The results are given in Figure 10-30. Fishbein et al. attempted to fit Gaussian curves to the data but were unable to achieve a satisfactory fit. A cubic fit was then tried with much better results, as shown in the figure. Later on, Hayes and Dyer suggested that the corner frequency of the cubic equation might be related to wind speed.[7] Currie et al. then determined that the spectral shapes for foliated trees were related to both wind speed and frequency.[12] They found that the spectrum had the shape

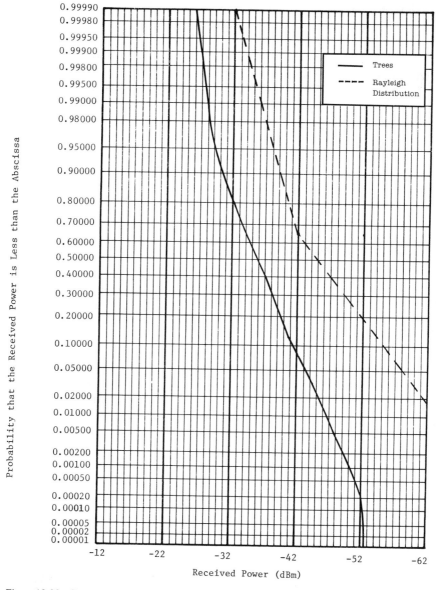

Figure 10-28. Cumulative probability distribution of the received power from deciduous trees; 9.5 GHz, horizontal polarization, and 4.1° depression angle. (From ref. 12)

Table 10-5. Summary of the Standard Deviation for Various Classes of Clutter.

Clutter Type	Polarization	Average Value of Standard Deviation			
		9.5 GHz	*16.5 GHz*	*35 GHz*	*95 GHz*
Deciduous trees summer	Vertical	3.9	—	4.7	—
	Horizontal	4.0	—	4.0	5.4
	Average	4.0	—	4.4	6.6
Deciduous trees fall	Vertical	3.9	4.2	4.4	6.4
	Horizontal	3.9	4.3	4.3	5.3
	Average	3.9	4.2	4.3	5.0
Pine trees	Vertical	3.5	3.7	3.7	6.8
	Horizontal	3.3	3.8	4.2	6.3
	Average	3.4	3.7	3.9	6.5
Mixed trees summer	Vertical	4.3	—	4.0	—
	Horizontal	4.6	—	4.2	—
	Average	4.4	—	4.1	—
Mixed trees fall	Vertical	4.1	4.1	4.7	6.3
	Horizontal	4.5	4.3	4.6	5.0
	Average	4.4	4.2	4.6	5.4
Tall grassy field	Vertical	1.5	—	1.7	2.0
	Horizontal	1.0	1.2	1.3	—
	Average	1.3	1.2	1.4	2.0
Rocky area	Vertical	1.1	2.2	1.8	1.6
	Horizontal	1.2	1.7	1.7	1.7
	Average	1.1	1.9	1.8	1.7
10-in. corner reflector (located in grassy field)		1.0	1.0	1.2	1.2

Source: Reference 12.

of a power function of the form given in equation 10-21, that is,

$$P(f) = \frac{A}{1 + (f/f_c)^n} \qquad (10\text{-}24)$$

The measured values for f_c and n are given in Table 10-6 as a function of transmitted frequency and wind speed. Thus, in a manner similar to that of rain, the spectral width increases with increasing frequency and wind speed. Once again, the use of a Gaussian rolloff characteristic to simulate the clutter spectrum at 95 GHz could result in underestimating the spectrum by a considerable amount.

The decorrelation time is of concern for land clutter, as it was in the case of

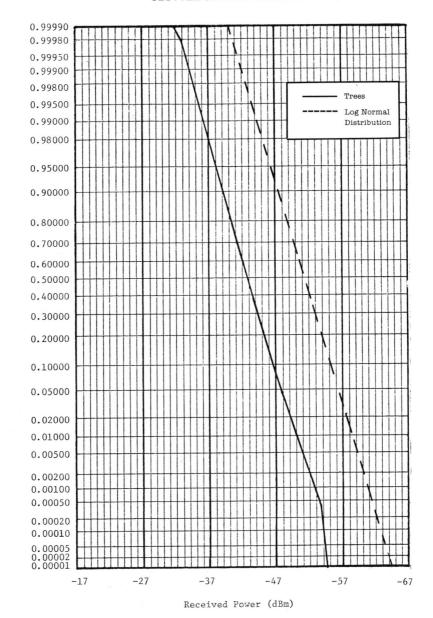

Figure 10-29. Cumulative probability distribution of the received power from deciduous trees; 95 GHz, horizontal polarization, and 4.1° depression angle. (From ref. 12)

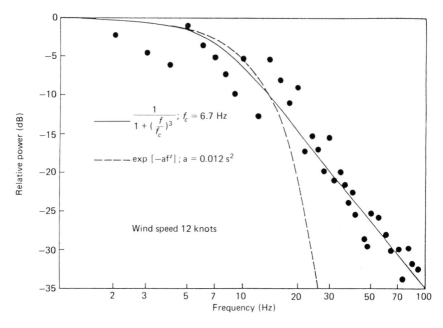

Figure 10-30. X-band power spectrum for deciduous trees compared with Gaussian and cubic power function curves. (From ref. 10)

rain. Decorrelation times were measured by Currie et al.,[12] and the results are tabulated in Table 10-7. At X-band and low wind speeds, very long decorrelation times were observed. As the wind speed or frequency was increased, the decorrelation times decreased. However, the maximum PRI rate for independent samples would be very low, as shown by Figure 10-31. Thus, wideband frequency agility would be imperative to decorrelate the clutter if one were trying to detect a target located near the clutter.

Table 10-6. Foliage Spectra Power Function Versus Frequency.

Power Function Parameter	Frequency			
	9.5 GHz	16 GHz	35 GHz	95 GHz
n(linear)	3	3	2.5	2
n(log)	4	3	3	3
f_c(Hz), 6–15 mph wind speed	9	16	21	35

Source: Reference 12.

Table 10-7. Measured Decorrelation Times for Deciduous Trees at 10, 16, 35, and 95 GHz.

Wind Speed (mph)	Decorrelation Time (ms)			
	10 GHz	16 GHz	35 GHz	95 GHz
2	210	150	65	30
3–5	92	55	27	12
6–10	45	28	23	8
11–14	21	11	2.8	0.8

Source: Reference 12.

10.3.3 Sea Clutter

Sea clutter is different from land clutter in two ways: the temporal variations tend to be larger in magnitude than those for land clutter, and the spatial variations tend to be smaller. With the exception of 90° depression angle, the reflectivity of the sea would be virtually nonexistent if it were not for the effects of wind-generated waves and capillaries. A smooth sea forward scatters all of the incident energy, so that none is reflected back toward the radar. However, the likelihood of having a smooth sea at any given time is relatively small, so that backscattered energy is normally received from the sea. Since backscattered energy from the sea is due to wave and capillary action, it is not too surprising that backscatter magnitude can be related to wave height or amplitude. Several quantities have been used to describe wave height including sea state, average wave height, significant wave height, and peak wave height.

Sea state. The sea state is a single number that attempts to describe the characteristics of the sea surface. The relationship between sea state and wave height is not clearly defined, although some relationship to average wave height is often described. Table 10-8 summarizes the Beaufort chart for sea states.

Average wave height. The average wave height is simply the mean value of the peak-to-trough height of a large number of waves. It is one of the hardest quantities to estimate visually and one of the easiest to calculate from an instrumented wave gauge.

Significant wave height. The significant wave height is defined as the average value of the peak-to-trough height of the one-third highest waves. In other words, it is the height of the most prominent waves. This is the height that is typically estimated by an observer.

Peak wave height. The peak wave height which is designated $h_{1/10}$ is the peak-to-trough height of the one-tenth highest waves. It has a value approximately twice the average wave height.

In general, sea waves are usually considered to be approximately sinusoidal

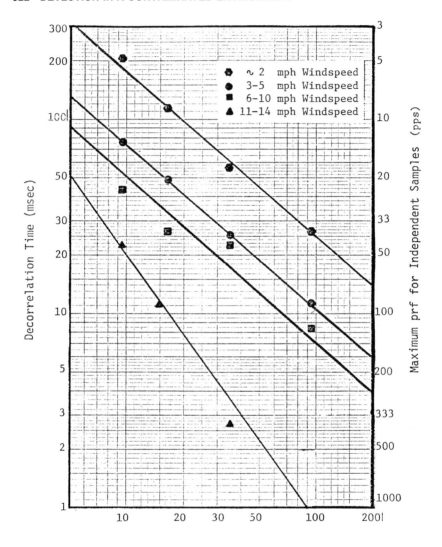

Figure 10-31. Decorrelation time versus frequency and wind speed for return from deciduous trees. (From ref. 12)

Table 10-8. Beaufort Chart for Sea States

Wind speed (knots)	4	5	6	7	8	9	10	15	20	25	30	40	50	60	70

Wind and description (Beaufort Scale)

Beaufort	Description
1	Light air
2	Light breeze
3	Gentle breeze
4	Moderate breeze
5	Fresh breeze
6	Strong breeze
7	Moderate gale
8	Fresh gale
9	Strong gale
10	Whole gale
11	Storm

Required fetch (miles): 50, 100, 200, 300, 400, 500, 600, 700

Fetch is the number of miles a given wind has been blowing over open water

Required wind duration (hours): 5, 20, 25, 30, 35

Duration is the time a given wind has been blowing over open water

If the fetch and duration are as great as those indicated above, the following wave conditions will exist. Wave heights may be up to 10% greater if fetch and duration are greater.

Wave height crest to trough (feet): 1, 2 — White caps form — 4

Sea state and description (Douglas sea number)

Douglas sea number	Description
1	Smooth
2	Slight
3	Moderate
4	Rough
5	Very rough
6	High
7	Very high
8	Precipitous

Note: The height of the waves is arbitrarily chosen as the height of the highest one-third of the waves (By permission, from Long, ref. 6; © 1983 Artech House, Inc.)

in shape and Gaussian distributed in amplitude and period. Radar backscatter occurs from the sides of the waves as well as from small facets superimposed on the waves which are generated by the wind. The primary characteristics of the reflectivity from the sea are discussed below.

Average Values

As in the case of land clutter, average values for sea clutter are usually expressed in terms of σ^0 (RCS per unit area). Generally, sea backscatter is divided into two regions: the low-angle region (0–30° depression angle) and the plateau region (30–90° depression angle). At low depression angles, the backscatter depends strongly on polarization for low to moderate sea states and is higher in average value for vertical than for horizontal polarization. The backscatter also depends strongly on the angle between the radar line of sight and the wind/wave direction, and is larger when the radar line of sight is upwind/upwave than when it is either down or cross wind/wave. An excellent discussion of sea backscatter dependencies is given by Long.[6]

In general, sea backscatter depends on wave height, wind speed, frequency, polarization (both transmit and receive), radar look direction relative to the wind, radar look direction relative to wave direction, and depression angle. The dependencies of sea backscatter changes as the frequency is increased from the microwave region to the millimeter region. These parametric dependencies are discussed below.

Microwave Region.   In the microwave region, σ^0 increases with increasing frequency. Also, for low to moderate sea states, σ_{VV} is greater than σ_{HH}. This is illustrated by Figures 10-32 through 10-35, which give the reflectivity of the sea as a function of depression angle for VV and HH polarizations, respectively, for 8910 and 438 MHz. The figures show an increase in reflectivity with frequency, depression angle, and wind speed. Also, VV is larger than HH, particularly for the lower depression angles and wind speeds. Figure 10-36 further illustrates the frequency dependence and indicates the dependence on radar look direction relative to the wind. The reflectivity is seen to increase from L-band up through K_a-band, and the returns from the upwind direction are greater than those from the downwind direction.

Millimeter Wave Region.   Some of the dependencies of sea reflectivity on the various parameters discussed above seem to be modified in the millimeter region. The dependence of reflectivity on frequency seems to change from a monotonically increasing relationship to a decreasing dependence. Figure 10-37 is a scatter diagram in which the reflectivity of the sea at 95 GHz is plotted on the vertical axis versus the reflectivity at 9.375 GHz measured simultaneously, with

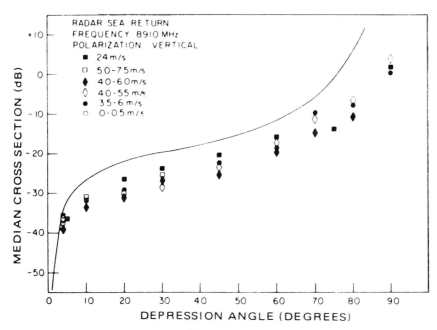

Figure 10-32. The variation of median σ_{VV}^0 with grazing angle and wind speed, X-band. (By permission, from Guinard et al., ref. 27; © 1970 IEEE)

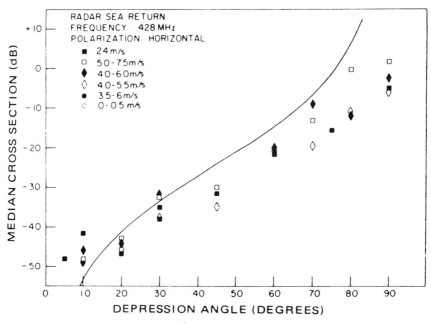

Figure 10-33. The variation of median σ_{HH}^0 with grazing angle and wind speed, X-band. (By permission, from Guinard et al., ref. 27; © 1970 IEEE)

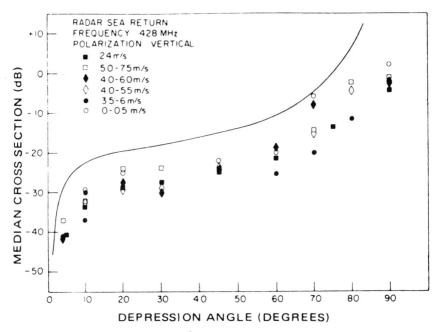

Figure 10-34. The variation of median σ^0_{VV} with grazing angle and wind speed, P-band. (By permission, from Guinard et al., ref. 27; © 1970 IEEE)

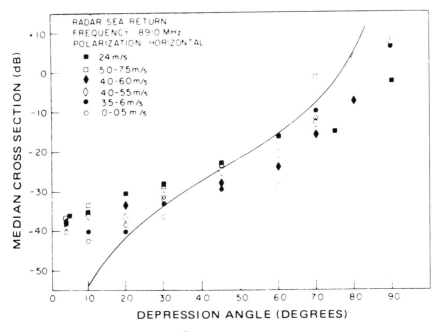

Figure 10-35. The variation of median σ^0_{HH} with grazing angle and wind speed, P-band. (By permission, from Guinard et al., ref. 27; © 1970 IEEE)

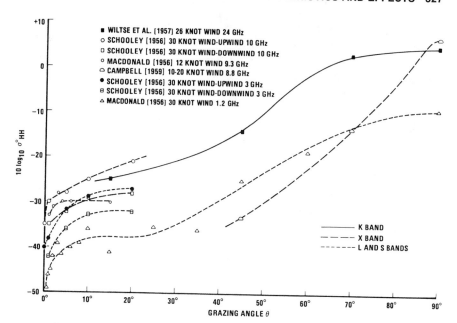

Figure 10-36. Measured values of σ_{HH}^0 at L-, S-, X-, and K-bands. (From ref. 28)

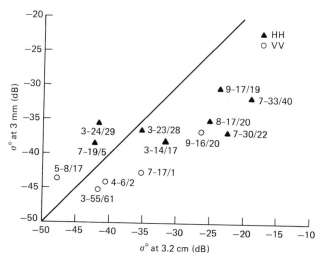

Figure 10-37. Scatter diagram of σ^0 at 3 mm and 3.2 cm. (From ref. 29)

the 9.375-GHz data on the horizontal axis. A 45° line indicates the points where $\sigma_{95} = \sigma_{9.375}$. As the figure shows, most of the points fall to the right of the 45° curve, indicating that $\sigma_{9.375}$ is greater than σ_{95}. The polarization dependence of the return also seems to be altered. Figure 10-38 gives a scatter diagram in which the reflectivity of the sea for VV polarization is plotted versus that for HH polarization for 95 GHz. Once again, a 45° line indicates the points for which $\sigma_{VV} = \sigma_{HH}$. Almost all of the points fall to the right of the line, indicating that the return for HH polarization is greater than that for VV polarization.

Low-Angle Region. At low angles as the Rayleigh roughness criterion becomes important, a very definite change is seen in the properties of sea reflectivity. In terms of a radar system, this is observed as a change in the range dependence of the sea backscatter.[30] Figure 10-39 illustrates this effect. Shown is the re-

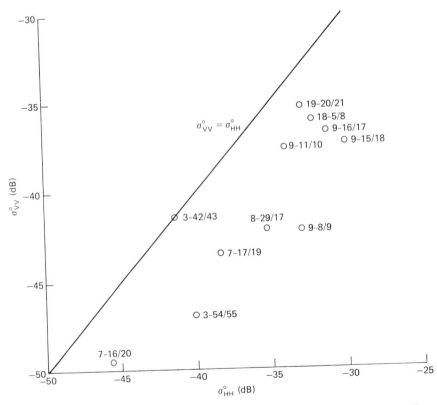

Figure 10-38. Scatter diagram of σ^0 for vertical and horizontal polarization; $\lambda = 3$ mm. (From ref. 29)

AVERAGE ECHO POWER FOR TWO SEA CONDITIONS, HH POLARIZATION

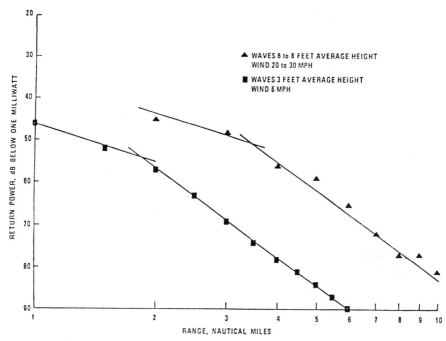

Figure 10-39. Average echo power for two sea conditions, HH polarization. (By permission, from Dyer et al., ref. 30; © 1974 IEEE)

ceived power versus range from sea clutter for two different wave heights. Each curve shows a transition from an R^{-3} to an R^{-7} range dependence. The range at which this transition occurs moves out with increasing wave height (or rms surface roughness), as predicted by the Rayleigh criterion. However, such dependencies can be modified by changes in multipath and atmospheric propagation conditions which can lead to anomalous propagation or ducting. Figure 10-40 illustrates this effect. The range dependence of sea backscatter is plotted for the morning and afternoon of the same day. The morning data show the effects of anomalous propagation, while the afternoon data show no abnormal effects. Abnormal propagations can occur quite often, as illustrated by Figure 10-41, which gives the frequency of occurrence of various range dependencies measured at Boca Raton, Florida, over a several-year period. During a significant portion of the time, range dependencies other than R^{-3} or R^{-7} were measured, indicating abnormal propagation conditions.

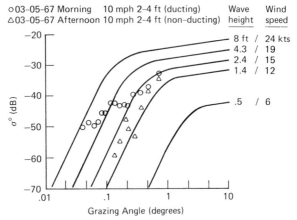

Figure 10-40. Comparisons of σ_{HH}^0 made under ducting and nonducting conditions. (By permission, from Dyer et al., ref. 30; © 1974 IEEE)

Fluctuations

Amplitude Variations. The amplitude variations of the sea were at first assumed to be Rayleigh distributed, and for VV polarizations this was not a bad assumption. However, for HH polarization and for smaller beam areas, the actual measured distributions appeared to be wider than would be predicted by Rayleigh statistics. Figure 10-42 illustrates this fact. It gives cumulative distribu-

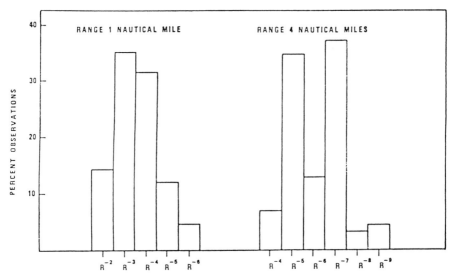

Figure 10-41. Percentage of observations versus range dependence of echo power. Data mixed with HH and VV polarizations. (By permission, from Dyer et al., ref. 30; © 1974 IEEE)

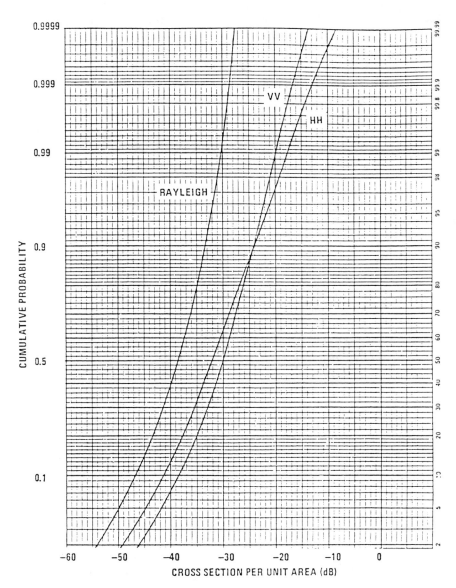

Figure 10-42. Comparison of the cumulative distribution of sea return at X-band for HH and VV polarization with a Rayleigh distribution. (By permission, from Dyer et al., ref. 30; © 1974 IEEE)

tions for sea return at X-band for VV and HH polarizations. A Rayleigh distribution is also plotted on the curves for comparison. The VV polarization curve is similar to the Rayleigh distribution, but the HH polarization is much wider, possessing a high-value *tail*. This high-value tail is usually associated with the phenomena called *spikes*. Spikes are isolated high-value reflections from the sea that appear to be associated with facets that form on the sea surface. Currently, there is a controversy regarding the exact causes of spikes, but it is known that they are small in area and are caused by wind effects on the sea. Spikes are present for both VV and HH polarizations but are much more prevalent for HH polarization, at least at microwave frequencies. Recent measurements indicate that this polarization difference is less noticeable at millimeter frequencies. Figure 10-43 compares the distributions measured for sea backscatter for VV and HH polarization at 35 and 95 GHz. At 35 GHz, the HH distribution is much wider than the VV distribution; at 95 GHz, the HH and VV distributions are almost identical, indicating a change in the mechanism causing spiking.

The amplitude statistics of the sea have been successfully modeled using lognormal or Weibull distributions. These distributions have an advantage over the Rayleigh distribution in that they are two-parameter distributions, thus allowing a better fit to existing data. Most current attempts to model the statistics of sea clutter use a two-parameter distribution which has been empirically fitted to actual sea clutter data.

Spectral Characteristics. The spectrum for sea clutter differs from that of land clutter in one fundamental way: since sea waves can have physical movement toward or away from a radar, the spectrum of the received signal can have a maximum value at a frequency other than zero. (Since most land clutter cannot move from its fixed position on the ground, its maximum return will always occur at zero frequency.) The frequency at which this maximum occurs is given by the familiar Doppler equation:

$$f_D = \frac{2V}{\lambda} \tag{10-25}$$

The width of the Doppler spectrum can be related to the width of the wave velocity spectrum by substituting F_V (the wave spectrum) for V and F_f (the Doppler Spectrum) for F_D and solving Eq. (10-25) to obtain

$$f_V = \frac{\lambda}{2} F_f \tag{10-26}$$

Figure 10-44 gives an example of the coherent spectrum of sea return at L-band, indicating the spectrum width for various amplitude levels. At higher

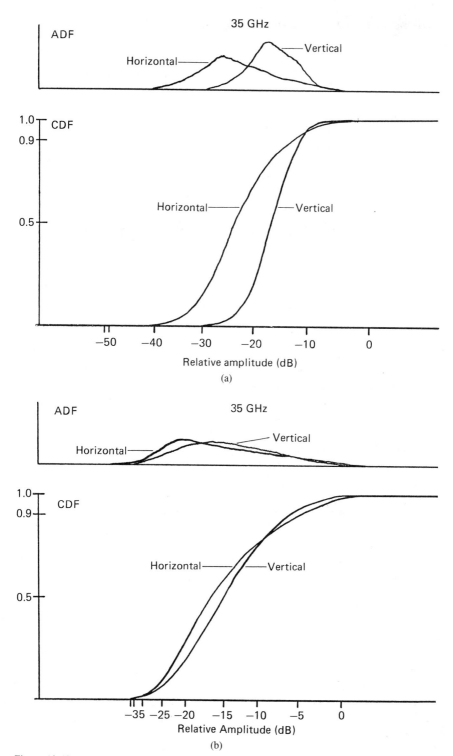

Figure 10-43. (a) 35-GHz sea return amplitude distributions. (b) 95-GHz sea return amplitude distributions. (From ref. 31)

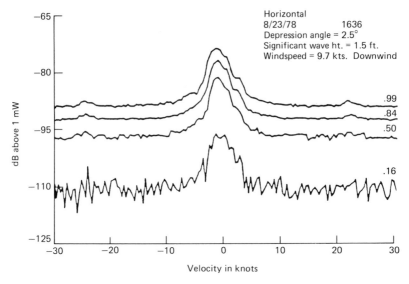

Figure 10-44. Percentile Doppler spectra of sea echo at L-band. (From ref. 32)

frequencies, the center frequency and spectral width would be expected to scale proportional to the transmitted frequency, as indicated by the Doppler equation. In general, the sea spectrum has been found to be approximately Gaussian distributed, as illustrated in Figure 10-44.

Another way of looking at sea spectrum is in terms of the autocorrelation function. Studies of the autocorrelation function for sea return have determined that it can be divided into two parts: a rapidly decaying part and a strongly correlated portion with a period of many seconds. Figure 10-45 illustrates this phenomenon. It shows the autocorrelation function for sea return at K_u-band for two time scales: on the short time scale, the autocorrelation function decays to approximately 0.85 of its DC value in about 2 ms, while the long time scale indicates that over 1 s is required for it to decay to nearly zero. The implication of Figure 10-45 is that frequency agility is required to obtain independent samples between radar pulses, and extremely wide bandwidths may be required to decorrelate the slowly varying part of the spectrum which is probably due to gross wave motion.

10.4 EXAMPLES OF CLUTTER MODELS

In this section, models of backscatter from rain and the earth's surface are summarized. These models have been developed by various experimenters based on data available in the unclassified literature and represent the best estimates of backscatter currently available. They are compared with representative data to illustrate their validity.

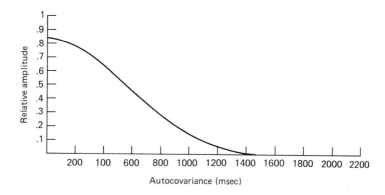

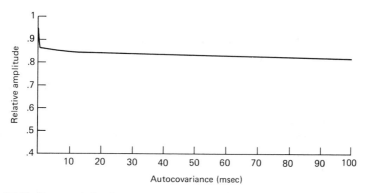

Figure 10-45. Autocorrelation function calculated for sea return at K_u-band, 3.9° depression and downwind/downwave look direction. (From ref. 34)

10.4.1   Rain Clutter Model

Calculation of rain backscatter effects requires an analytical expression for η, the radar backscatter per unit volume, and the backscatter cell size. Measured data for η exhibit rather large variations, as do those for σ^0. Data collected by several investigators were used to develop a clutter model of the form[15]

$$\eta = AR^B (m^2/m^3) \qquad (10\text{-}27)$$

where

η = backscatter cross section in square meters/cubic meters (sometimes the symbol σ_v is used)

R = rain rate (millimeters/hour)

A and B = parameters to be fitted to measured data for each frequency of interest

Values of A and B are given in Table 10-9.

Table 10-9. Rain Backscatter Model Coefficient.

Frequency (GHz)	A	B
9.4	1.3×10^{-8}	1.6
35	1.2×10^{-6}	1.6
70	4.2×10^{-5}	1.1
95	1.5×10^{-5}	1.0

Source: Reference 11.

10.4.2 Land Clutter Models

A large number of radar measurements have been performed on all types of clutter including soil, grass, crops, trees, sand, rocks, urban areas, and snow-covered ground. These data have been taken with a number of radar measurement systems and under many different environmental and physical conditions. Thus, there are large spreads in the available data. However, if one carefully picks data sets which appear to be consistent, models for the reflectivity of the various types of clutter as functions of frequency and depression angle can be developed. A model of average radar cross section per unit area was developed by the Georgia Institute of Technology.[16, 24] This model was developed by empirically fitting curves to a selected subset of available data. The equation for the model can be expressed as

$$\sigma^0 = A\left(\theta + C\right)^B \exp\left[-D/\left(1 + \frac{0.1\sigma_h}{\lambda}\right)\right] \qquad (10\text{-}28)$$

where

θ = depression angle in radians
σ_h = standard deviation of surface in centimeters
λ = radar wavelength
A, B, C, and D are empirically derived constants

Table 10-10 gives a summary of A, B, C, and D values for frequencies of 3 to 95 GHz. The validity of the model can be seen by comparing the model with actual measured data. Figures 10-46 through 10-48 give examples of σ^0 data for trees, urban clutter, and snow as a function of the incidence angle for X-band through 95 GHz. The model outputs plotted on each data set show good agreement between the model and the data sets. Note that since the data are plotted in logarithmic coordinates, the average of all of the values falls near the top of the set of data points, rather than in the middle of the data point cluster.

Table 10-10. Clutter Model Constants.

Constant	Frequency	Soil/Sand	Grass	Tall Grass Crops	Trees	Urban	Wet Snow	Dry Snow
A	3	0.0045	0.0071	0.0071	0.00054	0.362	—	—
	5	0.0096	0.015	0.015	0.0012	0.779	—	—
	10	0.25	0.023	0.006	0.002	2.0	0.0246	0.195
	15	0.05	0.079	0.079	0.019	2.0	—	—
	35	—	0.125	0.301	0.036	—	0.195	2.45
	95	—	—	—	3.6	—	1.138	3.6
B	3	0.83	1.5	1.5	0.64	1.8	—	—
	5	0.83	1.5	1.5	0.64	1.8	—	—
	10	0.83	1.5	1.5	0.64	1.8	1.7	1.7
	15	0.83	1.5	1.5	0.64	1.8	—	—
	35	—	1.5	1.5	0.64	—	1.7	1.7
	95	—	1.5	1.5	0.64	—	0.83	0.83
C	3	0.0013	0.012	0.012	0.002	0.015	—	—
	5	0.0013	0.012	0.012	0.002	0.015	—	—
	10	0.0013	0.012	0.012	0.002	0.015	0.0016	0.0016
	15	0.0013	0.012	0.012	0.002	0.015	—	—
	35	—	0.012	0.012	0.012	—	0.008	0.0016
	95	—	0.012	0.012	0.012	—	0.008	0.0016
D	3	2.3	0.0	0.0	0.0	0.0	—	—
	5	2.3	0.0	0.0	0.0	0.0	—	—
	10	2.3	0.0	0.0	0.0	0.0	0.0	0.0
	15	2.3	0.0	0.0	0.0	0.0	—	—
	35	—	0.0	0.0	0.0	—	0.0	0.0
	95	—	0.0	0.0	0.0	—	0.0	0.0

Note: These parameters correspond to dry conditions except for wet snow.
Source: Adapted from reference 24

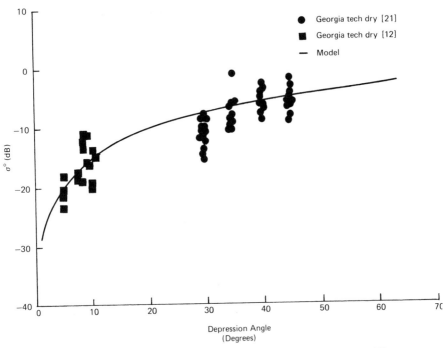

Figure 10-46. Comparison of radar reflectivity of trees and model at 95 GHz.

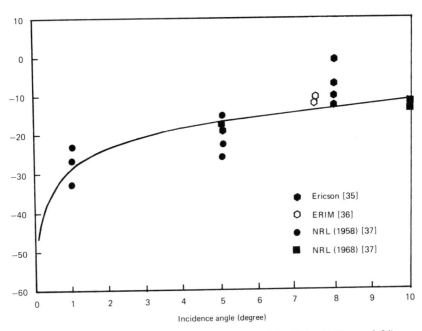

Figure 10-47. Comparison of urban clutter data to model at X-band. (From ref. 24)

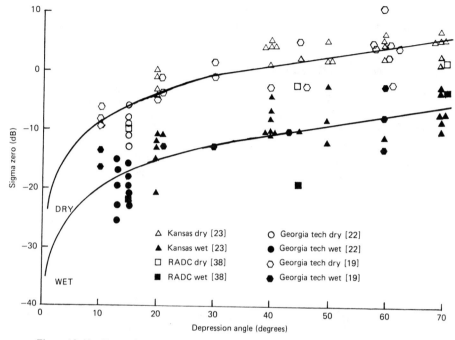

Figure 10-48. Comparison of wet/dry snow backscatter data with the model at 35 GHz.

The plotted data are time averages for different spatial looks so that the data variations can be attributed to spatial variations.

Many land clutter models (mostly concerned with incidence angles above 10°) have characterized σ^0 as decreasing to zero as the incidence angle goes to 0°. Although there is some theoretical basis for such behavior, actual σ^0 data from measurements at incidence angles below 5° have not all behaved in this way. The discrepancy might be explained by the ambiguity in defining the incidence angle. In most clutter measurements, the depression angle is used interchangeably with the grazing or incidence angle. Even if the experimenters are careful to account for terrain slope, curvature of the earth, and atmospheric influences, it is not possible to determine the exact incidence angle at every tree, bush, sand dune, or patch of grass. In an attempt to model experimental data more accurately and to prevent the sharp drop to $-\infty$ in decibels of σ^0 for grazing angles near 0°, a constant, C, was added to the equations as shown above. The C values prevent σ^0 from going to zero ($-\infty$ in decibel space).

The equations in Table 10-10 are given for dry conditions (except for snow). Although insufficient data were available to develop wet and dry models separately, previous clutter studies have determined that σ^0 for wet terrain tends to be 5 dB higher than for dry terrain.[21] This observation was used to develop the constants in Table 10-10. To obtain σ^0 values for wet terrain, one must multiply the equations in Table 10-10 by $10^{0.5}$ (or add 5 dB in decibel space).

10.5 CONCLUDING REMARKS

This chapter has discussed the use of standard descriptors of radar clutter such as σ^0 and η to describe clutter properties. However, it must be noted that the original purpose of defining σ^0 as the *average* cross section per unit area was to allow sea or ground return to be measured by one radar and have the normalized results applicable to other radars. It is important to understand that σ^0 is a fairly accurate descriptor *only* if the surface of the sea or land exhibits homogeneous properties across the area illuminated by the radar. Of course, this is more commonly the case for low-resolution radars than it is for moderate- to high-resolution systems. The fact that σ^0 is not a constant across large areas of land, or even of the sea, has been recognized since the early days of this terminology. A single value for σ^0 is often used in system analysis, but this can lead to incorrect results if the proper interpretation is not placed on the overall situation.

The key point is that there are physical reasons for the temporal and spatial variations in return as a radar views an extended area of the land, the sea, or a volume of space containing many scatterers. Therefore, any value of σ^0 which is based on averaging over a substantial spatial region will often not represent the important excursions of the return which are both higher and lower than this average value. This is true whether the average is obtained by viewing with a radar of extremely low resolution, so that the radar itself does the averaging (coherently), or whether a number of resolved cells are measured and these results subsequently averaged (noncoherently).

Clutter cross-section values such as those given by Nathanson[39] are useful for indicating trends and for limited system analysis. Their major deficiencies are that they are highly averaged values of radar returns which extend over a wide band of clutter amplitudes. This type of averaging further conceals both temporal variations of the return from any given region and the spatial variation from region to region of even apparently similar terrain. (As an aside, one of the difficulties in describing clutter is that the physical description of the land or sea is not precise enough to allow good intercomparison—that is, neither "State 3 sea" nor "heavy forest" is sufficiently descriptive to ensure agreement between values measured by different investigators observing different sites at different times.)

The clutter models presented in this chapter attempt to relate σ^0 to differences observed in the reflectivity due to frequency effects, polarization, depression angle, and environmental parameters. Although they also represent averaged values, they should prove more useful for real-world prediction of performance than previous models.

10.6 REFERENCES

1. H. A. Corriher et al., Chapter XVII, "Elements of Radar Clutter," of *Principals of Modern Radar,* H. Allen Ecker, ed., Georgia Institute of Technology, Atlanta, 1972.
2. Herbert Goldstein, "A Primer of Sea Echo," Report No. 157, U.S. Navy Electronics Laboratory, San Diego, 1950.
3. R. N. Trebits, "Radar Cross Section," Ch. 2, *Techniques of Radar Reflectivity Measurement,* Artech House, Dedham, Mass., 1984.
4. R. S. Berkowitz, *Modern Radar,* John Wiley & Sons, New York, 1965, p. 560.
5. J. D. Echard et al., "Discrimination Between Targets and Clutter by Radar," Final Technical Report on Contract DAAAG-29-780C-0044, Georgia Institute of Technology, Atlanta, December 1981.
6. M. W. Long, *Radar Reflectivity of Land and Sea,* Artech House, Dedham, Mass., 1983.
7. R. D. Hayes and F. B. Dyer, "Computer Modeling for Fire Control Radar Systems," Technical Report No. 1 on Contract DAA25-73-C-0256, Georgia Institute of Technology, Atlanta, December 1974.
8. Donald E. Kerr and Herbert Goldstein, "Radar Targets and Echoes," ch. 6, of *Propagation of Short Radio Waves,* vol. 13, MIT Radiation Laboratory Series, Boston Technical Publishers, Lexington, Mass., 1964, pp. 360–561.
9. E. J. Barlow, "Doppler Radar," *Proceedings of the IRE,* vol. 37, no. 4, April 1949, p. 351.
10. W. Fishbein et al., "Clutter Attenuation Analysis," Technical Report No. ECOM-2808, U.S. Army ECOM, AD665352 March 1967.
11. N. C. Currie, F. B. Dyer, and R. D. Hayes, "Analysis of Radar Rain Return at Frequencies of 9.5, 35, 70, and 95 GHz," Technical Report No. 2 on Contract DAA25-73-0256, Georgia Institute of Technology, Atlanta, February 1975.
12. N. C. Currie, F. B. Dyer, and R. D. Hayes, "Radar Land Clutter Measurements at 9.5, 16, 35, and 95 GHz," Technical Report No. 3 on Contract DAA25-73-0256, Georgia Institute of Technology, Atlanta, March 1975.
13. P. M. Austin, "Radar Measurements of the Distribution of Precipitation in New England Storms," *Proceedings of the 10th Weather Conference,* Boston, 1965, pp. 247–254.
14. Louis J. Battan, *Radar Characteristics of the Atmosphere,* University of Chicago Press, Chicago, 1973.
15. N. C. Currie, F. B. Dyer, and R. D. Hayes, "Radar Rain Return at Frequencies of 10, 35, 70, and 95 GHz," 1975 IEEE International Radar Conference, Washington, D.C., April 1975.
16. N. C. Currie, and S. P. Zehner, "MM Wave Land Clutter Model," IEE RADAR 82 International Symposium, London, September, 1982.
17. "Radar Terrain Return Study," Goodyear Aircraft Corp., Final Report on Contract NOAS-59-6186-C, GERA-463, Phoenix, September 1959.
18. W. H. Stiles, F. T. Ulaby, and E. Wilson, "Backscatter Response of Roads and Roadside Surfaces," Sandia Report No. SAND78-7069, University of Kansas Center for Research, Lawrence, March, 1979.
19. R. D. Hayes, N. C. Currie, and J. A. Scheer, "MMW Backscatter and Emissivity of Snow," IEEE 1980 International Radar Conference, Washington, D.C., April 1980.
20. Bush et al., "Seasonal Variations of the Microwave Scattering Properties of the Deciduous Trees as Measured in the 1–18 GHz Spectral Range," R.S.L. Report No. 177-60, University of Kansas Center for Research, Lawrence, June 1976.
21. N. C. Currie et al., "Radar Millimeter Wave Measurements: Part I, Snow and Vegetation," Report No. AFATL-TR-77-92 on Contract F08635-77-C-0221, Georgia Institute of Technology, Atlanta, July, 1977.
22. N. C. Currie, F. B. Dyer, and G. W. Ewell, "Radar Millimeter Measurements from Snow," Technical Report No. AFATL-TR-77-4 on Contract F08635-77-C-0221, Georgia Institute of Technology, Atlanta, January 1977.

23. William H. Stiles and F. T. Ulaby, "Microwave Sensing of Snowpacks," NASA Report No. 3263 on Contract NAS-5-73777, University of Kansas Center for Research, Lawrence, June 1980.

24. S. P. Zehner and M. T. Tuley, "Development and Validation of Multipath and Clutter Models for TAC Zinger in Low Altitude Scenarios," Final Technical Report on Contract No. F49620-78-C-0121, Georgia Institute of Technology, Atlanta, March 1979.

25. R. R. Boothe, "The Weibull Distribution Applied to Ground Clutter Backscatter Coefficient," Report No. RE-TR-69-15, U.S. Army Missile Command, Huntsville, Ala., June 1969.

26. Joseph E. Knox, "High Angle Backscatter from Snow on the Ground," Snow 1-A Symposium, Hanover N.H., October 1982.

27. N. W. Guinard and J. C. Daley, "An Experimental Study of a Sea Clutter Model," *Proceedings of the IEEE*, vol. 58, April 1970, pp. 534–550.

28. M. W. Long et al., "Wavelength Dependence of Sea Echo," Final Report on Contract N62269-3019, Georgia Institute of Technology, Atlanta, 1965.

29. W. K. Rivers, "Low Angle Sea Return at 3 mm Wavelength," Final Technical Report on Contract N62269-70-C-0489, Georgia Institute of Technology, Atlanta, November 1970.

30. F. B. Dyer and N. C. Currie, "Some Comments on the Characterization of Radar Sea Echo," *1974 IEEE AP-S International Symposium,* Atlanta, June 1974, pp. 323–326.

31. R. N. Trebits et al., "Millimeter Radar Sea Return Study," Interim Technical Report on Contract No. N60921-77-C-A168, Georgia Institute of Technology, Atlanta, July 1978.

32. D. K. Plummer et al., "Some Measured Statistics of Coherent Radar Sea Echo and Doppler," Final Technical Report on Contract NADC 78154-30, Georgia Institute of Technology, Atlanta, December 1979.

33. J. C. Wiltre, et al., "Backscattering characteristics of the Sea in the Region 10 to 50 KMC," *Proceedings of the IRE,* vol. 45, pp. 220–228, February, 1957.

34. R. N. Trebits and B. Perry, "Multifrequency Radar Sea Backscatter Data Reduction," Final Technical Report on Contract N60921-80-C-0176, Georgia Institute of Technology, Atlanta, September 1982.

35. L. O. Ericson, "Terrain Return Measurements with an Airborne X-band Radar Station," Sixth Conference of the Swedish National Committee on Scientific Radio, Stockholm, Sweden March 1963 (translated October 1966).

36. A. Maffet et al., *Radar Analysis Services,* vol. 2C: *Synthetic Aperture Radar Change Detection Model—Technical Reference Material,* Radar and Optics Division, Environmental Research Institute of Michigan (ERIM), Ann Arbor, June 1977.

37. F. C. McDonald, W. S. Ament, and D. L. Ringwalt, "Terrain Clutter Measurements," NRL Report 5057, Propagation Branch, Washington, DC, January 1958.

38. D. T. Hayes, "Radar Backscatter From Snow," 1978 IEEE AP/S International Symposium, Seattle, June 1978.

39. Fred E. Nathanson, *Radar Design Principles,* McGraw-Hill Book Co., New York, 1969, p. 229.

11
TARGET MODELS

Carl R. Barrett, Jr.

11.1 INTRODUCTION

When defining radar detection performance, the radar designer is normally confronted by the fact that the amplitude of the target return can be only statistically (nondeterministically) defined. That is, the target's instantaneous RCS fluctuates principally as a result of changes in the vector summation of the returns from the many scattering elements that contribute to the target's effective scattering cross section. These target-related amplitude fluctuations are forced by factors such as changes in illumination angle, frequency, and polarization; target motion and vibration; and kinematics associated with the sensor's host vehicle.

This chapter presents various statistical models used to describe target amplitude fluctuations and estimate radar performance. As will be demonstrated, target amplitude fluctuations can greatly modify the signal-to-noise ratio required to achieve a high single-look probability of detection.

Just as terrain backscatter can be one person's desired target (e.g., when ground mapping) and another's interference (e.g., when attempting to detect a stationary, small RCS target in the midst of terrain return), target amplitude fluctuations also may have a dual impact. That is, when one is attempting to achieve a high single-look detection probability, fluctuations may force the requirement for a higher signal-to-interference ratio. On the other hand, if the purpose is to classify returns resulting from specific target types, it may be found that target complexities leading to this fluctuation characteristic are, in fact, the source of the useful information. Because of the importance of target classification in today's radar technology, this subject is introduced in Chapter 21.

11.2 TARGET MODEL RELATIONSHIPS TO RADAR DETECTION THEORY

As stated above, the observed RCS of typical targets is produced by the complex vector summation of reflections from multiple scattering elements distributed over the entire extent of the target. Since these individual scatterers may,

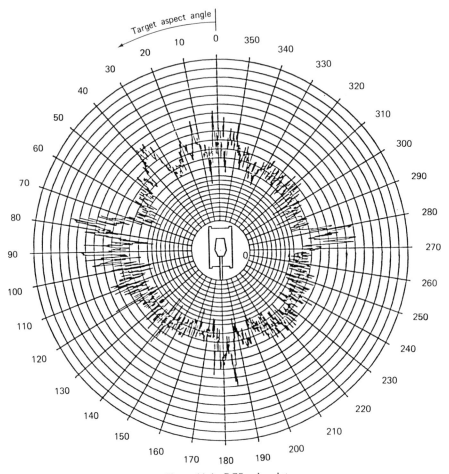

Figure 11-1. RCS polar plot.

in turn, be directive, frequency sensitive, and so on, it is easy to postulate how changes in target aspect angle or instantaneous illumination frequency and polarization can cause the observed RCS to fluctuate. An example of the measured RCS as a function of viewing aspect angle for a typical tactical target is given in Figure 11-1. This polar plot of RCS versus aspect angles was taken while the target was rotated through 360° of azimuth angle. A histogram of this observed RCS obviously differs greatly from that obtained from a nonfluctuating target.

Study of the RCS plot of Figure 11-1 suggests the difficulty of accurately predicting detection performance with mathematical tools and procedures that were developed based upon the assumption of a nonfluctuating target (e.g., as assumed in Chapter 9). Generally, the use of a nonfluctuating target model for predicting the detection of a fluctuating target leads to overestimated detection performance when high single-scan detection probabilities are desired. Conversely, the use of a nonfluctuating model leads to underestimated detection performance when low single-scan detection probabilities are desired.* Heuristically, this conclusion can be demonstrated by recalling that for a fluctuating target there is some probability that the observed RCS will be less than the mean value. The signal-to-interference ratio must therefore be increased to ensure that enough of the low RCS amplitudes cross the threshold such that the high-probability single-scan detection criterion is met. Conversely, associated with a target is a nonzero probability that the observed RCS will be greater than the mean value. In this case, the signal-to-interference ratio can be reduced until just enough of these high-amplitude returns cross the threshold to meet a low-probability single-scan detection criterion.

The radar analyst can develop and apply mathematical tools that permit accurate prediction or adapt nonfluctuating models to estimate the detection performance for fluctuating targets. The objective of this chapter is to present procedures and techniques that can be used to predict radar detection performance accurately when fluctuating targets are encountered. This chapter also introduces the concept of threshold loss and demonstrates the sensitivity of this loss to various target models, desired detection and false-alarm probabilities, and detection threshold criteria.

11.3 TARGET FLUCTUATION MODELS

Marcum[1] and Swerling[2] published classic works that describe the theory, methods, and procedures for predicting the detection performance of nonfluctuating and fluctuating targets, respectively. Methods and procedures for the nonfluctuating case are given in Chapter 9. Swerling extended Marcum's works to incorporate what has become known as the four Swerling models. His first two cases (cases 1 and 2) model targets that produce radar receiver output voltage envelopes that can be described by Rayleigh statistics. The target's RCS,[3] which is proportional to the voltage squared, is therefore distributed exponentially with the density function

$$p(\gamma) = \frac{1}{\bar{\gamma}} e^{-\gamma/\bar{\gamma}} U(\gamma) \qquad (11\text{-}1)$$

---

*This result leads to some very interesting conclusions when detection is defined by multiple-look criteria rather than single-look criteria.

where

γ = instantaneous RCS
$\bar{\gamma}$ = average RCS
$U(\gamma)$ = unit step function

This is a chi-square density function with a single duo-degree of freedom. The distinction between Swerling's first two cases can be summarized as follows:

Case 1 — Slow target fluctuation (correlated pulse-to-pulse but independent scan-to-scan).

Case 2 — Rapid target fluctuation (assumed independent pulse-to-pulse).

This target model represents targets composed of a number of independent scattering elements, none of which dominate.

Swerling's second two cases assume that the target RCS can be described by a chi-square density function with two duo-degrees of freedom (actually a gamma distribution resulting from normalization of the chi-square). The density function of the target RCS is therefore given by

$$p(\gamma) = \frac{4\gamma}{\bar{\gamma}^2} e^{-2\gamma/\bar{\gamma}} U(\gamma) \qquad (11\text{-}2)$$

This model represents targets that have a single dominant nonfluctuating scatterer, along with an assembly of many independent smaller scatterers. The differences between these second two cases are as follows:

Case 3 — Slow target fluctuation (correlated pulse-to-pulse but independent scan-to-scan).

Case 4 — Rapid target fluctuation (assumed independent pulse-to-pulse).

The radar analyst must realize that none of these four target models are universally valid under real-world conditions. An example of this situation can be given with the aid of Figure 11-2. This figure illustrates a family of cumulative probability density functions that might be used to represent target cross sections with varying duo-degrees of freedom chi-square distributions. The probability density function that represents a chi-square distributed target with k duo-degrees of freedom is given (after Mitchell and Walker)[5] as

$$p(\gamma) = \frac{1}{\Gamma(k)} \left(\frac{k}{\bar{\gamma}} \right)^k \gamma^{k-1} e^{-k\gamma/\bar{\gamma}} U(\gamma)$$

Incorporated into the figure is an example of empirical test data from an actual target. As demonstrated by these data, targets rarely behave in an easy-to-model manner. Knowledge of actual target fluctuation statistics, however, and statis-

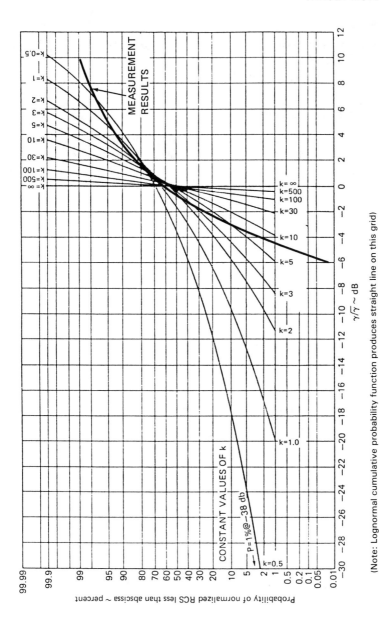

Figure 11-2. Cumulative RCS distribution of a target plotted against a family of chi-square distribution functions (k = number of duo-degrees of freedom). (By permission, from Meyer et al., ref. 4; © 1973 Academic Press, Inc.)

tical prediction techniques allow one to estimate expected performance and attach a degree of confidence to the estimate. Note that the curves for $k = \infty$, $k = 2$, and $k = 1$ in Figure 11-2 represent the nonfluctuating case, Swerling cases 3 and 4, and Swerling cases 1 and 2, respectively.

11.3.1   NONFLUCTUATING TARGET

The mathematical tools necessary for representing a fluctuating target RCS can be developed by considering standard techniques for estimating the probability of false alarm, P_{fa}, and the probability of detection, P_{d}. The estimation problem is illustrated in Figure 11-3. In this sketch, $p(V, \bar{x} = 0)$ represents the probability density of the receiver output voltage at the detection decision circuit when only thermal noise is present, and $p(V, \bar{x})$ represents the case when both a nonfluctuating target signal and noise are present.

From Figure 11-3, it is evident that

$$P_{\text{fa}} = \int_{V_T}^{\infty} p(V, \bar{x} = 0) \, dV \tag{11-3}$$

and

$$P_{\text{d}} = \int_{V_T}^{\infty} p(V, \bar{x}) \, dV \tag{11-4}$$

where V_T is the detection threshold. The performance prediction process then

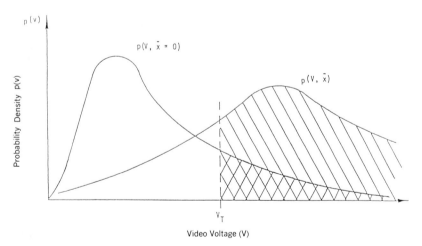

Figure 11-3. Relationship of the probability of detection and the probability of false alarm.

requires completion of the following serial steps:

1. Define the density function when noise alone exists at the detection decision circuit.
2. Define V_T that will produce the allowed P_{fa}.
3. Define the density function when signal-plus-noise exist at the detection circuit.
4. Calculate the detection probability that results with V_T.
5. Repeat these steps to include effects of threshold uncertainty resulting from an inadequate (errored) estimate of the interference level (the rms level or power of the interference and, hence, of V_T).

Threshold uncertainty applies to any system when the interference (level or distribution) is not known a priori and the threshold itself is based on an estimate of this interference. In such cases, Eq. (11-3) produces only an estimate of P_{fa}, or $\hat{P}_{fa}$, where the symbol ^ represents an estimate of the parameter. That is,

$$\hat{P}_{fa} = \int_{\hat{V}_T}^{\infty} p(V, \bar{x} = 0)\, dV \qquad (11\text{-}5)$$

Under such conditions, the P_{fa} is generally defined to mean the expected value of $\hat{P}_{fa}$. The false alarm probability is therefore represented as

$$P_{fa} = E[\hat{P}_{fa}] = \int_0^{\infty} p(\hat{V}_T) \int_{\hat{V}_T}^{\infty} p(V, \bar{x} = 0)\, dV\, d\hat{V}_T \qquad (11\text{-}6)$$

In this formulation, $p(\hat{V}_T)$ is the probability density function for $\hat{V}_T$. Similarly, the P_d is generally defined to mean the expected value of $\hat{P}_d$. The detection probability is therefore represented as

$$P_d = E[\hat{P}_d] = \int_0^{\infty} p(\hat{V}_T) \int_{\hat{V}_T}^{\infty} p(V, \bar{x})\, dV\, d\hat{V}_T \qquad (11\text{-}7)$$

Detection Threshold Estimation

Steps 1 and 2 defined in Section 11.3.1 lead to the definition of an unerrored detection threshold that produces an allowed false alarm probability. The analytical process associated with these two steps is defined here.

Early researchers[1] showed that square law detection and summing of N independent samples of thermal noise (noncoherent integration of N pulses), where the noise has Rayleigh envelope statistics, produces a chi-square density function with $2N$ degrees of freedom. That is, if z is the sum (at video) of N square law detected samples, as in

$$z = \sum_{i=0}^{N-1} x_i^2 \tag{11-8}$$

where the x_i values are zero mean Gaussian random variables (with variance σ^2), it has been shown that[1]

$$p(z) = \frac{z^{L/2-1}e^{-z/2\sigma^2}}{2^{L/2}\sigma^L \Gamma(L/2)} U(z) \tag{11-9}$$

where

$U(z) =$ unit step function
$\Gamma(L/2) =$ gamma function

By making the substitutions $L = 2N$ and $\lambda = 1/(2\sigma^2)$, one obtains the following results:

$$p(z) = \frac{z^{N-1}e^{-\lambda z}}{2^N \sigma^{2N} \Gamma(N)} U(z) \tag{11-10}$$

and

$$p(z) = \frac{\lambda(\lambda z)^{N-1}e^{-\lambda z}}{\Gamma(N)} U(z) \tag{11-11}$$

Now, with*

$$V = \lambda z = \sum_{i=0}^{N-1} \frac{x_i^2}{2\sigma^2}$$

so that

$$dV = \lambda\, dz \tag{11-12}$$

the form

$$p(V) = \frac{V^{N-1}e^{-V}}{(N-1)!} U(V)$$

results. This is the special case of the gamma distribution with $\beta = 1.0$.[6]

---

*This is equivalent to normalizing the noise power by its expected value. Similarly, by following the convention that the target power is the single-pulse signal-to-noise ratio, the entire predictive process is properly normalized.

As developed by Parzen,[7]

$$P_{fa} = \int_{Y_B}^{\infty} p(V) \, dV = g(N, Y_B) = \frac{\Gamma(N, Y_B)}{\Gamma(N)} \tag{11-13}$$

or

$$P_{fa} = e^{-Y_B} \sum_{m=0}^{N-1} \frac{Y_B^m}{m!} \tag{11-14}$$

where

$\Gamma(N, Y_B) =$ incomplete gamma function
$Y_B =$ normalized threshold value as found in the Pachares tables[8]

Note that the normalized threshold value, Y_B, is related to the threshold value, V_T, by

$$V_T = 2\sigma^2 Y_B \tag{11-15}$$

The expression for P_{fa} in Eq. (11-14) is very useful, since it relates the normalized detection threshold, Y_B, to the probability of a false alarm. Numerical techniques can use this expression to determine an unerrored detection threshold that will produce a given probability of a false alarm. Generation of the unerrored threshold completes step 2 of the serial process outlined above.

When applied to a system with Rayleigh envelope distributed noise and noncoherent integration of N independent samples, estimation of the mean of the integrated voltage, $E[z]$, is equivalent to an estimation of $2\sigma^2$. That is,

$$E[z] = N2\sigma^2 = \mu_z \tag{11-16}$$

and

$$\hat{\mu}_z = N2\hat{\sigma}^2 \tag{11-17}$$

Using Eq. (11-15), one can relate Eq. (11-17) to the threshold estimate as follows:

$$\hat{V}_T = 2\hat{\sigma}^2 Y_B \tag{11-18}$$

$$\hat{V}_T = \frac{Y_B}{N} \hat{\mu}_z \tag{11-19}$$

$$\hat{V}_T = k_T \hat{\mu}_z \tag{11-20}$$

as representative of a threshold that is based upon the level of the ambient interference. Also, since $\hat{\sigma}$ is only an estimate, the expected value of $\hat{P}_{fa}$ must be forced to the desired result before final selection of the threshold-defining constants (Y_B or k_T).

The Nonfluctuating Target

Marcum[1] has shown that for a nonfluctuating target the conditional voltage density function at video (after square law detection) for the sum of N pulses composed of signal-plus-noise is

$$p(V/x) = \left(\frac{V}{Nx}\right)^{(N-1)/2} e^{-(V+Nx)} I_{N-1}(2(NxV)^{1/2}) U(V) \qquad (11\text{-}21)$$

where

N = number of samples summed together
I_{N-1} = modified Bessel function of first kind with order $(N-1)$
x = signal-to-noise ratio (measured at *IF* prior to square law detection)

As discussed in Chapter 9, the expected probability of detection, P_d, can be obtained by entering the density function of Eq. (11-21) into Eq. (11-7) and performing the operations. That is,

$$P_d = \int_0^\infty p(\hat{V}_T) \int_{\hat{V}_T}^\infty p(V/x)\, dV\, d\hat{V}_T \qquad (11\text{-}22)$$

The estimated threshold, $\hat{V}_T$, is determined using the methods of the section "Detection Threshold Estimation." Mitchell and Walker[5] demonstrate that P_d can be calculated by the summation

$$P_d = \sum_{b=0}^\infty \frac{(Nx)^b}{b!} e^{-Nx} \sum_{m=0}^{N+b-1} \frac{Y_B^m}{m!} e^{-Y_B} \qquad (11\text{-}23)$$

when the signal level has been normalized to force $2\sigma^2 = 1.0$ and the variance of the threshold estimates (V_T and Y_B) equals zero. This is equivalent to stating that the ambient interference level is known exactly. The form of Eq. (11-23) is particularly useful, since it lends itself to recursive methods for machine calculation of detection performance. The references by Mitchell and Walker[5] and Shnidman[9] are particularly useful in defining these methods (and procedures of error testing to permit truncation of the infinite summation).

Sample detection results for situations in which the variance of the threshold estimate equals zero are given in Figure 11-4. A 90% single-scan P_d (for P_{fa}

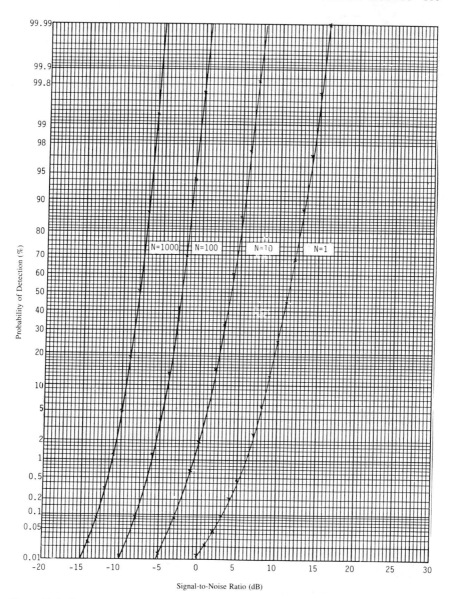

Figure 11-4. Detection probability as a function of signal-to-noise ratio and number of samples noncoherently integrated (nonfluctuating target and $P_{fa} = 10^{-6}$).

$= 10^{-6}$) is obtained for signal-to-interference ratios of approximately 13.2, 5.5, -1.3, and -7.0 dB when 1, 10, 100, and 1000 samples, respectively, are noncoherently integrated. Similar results and families of curves can be generated for different false alarm probabilities and number of samples noncoherently integrated.

When the theoretical threshold is not known exactly, a greater signal-to-interference ratio is required to produce the desired P_d for the allowed P_{fa}. This increase in the required signal-to-interference ratio is by definition the *constant false alarm rate (CFAR) loss*. That subject is discussed later in this chapter.

11.3.2 SWERLING CASES 1 AND 2

Swerling case 1 and 2 models apply to fluctuating targets whose RCS statistics can be modeled as single duo-degree of freedom chi-square functions. Returns from complex targets such as large aircraft, rain clutter, and terrain clutter are generally considered to meet this criterion. The distinction between case 1 and case 2 is based on the correlation interval, as discussed above. Calculation of the detection performance for a case 1 target can be initiated by rewriting the density function of the RCS to represent the single-sample signal-to-noise ratio (x). The density function of the signal-to-noise ratio is then

$$p(x) = \frac{1}{\bar{x}} e^{-x/\bar{x}} U(x) \qquad (11\text{-}24)$$

where $\bar{x}$ is the mean signal-to-noise ratio. This is a gamma distribution with a single degree of freedom. The unconditional normalized density function of the video voltage is then given as

$$P_1(\hat{V}, \bar{x}) = \int_0^\infty p(V/x)_{N=N} \, p(x) \, dx \qquad (11\text{-}25)$$

where

$p(V/x)_{N=N}$ is given by Eq. (11-21)
$\quad\quad p(x)$ is given by Eq. (11-24)

The probability of detection may be found by entering Eq. (11-25) into Eq. (11-4) or (11-7). Shnidman[9] gives the results

$$P_d = e^{-Y/(x+1)} \qquad \text{for } N = 1$$

and

$$P_{\mathrm{d}} = P(N - 1, Y) + \left(\frac{z + 1}{z}\right)^{N-1} e^{-Y(z+1)}$$

$$\cdot \left[1 - P\left(N - 1, \frac{Yz}{z + 1}\right)\right] \qquad \text{for } N \geq 2 \quad (11\text{-}26)$$

where

$$P(N, Y) = \sum_{m=0}^{N-1} \frac{Y^m}{m!} e^{-Y}$$

$z = N\bar{x}$

$\bar{x}$ = average signal-to-noise ratio on each sample

Y = the unerrored detection threshold (Y_{B} of the section "Detection Threshold Estimation")

Sample detection results for the case 1 target model when the desired threshold value is known precisely are given in Figure 11-5. A 90% single-scan P_{d} (for $P_{\mathrm{fa}} = 10^{-6}$) is obtained for signal-to-interference ratios of approximately 21.2, 13.8, 7.2, and 1.7 dB when 1, 10, 100, and 1000 samples, respectively, are integrated noncoherently. The corresponding signal-to-interference ratios necessary to achieve this detection performance are approximately 8.0, 8.3, 8.5, and 8.7 dB greater than for the nonfluctuating target (see the section "The Nonfluctuating Target").

For the case 2 (fast fluctuation) model, the unconditional density function for the signal-plus-interference video voltage to be inserted into Eq. (11-4) for P_{d} calculations is the N-fold self-convolution of

$$p_{2i}(V, x) = \int_0^{\infty} p(V/x)_{N=1}\, p(x)\, dx \qquad (11\text{-}27)$$

where

$p(V/x)_{N=1}$ is given by Eq. (11-21)

$p(x)$ is given by Eq. (11-24)

Performance of this N-fold convolution produces the unconditional density function

$$p_2(V, \bar{x}) = \frac{1}{\Gamma(N)} \frac{V^{N-1}}{(1 + \bar{x})^N} e^{-V/(1 + \bar{x})}\, U(V) \qquad (11\text{-}28)$$

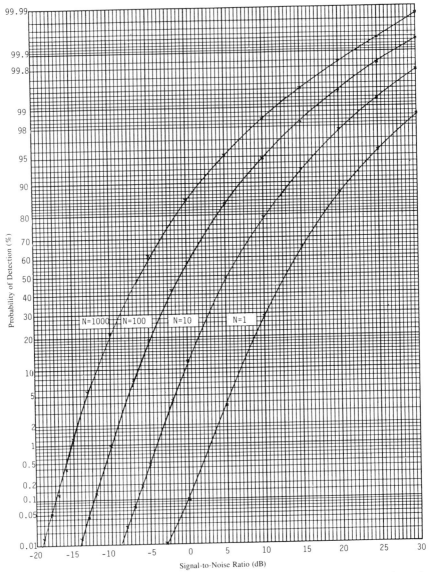

Figure 11-5. Detection probability as a function of signal-to-noise ratio and number of samples noncoherently integrated (Swerling case 1 and $P_{fa} = 10^{-6}$).

P_d may then be calculated by entering Eq. (11-28) into Eq. (11-4) or (11-7). Johnson[10] gives this result as

$$\hat{P}_d = \sum_{m=0}^{N-1} \frac{Y^m}{m!(1+\bar{x})^m} e^{-Y/(1+\bar{x})} \tag{11-29}$$

when the variance of the threshold estimate (V_T or Y) equals zero.

Sample detection results for the case 2 target model in situations where the desired threshold value is precisely known (zero variance) are given in Figure 11-6. A 90% single-scan P_d (for $P_{fa} = 10^{-6}$) is obtained for signal-to-interference ratios of approximately 21.2, 6.3, -1.2, and -6.8 dB when 1, 10, 100, and 1000 samples, respectively, are integrated noncoherently. Note that this target fluctuation model requires (for the same P_d and P_{fa}) signal-to-interference ratios approximately 8.0, 0.8, 0.4, and 0.2 dB, respectively, higher than for the nonfluctuating target for the same number of noncoherently integrated samples. Except for $N = 1$, the case 2 model produces a far smaller performance impact than does the case 1 model. It is easy to understand the value of frequency agility in forcing a case 1 target to fluctuate (to approximate the case 2 situation) during the noncoherent integration period.

11.3.3 SWERLING CASES 3 AND 4

Swerling case 3 and 4 models apply to targets whose RCS statistics can be fitted to a two duo-degree of freedom chi-square density function. Targets such as rockets, missiles, and space-based satellites are generally considered to have such scattering characteristics. Calculation of the detection performance of a case 3 target can be initiated by rewriting the density function of the RCS in a form that represents the single-sample signal-to-noise ratio (x). The density function for the signal-to-noise ratio is then given by

$$p(x) = \frac{4x}{\bar{x}^2} e^{-2x/\bar{x}} U(x). \tag{11-30}$$

Detection performance for cases 3 and 4 can be obtained by following the procedure laid out for cases 1 and 2. For case 3 when the threshold estimate has a zero variance, Johnson[10] gives the results

$$\hat{P}_d = e^{-2Y/(z+2)} \left[1 + \frac{2Yz}{(z+2)^2} \right] \quad \text{for } N = 1$$

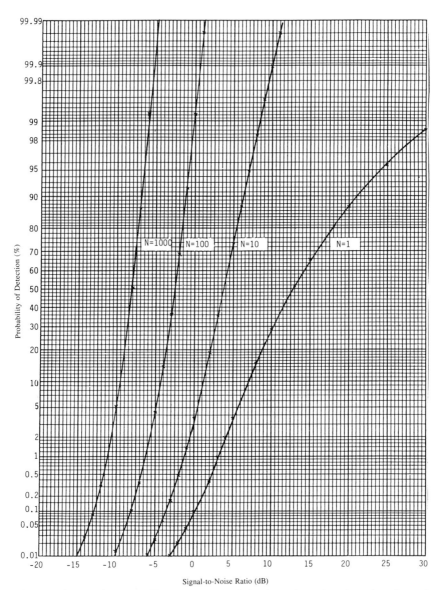

Figure 11-6. Detection probability as a function of signal-to-noise ratio and number of samples noncoherently integrated (Swerling case 2 and $P_{fa} = 10^{-6}$).

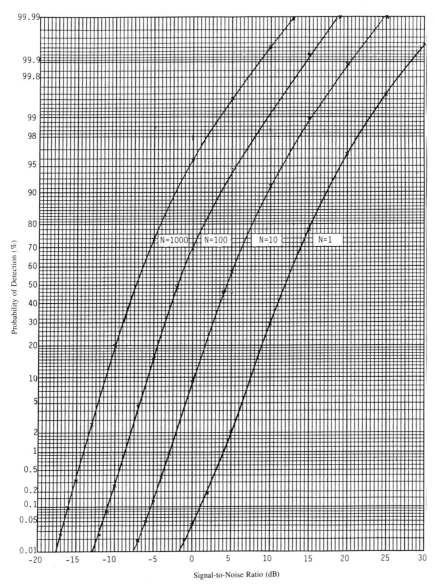

Figure 11-7. Detection probability as a function of signal-to-noise ratio and number of samples noncoherently integrated (Swerling case 3 and $P_{fa} = 10^{-6}$).

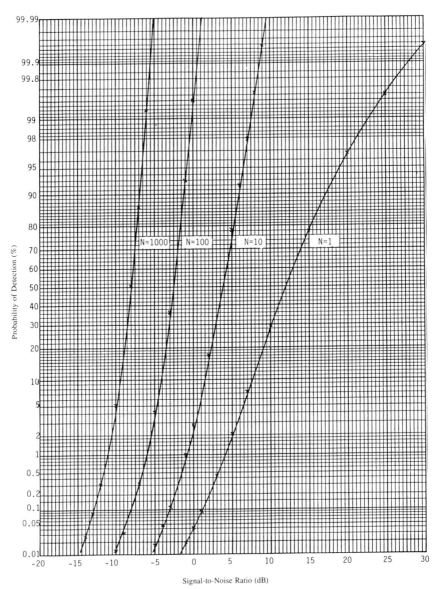

Figure 11-8. Detection probability as a function of signal-to-noise ratio and number of samples noncoherently integrated (Swerling case 4 and $P_{fa} = 10^{-6}$).

and

$$\hat{P}_{\mathrm{d}} = \left[\frac{Y^{N-2}\,e^{-Y}}{(N-2)!}\right]\left[\frac{2Y}{z+2}\right] + P(N-1,\,Y)$$
$$+ \left[\frac{z+2}{z}\right]^{N-2} [e^{-2Y/(z+2)}]\left[1 - \frac{2(N-2)}{z} + \frac{2Y}{z+2}\right]$$
$$\cdot \left[1 - P\!\left(N-1,\,\frac{Yz}{z+2}\right)\right] \qquad \text{for } N \ge 2 \quad (11\text{-}31)$$

where z and $P(N,\,Y)$ are as defined for Eq. (11-26).

For a case 4 fluctuation, Johnson gives the results

$$\hat{P}_{\mathrm{d}} = \sum_{m=0}^{N-1} \frac{V^m}{m!}\,e^{-V} + \left[\sum_{m=N}^{2N-1} \frac{V^m}{m!}\,e^{-V}\right]$$
$$\cdot \left[1 - \sum_{k=0}^{m-N} \frac{N!}{k!(N-k)!}\left(\frac{\bar{x}}{\bar{x}+2}\right)^k \left(\frac{2}{\bar{x}+2}\right)^{N-k}\right] \quad (11\text{-}32)$$

where $V = 2Y/(\bar{x}+2)$. Case 3 and case 4 results are the same for $N = 1$.

Sample detection results for cases 3 and 4 are given in Figures 11-7 and 11-8. Signal-to-noise ratios that are required to yield a 90% P_{d} and 10^{-6} P_{fa} are given in Table 11-1. As represented in this table, the performance impact of this fluctuation statistic is not as severe as for the case 1 and 2 statistics. As was true for cases 1 and 2, the use of frequency agility to force the target to fluctuate during the noncoherent integration period produces desirable detection performance enhancements.

Table 11-1. Signal-to-noise ratio (SNR) required for detection with case 3 and case 4 fluctuation models.

Swerling Case	Number of Samples Noncoherently Integrated	Approximate SNR Required for Detection ($P_{\mathrm{det}} = 0.9$, $P_{\mathrm{fa}} = 10^{-6}$) (dB)	Approximate SNR Loss Relative to Nonfluctuating Target (dB)
3	1	17.1	3.9
	10	9.6	4.1
	100	3.3	4.6
	1000	−2.2	4.8
4	1	17.1	3.9
	10	6.0	0.5
	100	−1.2	0.1
	1000	−7.0	0.0

11.3.4   OTHER MODELS

Comparison of the cumulative RCS distributions of sample chi-square functions and sample target observations leads one to question the validity of using only the four statistical models that have been discussed. In practice, other members of the chi-square family and other probability density functions are used. In addition to other numbers of duo-degree of freedom chi-square models, probability density functions such as the Weibull and the lognormal are commonly used to represent targets that tend to contain aspect angle–dependent specular scatterers. In addition, these two models have been used to represent the statistics of ground clutter when viewed at extremely low grazing angles or with high-resolution sensors.

Detection predictions for these alternative models can be obtained by inserting the appropriate probability density function for the single-pulse signal-to-noise ratio, x, into the process outlined above for cases 1 to 4. The probability density function for the general chi-square model is given as[5]

$$p(x) = \frac{1}{\Gamma(k)} \left(\frac{k}{\bar{x}} \right)^k x^{k-1} e^{-kx/\bar{x}} U(x) \tag{11-33}$$

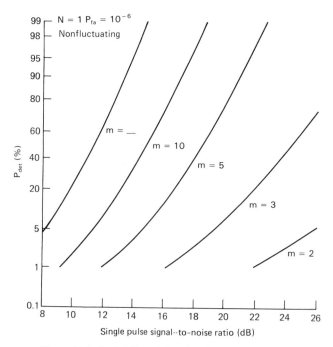

Figure 11-9. Probability of detection for the multicell test.

where

x = the single-pulse signal-to-noise ratio
k = the number of duo-degrees of freedom in the chi-square model
$\bar{x}$ = the mean single-sample signal-to-noise ratio

The Weibull probability density function for the single-sample signal-to-noise ratio is given as[6]

$$p_w(x) = \lambda \beta x^{\beta - 1}\, e^{-\lambda x^\beta}\, U(x) \tag{11-34}$$

where λ and β, the shape and magnitude parameters, are greater than zero. The lognormal probability density function for the single-sample signal-to-noise ratio is given as

$$p_{\ln}(x) = \frac{1}{\sqrt{2\pi}\,\sigma_s x} \exp\left[-\left(\frac{(\ln x - \ln x_m)^2}{2\sigma_s^2}\right)\right] \tag{11-35}$$

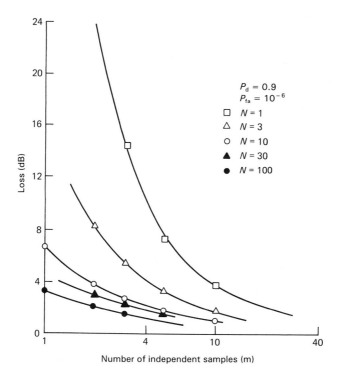

Figure 11-10. CFAR loss as a function of the number of cells (m) in a multicell CFAR test for integration of N samples into each cell.

where σ_s is the standard deviation of $\ln(x)$ and x is the median RCS. The references by Mitchel and Walker,[5] Shnidman[9], and Johnson[10] give useful recursive form equations that can be used to determine the P_d for the general chi-square model case. Efficient machine calculations for other fluctuation models will require developments that are similar to those presented in these references.

11.3.5 CAUTIONS AND COMMENTS ON TARGET MODELS

Plots of detection probability as a function of signal-to-noise ratio, as presented in Figures 11-4 through 11-8 when an exact value of the threshold ($\sigma_{\hat{V}_T} = 0$) can be generated, are available in the literature.[3,11] The procedure outlined above describes how an errored threshold estimate modifies the detection process and how the estimation procedure should be modified. These effects are examined in references 9, 12, 13, 14, and 15. Figure 11-9 demonstrates the effect of the number of cells used in a multicell threshold (CFAR) estimator on the expected detection performance.[5] It is assumed that the threshold is computed as K_T times

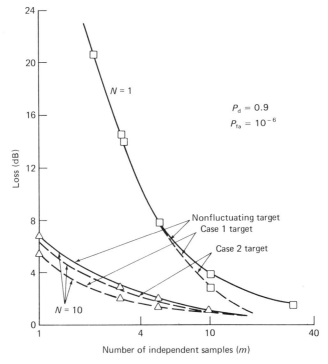

Figure 11-11. CFAR loss as a function of number of cells (m) in a multicell CFAR test after integration of N independent samples into each cell (for exponentially distributed noise power).

the mean power level estimate, as in Eq. (11-2) of Section 11.3. The change in signal-to-noise ratio as the number of cells decrease from infinity (an unerrored estimate) at a fixed detection percentage is the CFAR loss. Figures 11-10 and 11-11 demonstrate the variability of the effective detection loss with characteristics of the radar, the environment, and the target. No single number adequately describes this loss for all conditions.

The interference has been assumed to this point of discussion to have Rayleigh envelope statistics and to be independent sample-to-sample. In many cases, both of these assumptions are invalid. That is, the background clutter may be pulse-to-pulse correlated and may have other distributions (for example, Wei-

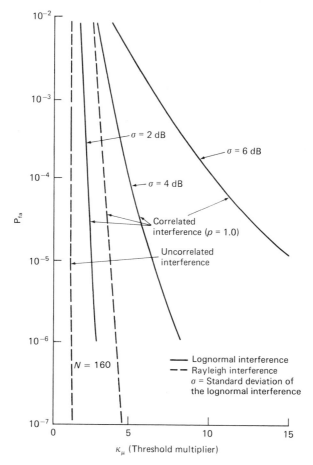

Figure 11-12. P_{fa} versus threshold multiplier K_μ where
$V_T = k_\mu \mu$
μ = mean of the voltage at the threshold device.

bull or lognormal). Also, as is true in many applications, these cases come mixed without an a priori knowledge of the mix.

An example of the magnitude of this difficulty may be presented with the aid of Figure 11-12. This figure shows that a system based upon a mean level threshold ($V_T = K_\mu\mu$) and the assumption of 160 independent samples (integrated) would have major P_{fa} problems with correlated clutter, either Rayleigh or lognormal.

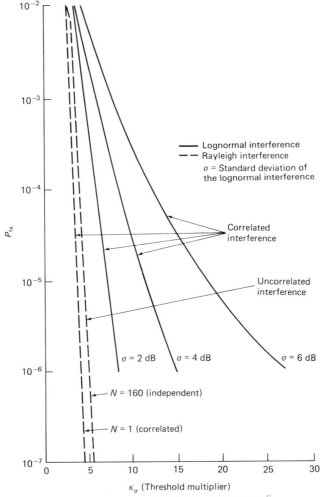

Figure 11-13. P_{fa} versus threshold multiplier K_σ where

$V_T = K_\sigma\sigma + \mu$.

σ = standard deviation of the voltage at the threshold device

μ = mean of the voltage at the threshold device

Similarly, a system designed for mean level estimation and correlated Rayleigh clutter would experience P_{fa} problems with the lognormal interference and sensitivity degradation when encountering thermal noise limits (uncorrelated interferences). Normally, neither of these situations is acceptable.

In an effort to "patch" the processor in order to survive these difficulties, some processors use a threshold value based upon estimates of both the mean and the standard deviation (a two-parameter method). Figure 11-13 illustrates P_{fa} results achievable for this case. The result produced is much better than that for the mean level estimator (single-parameter method), except that variations in the distribution of the interference continue to create large changes in P_{fa}. Since this processing form does not maintain a constant false alarm probability, it is called a *parametric* or *distribution-dependent process*. The definition of methods to overcome these problems effectively is the challenge that must be met if sensors with acceptable stationary target detection performance are to be developed.

11.4 REFERENCES

1. J. I. Marcum, "A Statistical Theory of Target Detection by Pulsed Radar: Mathematical Appendix," Research Memorandum RM-753, Rand Corp., July 1, 1948, republished in *IRE Transactions on Information Theory*, vol. IT-6, no. 2, April 1960, pp. 59–267.
2. Peter Swerling, "Detection of Fluctuating Pulsed Signals in the Presence of Noise," *IRE Transactions on Information Theory*, vol. IT-3, no. 3, September 1957, pp. 175–178.
3. M. I. Skolnik, *Radar Handbook*, McGraw-Hill Book Co., New York, 1970.
4. D. P. Meyer and H. A. Mayer, *Radar Target Detection Handbook of Theory and Practice*, Academic Press, New York, 1973.
5. R. L. Mitchell and J. F. Walker, "Recursive Methods for Computing Detection Probabilities," *IEEE Transactions on Aerospace and Electronic Systems*, vol. AES-7, no. 4, July 1971, pp. 671–676.
6. R. E. Walpole and R. H. Myers, *Probability and Statistics for Engineers and Scientists*, Macmillan Publishing Co., New York, 1972, p. 124.
7. E. Parzen, *Modern Probability Theory and Its Applications*, John Wiley & Sons, New York, 1960.
8. J. Pachares, "A Table of Bias Levels Useful in Radar Detection Problems," *IRE Transactions on Information Theory*, vol. IT-4, no. 1, March 1958, pp. 38–45.
9. D. A. Shnidman, "Evaluation of Probability of Detection for Several Target Fluctuation Models," Lincoln Laboratory Technical Note 1975-35, ADA013733, Lincoln Laboratory, Lexington, Mass., July 1975.
10. M. A. Johnson, "Radar Detection Calculations with the HP-65 and HP-67," Calculator Program Descriptions, IEEE International Radar Conference, April 1980.
11. M. I. Skolnik, *Introduction to Radar Systems*, McGraw-Hill Book Co., New York, 1962.
12. H. Rohling, "Radar CFAR Thresholding in Clutter and Multiple Target Situations," *Aerospace and Electronic Systems*, vol. 19, no. 4, July 1983, pp. 608–621.
13. F. E. Nathanson, *Radar Design Principles*, McGraw-Hill Book Co., New York, 1969.
14. J. S. Bird, "Calculating Detection Probabilities for Adaptive Thresholds," *Aerospace and Electronic Systems*, vol. 19, no. 4, July 1983, pp. 506–512.
15. G. V. Trunk, "Survey of Radar ADT," *Microwave Journal*, vol. 26, no. 7, July 1983, pp. 77–88.

12
ADAPTIVE THRESHOLDING AND AUTOMATIC DETECTION

Carl R. Barrett, Jr.

12.1 INTRODUCTION

The requirement for radar systems that can detect the presence of targets within background environments that are more complex and less known than thermal noise and maintain a controlled false alarm rate leads to increased emphasis on adaptive threshold automatic detection circuits. Systems that use automatic detection circuits to maintain, ideally, a constant false alarm rate by generating estimates of the receiver output when targets (of interest) are not present are called *constant false alarm rate (CFAR) systems*.

As described by Nitzberg,[1] automatic target detection for a search radar can be achieved by comparing the processed voltage in each range cell to:

1. A fixed threshold level.
2. Threshold levels based upon the mean amplitude of the ambient interference.
3. A level computed on the basis of partial (a priori) knowledge of the interference distribution.
4. Threshold levels determined by distribution-free statistical hypothesis testing that assumes no a priori knowledge of the statistical distribution of the interference.

In the first case, a detection decision is made if the processed signal, r_0, is equal to or greater than a preset threshold. That is, if

$$r_0 \geq T_p \tag{12-1}$$

a detection is declared. This form of automatic detector with a nonadaptive threshold produces large changes in false alarm rates, perhaps within different regions of the same search sector, when the interference (level, statistic, etc.) varies.

The second and third cases represent what are typically called *adaptive threshold CFAR* processors.[2] In these processors, estimates of the unknown parameters of the known distribution of the processed interference are formed. The second case can achieve CFAR when the distribution of the processed interference is completely described by its mean level. In this case, the detection threshold T_0 is adequately represented by

$$T_0 = K_0 \, \sigma_0^2 \qquad (12\text{-}2)$$

where K_0 is a constant defined by the desired probability of false alarm and the statistics of the processed interference, and σ_0^2 is the mean (power) level of the processed interference. A detection is declared when the processed signal r_0 is equal to or greater than T_0. That is, when

$$r_0 \geq T_0 \qquad (12\text{-}3)$$

The third case is similar to the second, except that knowledge of the mean (power) of the processed interference does not completely describe its statistics. The adaptive threshold CFAR processor in this third case forms estimates of the unknown parameters of the known (a priori) distribution. For example, it is found that the mean to standard deviation ratio of the processed interference is a function of the pulse-to-pulse correlation properties of the input signal. Changes in the correlation properties cause the mean to standard deviation ratio to change and hence dictate a need for an adjustment in the threshold constant [K_0 of Eq. (12-2)]. This third case can achieve CFAR when the form of the distribution of the processed interference is known and its unknown parameters can be adequately estimated.

The fourth case represents what are called *nonparametric CFAR* processors.[2] These distribution-free processes form a test variable whose statistics (in the no-signal case) are independent of the distribution of the input (nonprocessed) interference. (*Distribution free* implies that the false alarm probability is maintained at a constant value and is independent of the statistical distribution of the interference.) This class of detectors finds applications, for example, in scanning (search and track-while-scan) radars where the level and distribution of the background interference are unknown and are probably nonstationary over anything greater than a modestly sized search area. These conditions are generally found to exist, for example, when detection performance is background clutter (not thermal noise) limited. The data of Chapter 10 illustrate the fallacy associated with an a priori assumption of a particular describing distribution and observed amplitude when detection performance is background clutter limited.

This chapter presents an introduction to adaptive threshold CFAR processors and distribution-free (sometimes called *nonparametric*) CFAR processors.[3] It is intended to introduce the reader to this important subject. This introduction should provide insight into some of the problems that can restrict the usefulness of processors that are designed without full consideration (and knowledge) of key background (interference) characteristics.

12.2 ADAPTIVE THRESHOLD CFAR PROCESSORS

As stated above, the adaptive threshold CFAR processor[2] is applicable to situations where the distribution of the processed data (in the no-signal case) is known generally and unknown parameters associated with the distribution can be estimated. The adaptive threshold CFAR processor is a realization of the multicell test of Chapter 11 (Section 11.3.5). In practice, it is often implemented as a moving or sliding window through which estimates of the unknown parameters of the interference (without desired target response contamination, it is hoped) are formed.

An example of an adaptive threshold CFAR processor for noncoherent detection is given in Figure 12-1. If the optional processing blocks are ignored, this diagram represents a mean level detector of a type similar to that described by Finn and Johnson.[4] The tapped delay line forms the sliding window through which an estimate of the mean background in a region that leads and lags (in range) the cell under test can be formed. (It is assumed that measurement of the mean background is sufficient to describe the unknown parameters of its distribution such that CFAR can be realized.) As represented in Figure 12-1, the estimated mean level in the region surrounding the jth range cell (cell under evaluation), $\hat{\mu}_j$, is formed as

$$\hat{\mu}_j = \frac{1}{N} \left[\sum_{n=j-(L+N/2)}^{j-(L+1)} v_n + \sum_{n=j+(L+1)}^{j+(L+N/2)} v_n \right] \qquad (12\text{-}4)$$

where

N = number of range cells in the leading and lagging range regions of the sliding window

L = number of cells (on each side of the cell being evaluated) that are to be removed from the sliding window to prevent contamination of $\hat{\mu}_j$ by the signal in the region surrounding the jth cell

v_n = video amplitude in the nth range cell into the detection process

The detection threshold for the jth range cell $T_{\mu/j}$ is then formed [in the manner

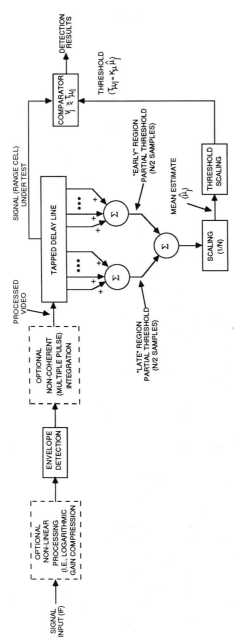

Figure 12-1. Mean-level adaptive threshold CFAR processor.

of Eq. 12-1] as

$$T_{\mu/j} = K_\mu \,\hat{\mu}_j \qquad (12\text{-}5)$$

where K_μ = desired scaling constant (to achieve a desired expected probability of a false alarm). A detection decision is reached when the signal in the jth range cell (the cell under test) exceeds this computed level.

Incorporation of the optional processing elements allows the mean-level adaptive threshold CFAR processor to be adapted to alternative specialized requirements. For example, use of the nonlinear processing block in the form suggested in Figure 12-1 converts this processor to a nonlinear receiver technique[2] that displays CFAR properties when the background (interference) is best represented by certain two-parameter distribution functions (such as log-normal and Weibull).[5] Incorporation of the optional noncoherent integration function modifies the processor of Figure 12-1 to represent the adaptive threshold CFAR process when multiple-pulse noncoherent integration is performed.

The data in Figure 11-12 illustrate the problems encountered in a mean-level adaptive threshold CFAR processor when the parameters of the distribution of the background (interference) are not those that were assumed for the specification of the threshold multiplier K_μ (see Figure 12-1). For example, if noncoherent integration of 160 independent envelope detected Rayleigh distributed noise samples is performed, a threshold multiplier (K_μ) of approximately 1.1 will produce a probability of false alarm (P_{fa}) of 10^{-6}. However, if these 160 samples were even partially correlated (as may be encountered when the interference originates, for example, from backscattered clutter returns), the P_{fa} will be much greater than 10^{-6}. In the limit, when all 160 samples are totally correlated, Figure 11-12 demonstrates that the P_{fa} increases by many orders of magnitude.

Factors such as these lead to the desirability of class 3 processors, as discussed above and defined by Nitzberg.[1] In the situation just described, the particular unknown is the pulse-to-pulse correlation characteristics of the return data (no signal case). A reasonable approach to solving this problem is to form an estimate of this unknown quantity and use the estimate to form a threshold value. The data in Figure 11-13 demonstrate this result when the threshold for the jth range cell $T_{\sigma/j}$ is formed based upon the assumption that the mean and variance of the processed return data (no signal case) are sufficient measures of the unknown characteristics. The threshold for the calculations leading to the data of Figure 11-13 is formed according to the equation

$$T_{\sigma/j} = \mu_j + K_\sigma \sigma_j \qquad (12\text{-}6)$$

where

μ_j = the mean level of the processed interference (at the threshold device)

σ_j = the standard deviation of the processed interference (at the threshold device)

K_σ = the threshold multiplier to produce the desired P_{fa}

It is seen that the change in the P_{fa} is much less disastrous than that observed for the class 2 processor. (K_σ was held constant for these calculations and, hence, was not influenced by μ/σ. This ratio could be used to vary K_σ and therefore to effectively achieve CFAR characteristics.) Figures 11-12 and 11-13 demonstrated the difficulty the adaptive threshold CFAR processor experiences while trying to produce CFAR characteristics when the distribution of the interference is not known a priori. Both figures demonstrate that adaptive threshold CFAR processors produce false alarm probabilities that are related to the distribution of the processed interference. These characteristics lead to the desirability of nonparametric (distribution-free) CFAR processors.

12.3 DISTRIBUTION-FREE (NONPARAMETRIC) CFAR PROCESSORS

Distribution-free (nonparametric)[3] CFAR processors are designed to provide CFAR characteristics when the background return (i.e., noise and clutter) has an unknown (and potentially varying) distribution. The purpose of this class of processors is to remain insensitive (rather than to adapt) to variations in the distribution. Therefore, no attempt is made to design for optimum performance. However, even though nonparametric processors generally experience an additional detection loss (relative to adaptive threshold CFAR processors), their CFAR properties make their application advantageous. This section presents a summary of selected detection techniques that are proposed and described in the literature.[4,6-8] Major approaches to be described include the following:

1. Double threshold detector
2. Modified double threshold detector
3. Rank order detector
4. Rank-sum double quantizer detector

(The double threshold detector is not a classic nonparametric processor. It does, however, have properties that justify its inclusion in this discussion.) Operation of this class of detectors will be examined through the use of examples. Note that there appears to be no general procedure, such as those based upon the Neyman-Pearson criteria, that allows the formation of an optimum distribution-free (nonparametric) processor. The process instead involves selection from among the many nonparametric procedures available in the general statistical literature and adaptation to the radar processor problem. The radar designer must determine an approach that adequately satisfies design objectives.

12.3.1 Double Threshold Detector Example

One of the interesting automatic (but nonadaptive) detection concepts is the double threshold method described by Swerling.[6] This concept is identified in the literature by several names, including the *M out of N (M/N) coincidence detector* and the *binary detector*. Among the several advantages that prompt its use are the following:

1. Its detection performance for a large number of samples (N) comes very close to that of an ideal N-pulse noncoherent integrator.
2. It is relatively simple to implement.
3. As a CFAR circuit (when the first threshold is set to yield the desired probability of threshold crossing when only interference is present), it is essentially distribution free.

In the manner described by Swerling[6] and as represented in Figure 12-2, the double threshold detector provides CFAR characteristics only when the input video level (no-signal presence) is fixed to provide a constant probability of exceeding the input threshold (V_T of Figure 12-2). As shown later, this processor can be modified to provide adaptive CFAR properties. Performance of the processor can be calculated as described in the following paragraphs. The notation used corresponds to that of Chapter 11.

When the input data come from noise alone, $\bar{x} = 0$, the input signal [that is described by the density function $p(V, \bar{x} = 0)$] is compared with the threshold voltage, V_T, to form the quantized video. The quantized video has two states—

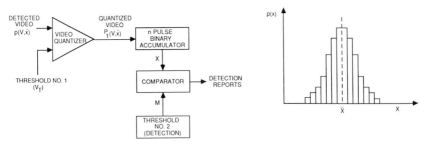

Figure 12-2. Double threshold detector.

success or failure. Success is defined when the input video is equal to or greater than V_T. Conversely, failure is defined when the input video is less than V_T. This repeated process is called a *binomial experiment*. The binary accumulator determines the number of successes X over a series of n trials. The number of successes is compared with a second threshold M to determine the outcome of the detection test. A detection is declared if $X \geq M$.

Since X represents the number of successes out of n repeated trials of a binomial experiment, the binomial variable X can be represented by the binomial distribution[9]

$$b(X; n, P_1(V, \bar{x})) = \binom{n}{X} P_1(V, \bar{x})^X (1 - P_1(V, \bar{x})^{n-X} \qquad (12\text{-}7)$$

By writing $P_1(V, \bar{x})$ (the probability of success of the binomial experiment) as P_1, this can be rewritten as

$$b(X; n, P_1) = \binom{n}{X} P_1^X (1 - P_1)^{n-X} \qquad (12\text{-}8)$$

where

$$\binom{n}{X} = \frac{n!}{X!(n - X)!}$$

Since a detection is declared when $X \geq M$, the probability of detection, P_d, is given as

$$P_d = \sum_{X=m}^{n} \binom{n}{X} P_1^X (1 - P_1)^{n-X} \qquad (12\text{-}9)$$

and the performance of the circuit can be determined. The probability of success of the binomial experiment, P_1, is given as

$$P_1 = \int_{V_T}^{\infty} p(V, \bar{x}) \, dV \qquad (12\text{-}10)$$

By knowing these quantities and the number of independent samples in the binomial experiment, thresholds may be determined in the no-signal case ($\bar{x} = 0$) to establish a desired false alarm probability P_{fa}. Similarly, given these thresholds and the input video density function when the signal is present ($\bar{x} \neq$

0), the probability of detection may be determined. These relationships are sketched in Figure 12-3.

From Figure 12-3 and this discussion, it is seen that for a known input interference level and distribution:

1. A fixed V_T produces a known probability of success of the binomial experiment.
2. A known density function is established after summing n independent outcomes of the binomial experiment.
3. A value for the second threshold M can be determined to establish a desired P_{fa}.

When a signal is added to the input (signal-plus-noise), the new probability

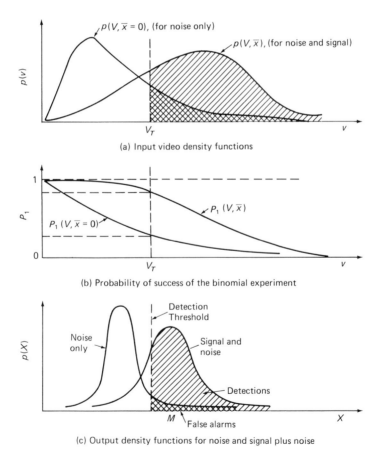

(a) Input video density functions

(b) Probability of success of the binomial experiment

(c) Output density functions for noise and signal plus noise

Figure 12-3. Sample functions through the double threshold detector.

of success of the binomial experiment can be determined, the density function of X can be determined, and the probability of detection can be calculated.

If this process is repeated for different values of the first threshold (input signals fixed), an optimization curve such as the one in Figure 12-4 is obtained. It is seen that there are optimum settings of M (and hence of P_1 and V_T) that maximize the detection performance, and that these settings vary with target (and background) fluctuation characteristics.

The most significant difficulties with the double threshold detector are as follows:

1. A change in the noise level (or distribution) that affects the probability of success of the binomial experiment destroys the ability of the detector to control the false alarm probability.

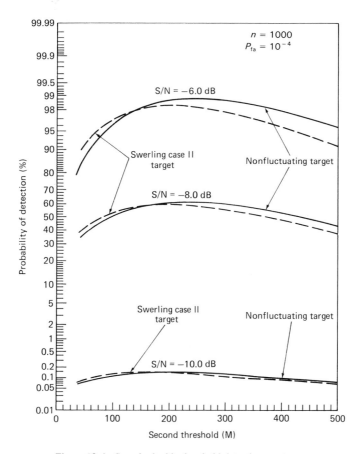

Figure 12-4. Sample double threshold detection contours.

2. The interference is not always constant with range, and therefore, a fixed threshold level applied across the entire instrumented range of a radar system is by the problem stated above inadequate.
3. Since the interference is not always pulse-to-pulse independent, CFAR performance may be severely degraded.

[In the worst case, the interference has a sample-to-sample normalized correlation coefficient of unity. In this case, P_{fa} equals $P_1(V, \bar{x} = 0)$ (approximately 0.35).]

Figure 12-5 compares the detection performance of the double threshold detector to that which can be achieved by an ideal noncoherent integrator. It is seen that for $n = 100$ the relative detection loss ranges from approximately 1.0 to 1.2 dB (and decreases as n becomes larger) when the background has Rayleigh envelope statistics.

12.3.2 Modified Double Threshold Detector Example

In an effort to overcome the limitations of the double threshold detector, it is sometimes used in a modified form. Modifications typically include:

1. A monitor loop to set the first threshold such that a desired success probability of the binomial (noise-only) experiment is established.

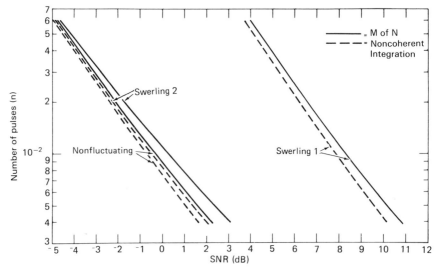

Figure 12-5. Detection performance of the double threshold detector compared to an ideal noncoherent integrator (Rayleigh envelope distributed interference, $P_{fa} = 10^{-6}$ and $P_d = 0.9$). (By permission, from Walker, ref. 10; © 1971 IEEE)

2. An adaptive second threshold to allow for errors in the first threshold and for nonzero correlation coefficients (when the interference is not pulse-to-pulse independent).

An example of a modified double threshold detector is given in Figure 12-6. In this case, the adaptive first threshold regulates the probability of success of the binomial experiment to make it independent of the statistics and level of the input noise. Provided that the level and statistics of the noise (interference) are constant over the extent of the sliding window that surrounds that cell under test, this circuit functions well to accomplish its CFAR objective. The adaptive second threshold compensates (within limits) for errors in the first threshold settings and for single-lag correlation coefficients (of interference) of values other than zero. The regions over which $\hat{\mu}_x$ and $\hat{\sigma}_x$ are estimated will be discussed later in this chapter. (It is assumed that $\hat{\mu}_x$ and $\hat{\sigma}_x$ can be used to represent adequately the unknown parameters in the distribution that describes the test statistic, X.) Since the density function of X is binomially distributed and independent of the input video density function, the processor is (under most situations) distribution free.

The detection efficiency of the double threshold detector exceeds that of the modified double threshold detector and would, therefore, be used if its non-adaptive nature could be tolerated. When adaptive characteristics are required, however, the added detection loss associated with the modified technique must be accepted (since the number of independent samples that form the threshold estimate is not infinite, a finite loss is encountered). This detector, therefore, has a detection loss relative to an ideal integrator whose output can be tested against a precisely known (a priori) threshold that is at least equal to the sum of the losses of the multicell test (Section 11.3.5) and the double threshold detector (Section 12.3.1).

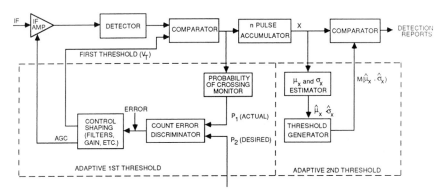

Figure 12-6. Modified double threshold detector.

12.3.3 Rank Order Detector Example

Another example of major importance is the rank order detector.[8,11] This detector is based on a binary detection concept which incorporates a nonparametric statistical decision process to regulate the probability of success of the binomial experiment. This detector, shown in block diagram form in Figure 12-7, uses a sliding window technique to generate efficiently the quantization threshold for all radar ranges. The performance of the rank order detector is not as good (from a signal detection probability point of view) as that of the double threshold detector. With the proper choice of the rank threshold, T_R, and the number of independent cells, N_R, used in the ranking process, however, the performance of the rank order detection technique can be shown to approach that of the modified double threshold detector. Because of its simple mechanization and adaptive properties, the rank order detector has many application advantages.

The rank order detector shown in Figure 12-7 relies on n independent noise samples for CFAR operation. Single-lag correlation coefficients greater than zero require that the second threshold M_R be made adaptive. Also, single-lag correlation coefficients approaching unity violate the premises on which the concept is based (thereby voiding detector CFAR operation).

Reid et al.[11] analyzed this processor for the case in which there were 24 cells in the sliding window (24 local samples are compared as indicated in Figure 12-7, with each test sample), a rank threshold (T_R) of 23 (the test cell amplitude

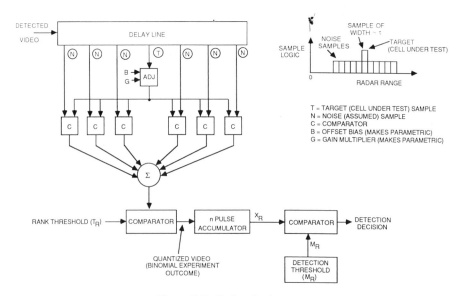

Figure 12-7. Rank order detector.

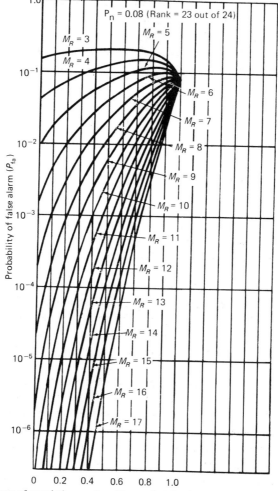

Figure 12-8. Effects of correlation on P_{fa} with a rank order detector. (By permission, from Reid, et al., ref 11; © 1975 IEEE)

must exceed on a per range sweep basis 23 of the 24 local samples) was used, and the results of the ranking process were accumulated in each test cell over 17 pulse repetition intervals. Figure 12-8 illustrates the results of their study. It demonstrates (for the selected rank threshold and 17 sample integrations) the relationship between the second threshold (M_R), the single-lag correlation coefficient (ρ), and the probability of false alarm (P_{fa}). It is seen that a P_{fa} of less than 10^{-6} is achievable with a second threshold (M_R) of 10 when ρ equals zero (a P_{fa} of exactly 10^{-6} can be achieved only at unique values of ρ and M_R). As ρ increases from zero, the second threshold (M_R) must increase to maintain the desired P_{fa}. When ρ increases beyond approximately 0.45 and M_R reaches its maximum value of 17 (defined by the total number of pulse repetition periods in the processing interval), P_{fa} must begin to increase from the desired value of 10^{-6}. At the limit, when ρ equals 1.0, P_{fa} must equal the rank quantizer crossing probability (P_n). P_n is given by[8]

$$P_n = \frac{N - T_R + 1}{N + 1} \tag{12-11}$$

where

$N =$ number of cells in the ranking process
$T_R =$ ranking threshold

This detector can therefore be made to function to its limits by sensing the correlation coefficient.[12] Knowledge of this coefficient can then be used to select the desired second threshold (M_R). Techniques that are useful for estimating this correlation coefficient have been presented by Reid et al.[11] and Lefferts.[13]

Figure 12-8 demonstrates the limits of ρ over which a nearly constant P_{fa} can be achieved. It can be demonstrated (at the expense of making the process non-distribution-free) that the range of ρ over which P_{fa} can be held constant can be extended by introducing, prior to the ranking process, gain and DC level shifts into the test cell video. Figure 12-7 illustrates where these compensations might be inserted.

Performance results for this processor when the rank threshold, T_R, equals the number of cells in the reference window, N, are given in Figures 12-9 and 12-10. Figure 12-9 demonstrates that when the number of pulses integrated n equals 50 use of approximately 10 cells ($N_c = 10$) in the reference window produces near maximum detection performance (for a fixed signal-to-noise ratio). Figure 12-10 demonstrates for this same case ($N_c \approx 10$) the relative performance of an ideal noncoherent integrator, the double threshold detector, and the rank order detector. It is seen that the double threshold detector and the rank sum detector experience (when $n = 50$, $P_d = 0.9$, $P_{fa} = 10^{-6}$) a loss of approximately 1.0 and 1.8 dB, respectively, relative to the ideal noncoherent integrator. (The interference is assumed to have Rayleigh envelope statistics and to have a pulse-to-pulse correlation coefficient of zero.) Because of the distribution-free characteristics of this processor (and the relative simplicity of its implementation), this small added loss is acceptable in many applications.

12.3.4   Rank-Sum Double Quantizer Example

Several techniques have been proposed to reduce the sensitivity of automatic detection (nonparametric) detectors to interference with single-lag (pulse-to-

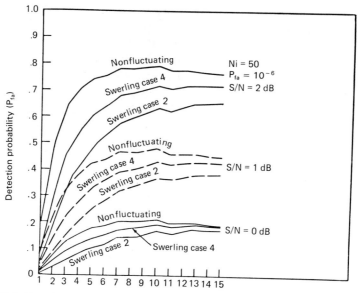

Figure 12-9. Detection performance of the rank order processor as a function of the number of cells in the local reference window. (By permission, from Dillard, et al., ref. 12; © 1970 IEEE)

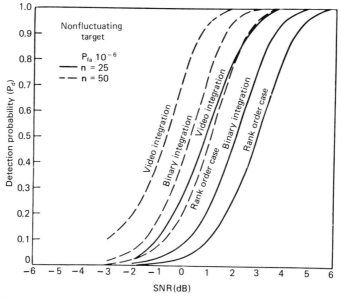

Figure 12-10. Representative detection performance of the rank order processor relative to video integration and the double threshold (binary integration) processor. (By permission, from Dillard, et al., ref. 12; © 1970 IEEE)

pulse or azimuthal) correlation coefficients greater than zero. One such technique is the rank-sum double quantizer, as presented in Figure 12-11 and as described by Barnes et al.[14] As shown in the figure, the radar video is passed into a quantizer that functions like that of the rank quantizer, except that the rank threshold (to develop the binomial decision) is not evoked. This number, which represents the number of times the video exceeds that of the range bins that surround it, is passed into a sliding window integrator. After integrating

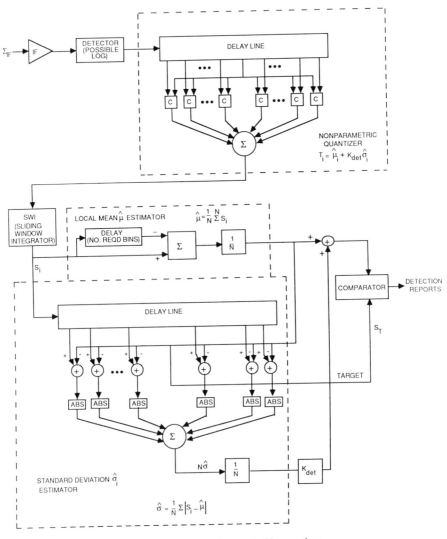

Figure 12-11. Rank-sum double quantizer.

the desired number of sweeps (pulse repetition intervals), the resultant signal in each range bin has a distribution dependent only upon the single-lag (pulse-to-pulse) correlation coefficient (or azimuthal correlation). A threshold and a detection decision can be computed by estimating the mean (in the local mean estimator) and standard deviation (in the standard deviation estimator) in the range region surrounding the cell under test.

12.4 SCANNING RADAR CFAR APPLICATIONS

Techniques that were presented in Chapter 11 for the prediction of false alarm and target detection performance and in the preceding sections of this chapter are generally based on the assumption that the target return can be represented by a uniform amplitude (constant mean power) train of N pulses [see Eq. (11-8)]. In practical terms, since the antenna spatial pattern is not adequately represented by a gain that is equal to the peak (on-axis) gain and constant over its beamwidth (and near zero elsewhere), it is not accurate to assume that a continuously scanning beam produces a mean target return that is constant over the 3-dB beam dwell interval. Some adjustments must, therefore, be made to incorporate the fact that the mean signal-to-noise ratio of the return varies as a function of the line of sight of the target within the beam (and therefore as a function of position within the n pulse integration period). Also, note that the relative phasing of the n pulse integration period to the amplitude versus time history of the target response created by the antenna scan across the target affects the predicted result (and must therefore be considered). It is accepted[2] that the optimum processor for a pulsed, noncoherent waveform of n pulses is a square law detector followed by an n pulse noncoherent integrator that uses equal weighting of each detected pulse [Eq. (11-8)]. However, the same theory that predicts this result[2] suggests that the optimum output signal-to-noise ratio from an integrator for a weighted input pulse train is obtained when the integrator itself uses weights, w_i, that are proportional to the per sample weights of the input waveform. Eq. (11-8) should therefore be modified to the form

$$z = \sum_{i=0}^{n-1} w_i x_i^2 \qquad (12\text{-}12)$$

where the w_i values are defined as above. In a practical situation, the w_i values of Eq. (12-12) may be determined as a result of the implementation used for the noncoherent integrator and, as such, may yield only an approximation to the matched integrator.

In a scanning radar application, the integrator must not only be realizable in a practical sense but also (1) provide as small a detection (matching) loss as possible, (2) provide a means of minimizing losses associated with integration

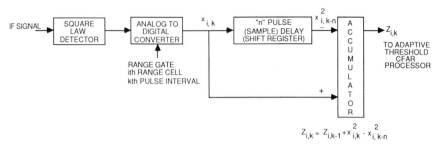

Figure 12-12. Sliding window integrator.

sample window and scanning beam straddle of the target, and (3) in track-while-scan applications, permit accurate measurements of the target angular position. Because of the unknown angular position of the target, there is no a priori knowledge that defines when to initiate gathering of an n pulse sample set that maximizes the output signal-to-noise ratio. Integrators are typically configured as sliding (moving) window detectors or as feedback integrators to overcome this difficulty. The output from the integrator appears in each sample period (or desired multiple of the sample period) such that samples that approach the maximum processed signal-to-noise ratio (best phase match) can be obtained. A typical (digital) sliding window integrator that accomplishes the function represented by Eq. (11-8) is given in Figure 12-12. Figure 12-13 represents a single-loop feedback integrator. It is seen that the sliding window integrator requires the storage of data (for the processed range gates) for n interpulse (pulse repetition) periods. The single-loop processor requires storage of the data for a single interpulse period (one data word for each processed range cell). The conclusion, of course, is that if data memory is somehow restricted and if performance is acceptable, the feedback approach is preferred.

Various references[15-17] compare the detection performance of the moving

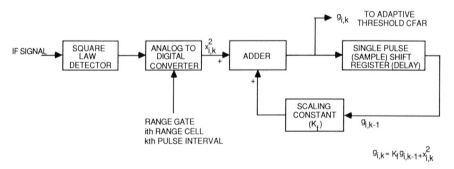

Figure 12-13. Single-loop feedback integrator.

weighted N pulse process. When the feedback constant (K_f of Figure 12-13) is adjusted in accordance with

$$n \approx \frac{1}{0.63} \left(\frac{1}{1 - K_f} \right)$$

(12-13)

and n (assumed large) equals the number of pulses that occur during the beam 3-dB one-way dwell time[18] (Gaussian-shaped pattern), Trunk[16] gives the relative detection loss (relative to an antenna gain that is constant and equal to the beam on-axis gain and a uniformly weighted integrator) for the single-loop integrator as 1.6 or 0.3 dB when single-look or cumulative (per scan) detection laws, respectively, are used. Palmer and Cooper[17] present data demonstrating that a sliding window integrator and an optimum double loop feedback integrator provide, for single-look detection laws, approximately a 0.5-dB sensitivity improvement relative to the single-loop feedback integrator. The designer must determine the merits of this increased performance in light of the implementation costs.

The remaining important factor in track-while-scan application is the impact of the integrator on angle sensing accuracy. Skolnik,[19] in his section on "Angle Accuracy and Parameter Estimation," gives the minimum bound for angle estimation accuracy as

$$\sigma_{min} \approx \frac{1.26 \, \theta_B}{(S/N)_c \, \sqrt{n}} \qquad \text{for } (S/N)_c \ll 1$$

(12-14)

$$\sigma_{min} \approx \frac{1.06 \, \theta_B}{\sqrt{N(S/N)_c}} \qquad \text{for } (S/N)_c \gg 1$$

(12-15)

where

$(S/N)_c$ = signal-to-noise ratio at peak antenna gain (target is on-axis)
θ_B = antenna 3-dB beamwidth

These bounds, therefore, become the yardstick by which angle-sensing accuracy can be gauged. Trunk[15] demonstrates that the sliding window detector provides better (by approximately 8%) angle-sensing accuracy than the feedback integrator. One of the interesting results from Trunk[15] is that the measured value for the mean of the sensed position depends on $(S/N)_c$ in the feedback integrator. Because of this sensitivity in the feedback integrator, some method[15] must be used to minimize this effect. Overall ease of implementation must also be considered in view of achieved angle position measurement results.

The topic of search radar thresholding should include some discussion of

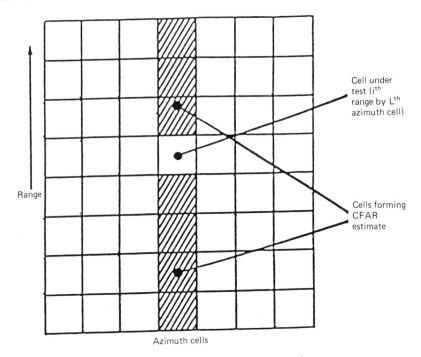

Range

Azimuth cells

Cell under test (i^{th} range by L^{th} azimuth cell)

Cells forming CFAR estimate

Figure 12-14. Region for threshold generation.

regions over which the detection threshold is formed. Some sample regions over which the threshold may be computed are shown in Figures 12-14 through 12-18. Figure 12-14 illustrates the *line technique* implied in the detectors that have been described. The results in the ith range by Lth azimuth cell represent the values computed (for a particular pulse repetition interval) by the sliding window or feedback integrator. The number of cells used to form the CFAR estimate is selected to minimize the multicell CFAR loss (see Chapter 11) while minimizing the sensitivity of the computed results to nonstationary characteristics of the background (interference). Figure 12-15 shows an area technique that sometimes has advantages in more discontinuous (nonhomogeneous) background environments. The values in the surrounding azimuth cells represent the result of integrating in a search radar concoherent data sets that are spaced (separated) n pulse repetition intervals (center-to-center). Figure 12-16 shows how the previous two threshold techniques may be modified to function with an MTI radar that uses a clutter rejection (or stop band) filtering technique. Figure 12-17 shows the logical extension of Figure 12-16 when the radar uses a contiguous Doppler filter passband structure.

Figure 12-18 shows how the line CFAR technique of Figure 12-14 can be established in the Doppler dimension for a radar that uses a contiguous Doppler

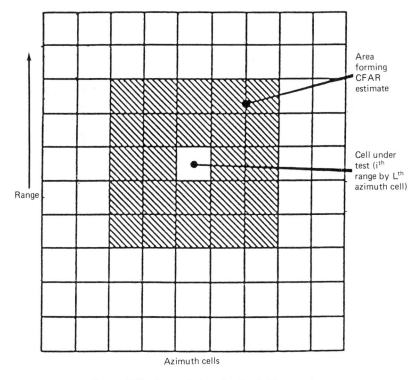

Figure 12-15. Area technique for threshold generation.

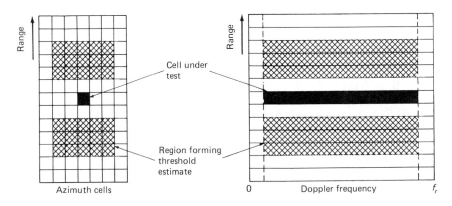

Figure 12-16. Area (range by azimuth) lead and lag range cell technique for threshold generation.

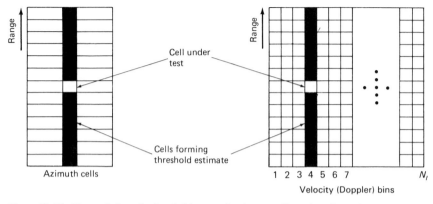

Figure 12-17. Line technique for threshold generation in range dimension, for each range by Doppler cell.

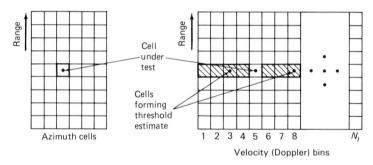

Figure 12-18. Line technique for threshold generation in velocity (Doppler) dimension for each range by Doppler cell.

filter passband structure. In this case, the independent samples for the threshold estimation are the outputs of Doppler filters that lead and lag the target in Doppler frequency within each range by azimuth cell.

12.5 SCANNING RADAR EXAMPLE

It is instructive to consider the detection performance of a frequency agile, noncoherent, pulsed radar that detects, for example, a nonfluctuating target in a clutter-limited environment. The goal is to determine the single pulse signal-to-clutter ratio that is referenced to the on-axis signal-to-clutter ratio $(S/C)_c$ and required to achieve a 90% single-look probability of detection at a 10^{-6} probability of false alarm. The sensor pulse repetition frequency and antenna scan rate are such that 100 pulses ($n = 100$) are emitted between the antenna 3-dB gain points. It is so defined that the background clutter has Rayleigh envelope statistics and that the system has an adequate frequency-agile bandwidth to satisfy the requirement that all 100 background samples are independent. Because of the desire to obtain efficient detection, good track-while-scan angle sensing performance, a hardware-efficient design, and limited CFAR loss, a single-loop feedback integrator has been selected. A line-type (lead/lag) multicell adaptive threshold processor with 10 cells has also been selected.

The data of Figure 11-4 define that an effective single pulse (S/C) of approximately -1.3 dB is required for detection. Figure 11-10 can be used to estimate the CFAR loss as approximately 0.6 dB. The results of the preceding section define a loss of approximately 1.6 dB for the feedback integrator in the scanning radar application. The required $(S/C)_c$ at the IF input when range gate straddle loss is neglected is therefore seen to equal

$$(S/C)_c \approx (-1.3 + 0.6 + 1.6) \text{ dB}$$
$$\approx 0.9 \text{ dB}.$$

Other system losses, propagation effects, and so on can then be used to extend this result to the synthesis of desirable remaining system parameters and characteristics.

This process can be performed for other examples using the techniques outlined in the referenced sections to expand the data base beyond that which is included.

12.6 SUMMARY

Basic concepts of automatic detection have been presented. The previous discussions are not complete, but they do provide insight into the problem and a guide for developing the detector (detection) circuit that best fits the intended application.

The nonparametric techniques that were discussed use multiple independent samples for forms of effective noncoherent integration prior to generation of a detection decision. In some cases, other system considerations do not allow noncoherent integration. In these cases, more conventional techniques that use regions about the cell under test (line, area, etc., as discussed previously) or samples obtained in the time$^{1.5}$ dimension within each cell under test must be used to arrive at a value for the detection threshold.

12.7 REFERENCES

12.7.1 Cited References

1. R. Nitzberg, "Constant-False-Alarm-Rate Signal Processors for Several Types of Interferences," *IEEE Transactions on Aerospace and Electronic Systems*, vol. AES-8, no. 1, January 1972, pp. 27–34.
2. D. Curtis Schleher, *Automatic Detection and Radar Data Processing*, Artech House, Dedham, Mass., 1980.

392 DETECTION IN A CONTAMINATED ENVIRONMENT

3. J. B. Thomas, "Nonparametric Detection," *Proceedings of the IEEE*, vol. 58, no. 5, May 1970, pp. 623–631.

4. R. S. Finn, and R. S. Johnson, "Adaptive Detection Mode with Threshold Control as a Function of Spatially Sampled Clutter Level Estimates," *RCA Review*, vol. 29, no. 3, September 1968, pp. 414–464.

5. G. B. Goldstein, "False-Alarm Regulation in Log-Normal and Weibull Clutter," *IEEE Transactions on Aerospace and Electronic Systems*, vol. AES-9, no. 1, January 1973, pp. 84–91.

6. Peter Swerling, "The 'Double Threshold' Method of Detection," RM-1008, Rand Corp., Santa Monica, CA., December 17, 1952.

7. L. D. Davisson, et al., "The Effects of Dependence on Nonparametric Detection," *IEEE Transactions on Information Theory*, vol. IT-16, no. 1, January 1970, pp. 32–41.

8. L. E. Vogel, "An Examination of Radar Signal Processing via Non-Parametric Techniques," *Proceedings of the IEEE International Radar Conference*, April 1975, pp. 533–537.

9. R. E. Walpole, and R. H. Myers, *Probability and Statistics for Engineers and Scientists*, Macmillan Publishing Co., New York, 1972, p. 83.

10. J. F. Walker, "Performance Data for a Double-Threshold Detection Radar," *IEEE Transactions on Aerospace and Electronic Systems*, vol. AES-7, no. 1, January 1971, pp. 142–146.

11. W. S. Reid, et al, "Techniques for Clutter False Alarm Control in Automated Radar Detection Systems," *Proceedings of the IEEE International Radar Conference*, April 1975, pp. 294–299.

12. G. M. Dillard, et al., "A Practical Distribution-Free Detection Procedure for Multiple-Range-Bin Radars," *IEEE Transactions on Aerospace and Electronic Systems*, vol. AES-6, no. 5, September 1970, pp. 629–635.

13. Robert E. Lefferts, "Adaptive False Alarm Control in Double Threshold Radar Detection," *IEEE Transactions on Aerospace and Electronic Systems*, vol. AES-17, no. 5, September 1981, pp. 666–675.

14. R. M. Barnes, et al., "ARTS Enhancement Support Program Multisensor System Study," MSD-F-183, John Hopkins University Applied Physics Laboratory, Baltimore, January 31, 1973.

15. G. V. Trunk, "Comparison of Two Scanning Radar Detectors: The 'Moving Window' and the Feedback Integrator," *IEEE Transactions on Aerospace and Electronic Systems*, vol. AES-7, no. 2, March 1971, pp. 395–398.

16. G. V. Trunk, "Detection Results for Scanning Radars Employing Feedback Integration," *IEEE Transactions on Aerospace and Electronic Systems*, vol. AES-6, no. 4, July 1970, pp. 522–527.

17. D. S. Palmer, and D. C. Cooper, "An Analysis of the Performance of Weighted Integrators," *IEEE Transactions on Information Theory*, vol IT-10, no. 4, October 1964, pp. 296–302.

18. T. P. Kabaservice, and S. M. Newman, "Recent Technical Developments in Ground-Based Air Surveillance Radar Systems," *Proceedings of Military Microwaves '80*, October 1980, pp. 683-700.

19. M. I. Skolnik, *Introduction to Radar Systems*, McGraw-Hill Book Co., New York, 1962.

12.7.2 Other References

20. G. V. Trunk, et al., "Modified Generalized Sign Test Processor For 2-D Radar," *IEEE Transactions on Aerospace and Electronic Systems*, vol. AES-10, no. 5, September 1974, pp. 574–582.

21. Eli Brookner, *Radar Technology*, Artech House, Dedham, Mass., 1978.

22. M. I. Skolnik, *Radar Handbook*, McGraw-Hill Book Co. New York, 1970.

23. V. G. Hansen, "Comments on 'Performance Data for a Double-Threshold Detection Radar',"

IEEE Transactions on Aerospace and Electronic Systems, vol. AES-7, no. 1, May 1971, p. 561.

24. J. F. Walker, "Performance Data for a Double-Threshold Detection Radar," *IEEE Transactions on Aerospace and Electronic Systems*, vol. AES-7, no. 1, January 1971, pp. 141–146.

25. G. M. Dillard, "A Moving-Window Detector for Binary Integration," *IEEE Transactions on Information Theory*, vol. IT-13, no. 1, January 1967, pp. 2–6.

26. G. R. Martin, "Adaptive Threshold Device Employing Spatial Integration in Two Dimensions," *Electronics Letters*, vol. 12, no. 12, June 1976, pp. 307–308.

27. Lesli M. Novak, "Optimal Target Designation Techniques," *IEEE Transactions on Aerospace and Electronic Systems*, vol. AES-17, no. 5, September 1981, pp. 676–684.

28. V. Gregers Hansen, and Harold R. Ward, "Detection Performance of the Cell Averaging Log/ CFAR Receiver," *IEEE Transactions on Aerospace and Electronic Systems*, vol. AES-8, no. 5, September 1972, pp. 648–652.

29. Hermann Rolling, "Radar CFAR Thresholding in Clutter and Multiple Target Situations," *IEEE Transactions on Aerospace and Electronic Systems*, vol. AES-19, no. 4, July 1983, pp. 608–621.

30. R. Nitzberg, "Constant False Alarm Rate Processors for Locally Nonstationary Clutter," *IEEE Transactions on Aerospace and Electronic Systems*, vol. AES-9, no. 3, May 1973, pp. 399–405.

PART 4
RADAR WAVEFORMS AND
APPLICATIONS

- Continuous Wave Radar—Chapter 13
- MTI and Pulsed Doppler Radar—Chapter 14
- Pulse Compression Radar—Chapter 15
- Synthetic Aperture Radar—Chapter 16

Preceding parts of this book have covered propagation, basic elements of the radar, and detection in some detail. We are now ready to consider specific implementations and applications of radar. In particular, the type of radar waveform selected, and the intended utilization of the radar will determine, to a large extent, many of the operating characteristics and performance parameters. For example, a radar application requiring only detection and, perhaps, target velocity measurement certainly implies significantly different radar characteristics and specific system configuration than an application requiring highly accurate target position location and spatial tracking.

Starting with a discussion of continuous wave (CW) radar fundamentals, specific applications and advantages and disadvantages as compared with pulsed radar, Chapter 13 then reviews the basis for CW radar detection—the Doppler effect. The spectral features of a CW radar's received signal are considered and related to measurement resolution. Also presented in Chapter 13 are techniques for predicting CW radar range performance and several waveform modulation techniques, including frequency modulation, phase modulation, and multiple frequency CW radar.

Many target detection scenarios require the detection of a very small moving target in the presence of a very large stationary or almost stationary background of clutter. For example, the detection of a low flying aircraft against a background of fixed ground returns presents a classical, albeit difficult, radar detection problem that has been investigated extensively since the early days of radar development. Several specific radar implementations and signal processing techniques have been developed to solve this problem. Chapter 14 reviews the limitations of radar in a moving target detection application and then presents an overview of various moving target indication (MTI) and pulsed Doppler radar techniques, including a discussion of the differences between these to moving target radar approaches.

Pulse compression allows a radar to transmit a pulse of relatively long duration and low peak power to attain the range resolution and detection performance of a short pulse, high peak power waveform. In many operational situations, it is important, especially from an electronic warfare vulnerability viewpoint, to reduce the peak power transmitted by the radar. Chapter 15 presents techniques for coding the transmitted waveform and then detecting the reflected signal in a manner that produces signal compression (pulse compression) on reception.

Prior to development of the synthetic aperture radar (SAR) concepts described in Chapter 16, achieving high angular, or cross-range, resolution in a long range airborne search radar was virtually impossible. By processing stored backscatter amplitude and phase information obtained by coherently transmitting and receiving electromagnetic energy along a linear flight path, a synthetic array radar provides high cross-range resolution imagery, in effect recreating in time a long, linear antenna array. The basic mathematical relationships which allow the design of a SAR and prediction of performance are developed in Chapter 16. Application of SAR technology and examples of SAR imagery are also given in this chapter.

13
CONTINUOUS WAVE RADAR

William A. Holm

13.1 INTRODUCTION

A radar system transmits EM waves and receives them upon scattering from a target. The extent of the information about the target derived by the radar depends on the transmitted and received waveform parameters, i.e., amplitude, frequency, and phase, and on the vector nature of EM waves, i.e., polarization.

The target imparts information to the radar by interacting with the waves and thus changing their waveform. The most important result of this interaction is the change in direction of propagation (scattering) of the waves, because the target must direct a portion of the wave energy toward the radar receiver so that the radar can determine the presence and bearing of the target. The interaction of the target with the waves can also affect the waveform parameters. Knowledge of the waveform parameters of the transmitted signal, measurement of the waveform parameters of the received signal, and correlation of these two sets of parameters reveal all the target information that was imparted to the waves. This is the classic inverse scattering problem: to configure the transmit waveform and to measure the receive waveform in order to extract optimally scatterer (target) information. Some of the target parameters that can be determined by a radar and the waveform parameters from which these target parameters are obtained are shown in Table 13-1.

The amount of target information that can be extracted from the received waves is a function of the transmitted wave characteristics, that is, of waveform design. For example, target range can be ascertained by modulating either the amplitude, frequency, or phase of the transmitter carrier signal and observing the time T required for this modulation to go from the radar transmitter to the target to the radar receiver. If the radar transmitter and receiver are colocated, the target range R can be determined from

$$R = \frac{c}{2} T \qquad (13\text{-}1)$$

where c is the propagation speed of the EM waves in free space. Conventional

Table 13-1. Target Parameters and the Waveform Parameters From Which They are Determined.

Target Parameters	Waveform Parameters
Presence and bearing	Amplitude
Range	Amplitude, frequency, phase
Size (RCS)	Amplitude
Classification	Amplitude, frequency, polarization
Radial speed	Frequency
Discrimination from clutter	Frequency, amplitude, polarization

pulsed radar is an example of a radar that uses amplitude modulation (AM) for range determination. Frequency modulation (FM) or phase modulation (PM) techniques can be used in CW radar for range determination.

A CW radar continuously transmits EM waves; this is the simplest type of radar. Simple CW radars transmit unmodulated waves and were the first experimental radars developed.[1] This type of radar is useful for detecting targets when the range between the radar and the target is changing, but it cannot determine target range. Detection is based on the Doppler effect, which is discussed later in this chapter. If the CW radar transmits waves that are modulated, then target range determination is possible.

A "timing mark" or modulation must be applied to the transmitted wave to measure target range with a CW radar. FM and PM techniques applied to CW radars for target range determination are also discussed in this chapter.

Particular applications dictate whether CW or pulsed radars are utilized. There are several advantages, however, in choosing a CW radar over a pulsed radar. CW radar hardware tends to be simpler. For example, no high-voltage modulators are required for simple CW radars, and CW transmitters tend to be smaller and lighter.

A radar's ability to detect targets is determined by (among other factors) the average power it transmits. Since a CW radar's duty factor is unity, its transmitted power level is lower than the peak power level of a pulsed radar with the same average power. Thus, high-power electrical breakdown is less of a problem with CW radars.

CW radars can also detect targets at shorter ranges than many pulsed radars. Pulsed radar systems have duplexers that use TR switches or tubes to protect the receiver during transmit. Because of the finite recovery time of the TR tube after transmit, echo returns from short-range targets will not reach the receiver and, thus, these targets will not be detected. CW radars do not use TR tubes; therefore, they do not have this short-range problem. Transmitter/receiver isolation is also necessary in CW radars, but this is achieved by using other types of duplexers (such as ferrite circulators) or by implementing frequency modulation techniques, which are discussed in this chapter.

Table 13-2. Comparison of CW Radar and Pulsed Radar.

Requirements	CW Radar Advantage	Pulsed Radar Advantage
Low hardware complexity	X	
High average power	X	
Short-range target detection	X	
Moving target discrimination	X	
Target range determination		X
High transmitter/receiver isolation		X

The simplest CW radars discriminate between moving targets and stationary clutter. The same capability in a pulsed radar requires additional signal-processing equipment such as a range-gated filter, delay-line canceller, or FFT processor. Even with this additional signal processing, pulsed radars may still be unable to detect targets with Doppler frequencies equal to an integer multiple of the pulse repetition frequency (PRF) of the pulsed radar. This limitation is not present in a CW radar.

Although the main disadvantages of a simple CW radar system—no target range information and little transmitter/receiver isolation—led to the development of the pulsed radar system, these disadvantages can be overcome by using the proper CW waveform design. A comparison of CW radar and pulsed radar advantages is given in Table 13-2.

Some of the more common CW radar applications are in weapon fuzes, weapon seekers, lightweight portable personnel detectors, aircraft altimeters, and police radars. Simple (uncoded waveform) CW radar systems can be used for proximity fuzes (where weapon detonation is based on the power level of the echo return from the target), seekers, personnel detectors, and police radars, since no range information is necessary for these applications. Radar aircraft altimeters usually are frequency modulated CW radars capable of aircraft-to-terrain range determination.

13.2 THE DOPPLER EFFECT

Moving target discrimination and resolution in simple CW radar systems are based on the Doppler effect, which is now discussed.

Consider a stationary radar with a frequency of $f = 1/T_0$, where T_0 is the period of the transmitted wave, and a target moving at a constant speed, v, toward the radar, as shown in Figure 13-1. (The speed v will be assumed to be positive if the target is approaching the radar, negative if receding.) At time $t = t_0$, let the target be at range $R = R_0$, and assume that at that time a peak or crest of the wave (point A in Figure 13-1) is emerging from the radar's antenna. At time $t = t_0 + T_0$, the next crest of the wave (point B) is emerging; at this

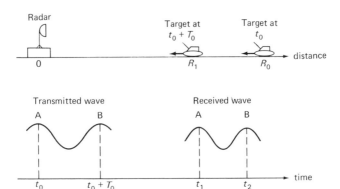

Figure 13-1. Target geometry and transmitted and received waveforms for Doppler effect derivation.

time, let the target's range be $R = R_1$. The time, Δt, necessary for point A on the wave to travel from the radar to the target is the distance traveled divided by the speed of the wave:

$$\Delta t = \frac{R_0 - v\Delta t}{c}$$

or   (13-2)

$$\Delta t = \frac{R_0}{c + v}$$

where c is the wave's speed, that is, the speed of light. The time necessary for point A to return to the radar is again Δt. Therefore, the total round trip takes $2\Delta t$, and point A returns to the radar at time

$$t_1 = t_0 + \frac{2R_0}{c + v} \tag{13-3}$$

Similarly, point B returns to the radar at time

$$t_2 = t_0 + T_0 + \frac{2R_1}{c + v} \tag{13-4}$$

Since t_1 and t_2 are the times of arrival of two adjacent peaks of the wave (points A and B), the period of the received wave, T_0', is $t_2 - t_1$, or

$$T_0' = T_0 - \frac{2(R_0 - R_1)}{c + v} \tag{13-5}$$

As can be seen from Eq. (13-5), the period (frequency) of the received wave decreases (increases) from the period (frequency) of the transmitted wave due to scattering from a moving target. This is the Doppler effect.

Now, since $vT_0 = R_0 - R_1$,

$$T_0' = T_0 \left(\frac{c - v}{c + v} \right) = T_0 \left(\frac{1 - v/c}{1 + v/c} \right) \tag{13-6}$$

or, in terms of the received frequency, $f' = 1/T_0'$,

$$f' = f \left(\frac{1 + v/c}{1 - v/c} \right) \tag{13-7}$$

For most cases of interest, $v/c \ll\ \ll 1$. Thus,

$$\begin{aligned} f' &= f \left(1 + \frac{v}{c} \right) \left(1 + \frac{v}{c} + \frac{v^2}{c^2} + \ldots \right) \\ &= f \left(1 + \frac{2v}{c} + \frac{2v^2}{c^2} + \ldots \right) \end{aligned} \tag{13-8}$$

is well approximated by

$$f' \simeq f \left(1 + \frac{2v}{c} \right) = f + \frac{2v}{\lambda} \tag{13-9}$$

where the relationship $c = f\lambda$ has been used. Therefore, the received wave has been shifted in frequency from the transmit wave by the amount

$$f_d = \frac{2v}{\lambda} \tag{13-10}$$

This is the Doppler frequency shift.

As a rule of thumb, at X-band (10 GHz), the Doppler frequency shift is approximately 30 Hz per statute mile per hour of target radial speed. Doppler frequency shifts at other radar bands can be determined by using this rule of thumb and a linear extrapolation in frequency, for example, 3 Hz/mph at L-band (1 GHz), 105 Hz/mph at K_a-band (35 GHz), and so on. Doppler frequency shifts for various radar frequency bands and target speeds are shown in Table 13-3.

Table 13-3. Doppler Frequency Shifts (Hz) for Various Radar Frequency Bands and Target Speeds.

	Radial Target Speed		
Radar Frequency Band	1 m/s	1 knot	1 mph
L (1 GHz)	6.67	3.43	2.98
S (3 GHz)	20.0	10.3	8.94
C (5 GHz)	33.3	17.1	14.9
X (10 GHz)	66.7	34.3	29.8
K_u (16 GHz)	107	54.9	47.7
K_a (35 GHz)	233	120	104
mm (95 GHz)	633	326	283

13.3 SIMPLE CW RADAR SYSTEMS[2–4]

A block diagram of the simplest CW radar possible is shown in Figure 13-2. Here the output of a CW transmitter at frequency f is routed through a circulator to the antenna. The wave transmitted by the antenna propagates to, and is scattered from, a moving target and is received back at the antenna. The wave now has a frequency of $f + f_d$. The wave passes through the circulator to the receiver. Since the circulator does not provide perfect isolation between the transmitter and receiver, some of the transmitted signal at frequency f leaks through to the receiver.

At the front end of the receiver is a mixer that heterodynes the two signals together to produce a beat signal with frequency f_d. This signal is then amplified by a low-frequency amplifier and displayed. This type of receiver, which beats the received signal with the transmitted signal, is called a *homodyne*, or *zero-IF, receiver*.

The amount of transmitter/receiver isolation required of the circulator in the simple CW radar shown in Figure 13-2 depends on the power level of the trans-

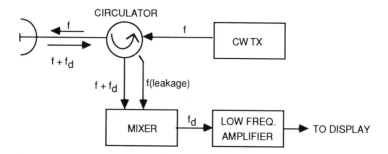

Figure 13-2. Simple CW radar system with a homodyne receiver.

mitted signal and the sensitivity of the receiver. For example, in Hewlett Packard's Series 35200 low-power, solid-state, X-band, Doppler radar TR modules,[5] only 18 dB of isolation is required of the circulator. (Circulators with 30 to 40 dB of isolation and higher have also been built.) In high-power radars, however, more isolation than the circulator can provide may be required. In these radars, other isolation techniques, such as dual antennas, may have to be employed.

A major shortcoming of the simple homodyne CW radar is its lack of sensitivity. This is due to flicker noise (from electronic devices within the radar), which varies with frequency as $1/f$. Therefore, at low Doppler frequencies, flicker noise is very strong and is amplified with the Doppler signal in the low-frequency amplifier. One way to avoid this problem is to amplify the received signal at a higher frequency where flicker noise is negligible and then heterodyne the signal down to lower frequencies. This is done in the simple dual-antenna configuration CW radar system shown in Figure 13-3. The receiver of this system is called a *superheterodyne receiver*.

Instead of mixing a portion of the transmitted signal directly with the received signal, as is done in a homodyne receiver, in the superheterodyne receiver a portion of the transmitted signal is shifted in frequency by an amount equal to the IF before it is mixed with the received signal. The IF signal is provided by an LO and is first mixed with a portion of the transmitted signal. The signal out of this mixer has many frequency components; three occur at frequencies around the transmit frequency: f, $f - f_{IF}$, and $f + f_{IF}$. A bandpass filter removes

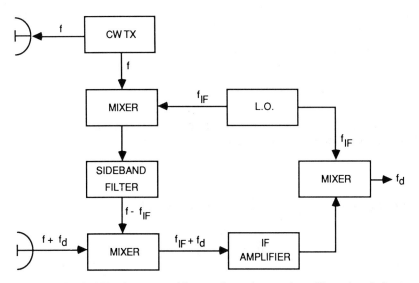

Figure 13-3. Simple CW radar system with a superheterodyne receiver. (Shown in a dual antenna configuration.)

all of the components except the lower sideband at $f - f_{IF}$. This signal is then mixed with the received signal, and the output of this mixer is then amplified by an IF amplifier. The IF signal is at a high enough frequency (e.g., 60 MHz) so that flicker noise is negligible. The output signal of the IF amplifier is then mixed with the IF signal from the LO to obtain the baseband (Doppler frequency) signal.

Target radial speed can be determined by appropriately filtering the baseband signal to determine the Doppler frequency. This filtering can be accomplished by using a tracking or sweeping LO followed by a single filter,[6] or by using a set of contiguous (in frequency) analog filters, or by digitizing the baseband signal by an A/D converter and then using a set of contiguous digital filters or an FFT processor. The bandwidth of the individual Doppler filters should be matched to the bandwidth of the target signal to maximize the signal-to-noise ratio and, thus, the probability of detection. The spectrum of the CW signal, therefore, must be known.

13.4 CW RADAR SPECTRUM AND RESOLUTION

If a CW radar illuminates a target moving with constant radial velocity toward the radar for an infinite period of time, then the received signal (ignoring amplitude factors, and fixed-phase terms) is

$$s(t) = \cos(\omega_0 + \omega_d)t \qquad (13\text{-}11)$$

where

$$f_0 = \frac{\omega_0}{2\pi} = \text{transmitted frequency}$$

$$f_d = \frac{\omega_d}{2\pi} = \text{Doppler frequency shift}$$

After this signal is heterodyned down to baseband (video band), it becomes (again ignoring amplitude factors and fixed-phase terms)

$$s(t) = \cos(\omega_d t) \qquad (13\text{-}12)$$

The Fourier transform of the signal given by Eq. (13-12) is defined to be[7]

$$F[s(t)] \equiv S(\omega) = \int_{-\infty}^{\infty} \cos(\omega_d t)\, e^{-j\omega t}\, dt \qquad (13\text{-}13)$$

or

$$S(\omega) = \pi\delta(\omega - \omega_d) + \pi\delta(\omega + \omega_d) \qquad (13\text{-}14)$$

where $\delta(x)$ is the impulse, or (Dirac) delta, function.[8] The delta function, $\delta(x)$, is an infinitely high, infinitely narrow peak at $x = 0$ with the following properties:

$$\delta(x) = 0 \qquad x \neq 0 \tag{13-15}$$

$$\int_{-a}^{b} \delta(x) \, dx = 1 \qquad a, b > 0 \tag{13-16}$$

The spectrum of the received signal after conversion to baseband is plotted in Figure 13-4.

First, note that the two delta functions appear at plus and minus ω_d. This is a general result of taking the Fourier transform of a real function, that is, the negative frequency spectrum is a mirror image of the positive frequency spectrum. Therefore, if nothing additional is done in the receiver in the down conversion of the received signal from RF to baseband, then the sign of the Doppler frequency shift will be lost and the relative target motion (approaching or receding) will be indeterminate. This problem of ambiguous relative target motion can be eliminated by splitting the IF signal into two channels and phase shifting the signal in one (quadrature-phase, or Q) channel 90° with respect to the signal in the other (in-phase, or I) channel. Both channels are then down converted to baseband. If the I channel and Q channels are thought of as the real and imaginary components, respectively, of a complex target signal, then the baseband signal is given by

$$s(t) = \frac{1}{\sqrt{2}} \left[\cos(\omega_d t) + j \sin(\omega_d t) \right] = \frac{1}{\sqrt{2}} e^{j\omega_d t} \tag{13-17}$$

The Fourier transform of Eq. (13-17) is given by

$$S(\omega) = \sqrt{2} \, \pi \delta(\omega - \omega_d) \tag{13-18}$$

Now the sign of the Doppler frequency shift is unambiguous. A block diagram

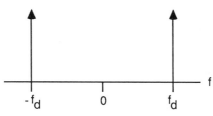

Figure 13-4. Baseband spectrum of a signal from a moving target illuminated by a CW radar for an infinite amount of time.

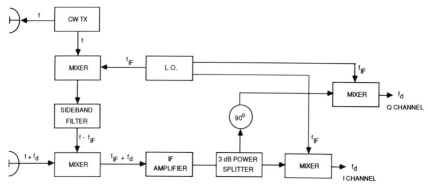

Figure 13-5. CW radar system with an I/Q detector.

of a CW radar system with an I and Q channel (synchronous detector) receiver is shown in Figure 13-5.

The second thing to note about the spectrum in Figure 13-4 is that the target signature bandwidths are zero. This is due to the unrealistic assumption that the target was illuminated for an infinitely long time. Assuming that the target is illuminated for a finite time, T_d, then the received target signal after conversion to baseband (assuming I/Q detection) is given by

$$s_{T_d}(t) = p_{T_d}(t) \frac{e^{j\omega_d t}}{\sqrt{2}} \qquad (13\text{-}19)$$

where

$$p_{T_d}(t) = \begin{cases} 1 & |t| < T_d/2 \\ 0 & |t| > T_d/2 \end{cases} \qquad (13\text{-}20)$$

The Fourier transform of this signal is

$$F[s_{T_d}(t)] \equiv S_{T_d}(\omega) = \frac{1}{2\pi} \int_{-\infty}^{\infty} P_{T_d}(s) S(\omega - s) \, ds \qquad (13\text{-}21)$$

where

$$P_{T_d}(s) = \frac{T_d \sin(sT_d/2)}{sT_d/2} \qquad (13\text{-}22)$$

and

$$S(\omega - s) = \sqrt{2} \, \pi \delta(\omega - s - \omega_d) \qquad (13\text{-}23)$$

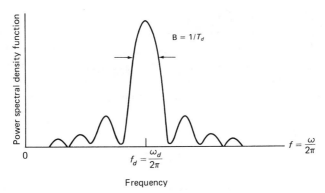

Figure 13-6. Spectral power density function of a signal from a moving target illuminated by a CW radar for a time T_0'.

Thus,   (13-24)

$$S_{T_d}(\omega) = \frac{T_d}{\sqrt{2}} \frac{\sin [(\omega - \omega_d)T_d/2]}{(\omega - \omega_d)T_d/2}$$

The resulting power spectral density function is proportional to $|S_{T_d}(\omega)|^2$ and is shown in Figure 13-6. The 3-dB bandwidth, B, of the main lobe of this function is approximately $B \approx 1/T_d$ Hz. Assuming no other contributions to the spectral broadening, for example, target cross section fluctuations and target radial accelerations, then the bandwidths of the Doppler filters should be $1/T_d$ Hz. For example, for an antenna with a $2°$ beamwidth scanning at a rate of 6 rpm ($36°/$s), the target will be illuminated by the radar for $T_d = 1/18$ s. Therefore, the bandwidth of a matched Doppler filter should be $B = 18$ Hz.

13.5   THE CW RADAR RANGE EQUATION

The radar range equation[9] was given in Chapter 1:

$$\text{SNR} = \frac{PG^2\lambda^2\sigma}{(4\pi)^3 R^4 kTBFL}$$ (13-25)

with the various terms defined there. Depending on the definitions of P and B, Eq. (13-25) can represent the single-pulse or integrated (over a dwell period) signal-to-noise ratio for a coherent (or noncoherent) pulsed radar, or the signal-to-noise ratio for a CW radar. To determine the appropriate definitions of P and B that make Eq. (13-25) applicable to a CW radar, first consider the single-pulse signal-to-noise ratio of a pulsed radar. In what follows, all radars are assumed to have coherent, synchronous detectors unless otherwise stated.

Eq. (13-25) represents the single-pulse signal-to-noise ratio of a pulsed radar when P is the peak power, P_t, and B is the IF bandwidth, B_{IF}. The integrated signal-to-noise ratio obtained during a target dwell period, T_d, is then given by multiplying this resulting equation by the number of pulses, n, transmitted during T_d. Now since

$$n = f_r T_d \tag{13-26}$$

where f_r is the pulse repetition frequency of the pulsed radar and $B_{IF} = 1/\tau$ for a matched-filter receiver, the integrated signal-to-noise ratio is given again by Eq. (13-25) when the following identifications are made:

$$P = P_{avg} = P_t f_r \tau \tag{13-27}$$

and

$$B = B_{vid} = 1/T_d \tag{13-28}$$

where P_{avg} is the average power transmitted and B_{vid} is the video bandwidth. The loss parameter, L, will have to be adjusted for any additional losses, such as antenna beamshape losses, signal processing losses, that may occur. (In fact, noncoherent detection can be handled by including a noncoherent integration loss in L.)

The integrated signal-to-noise ratio for a pulsed radar given by Eq. (13-25) by making the identifications given in Eqs. (13-27) and (13-28) is precisely the same signal-to-noise ratio equation for a CW radar. Of course, P_{avg} in a CW radar is the same as the transmitted power, P.

For a single (I) channel CW radar, an additional 3 dB loss relative to a CW radar with a synchronous detector is incurred due to the presence of noise from the image sideband.

As an example of an application of the CW radar range equation, consider a CW police radar with a single-channel, superheterodyne receiver. A typical block diagram of such a radar is shown in Figure 13-3, with the dual antenna configuration replaced with a single antenna and a circulator. Typical parameters for this radar are: $P = 100$ mW, $G = 20$ dB, $f = 10.525$ GHz ($\lambda = 2.85$ cm), $F = 6$ dB, $L = 9$ dB, and $\delta V = 1$ mph. The velocity resolution of 1 mph is equivalent to approximately 30-Hz resoltuion at X-band, resulting in a bandwidth of $B = 30$ Hz. For a car with an RCS of 30 m^2 and a required signal-to-noise ratio of 10 dB for detection, this police radar would, from the radar range equation, have a detection range of 4.2 km or 2.6 statute miles.

13.6 FREQUENCY MODULATED CONTINUOUS WAVE RADAR[10-15]

One main disadvantage of simple unmodulated CW radar systems is their inability to measure target range. This limitation can be overcome by modulating the CW signal, thus, essentially applying a timing mark to the carrier. As mentioned previously, the carrier signal can be modulated in amplitude, frequency, or phase. Pulsed radar is an example of a radar utilizing AM to determine range. Frequency-modulated continuous wave (FMCW) radar techniques are now considered, and multiple-frequency and phase-coded CW radar techniques are discussed in the following subsections.

FMCW radar systems measure *target range* by functionally varying the frequency of the transmitted signal in time and measuring the frequency of the received signal. The correlation of the frequencies of transmitted and received signals at any given instant of time is a measure of not only the target's range 'ut also its radial speed, v. This correlation can be illustrated by considering the linear frequency modulation of the transmitted signal and the frequency history of the received signal scattered from a moving target, as shown in Figure 13-7. The difference in frequency between the peaks of the two curves is the Doppler shift, $f_d = 2v/\lambda$, imposed on the scattered signal by the moving target, and the time difference, T, between the occurrence of two successive peaks is the signal transit time from the radar transmitter to the target to the radar receiver. Thus, if the transmitter and receiver are colocated, then the target's range is given by $R = cT/2$.

Simple geometrical considerations reveal that the instantaneous frequency

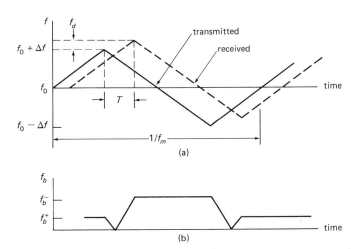

Figure 13-7. (a) Linear triangular frequency modulation and (b) frequence difference f_b, between transmitted and received signals.

difference, f_b, between the two curves when they both have positive slopes is

$$f_b = f_b^+ = \frac{8\Delta ff_m R}{c} - f_d \tag{13-29}$$

where

Δf = amplitude of the frequency modulation
f_m = frequency modulation rate

Similarly, the instantaneous frequency difference between the two curves when they both have negative slopes is

$$f_b = f_b^- = \frac{8\Delta ff_m R}{c} + f_d \tag{13-30}$$

Thus,

$$R = \frac{c}{8\Delta ff_m} \langle f_b \rangle \tag{13-31}$$

where $\langle f_b \rangle$ is the average frequency difference
and

$$v = \frac{\lambda}{4} (f_b^- - f_b^+) \tag{13-32}$$

Therefore, the target range R and the radial speed v can be determined by measuring the instantaneous frequency difference, f_b, between the transmitted and received signals. A simple FMCW radar system configured to measure both R and v is shown in Figure 13-8.

Some FMCW radar systems employing *linear frequency modulation* use a modulation like the one depicted in Figure 13-9. This modulation scheme provides an independent measure of v in the unmodulated portion ($df/dt = 0$) of the curve. The remainder of the frequency versus time curve is identical to the linear FM just discussed; thus, the differences in frequencies, f_b^+ and f_b^-, between the transmitted and received curves are the same as those given above. Since f_d has already been determined from the unmodulated portion of the curve, only one of these differences in frequencies, f_b^+ or f_b^-, is needed to determine R; the other can be used as a check on the computed value of R.

The range resolution, δR, obtainable with an FMCW radar system will depend on the resolution, or accuracy, with which the frequency difference, f_b, can be measured. This accuracy, δf_b, will depend on the bandwidth, $B = 2\Delta f$,

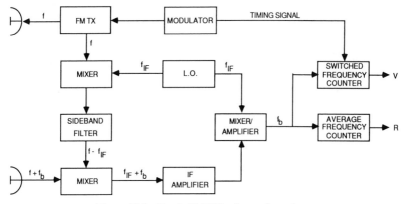

Figure 13-8. Simple FMCW radar configuration.

of the FM and the accuracy with which the FM waveform can be maintained. For example, for a linear FM, δf_b (and thus δR) will depend on the bandwidth and the *linearity* of the modulation. The nonlinearity of the modulation is given by $\delta f/B$, where δf is the deviation in the modulation from linear. The nonlinearity of the modulation must be much less than the inverse of the time–bandwidth product, that is, f_m/B, of the waveform not to adversely affect the maximum range resolution obtainable. Since B determines the maximum range resolution, $\delta R = c/2B$ ($T << 1/f_m$) obtainable, and f_m determines the maximum unambiguous range, $R_m = c/2f_m$, the constraints on linearity for a given δR and R_m are given by nonlinearity $<< f_m/B = \delta R/R_m$. Thus, for a range resolution of 1 ft and a maximum unambiguous range of 3000 ft, the nonlinearity of the FM waveform must be less than 0.03%.

One of the easier frequency modulations to obtain from a practical point of view (and one that is more mathematically tractable) is the *sinusoidal FM*. This FM for a transmitted and received signal is shown in Figure 13-10 for a stationary target. The frequency of the transmitted signal can be written as

$$f(t) = f_0 + \Delta f \cos (2\pi f_m t) \tag{13-33}$$

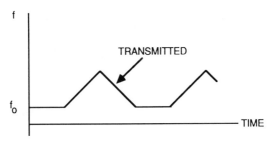

Figure 13-9. Segmented linear FM.

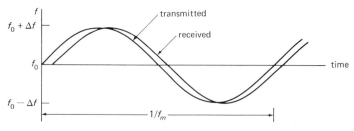

Figure 13-10. Sinusoidal FM.

The transmitted signal can be written as

$$s(t) = A_1 \sin [\phi(t)] \qquad (13\text{-}34)$$

where

$$\phi(t) = 2\pi \int f(t)\, dt = 2\pi f_0 t + \frac{\Delta f}{f_m} \sin (2\pi f_m t) \qquad (13\text{-}35)$$

Thus,

$$s(t) = A_1 \sin \left[2\pi f_0 t + \frac{\Delta f}{f_m} \sin (2\pi f_m t) \right] \qquad (13\text{-}36)$$

The received signal (from a stationary target) will be a replica of the transmitted signal but displaced in time by the transit time $T = 2R/c$;

$$r(t) = A_2 \sin \left[2\pi f_0 (t - T) + \frac{\Delta f}{f_m} \sin [2\pi f_m (t - T)] \right]$$

$$= A_2 \sin [\phi(t - T)] \qquad (13\text{-}37)$$

After mixing and filtering of the two signals in the receiver, the output from the lower sideband filter will be

$$s(t) = A_3 \sin [\phi(t) - \phi(t - T)]$$

$$= A_3 \sin \left(2\pi f_0 T + \frac{2\Delta f}{f_m} \sin (\pi f_m T) \cos \left[2\pi f_m \left(t - \frac{T}{2} \right) \right] \right) \qquad (13\text{-}38)$$

If one assumes that $T \ll 1/f_m$, then it is easy to show that the frequency, f_b, of the above beat signal averaged over one-half of a modulation cycle is

$$\langle f_b \rangle = \frac{8R f_m \Delta f}{c} \qquad (13\text{-}39)$$

or

$$R = \frac{c}{8\Delta f f_m} \langle f_b \rangle \tag{13.40}$$

which is the same result obtained in Eq. (13-31) for the linear FM case.

Transmitter/receiver isolation is the other major problem with CW radar systems. Separate transmitting and receiving antennas increase isolation, but in high-power, long-range CW radar systems, transmitter leakage into the receiver can still be a problem, as can unwanted return from close-in clutter.

FM techniques can be used to increase transmitter/receiver isolation. For a sinusoidal modulation, the transmitted signal is given above. The video spectrum of the received signal from a moving target consists of a pair of sidebands displaced by the Doppler frequency, f_d, about the modulation frequency, f_m, and each of its harmonics, plus a spectral line at f_d above DC. This spectrum is shown in Figure 13-11. The amplitude of the pair of spectral lines surrounding the nth harmonic is proportional to $J_n(D)$, the nth order Bessel function of the first kind, where

$$D = \frac{2\Delta f}{f_m} \sin (\pi f_m T) \tag{13-41}$$

Since $T = 2R/c$, where R is the range of the target, D is a function of target range.

Any one of the spectral components can be used to determine f_d. However, if a low response is desired when the received signals are from close-in clutter, the spectral components around higher-order harmonics should be chosen. This is due to the nature of the Bessel function, $J_n(D)$. For small arguments,

$$J_n(D) \cong D^n \sim R^n \quad (D \ll 1) \tag{13-42}$$

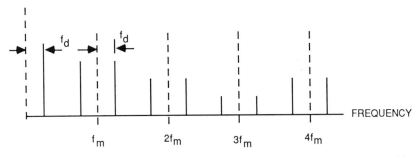

Figure 13-11. Spectrum of the video signal resulting from the detection of a moving target with a sinusoidal FMCW radar.

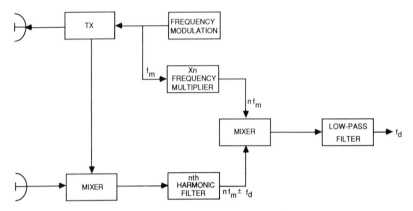

Figure 13-12. FMCW radar that extracts the nth harmonic from the received signal.

Therefore, spectral components around higher-order harmonics of f_m are greatly attenuated for received signals from close-in (small R) clutter and, indeed, theoretically, zero range leakage can be totally eliminated. A block diagram of an FMCW radar system which extracts the nth harmonic (J_n Bessel spectral component) is shown in Figure 13-12.

13.7 MULTIPLE-FREQUENCY CW RADAR[16,17]

Since the phase of an EM wave is (among other things) a function of the distance, d, traveled by the wave, then target range, R, should be determinable in a CW radar by measuring the relative phase difference between the transmitted and received waves. This relative phase difference, $\Delta\phi$, is given by

$$\Delta\phi = \frac{2\pi d}{\lambda} = \frac{4\pi R}{\lambda} \tag{13-43}$$

Thus, the target range is given by

$$R = \frac{\lambda\Delta\phi}{4\pi} \tag{13-44}$$

The maximum unambiguous range, R_{max}, is that range for which $\Delta\phi = 2\pi$; thus,

$$R_{max} = \lambda/2 \tag{13-45}$$

which varies from 15 cm at L-band (1 GHz) to 1.6 mm at 95 GHz. This maximum range is too small for most practical applications.

The maximum unambiguous range can be increased by transmitting multiple-frequency signals. Consider the case where two CW signals with frequencies f_1 and f_2 are transmitted. The signals received during the observation time (again ignoring amplitude factors) from a target moving with constant radial speed, v, are given by

$$s_i(t) = \cos(\omega_i t + \omega_{d_i} t + \Delta\phi_i) \qquad (13\text{-}46)$$

where

$$i = 1, 2$$

$$\omega_i = 2\pi f_i$$

$$\omega_{d_i} = 2\pi f_{d_i} = \frac{4\pi v}{\lambda_i} = \frac{4\pi v f_i}{c}$$

$$\Delta\phi_i = \frac{4\pi R}{\lambda_i} = \frac{4\pi R f_i}{c} \qquad (13\text{-}47)$$

After these signals are heterodyned down to video band, they become (again ignoring amplitude factors)

$$s_i(t) = \cos(\omega_{d_i} t + \Delta\phi_i) \qquad (13\text{-}48)$$

The relative phase difference between the two signals measured for a time T_d is

$$\Delta\phi = \frac{4\pi v}{c}(f_2 - f_1)T_d + \frac{4\pi R}{c}(f_2 - f_1)$$

or

$$\Delta\phi = \frac{4\pi \Delta f}{c}(R + vT_d) \qquad (13\text{-}49)$$

where $\Delta f = f_2 - f_1$. In most cases, the target range, R, is much greater than the target's radial displacement, vT_d, during the observation time. Therefore,

$$\Delta\phi \simeq \frac{4\pi \Delta f R}{c} \qquad (13\text{-}50)$$

or

$$R = \frac{c\Delta\phi}{4\pi \Delta f} \qquad (13\text{-}51)$$

The maximum unambiguous range ($\Delta\phi = 2\pi$) now is given by

$$R_{max} = \frac{c}{2\Delta f} \tag{13-52}$$

where, for example, a 50-kHz frequency difference results in a 3-km unambiguous range.

13.8 PHASE MODULATED CW RADAR[18]

Another method by which a timing mark is applied to the CW carrier to provide
target range is phase modulation (PM). PM CW radar systems measure target
range by applying a discrete phase shift every τ seconds to the transmitted CW
signal, thus producing a phase-coded waveform. The returning waveform is
correlated with a stored version of the transmitted waveform. The time delay
between the transmission of the coded waveform and the occurrence of the
maximum correlation between the received waveform and the stored version
provides target range information. For high-range resolution, the autocorrelation function of the transmitted signal must have a narrow peak and low sidelobes.

Basically, there are two types of phase-coding techniques: binary phase codes
and polyphase codes, as discussed in Chapter 15. Since the principle of range
determination is the same for both coding techniques, only the simpler binary
phase codes are considered here. A binary phase code is generated by dividing
the CW carrier into N time segments of equal duration, τ, and coding each
segment with either +1 or 0. A +1 corresponds to no phase shift (in-phase)
with respect to the unmodulated carrier, while a 0 corresponds to a 180° phase
shift (out-of-phase). A binary phase code and the resulting modulated CW
waveform are shown in Figure 13-13. The codes in CW radar systems are periodic, that is, they are continuously repeated.

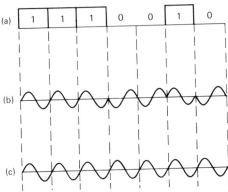

Figure 13-13. (A) Binary phase code and (B) resulting modulated CW waveform from (C) an
unmodulated CW waveform.

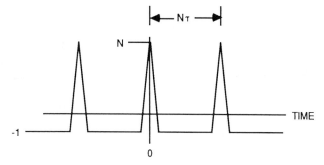

Figure 13-14. Autocorrelation function of a periodic maximal-length binary code of length N. Each time segment of the code is τ s.

Maximal length (or pseudorandom or pseudonoise) binary phase codes have autocorrelation functions which possess the desired properties mentioned above for good range resolution. These codes are most easily generated by linear feedback shift registers. A maximal length code has a length of $N = 2^n - 1$, where n is the number of stages in the shift register generating the code. Since the code is periodic, its autocorrelation function has a peak value of N at $t = 0$ and integer multiples of $N\tau$, and a value of -1 everywhere else, as shown in Figure 13-14.

An example of a three-stage linear feedback shift register that will generate a maximal length code is shown in Figure 13-15. The outputs from stages 2 and 3 are summed by modulo 2 (carry-free) arithmetic, and the sum is fed into stage 1. The three stages of the shift register are initially set to all 1's or a combination of 1's and 0's. An all-0 initial state results in an all-0 sequence; therefore, it is excluded. The output is a sequence of length $N = 2^3 - 1 = 7$. This is the maximum-length sequence code that can be generated before the code is repeated. Feedback connections to different stages of the shift register generate different code sequences which may or may not be of maximal length. However, for a given n-stage shift register, there are several different possible

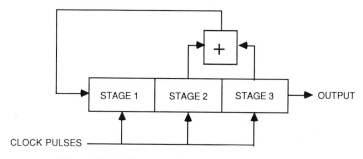

Figure 13-15. Three-stage linear feedback shift register.

feedback connections that will generate maximal length codes, that is, codes of length $N = 2^n - 1$. For example, for a seven-stage shift register, a maximum of 18 different maximal length codes, each of length $N = 127$, can be generated.

As mentioned above, a PM CW radar measures target range by correlating the received waveform with a stored version of the transmitted code. Let $t = 0$ correspond to the time at which the last segment (before repetition) of the N-segment coded waveform is transmitted, and let T be the round-trip time of the signal between the radar and the target. The return signal is decoded and sent through a sequenced delay line consisting of N serial delays each τ seconds long (see Figure 13-16). By sampling the signal before each delay, the entire code of the return signal can be simultaneously sampled and correlated with the stored version of the transmitted code. The result of this correlation is the value -1 (for a pseudorandom code) at all times, except for $T = T + kN\tau$, where k is an integer. When $T = T + kN\tau$, the result of the correlation is the value N.

Figure 13-16 shows a seven-segment delay line and correlator processing a received code at times $T - 3\tau$, T, and $T + 2\tau$. At time T, a maximum correlation value of 7 is obtained. At all other times, for example, $T - 3\tau$ or $T + 2\tau$, a correlation value of -1 is obtained.

An alternative method of implementing a PM CW radar is by correlating the received waveform with a *delayed* version of the transmitted code. If the delay is equal to the transit time of the transmitted signal, then the autocorrelation function will be a maximum; otherwise, it will be a minimum. Therefore, adjusting the delay time to maximize the autocorrelation function will yield target range. An implementation of this method is shown in Figure 13-17. In a radar system utilizing a shift register to generate the code, this delay is easily accomplished. The output of the kth stage of a shift register is delayed by one clock period from the output of the $(k - 1)$th stage. Therefore, any delay desired can be obtained by cascading additional stages. Alternatively, a delay can also be obtained by using a second shift register and delaying its clock pulses with respect to those of the first.

The transmitted waveform must return to the radar in $N\tau$ seconds or less in order to eliminate ambiguities in range measurements in a phase-modulated CW radar. Thus, the maximum unambiguous range of a PM CW radar is

$$R_{max} = \frac{cT_{max}}{2} = \frac{cN\tau}{2} \tag{13-53}$$

Moving target Doppler frequency shifts can be measured by using a set of contiguous (in frequency) filters to filter the video signal. The main effect of the phase coding is to cause an increase in the frequency sidelobe level over that of an uncoded signal. Thus, there is an increased likelihood of incorrect Doppler frequency measurements.

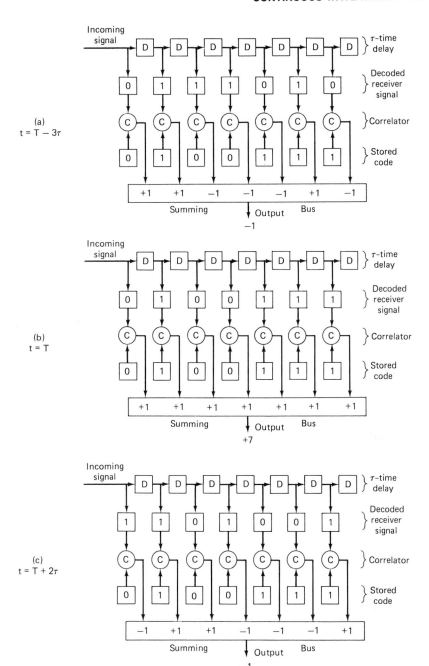

Figure 13-16. Seven-segment delay line and correlator processing a received signal at times (a) T-3τ, (b) T, and (c) $T + 2\tau$.

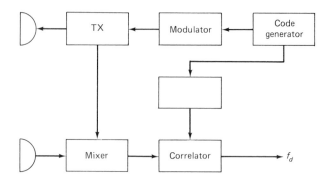

Figure 13-17. Pseudorandom PM CW radar.

13.9 SUMMARY

CW radar systems are more suitable than pulsed radar systems for many applications, especially in situations where a small, lightweight radar sensor is required. A basic homodyne CW radar is the simplest radar imaginable and can be made to fit into a package commensurate with its simplicity. For example, the Hewlett Packard Series 35200 X-band CW radar mentioned earlier in this chapter has a transmitter and receiver that together fit into a module approximately 2.25 in.^{3} in volume.

Methods of overcoming the simple CW radar's basic disadvantages—no target range information and little transmitter/receiver isolation—were discussed. FM techniques were shown to help provide increased isolation, while FM, multiple-frequency, and PM techniques were shown to provide target information. The interested reader is referred to references 19 to 24 for additional information on CW radar systems.

13.10 REFERENCES

1. M. I. Skolnik, *Introduction to Radar Systems,* 2nd ed., McGraw-Hill Book Co., New York, 1980, p. 9.
2. E. J. Barlow, "Doppler Radar," *Proceedings of the IRE,* vol. 37, no. 4, April 1949, pp. 340–355.
3. Skolnik, *Radar Systems,* pp. 70–81.
4. S. A. Hovanessian, *Radar Detection and Tracking Systems,* Artech House, Dedham, Mass., 1978, pp. 4–6.
5. "X-Band Doppler Radar Modules, 35200 Series," Hewlett-Packard Technical Data Notes, Hewlett-Packard Co., Palo Alto, November 1971.
6. K. E. Barton, *Radar Systems Analysis,* Artech House, Dedham, Mass., 1976, p. 382.
7. J. Mathews and R. L. Walker, *Mathematical Methods of Physics,* 2nd ed., W. A. Benjamin, Inc., Menlo Park, Calif., 1970, p. 102.

8. P. A. M. Dirac, *The Principles of Quantum Mechanics*, 3rd ed., Oxford University Press, Oxford, 1944.

9. Barton, *Radar Systems Analysis*, p. 113.

10. J. C. Toomay, *Radar Principles for the Non-Specialists*, Lifetime Learning, Belmont, Calif., 1982, pp. 102–106.

11. Hovanessian, *Radar Detection*, pp. 4-8 to 4-15.

12. Skolnik, *Radar Systems*, pp. 81–92.

13. F. E. Nathanson, *Radar Design Principles Signal Processing and the Environment*, McGraw-Hill Book Co., New York, 1969, pp. 365–367.

14. W. K. Saunders, "Post-War Developments in Continuous-Wave and Frequency-Modulated Radar," *IRE Transactions in Aerospace and Navigational Electronics*, vol. ANE-8, no. 1, March 1961, pp. 7–19.

15. R. D. Tollefson, "Application of Frequency-Modulation Techniques to Doppler Radar Sensors," *IRE NAECON Record*, Dayton, 1959, pp. 683–687.

16. Hovanessian, *Radar Detection*, pp. 4-15 to 4-17.

17. Skolnik, *Radar Systems*, pp. 95–98.

18. S. E. Craig, W. Fishbein, and O. E. Rittenbach, "Continuous-Wave Radar with High Range Resolution and Unambiguous Velocity Determination," *IRE Transactions on Military Electronics*, vol. MIL-6, no. 2, April 1962, pp. 153–161.

19. Toomay, *Radar Principles*, chaps. 5, 6.

20. Hovanessian, *Radar Detection*, chap. 4.

21. Skolnik, *Radar Systems*, chap. 3.

22. Nathanson, *Radar Design Principles*, chaps. 10, 12, and 13.

23. D. K. Barton (Ed.), *Radars*, Vol. 7: CW and Doppler Radar, Artech House, Dedham, Mass., 1980.

24. M. I. Skolnik (Ed.), *Radar Handbook*, McGraw-Hill Book Co., New York, 1970, chap. 16.

14
MTI AND PULSED DOPPLER RADAR

Carl R. Barrett, Jr.

14.1 INTRODUCTION

Relative motion between a signal source and a receiver creates a Doppler shift of the source frequency. Likewise, when a radar system illuminates a moving target that has a radial velocity component relative to the radar, the signal reflected from the target and received by the radar is also frequency shifted. This Doppler shift can be used in radar system applications for several advantageous purposes. For this discussion, applications of this effect are limited to (1) separating desired target returns from those of nonmoving targets and background clutter and (2) extracting information concerning the target's radial velocity.

In practice, the term *moving target indication (MTI) radar* usually refers to a system that produces ambiguous velocity and unambiguous range target data and uses a delay line canceller filter to isolate moving targets from the nonmoving background. Also, in practice, the term *pulsed Doppler radar* means a pulsed radar in which the Doppler measurement is unambiguous, the range measurement can be either ambiguous or unambiguous, and the Doppler data are extracted by the use of range gates and Doppler filters.[1-4]

The following section describes processes by which Doppler spectra are produced, defines the differences between CW and low-, medium-, and high-pulse repetition frequency (PRF) systems, and describes spectra that are typically encountered when these waveform types are employed.

14.2 PULSED RADAR WAVEFORMS

Pulsed waveforms are sometimes used instead of CW waveforms to circumvent problems related to transmitter signal leakage into the receiver, large-amplitude signals reflected from nearby clutter, and the summation (integral) of clutter extending from short range to the radar horizon. These classes of pulsed waveforms and their salient properties may be described as follows:

1. A high-PRF waveform normally has unambiguous Doppler frequency and highly ambiguous range* measurement characteristics which
   a. effectively solve the transmitter-to-receiver coupling problem of CW systems but
   b. introduce range blind zones during time periods when the transmitter is radiating.
2. A low-PRF waveform permits unambiguous range measurements but, in general, has highly ambiguous Doppler measurements. A low-PRF waveform
   a. circumvents the transmitter-to-receiver coupling problems and
   b. introduces Doppler blind zones.
3. A medium-PRF waveform may have range and Doppler frequency ambiguities but does
   a. circumvent the transmitter-to-receiver leakage problem of the CW system
   b. improve noise-limited detection performance relative to the low-PRF waveform and
   c. minimize the number of introduced Doppler blind zones (relative to the low-PRF system).

These characteristics imply the generic definitions of high-, medium-, and low-PRF radars.[1] Selection of the waveform that best fits a particular application can be made based only upon system requirements and operating conditions, and upon the target parameters and its environment. The radar system designer must balance performance requirements with waveform choice, system complexity, and achievable performance. The following discussions present a foundation upon which this choice can be based.

The performance of MTI and pulsed Doppler radars is a function of the target-to-clutter (plus noise) ratio** that can be achieved at the output of the radar processor. This ratio is influenced by both the passband characteristics of the filter and the spectral characteristics of the input interference term. Specifically, the spectral characteristics of the received clutter signal are defined by the following components:

1. Clutter internal motion.
2. Platform motion.
3. Antenna scanning modulation (or finite record lengths created by antenna "step scan" algorithms).

---

*An FM ranging code is assumed to permit range measurement with the CW system. A code such as FM ranging is also required with the high PRF waveform to achieve range ambiguity resolution.
**For this discussion, it is assumed that the term of interest is the background clutter (the thermal noise term is ignored).

4. Baseline RF system spectral stability.
5. Radar waveform.

Power spectral densities of the received waveforms from a moving platform-mounted CW and pulsed Doppler radars are illustrated in Figure 14-1. It is seen for the CW system that the sidelobe spectrum extends from $(f_c - 2V_c/\lambda)$ to $(f_c + 2V_c/\lambda)$, where f_c is the carrier frequency, V_c is the platform velocity, and λ is the RF wavelength. The mainlobe portion of the spectrum is centered on $(f_c + 2V_c \cos \phi/\lambda)$, where ϕ is the beam scan angle away from the platform velocity vector. By defining the relative velocity of the target in relation to the platform (radar) as V_R, it is seen that the target frequency is $(f_c + 2V_R/\lambda)$. For the case illustrated, the target is located in a frequency region that is clear of clutter when $V_R > V_c$, and the target velocity can be measured unambiguously. Ranging codes as discussed in Chapter 13 are necessary to determine target range. In the pulsed Doppler example of Figure 14-1, the radar PRF (f_R) is high enough to maintain a velocity-clear region and to allow unambiguous Doppler measurements. Techniques for removing velocity and/or range ambiguities are presented later in the chapter.

The platform motion spectrum results from the interaction of the radar beam

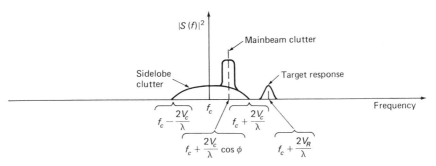

(a) CW radar received expected power spectrum

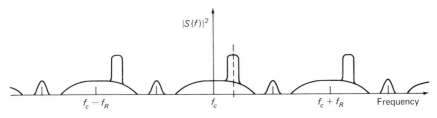

(b) Pulsed Doppler received expected power spectrum

Figure 14-1. Sample received expected power spectrums for a CW and a pulsed Doppler radar.

with the terrain beneath the moving platform. Consider, for example, the relationship

$$f_{\mathrm{d}} = \frac{2V_c}{\lambda} \cos \phi \tag{14-1}$$

where

f_{d} = Doppler shift
V_c = platform velocity
λ = RF wavelength
ϕ = angle between the platform velocity vector and the line of sight (LOS) to the region of interest

which defines the Doppler shift created by platform motion along a particular LOS. The appropriate geometry is pictured in Figure 14-2. By rotating the LOS about the velocity vector at a constant angle, a cone of revolution is generated. This cone defines points in space that correspond to the platform motion–induced Doppler shift described by Eq. (14-1). It can be shown (Figure 14-3)

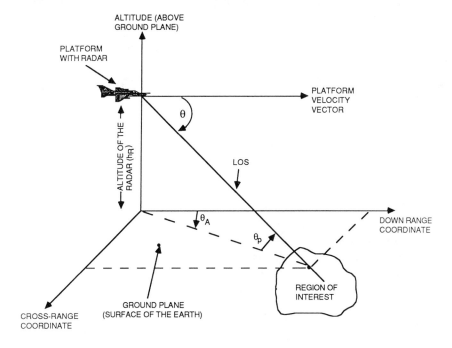

Figure 14-2. Radar/terrain geometry.

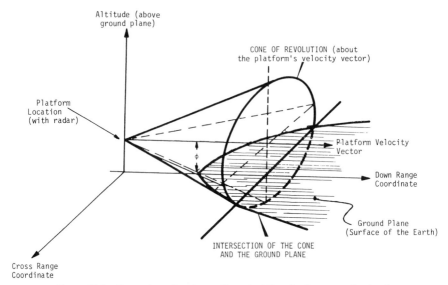

Figure 14-3. Generation of contours of constant Doppler frequency (isodops).

that the intersection of this cone with the stationary terrain beneath the moving platform defines a locus of constant Doppler shift (an isodop). These intersections, as sketched in Figure 14-4, define, for level flight, hyperbolas. For the special case of level flight (velocity vector parallel to the positive X-axis), constant range and velocity contours as represented in Figure 14-5 are obtained. If nonlevel flight is to be considered, the isodops must be appropriately modified to appear as different conic sections.

If an illuminated area defined between two isodops Δf_d apart (Figure 14-5) within a range cell is determined and input into the radar range equation (together with the appropriate two-way antenna gain at the observation LOS), then that region's contribution to the platform motion spectrum can be calculated. This process is illustrated in Figure 14-6. The complete platform motion spectrum can be completed by following the lead laid out in Figure 14-6 and extending the calculation to include all regions that produce return signals during the observation interval. Figure 14-7 illustrates the platform motion clutter power spectral density that would be produced by ground clutter in an airborne radar. Note that in this figure the maximum Doppler shift asymptotically approaches the extremes of permitted Doppler shifts ($\pm 2 V_c/\lambda$) as the slant range increases from the radar altitude (h_R) toward the radar horizon. It should also be noted that when the platform is in level flight, the maximum Doppler shift at each slant range corresponds to a point on the earth's surface that is coincident with the down-range projection (in a plane that is normal to the earth's surface) of the platform's velocity vector. Positive Doppler shifts correspond

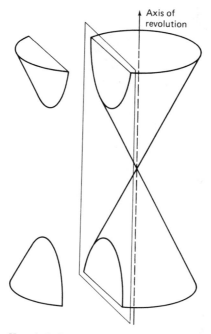

Figure 14-4. Hyperbola formed by cutting a cone with a plane surface.

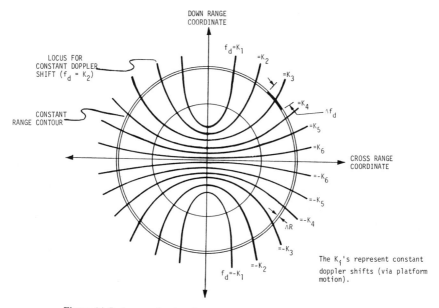

Figure 14-5. Locus of points for constant range and Doppler (isodops).

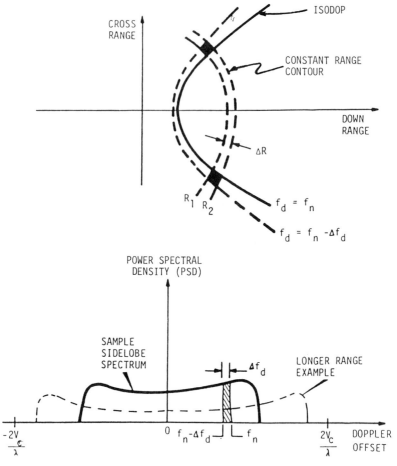

Figure 14-6. Sample process for generating the platform motion contribution to the resultant power spectral density.

to forward hemisphere closing situations, while negative Doppler shifts correspond to rear hemisphere opening situations. Figure 14-7 illustrates the large densities associated with the altitude line, in near-range sidelobe regions, and in the mainlobe region.

After data of the form of Figure 14-7 are generated and the target response is placed, the ambiguity function of the radar waveform can be used to determine the resultant radar composite response (output signal-to-interference ratio).[5] The ambiguity function $\chi(\tau, f_d)$ which is given by[1]

$$\chi(\tau, f_d) = K \int_{-\infty}^{\infty} u(t)\, u^*(t + \tau) e^{j 2\pi f_d t}\, dt \qquad (14\text{-}2)$$

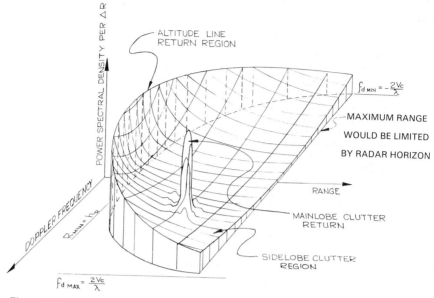

Figure 14-7. Representative background clutter distribution as a result of platform motion.

where

K = constant that is selected to force $\chi(0, 0) = 1.0$

$u(t)$ = complex envelope of the transmitted waveform

τ = delay of the reflecting element beyond a reference delay

f_d = closing Doppler shift of the reflecting element beyond a reference Doppler shift

describes the response of the radar to reflecting elements that are displaced (by amounts τ and f_d) from the reference values. Therefore, if it is assumed that the radar is matched to the target response and the squared magnitude of $\chi(\tau, f_d)$ is used as a multiplier of the clutter distribution (Figure 14-7), an integral over the entire resulting "weighted" distribution describes the effective clutter power, referenced to the output terminal of the receive antenna, that will appear at the output of the matched filter. That is, if the amplitude of the clutter distribution (Figure 14-7) is given* as $W(\tau, f_d)$, the output clutter power, C_o, from the matched filter, when referenced to the terminals of the receive antenna, is given as

$$C_o = \int_{\tau_{min}}^{\tau_{max}} \int_{f_{d_{min}}}^{f_{d_{max}}} |\chi(\tau, f_d)|^2 \, W(\tau, f_d) \, df_d \, d\tau \qquad (14\text{-}3)$$

---

*The range dimension has been transformed into a time delay (τ) dimension.

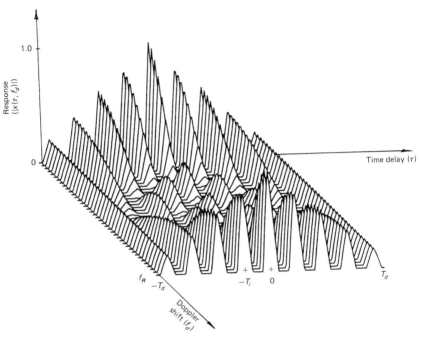

Figure 14-8. Central region of the ambiguity function for a coherent train of five equally spaced uncoded pulses.

when the condition $\chi(0, 0)$ is defined to match the target position. The ambiguity function illustrates the effect of both ambiguous range and Doppler-shifted returns that compete with the desired response.

Figure 14-8 demonstrates the central region of the ambiguity function for a coherent group of five ($N = 5$) equal-amplitude, equally spaced, uncoded pulses. In the case plotted, the PRF (f_R) was 30 kHz, the interpulse period (T_i) was $1/f_R$, and the pulse length (τ_p) was 10 μs. The one-half duration T_d as illustrated in Figure 14-8 can be shown to equal

$$T_d = (N - 1)/f_R + \tau_p \qquad (14\text{-}4)$$

It is convenient to represent the clutter distribution and ambiguity function as contours in a time delay (range) by Doppler shift space.[5,6] Figures 14-9 and 14-10 illustrate this representation for the clutter distribution and ambiguity function, respectively. The waveform represented in Figure 14-10 has a much greater number of pulses than the $N = 5$ case of Figure 14-8. The crosshatched regions in Figure 14-10 represent the location of the "ridges" that are evident in the plot of the ambiguity function of Figure 14-8. The solid dots represent the peaks of the ambiguity function that are greater in amplitude than a threshold placed 3 dB below the peak response (3 dB below $\chi(0, 0)$).

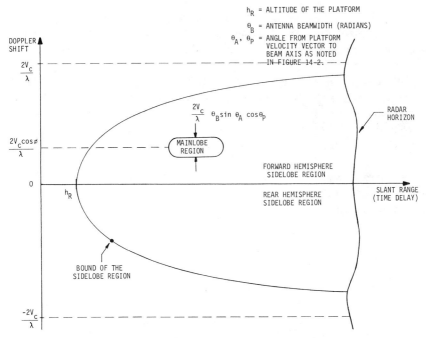

Figure 14-9. Slant range (time delay) by Doppler shift representation of the clutter distribution.

Note that Figure 14-9 contains a representation of the position and width of the main beam clutter return. This width will be demonstrated later to be an important factor in the choice of PRF. The information contained in Figures 14-9 and 14-10 provides a method for describing portions of the background return that contribute to the matched-filter output. Included in these two figures are data that describe the relative weighting that is appropriate for ambiguous

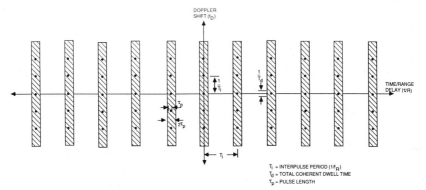

Figure 14-10. Time delay (slant range) by Doppler shift representation of the central region of the ambiguity function for a long sequence of equal-amplitude, equally spaced pulses (ambiguity diagram).

returns within various portions of the slant range by Doppler shift space. For the case of the nonrange ambiguous, low-PRF waveform, single-range ridges in the ambiguity diagram produce returns that contribute to the resultant platform spectrum.

In the case of the medium- and high-PRF waveforms, multiple (ambiguous)-range rings (ridges) produce returns that must be summed to determine the platform motion power spectral density (PSD) in each effective range bin. Therefore, to determine the complete platform motion spectrum in each range bin, the technique outlined in Figure 14-6 must be expanded to permit summation over all affected range rings. In the limit case, the CW waveform requires summation over all ranges. This process is represented pictorially in Figure 14-11. The right-hand section of the figure represents the clutter density function described in Figures 14-7 and 14-9. The resultant PSD is given as the integral of this density function $W(R, f_d)$ over the range region defined by the waveform. Thus,

$$\text{PSD}(f_d) = \int_R W(R, f_d) \, dR \qquad (14\text{-}5)$$

This result is valid in situations where the length of the waveform is long compared with the length of the scattering space (generally representative of airborne CW and low-, medium-, and high-PRF radar applications). The result of

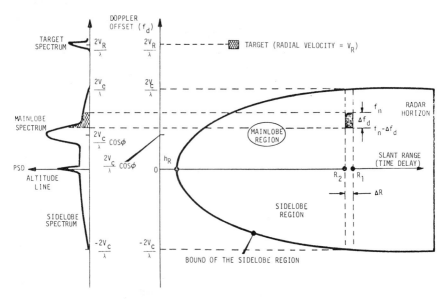

Figure 14-11. Propagation of the platform motion spectrum for a CW radar.

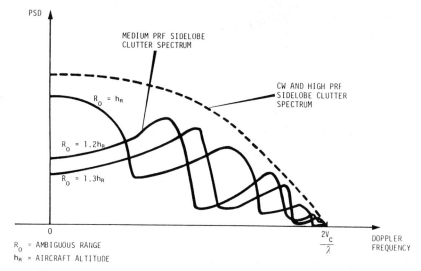

Figure 14-12. Sample medium- and high-PRF and CW platform motion spectra.

accomplishing this integration for the CW case is represented by the left-hand section of Figure 14-11.

This process can be used to demonstrate that the platform motion spectral shape achieved with a high-PRF waveform appears essentially as sketched for the CW case. However, for the medium-PRF case, the platform motion contribution to the spectral shape varies dramatically from range bin to range bin. An example of this is sketched in Figure 14-12. In this figure the ambiguous or indicated range (R_0) is given by

$$R_0 = R_t - nR_{AMB} \tag{14-6}$$

where

R_t = actual range of the element creating the response
n = positive integer (including zero) chosen to make $0 \leq R_0 < R_{AMB}$
$R_{AMB} = c/(2f_R)$ with c the velocity of light

It is interesting to observe that FM ranging codes that are used with CW and high-PRF waveforms* for range estimation distort the platform motion spectrum. One such code is the three-phase code suggested in Figure 14-13. By using the constant frequency, linear-up and linear-down FM phases, it can be shown that the change in apparent Doppler frequency between the three phases

---

*Various codes are also used with medium-PRF systems to resolve their ambiguities (for example, some codes use PRF programming with types of coincidence logic).

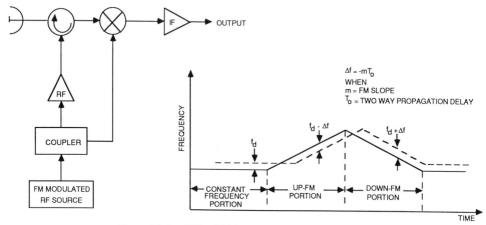

Figure 14-13. CW and high-PRF ranging code example.

permits (1) range estimation and (2) resolution of ambiguities between multiple targets. Figure 14-13 illustrates that during the up-FM cycle the apparent Doppler offset decreases. The apparent Doppler increases during the down-FM cycle. Thus, by reference to Figure 14-6, it can be shown that the platform motion power spectral density, $PSD(f_d)$, during either the up- or down-FM can be obtained, conceptually, by integration of $W(R, f_d)$ along the slopes, as represented in Figures 14-14 and 14-15. Thus,

$$PSD(f_d)_{FM} = \int_R W(R, f_d - mR) \, dR \qquad (14\text{-}7)$$

where m is the effective slant range to apparent Doppler shift scale factor of the FM ranging code.

By using this technique, it is possible to visualize the impact such a code sequence will have upon the platform motion spectrum produced at the output of the radar receiver. Figures 14-13, 14-14, and 14-15 illustrate how a target that was initially located within the so-called velocity-clear region can end up competing during different phases of the ranging code with background clutter in a non-velocity-clear region.

Various models in the literature[6-11] describe the characteristics of background clutter temporal fluctuations (internal motion). Similarly, radar system instabilities[12,13] (master oscillator AM and FM noise, transmitter-induced modulation, timing jitter, etc.) must be modeled and used to broaden* the spectrum

---

*Analytically, the mathematical process is the convolution of the platform motion voltage spectrum, $S_{pm}(f)$, with the internal motion, $S_{IM}(f)$, and radar stability, $S_R(f)$, voltage spectra. However, since we have predicted only expected PSDs, this process can be performed by Monte Carlo simulation techniques. See reference 14 for further study of this subject.

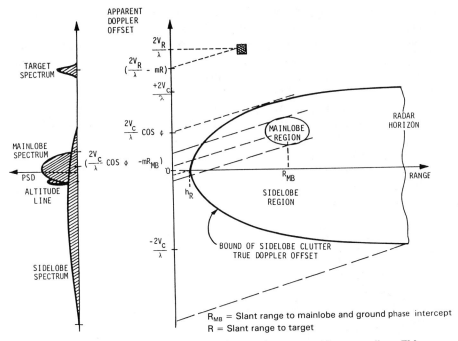

Figure 14-14. Modified CW and high-PRF platform motion spectra with up-ramp linear FM.

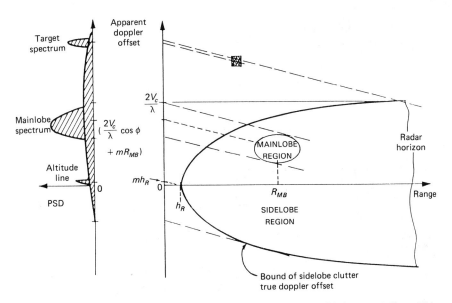

Figure 14-15. Modified CW and high-PRF platform motion spectrum with down-ramp linear FM.

produced by platform motion alone. Incorporation of these processes produces an effective broadened platform-expected PSD.

Since the radar uses a pulsed waveform with a fixed pulse length, τ_p, and interpulse period, T_i, its effect on the resulting received spectrum must be considered. In cases where the antenna scans, the beam dwell time is also finite. Limited (finite) beam dwell may result by scanning the antenna across the target so that a scanning modulation equivalent to the two-way voltage pattern of the antenna is impressed upon the return signal. Finite beam dwell may also result from beam step scan algorithms.

Step-scan techniques typically lead to a scanning modulation spectrum that can be characterized by a group of $N + 1$ uniformly weighted pulses. Burdic[15] has shown that such a group of $N + 1$ pulses (with the center of the central pulse located at $t = 0$) can be represented in a contracted notation as

$$s(t) = \sum_{m = -N/2}^{N/2} s_1 (t - mT_i) \tag{14-8}$$

where

$$s(t) = \text{rect } (t/\tau_p)$$
$$\text{rect } (t/\tau_p) = 1.0 \text{ for } -\tau_p/2 \leq t \leq \tau_p/2$$
$$= 0.0 \text{ elsewhere}$$
$$s_1(t - mT_i) = s_1(t) \text{ replicated at time centers } mT_i$$

Since the amplitude spectrum $S(f)$ is defined as the Fourier transform of the time waveform, $s(t)$, it can be shown[15] that

$$S(f) = S_1(f) \sum_{m = -N/2}^{N/2} \exp \left(-j2\pi fnt_i \right)$$

$$= S_1(f) \frac{\sin((N + 1)\pi f T_i)}{\sin(\pi f T_i)} \tag{14-9}$$

Since the transform* of $s_1(t)$ is given as

$$S_1(f) = \tau_p \text{ sinc } (f\tau_p) \tag{14-10}$$

the transform of the train of $N + 1$ pulses is then given as

$$S(f) = \tau_p \text{ sinc } (f\tau_p) \frac{\sin ((N + 1)\pi f T_i)}{\sin (\pi f T_i)} \tag{14-11}$$

---

*The sinc(x) function is defined in reference 15 (p. 48) as sinc(x) = $\sin (\pi x)/(\pi x)$.

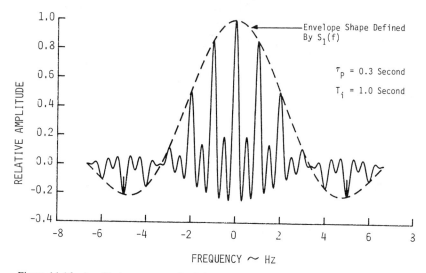

Figure 14-16. Amplitude spectrum of a finite number of rectangular pulses $[(N + 1) = 5]$.

This function is sketched in Figures 14-16, 14-17, and 14-18. It should be noted that this result describes the shape of the cut through the ambiguity function for the situation when $\tau = 0$.

Figure 14-8 illustrates that the shape of a cut through the ambiguity function for constant offset delay (τ) changes as a function of the delay. A rationale for this change is illustrated in Figure 14-19. This figure shows how ambiguous

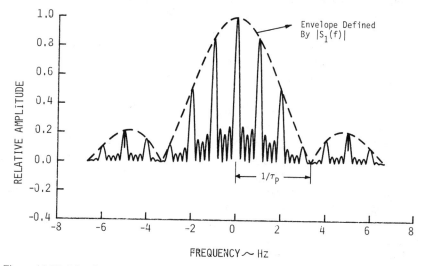

Figure 14-17. Magnitude of the amplitude spectrum for a finite number of rectangular pulses $[(N + 1) = 5]$.

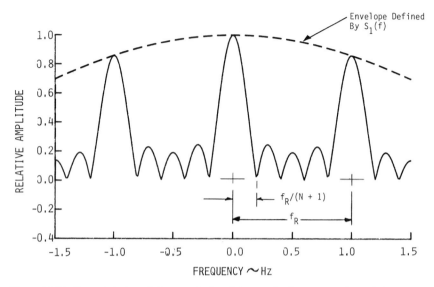

Figure 14-18. Expanded plot of the magnitude of the amplitude spectrum for a finite number $[(N + 1) = 5]$ of rectangular pulses.

range segments contribute to the composite signal return. As demonstrated in the figure, a five-pulse burst waveform with interpulse period T_i and pulse width τ_p is used with a range-gated receiver to sample the return from a scatterer at slant range R_T. The receiver is gated on in a burst fashion that is delayed from the envelope of the transmit waveform by a time interval given by $2R_T/c$ (c equals the velocity of propagation of light). The pulse repetition rate (f_R) defines a maximum unambiguous range (R_{AMB}) of $c/(2f_R)$. As seen in the figure, a scatterer at a slant range of $R_T - R_{AMB}$ contributes four pulse samples to the total sampled output. Similarly, it is seen that a scatterer at slant range $R_T + 4R_{AMB}$ contributes only one pulse sample to the total sampled output. Intermediate-range positions are also indicated.

From this result, it can be demonstrated that the shape of a constant delay cut through the ambiguity function is controlled by the time delay away from the matched filter design time reference. In the case described in Figure 14-19, scatterers at ranges ($R_T - R_{AMB}$) and ($R_T + R_{AMB}$) contribute only four pulse samples $[(N + 1) = 4]$ to the gated received signal. This produces an amplitude spectrum as represented in Figure 14-20. This spectrum also represents the shape of cuts through the ambiguity function (Figure 14-8) for $\tau = \pm T_i$. The effect on the scanning modulation spectrum can be seen by comparing the result with Figure 14-17.

The shape of both the spectrum and the ambiguity function for τ equal to nonzero multiples of the interpulse period is important, since it is key to the determination of the performance of MTI and pulsed Doppler radars. An ex-

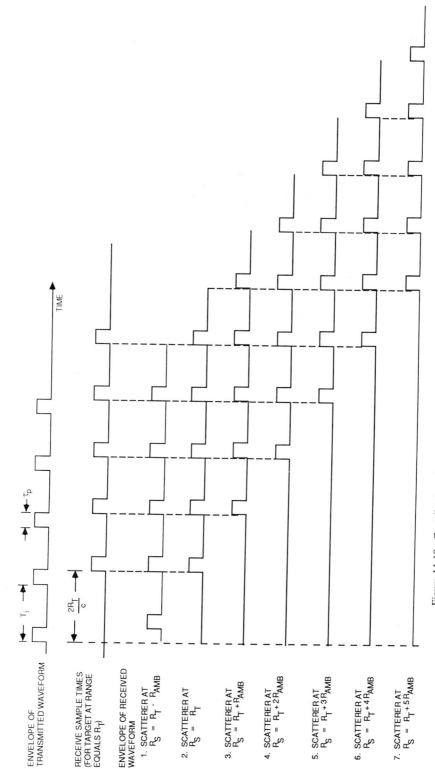

ENVELOPE OF
TRANSMITTED WAVEFORM

RECEIVE SAMPLE TIMES
(FOR TARGET AT RANGE
EQUALS R_T)

ENVELOPE OF RECEIVED
WAVEFORM

1. SCATTERER AT
   $R_S = R_T - R_{AMB}$

2. SCATTERER AT
   $R_S = R_T$

3. SCATTERER AT
   $R_S = R_T + R_{AMB}$

4. SCATTERER AT
   $R_S = R_T + 2R_{AMB}$

5. SCATTERER AT
   $R_S = R_T + 3R_{AMB}$

6. SCATTERER AT
   $R_S = R_T + 4R_{AMB}$

7. SCATTERER AT
   $R_S = R_T + 5R_{AMB}$

TIME

τ_p

τ_i

$\dfrac{2R_T}{c}$

Figure 14-19. Contribution of ambiguous range segments to the sampled return by a burst waveform of five pulses.

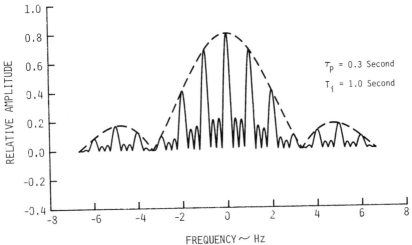

Figure 14-20. Magnitude of the amplitude spectrum for a finite number $[(N + 1) = 4]$ of rectangular pulses.

cellent treatment of the impact of this result on a practical system has been described by Ward.[16]

If the modulation function, $s(t)$, is used to modulate a carrier term in the radar transmitter as represented in the following sketch

$$\cos (2\pi f_0 t) \longrightarrow \boxed{\text{MODULATOR}} \xrightarrow{s_0(t)}$$
$$\uparrow$$
$$s(t)$$

so that

$$s_0(t) = s(t) \cos (2\pi f_0 t) \tag{14-12}$$

it can be shown[17] that

$$S_0(f) = S(f) * [\tfrac{1}{2}(\delta(f - f_0) + \delta(f + f_0))] \tag{14-13}$$

where $*$ denotes convolution and $\delta(f)$ is the delta function as defined in Chapter 13, or that

$$S_0(f) = \frac{\tau_p}{2} \left[\text{sinc} \left((f - f_0)\tau_p \right) \frac{\sin \left((N + 1)\pi(f - f_0)T_i \right)}{\sin \left(\pi(f - f_0)T_i \right)} \right.$$
$$\left. + \text{sinc}((f + f_0)\tau_p) \frac{\sin \left((N + 1)\pi(f + f_0)T_i \right)}{\sin \left(\pi(f + f_0)T_i \right)} \right] \tag{14-14}$$

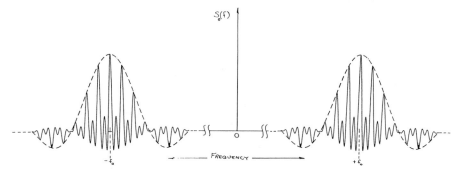

Figure 14-21. Voltage spectrum of a five-pulse burst RF waveform.

This function, which is in fact the voltage spectrum of the truncated ($N + 1$ pulse) transmitted waveform, is sketched in Figure 14-21. When this function, $S_0(f)$, is convolved with the voltage spectrum that incorporates results which include the effects of platform motion, internal motion, and system stability, the voltage spectrum that the Doppler filter must contend with is obtained. Sample received high-, medium-, and low-PRF expected power spectra are sketched in Figure 14-22.

These results allow evaluation of the signal-to-clutter ratio performance of the system or definition of the MTI filter that best fits the system application. Also, the characteristics of the filtered clutter residue define the requirements imposed on an automatic detection thresholding circuit, if a desired CFAR is to be retained. Selected MTI filter and CFAR topics are discussed in other sections.

14.3   NONCOHERENT PULSED RADAR

A simplified block diagram of a noncoherent search radar* is given in Figure 14-23. As this figure shows, no reference signal used by the receiver is phase coherent to the output phase of the transmitter. The output of a free-running pulsed transmitter is propagated through a coupler and a duplexing circulator and radiated. The signal sample taken by the coupler is fed to an automatic frequency control (AFC) circuit such that the local oscillator is made to track the transmitter frequency with suitable long time constants. After duplexing, the returned signal is mixed with the local oscillator to generate an IF signal. This signal is bandpass filtered and amplified by the IF amplifier. After square law (noncoherent) detection, the signal is processed by the signal processor [in this case, noncoherently integrated (see Chapter 1) and compared (see Chapter 12) with an automatically computed adaptive threshold] in preparation for presenting threshold crossings, via the display, to an operator.

---

*A noncoherent radar is one in which the receiver does not have or does not make use of a reference signal that is phase coherent with that of the transmitted waveform.[1]

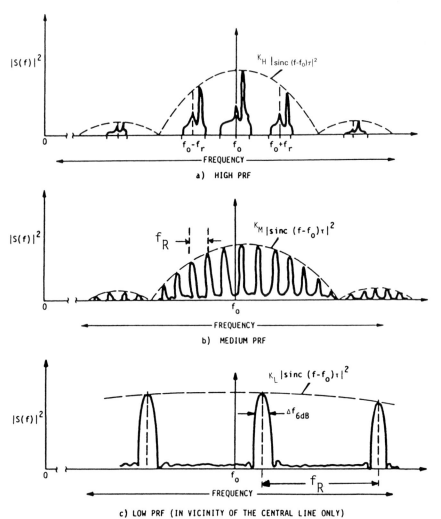

Figure 14-22. Sample received spectra.

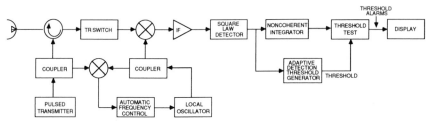

Figure 14-23. Block diagram of noncoherent pulsed search radar.

Table 14-1. Noncoherent Millimeter Wave Radar Application Parameters (For Clutter-Limited Detection).

σ_t	5 m^2	Target RCS
SW#	3	Target Swerling fluctuation number
σ°	-20 and -30 dBm^2/m^2	Clutter backscatter coefficient
τ_p	70×10^{-9} s	Effective pulse length
θ_B	1.0°	Antenna 3-dB (one-way) beamwidth
h_R	200 ft	Flight altitude of the platform (helicopter)
$\dot{V}_c$	0.0 mph	Platform flight velocity
$\dot{\theta}$	120°/s	Antenna scan rate (yaw)
f_r	10 kHz	Transmitter PRF
P_{fa}	10^{-6}	Design probability of false alarm
P_d	0.9	Design probability of single-scan detection (at R_d)
R_d	10 km	Desired detection range (P_{fa} and P_d)
$L_{S/C}$	4.0 dB	Total radar system losses (applicable to clutter-limited performance)
λ	3.2 mm	RF wavelength

Problems encountered in detecting small-RCS targets in expected background clutter environments may be demonstrated by analysis of this system. With system design restrictions that constrain parameters such as operating frequency, maximum permitted antenna dimensions, maximum processing dwell time, maximum pulse length, and operating geometry, it may be shown that the radar designer is left with only a few techniques to minimize the performance limits imposed by return from background clutter.

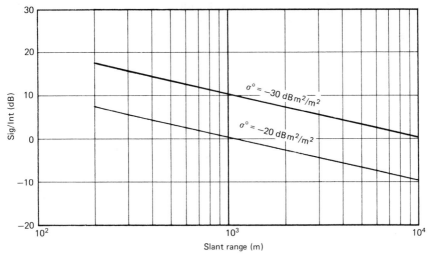

Figure 14-24. Signal-to-clutter ratio versus slant range for the parameters of Table 14-1.

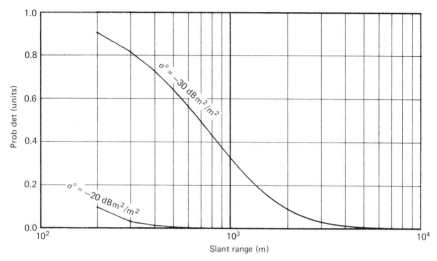

Figure 14-25. Single-scan probability of detection for monochromatic, noncoherent pulsed radar with parameters and conditions of Table 14-1.

As an example of these problems, consider the clutter-limited performance of a noncoherent pulsed millimeter wave radar that operates from a hovering helicopter and is intended to detect small-RCS surface targets. Significant parameters are given in Table 14-1. The signal-to-clutter ratio as a function of range for this system is given in Figure 14-24. The achievable detection performance for this noncoherent radar, when operated at a fixed frequency, is given in Figure 14-25. The difficulty encountered in achieving the desired 10-km range (0.9 P_d and 10^{-6} P_{fa}) is evident.

As an example to illustrate the procedures used to calculate this performance, consider the equation

$$\frac{S}{C} = \frac{\gamma_t}{R\theta_B \dfrac{c\tau}{2} \sigma^0 \sec \theta_G L_{s/c}}$$

for the single pulse signal-to-clutter ratio. The grazing angle of the radar LOS with the terrain, θ_G, is given as

$$\theta_G = \sin^{-1}\left(\frac{h_R}{R}\right)$$

At 1000 meters slant range, using the parameters of Table 14-1 and for $\sigma^0 =$

$-30 \text{ dBm}^2/\text{m}^2 \ (10^{-3} \text{ m}^2/\text{m}^2),$

$$\left.\frac{S}{C}\right|_{\text{dB}} \cong 10 \log$$

$$\cdot \left[\frac{5}{(1000)\left(\dfrac{1.0}{57.3}\right) \dfrac{(3 \times 10^8)(70 \times 10^{-9})}{2} (10^{-3}) \sec\left(\sin^{-1}\left(\dfrac{200}{3.27 \times 10^3}\right)\right) 2.51} \right]$$

$\cong 10.4 \text{ dB}.$

The data of Figure 11-7 for single pulse detection (no independent samples per beam dwell) yield a $P_d \cong 32$ percent. These results may be observed in Figures 14-24 and 14-25, respectively.

One potential way of minimizing this performance handicap is to incorporate frequency agility.* Figure 14-26 illustrates the single scan detection performance of this system when a frequency-agile bandwidth of 500 MHz is used. These results are calculated in the manner illustrated above except that the number of independent clutter samples created by frequency agility (or other means) defines the particular curve (or extrapolated curve) of Figure 11-7 that relates the achievable detection performance. The performance enhancement relative

---

*The transmitted frequency is changed pulse-to-pulse to force the background (and potentially the target) to fluctuate pulse-to-pulse. This leads to a minimum detection threshold and hence enhanced performance (see Chapter 11).

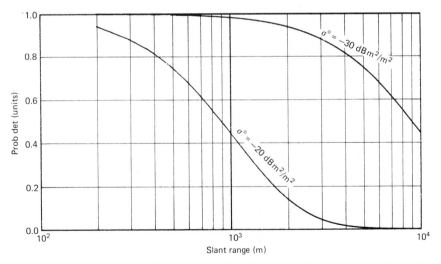

Figure 14-26. Single-scan probability of detection for a frequency-agile, noncoherent pulsed radar with parameters and conditions of Table 14-1 (500 MHz frequency-agile bandwidth).

to the monochromatic case is evident. The desired detection range goal (even in clutter represented by the smaller backscatter ratio) is still not met. The need for performance beyond that achievable by this technique (and parameter changes) leads to the use of processes such as polarization diversity (and agility) and to the use of the Doppler shift to allow detection of targets that have a radial velocity component relative to the background return. Adaptations of a non-coherent pulsed radar to exploit the Doppler shift so as to provide detection of targets that are moving relative to their background are discussed in Section 14.5.

14.4   COHERENT PULSED RADAR

As described in Section 14.3, the signal processing in a noncoherent pulsed radar is performed independently of the pulse-to-pulse phase of the transmitted waveform. In a coherent pulsed radar, however, the phase of the transmitted waveform is preserved in a reference signal that is used within the receiver for signal demodulation.[18] A simplified block diagram for a coherent pulsed radar is shown in Figure 14-27. As illustrated in this figure, the output of a coherent reference oscillator (COHO) is propagated through a coupler and into a pulse-forming gate. This gate forms, during each pulse repetition interval, a burst of the IF carrier that is equal in length to and timed to be in time coincidence with the "on time" of the pulse-modulated transmitter (RF amplifier). The gated COHO signal is used as a reference in a single sideband modulator such that a gated RF carrier with a known frequency and phase is generated. This pulse

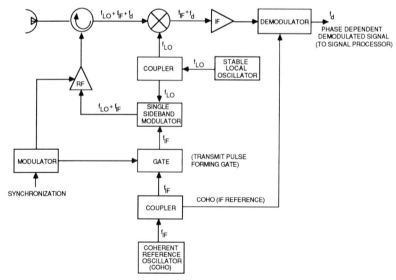

Figure 14-27. Simplified diagram of coherent pulsed radar.

burst is amplified in the transmitter (composed of TWTs, CFAs, or similar devices) and transmitted.

The second input into the single sideband modulator is derived from the output of the STALO. The STALO is also used as an LO reference in the receiver to translate the received signal to a convenient IF region. After amplification in the IF amplifier, the received signal is routed into a phase-sensitive demodulator (which uses a sample of the COHO as a frequency and phase reference). The resulting video signal is routed into the signal processor for desired processing.

The use of the STALO and COHO reference signals to store the phase of the transmitted waveform and to allow incorporation of this phase information into later signal processing identifies the radar described in Figure 14-27 as being coherent. On the other hand, the absence of such references (and the use of the square law detector as a demodulator at the IF output) identifies the radar that is described in Figure 14-23 as being noncoherent.

A comparison of Figures 14-23 and 14-27 demonstrates the relative complexity of the RF/IF sections of coherent and noncoherent systems. It can generally be demonstrated that this relative complexity is also carried into the signal processor. Therefore, if it were not for performance advantages, the noncoherent configuration would be exclusively used in search radar applications. As illustrated in Section 14.3, there are many situations in which the relative motion between the desired target and its background permits the exploitation of differential Doppler shifts to isolate desired target responses from large dominating (in amplitude) background returns.

As will be demonstrated in Section 14.5, there are techniques through which the noncoherent radar form (Figure 14-23) can be used to accomplish Doppler shift–aided detection of targets (as desired when small-RCS returns are otherwise obscured by background clutter or when the presence of a Doppler shift is useful as a means of target discrimination). However, it will be seen that these techniques have definite limitations and that their performance, at best, can only approach that of the coherent system. The radar designer, therefore, has the challenge of determining for the application being considered the relative performance versus cost merits of noncoherent versus coherent radar concepts.

14.5 MTI RADAR

As discussed in Section 14.1, the IEEE[1] defines an MTI radar as one in which the detection and display of moving targets are improved by the suppression of fixed targets.[19-21] This definition is expanded to incorporate Doppler processing as one possible form of MTI implementation. Barton[22] states that simple band rejection processing (filtering) is characteristic of MTI radars. The discussion in Section 14.1 illustrates the confusion relative to the definition by introducing alternative definitions that are summarized by Skolnik.[4] Without attempting to

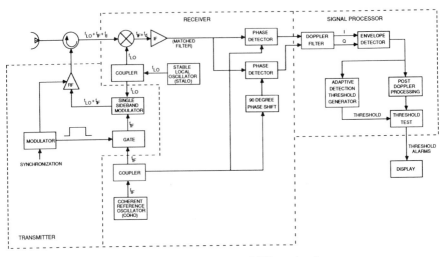

Figure 14-28. Pulsed coherent MTI search radar.

resolve this confusion, an MTI radar is defined, for purposes of this discussion, as one that uses simple band reject filtering to reject the return from fixed (stationary) targets. Enhanced detection and display of moving target responses are thereby achieved.

Figure 14-28 is a block diagram for a sample pulsed coherent MTI search radar. It uses a STALO to generate a reference signal for the receiver mixers. Similarly, a COHO is used as a reference to offset, in a single sideband mixer, a sample of the STALO output. This forms the transmitter excitation reference. By mixing the return signal with the LO and the output of the IF amplifier with a sample of the COHO, both phase and amplitude effects introduced into the signal by propagation and the target (i.e., Doppler and relativistic effects) may be recovered. The Doppler filter and remaining signal-processing circuits implement functions required for signal-to-interference ratio enhancement and desired detection performance.

The Doppler filter is structured to have a desirable passband such that relatively narrow bandwidth clutter is rejected. Since the target velocity is unknown a priori, the filter has a broad passband (in the Doppler shift or frequency dimension). The outputs of the filter are envelope detected and supplied as inputs to the post-Doppler processing stages such that noncoherent integration (see Chapter 11) and automatic detection using, potentially, adaptive thresholding techniques (see Chapter 12) can be performed.

The desired response of the MTI filter relative to the spectrum of the signal-plus-noise-plus-background clutter (fixed response) is illustrated in Figure 14-29. The rejection notches in the passband should be placed in frequency

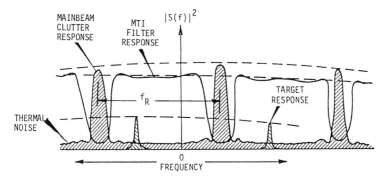

Figure 14-29. Sample MTI filter frequency response.

about the responses that are to be rejected and should only be as wide as required to achieve the desired clutter cancellation.* Placement of the notch must, in airborne MTI applications, be programmable such that it matches the Doppler shift associated with the centroid of the mainbeam clutter. The Doppler shift of the mainbeam clutter is given as

$$f_{d/MB} \cong \frac{2V_c}{\lambda} \cos \phi \qquad (14\text{-}15)$$

where V_c equals the platform velocity, ϕ is the angle between the velocity vector of the aircraft and the beam axis, and λ is the RF wavelength. This process is accomplished either by appropriate frequency shifts of the LO or COHO receiver reference signals or by programmable features of the MTI filter.

Many techniques are available for realizing the MTI filter.[24-27] Two particularly advantageous implementations are the delay line canceller and the range gate and filter. A block diagram of a generalized digital filter (multipulse canceller) is given in Figure 14-30. This figure illustrates a filter that uses both recursive and nonrecursive parts.** The c and d values are coefficients that are selected to obtain, within each instrumented range cell, the desired passband characteristics. The blocks labeled T_d represent a real-time delay equal to the delta time between input samples (taken as the pulse repetition interval).

The MTI processor for a range gate and filter MTI system is illustrated in Figure 14-31. The output of the phase detector (either the I or Q channel of Figure 14-28) is provided as the input to N sample-and-hold circuits. Each sample-and-hold circuit receives a sample gate (R_g #n) that is used to take a range

---

*Clutter cancellation is defined[23] as the ratio of the clutter output after cancellation to source clutter output without cancellation.
**Note: A recursive filter makes use of feedback paths while the nonrecursive filter uses only feedforward paths.

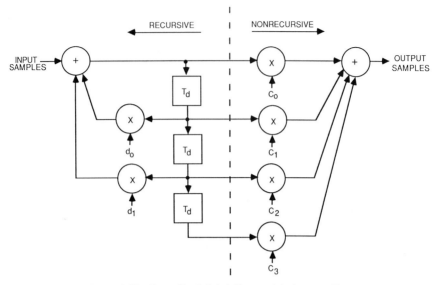

Figure 14-30. Generalized digital filter (multipulse canceller).

sample for the nth range cell. The output of each sample-and-hold circuit is filtered by a lumped constant (or active) bandpass filter to establish the desired passband. The effective achieved passband for a single range gate is illustrated in Figure 14-32. In this figure, f_L and f_H represent the lower and upper corner frequencies for the bandpass filter, respectively (f_R is the PRF).

In practice, modern digital technology leads to very package-efficient, reproducible-delay line canceller configurations. Further, because digital implemen-

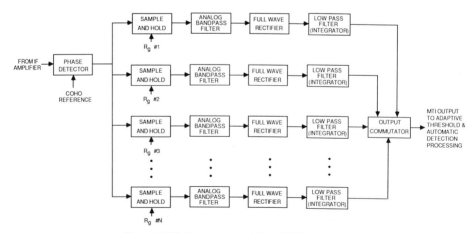

Figure 14-31. Range gate and filter MTI processor.

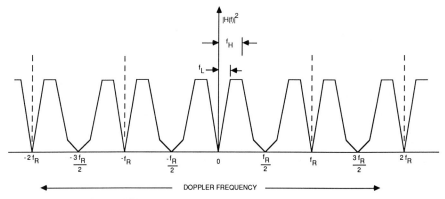

Figure 14-32. Sample passband for a range gate and filter MTI.

tations of the delay line canceller can provide the desired filter passband characteristics with the flexibility of passband programmability, they are the preferred choice. This conclusion is, therefore, quite different from that reached prior to the existence of current digital technology.[4]

Radars that are classically configured as noncoherent systems are sometimes used to detect and display targets that are in motion relative to their background. Figure 14-33 gives a block diagram of a simple, pulsed, noncoherent MTI search radar. Its principle of operation can be described with the aid of Figure 14-34, which demonstrates that A-scope range video presentations contain "butterflies" at the slant range (R_T) of the moving target. Each butterfly is created (over several interpulse periods) by the fluctuating amplitude of the sum of the return from both the background and the target.[18,19] A requirement for its operation is that an adequate background return exist in the range cell with the target to provide this referencing function.

The noncoherent MTI technique has, for many applications, been demonstrated to extend adequately the background clutter constrained detection limit.

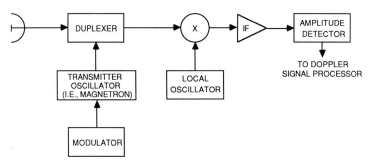

Figure 14-33. Simplified diagram of noncoherent MTI radar.

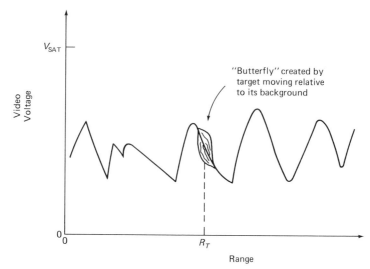

Figure 14-34. A-scope presentation of video from a monochromatic, noncoherent pulsed radar.

It does, however, have definite constraints. Some are as follows:

1. Ambiguous range responses seriously degrade clutter rejection performance.
2. Adequate MTI performance is dependent upon acceptable ranges of signal-to-clutter and clutter-to-noise ratios.
3. System interpulse amplitude stability is important.
4. Greater processing losses (relative to coherent) exist.

Certain of these restrictions can be visualized by understanding that the relative phase between the signals reflected from the background and the target on a pulse-by-pulse basis is the useful piece of information. Thus, various cross-products that are produced by the detector between the target and clutter contain the useful data. If the resulting desired cross-products are of sufficient amplitude relative to the random noise components (i.e., as a result of cross-products between noise and clutter and noise and target), then the MTI filter can perform its desired filtering function and detection can occur.

Noncoherent MTI systems such as those described here are sometimes called *clutter-referenced* or *externally coherent MTI systems*.

Two other interesting classes of noncoherent MTI radars are called *phase-sensitive* and *coherent-on-receive*.[28] The phase-sensitive noncoherent MTI is represented by Figure 14-35. As presented in this figure, a signal displaced one pulse length from the processed cell is amplitude limited and used as a phase demodulation reference. It can be shown[18,19] that the restrictions necessary for

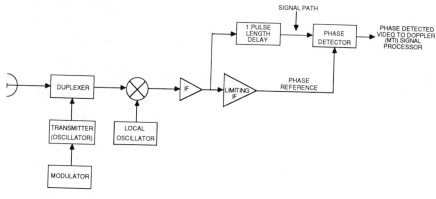

Figure 14-35. Block diagram of phase-sensitive, noncoherent MTI search radar.

effective operation are similar to those described for the noncoherent MTI approach described above (except that the limits on signal-to-clutter levels are not as stringent). Its performance (when there is both an acceptable amplitude stability and clutter-to-noise ratio in the reference cell) approaches that of a coherent system.

A simplified block diagram for a coherent-on-receive MTI radar is shown in Figure 14-36. In this concept, a sample of the transmitted phase is used to phase shift, on each pulse, the COHO phase detection reference. Useful phase information is then available, and signals are processed as though the system were coherent. Other realizations of this concept are possible. If the implementation is adequate and the constraints stated above (and following) are met, its performance approaches that of fully coherent systems. It is found that second time around echos (STAE) result in noise-like phase-detected signals (since the transmitted phase is not controlled pulse-to-pulse). STAE (or ambiguous range responses) and clutter responses, therefore, are not adequately rejected by the MTI filter. Degraded performance results. If the PRF can be set such that only

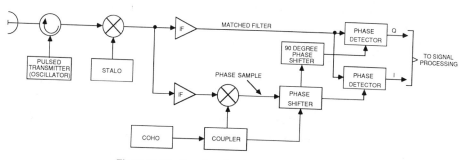

Figure 14-36. Sample coherent-on-receive MTI radar.

insignificant STAE clutter responses are produced, the mechanization errors can be held within necessary bounds, and the clutter spectral width (and shape) is supportive of adequate clutter rejection, this can be a practical approach.

14.6   CLUTTER REJECTION, TARGET ENHANCEMENT, IMPROVEMENT FACTOR, AND SUBCLUTTER VISIBILITY

Four quantities that are useful as gauges for quantifying the performance of MTI radars are (1) clutter rejection, (2) target enhancement, (3) improvement factor, and (4) subclutter visibility. *Subclutter visibility* is defined by the IEEE[1] as "the ratio by which the target echo power may be weaker than the coincident clutter echo power and still be detected with specified detection and false alarm probabilities. Target and clutter powers are measured on a single pulse return and all target radial velocities are assumed equally likely." Since the signal-to-clutter ratio that is required upon completion of MTI processing is a function of the desired detection performance, the remaining three quantities have more general (less specific) application.

As stated in Section 14.5, *clutter rejection* (taken to be the reciprocal of the cancellation ratio) is a measure of the ratio of the output clutter power to the input clutter power. The target enhancement factor defines the average target gain of the canceller when the probability density of the target velocity is assumed to be uniformly distributed over all possible radial velocities.[29] The target enhancement factor (T_E) can therefore be written as[20]

$$T_E \triangleq \left(\frac{T_0}{T_{in}}\right) \qquad (14\text{-}16)$$

where

T_0 = output target power averaged over all possible radial velocities
T_{in} = input target power

The *MTI improvement factor* is defined as the ratio of the output signal-to-clutter ratio to the input signal-to-clutter ratio when averaged over all target velocities of interest.[1] Therefore, if clutter rejection and the MTI improvement factor are designated by CR and I, respectively, then

$$I = T_E CR \qquad (14\text{-}17)$$

These definitions state the rather specific meaning of each of the four quantities. The reader is cautioned that knowledge of specific meanings is required if published results are to be properly applied.

Performance results for a two-pulse canceller MTI as given by Barton[29] are

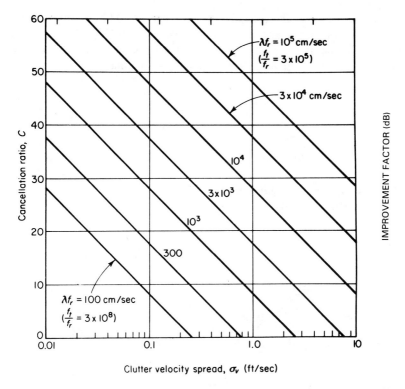

Figure 14-37. Clutter rejection and improvement factor for a two-pulse canceller MTI. (By permission, from Barton, ref. 29; © 1976 Artech House, Inc.)

reproduced in Figure 14-37. These results are based upon the simple canceller form of Figure 14-38 and the assumption of coherent operation. They are applicable when clutter can be adequately represented as having a Gaussian spectrum. The standard deviation of the power spectrum (σ_f) is transformed into an equivalent velocity spread (σ_v). As used in the figure, λ is the transmitted wavelength, f_r is the PRF of the waveform, and f_t is the RF carrier frequency. A 3-

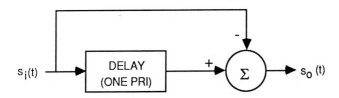

Figure 14-38. Simple two-pulse canceller (single delay).

dB difference between the cancellation ratio and the MTI improvement factor that accounts for the average target gain of the canceller (Eq. 14-16) is evident.

As an example of achievable MTI performance, consider a millimeter wave radar with parameters given in Table 14-1. It is seen that

$$\lambda f_r = (0.32)10^4$$

$$= 3.2 \times 10^3 \text{ cm/s}$$

If the clutter spectral spread (σ_v) is 1.0 ft/s, the achievable MTI improvement factor (from Figure 14-37) is approximately 20 dB. The sensitivity of this result to PRF and σ_v can be readily observed.

Similar results for both coherent and noncoherent MTI systems, as presented

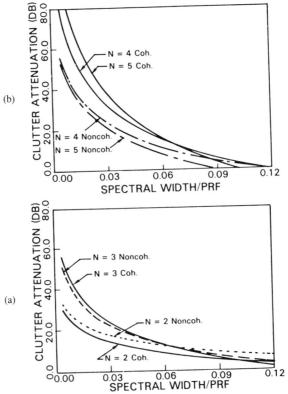

Figure 14-39. Clutter attenuation of coherent and noncoherent (linear detector) MTIs. (a) Two- and three-pulse canceller. (b) Four- and five-pulse canceller. (By permission, from Kretachner, et al., ref. 20; © 1983 IEEE)

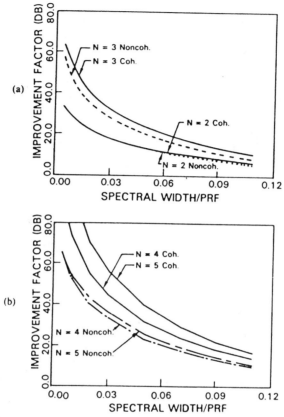

Figure 14-40. Improvement factors of coherent and noncoherent (linear detector) MTIs. (a) Two- and three-pulse canceller. (b) Four- and five-pulse canceller. (By permission, from Kretachner, et al., ref. 20; © 1983 IEEE)

by Kretschner et al.,[20] are given in Figures 14-39 through 14-41. These results demonstrate the clutter rejection (labeled *clutter attenuation* in Figure 14-39) and the MTI improvement factor that can be achieved as the number of pulses used in the MTI canceller increases. The results are given as a function of the ratio of the one sigma clutter spectral width to the PRF of the waveform and demonstrate the effect of both square law and linear detector functions on the achievable noncoherent MTI performance when the clutter-to-target ratio is 20 dB (the results were stated to be approximately the same for clutter-to-target ratios of 10 dB or greater). Consistent with the discussion of Section 14.5, the clutter-to-noise ratio (when considering the nonlinear, noncoherent MTI mode) is assumed to be large.*

---

*Achievable noncoherent MTI performance is a function of the clutter-to-noise ratio into a noncoherent (square law or linear) detector.

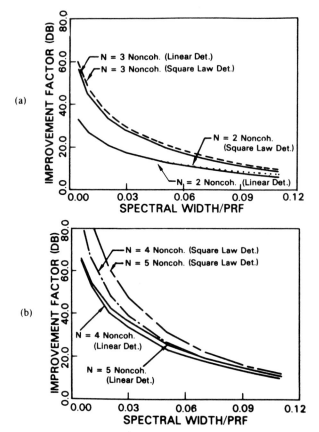

Figure 14-41. Improvement factors for noncoherent MTIs using a linear and a square law detector. (a) Two- and three-pulse canceller. (b) Four- and five-pulse canceller. (By permission, from Kretachner, et al., ref. 20; © 1983 IEEE)

The multipulse canceller assumed by Kretschner was a simple, nonrecursive, finite-impulse response (FIR) filter of the form given in Figure 14-42. If the data in Figures 14-39 through 14-41 are used as the only measure of performance, the designer would conclude that the order of the canceller should be increased until the desired MTI improvement factor is reached. A necessary additional factor that must be considered is the shape of the Doppler passband that is being formed. The key result for Doppler frequencies in the range of 0 to f_R is illustrated in Figure 14-43. This figure shows that if, for example, the -10 dB relative bandwidth for each response is used as the measuring point, the two-, three-, and five-pulse cancellers will reject approximately 20, 38, and 54% of the possible range of target Dopplers, respectively. It is this consideration that leads to the interest in, for example, recursive (feedback) filters that allow the realization of passbands that approach the desired result illustrated in

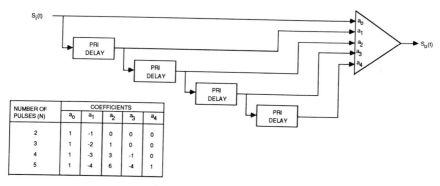

Figure 14-42. *N* pulse canceller.

Figure 14-29. The reader is encouraged to review books such as that by Stanley et al.[26] for greater depth in this subject.

The results as presented in this section, when used together with the methods presented in Section 14.2 for determining the shape (width) of the competing clutter spectrum, provide the mechanisms for predicting the MTI improvement factor for candidate radar systems.

14.7 PULSED DOPPLER RADAR

Consistent with the definition established in Section 14.5, a pulsed Doppler radar is defined for purposes of this section as one that uses a pulsed waveform and exploits the Doppler shift to obtain velocity information concerning the

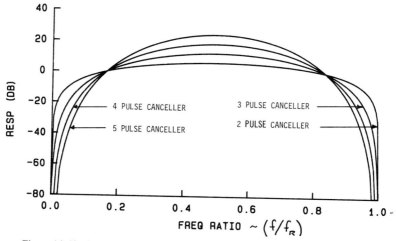

Figure 14-43. Sample passband for simple feed-forward, multipulse cancellers.

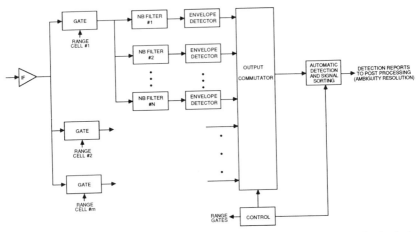

Figure 14-44. Block diagram of a sample signal processor for pulsed Doppler radar (analog).

target. Reference 2 traces the early history of the development of high-PRF pulsed Doppler airborne radars, and reference 30 presents the measured performance of a medium-PRF pulsed Doppler airborne radar.

Figures 14-44 and 14-45 are block diagrams of two signal processor forms that can be used with pulsed Doppler radars. In Figure 14-44, the output of the IF amplifier is gated at times that correspond to desired range cells. The gated output is then filtered by a bank of contiguous narrowband (NB) filters that are placed in frequency to span collectively the range of expected target Doppler shifts. The outputs of these filters are sampled via an output commutator and provided as inputs to an automatic detection and signal-sorting function. Detection reports are supplied to the postprocessor so that techniques such as *m*-out-of-*n* detection logic can be used to resolve potential range and velocity ambiguities.

Figure 14-45 is the digital equivalent of the analog processor in Figure 14-44. In this approach, the complex signal envelope is demodulated as in-phase

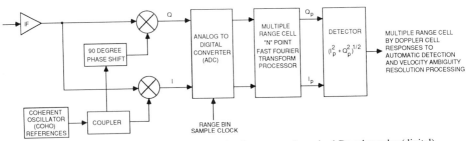

Figure 14-45. Block diagram of a sample signal processor for pulsed Doppler radar (digital).

(I) and quadrature (Q) video. After sampling at times corresponding to desired range cell placement and analog-to-digital conversion, an N-point fast Fourier transform (FFT) (that corresponds to the N complex data samples at each range cell position) is completed. The FFT produces N contiguous filter passbands that span the frequency region from zero to the PRF. After completion of the FFT for each range cell, the outputs are envelope detected and supplied as inputs to the automatic detection and postdetection processors.

As the number of range cells (m) and Doppler cells (N) increases, the amount of hardware necessary to support the approach of Figure 14-44 grows dramatically. In contrast, modern digital technology leads to hardware efficient realizations of the approach given in Figure 14-45. It can also be demonstrated that the N-point FFT which produces a finite impulse response filter is the desired matched-filter characteristic for a finite sequence of N equally weighted pulses (as discussed in Section 14.2 for step scan algorithms). The NB filter in the analog approach can at best be only an approximation of the desired matched-filter characteristic. Further, since the filters have infinite impulse response properties, they lead (in a scanning or search radar) to difficult transient response problems. Considerations such as these have led to selection of the digital approach (Figure 14-45) for most modern pulsed Doppler radar designs.

One of the performance limitations that must be resolved in a pulsed Doppler radar is the spectral leakage that results when a uniformly weighted set of N samples is processed through the FFT. This result was demonstrated in Section 14.2 to result in a zero range (time delay) cut through the ambiguity function with a $\sin(nx)/\sin(x)$ characteristic. The high sidelobes of this function permit responses in the same range cell, but with other Doppler shifts to compete with the target return. To minimize this spectral leakage problem, aperture weighting over the N sample data set is used. The article by Harris[31] presents typical weighting functions and techniques. Similarly, the article by Ziemer et al.[32] describes the resulting improvement factors as a function of the clutter spectral width for various FFT coefficients and aperture weighting functions. Results such as these demonstrate achievable improvement factors and associated blind zone widths at Doppler shifts approaching integer multiples of the PRF.

14.8   SUMMARY

As a final example, consider a 16-GHz radar to be operated from a moving platform with a velocity (V_c) of 1000 fps. The radar antenna has an azimuth scan rate ($\dot{\theta}$) of 100°/s, a 3-dB beamwidth (θ_B) of 2.6°, and a scan sector, θ_s, up to 45° off the aircraft velocity vector. A minimum clutter rejection of 25 dB is required to achieve the desired performance goals. The RMS clutter internal motion spectral width (1σ) is 20 Hz. Assume that the PRF is set at 2 kHz and that the antenna can be adequately represented as having a Gaussian pattern.

The problem is to determine the improvement factor possible and to assess the choice of the PRF.

Barton[25] gives the RMS width of the main beam platform motion spectrum (σ_p) as

$$\sigma_p = \frac{\theta_B V_c}{1.18\lambda} \sin(\theta_s) \tag{14-18}$$

where λ is the RF wavelength and the beamwidth is in radians. When evaluated at $\theta_s = 45°$, $\sigma_f = 443.5$ Hz. The width of the main beam scan modulation spectrum (σ_s) is given[25] as

$$\sigma_s = \frac{\dot\theta}{3.78\theta_B} \tag{14-19}$$

When evaluated, this contribution is 10.2 Hz. If the contribution to the clutter spectral width by system stability can be considered to equal zero, the overall main beam spectral width (σ_t) is given as

$$\sigma_t^2 = \sigma_p^2 + \sigma_s^2 + \sigma_i^2 \tag{14-20}$$

when the contributors are assumed to have Gaussian shapes. Evaluation of σ_t yields 444 Hz. The ratio of the spectral width to the PRF is then approximately 0.22. The data of Figure 14-40 can then be used to determine that none of the illustrated coherent and noncoherent multiple-pulse canceller MTI systems can achieve the required 25 dB of improvement.

In the situation shown, the PRF must be raised to produce an adequate result. The choice between a medium- and a high-PRF waveform must then be made based upon achievable performance (and cost), using the techniques introduced in this chapter.

This chapter provides insight into the mechanisms by which platform motion, scan modulation, clutter internal motion, and system instabilities interact with the radar waveform to produce a particular spectrum at the output of the receiver. The ambiguity function provides a method of describing ambiguous range and Doppler regions that contribute to the observed output from a matched filter corresponding to a desired range by Doppler coordinate. The chapter also deals with definitions for low-, medium-, and high-PRF waveforms and with typical noncoherent and coherent pulsed radar configurations. The confusion of the meaning of MTI and pulsed Doppler radars is discussed, and a definition is stated. Typical MTI and pulsed Doppler radar performance issues and block diagrams are presented.

14.9 REFERENCES

1. *IEEE Standard Radar Definitions*, IEEE Std. 686-1982, IEEE, New York, 1982.
2. Leroy C. Perkins, et al., "The Development of Airborne Pulse Doppler Radar," IEEE Transactions on *Aerospace and Electronic Systems*, vol AES-20, no. 3, May 1984, pp. 292–303.
3. Alex Ivanov, "Semi-Active Radar Guidance," *Microwave Journal*, vol. 26, no. 9, September 1983, pp. 105–120.
4. Merril I. Skolnik, *Introduction to Radar Systems*, McGraw-Hill Book Co., New York, 1962, pp. 113, 153.
5. F. C. Williams and M. E. Radant, "Airborne Radar and the Three PRF's," *Microwave Journal*, vol. 26, no. 7, July 1983, pp. 129–135.
6. Fred E. Nathanson, *Radar Design Principles*, McGraw-Hill Book Co., New York, 1969, pp. 283–292.
7. R. D. Hayes, "Fluctuation in Radar Backscatter from Rain and Trees," Report of the ARPA/NELC Tri-Service MMW Workshop, December 1976.
8. W. Fishbein, "Clutter Attenuation Analysis," ECOM-2808, Army Electronics Command, Fort Monmouth, NJ, March 1967, AD665352.
9. R. D. Hayes et al., "Backscatter from Ground Vegetation at Frequencies between 10 and 100 GHz," AP-S/URSI Symposium, October 1976.
10. V. W. Richard, "Millimeter Wave Radar Applications to Weapon System," BRL Memorandum Report No. 2631, BRL, Aberdeen, June 1976.
11. L. C. Bomar, et al., "MM-Wave Reflectivity of Land and Sea," *Microwave Journal*, vol. 21, no. 8, August 1978, pp. 49–53, 83.
12. R. S. Raven, "Requirements on Master Oscillators for Coherent Radar," *Proceedings of the IEEE*, vol. 54, no. 2, February 1966, pp. 237–243.
13. D. B. Laeson, et al., "Short-Term Stability for a Doppler Radar: Requirements, Measurements, and Techniques," *Proceedings of the IEEE*, vol. 54, no. 2, February 1966, pp. 244–248.
14. Richard L. Mitchell, *Radar Signal Simulation*, Artech House, Dedham, Mass., 1976.
15. W. S. Burdic, *Radar Signal Analysis*, Prentice-Hall, Inc., Englewood Cliffs, N.J., 1968, pp. 73–74.
16. H. R. Ward, "Doppler Processor Rejection of Range Ambiguous Clutter," *IEEE Transactions on Aerospace and Electronic Systems*, vol. AES-11, no. 4, July 1975, pp. 519–522.
17. Burdic, *Radar Signal Analysis*, chap. 2.
18. Skolnik, *Radar Systems*, p. 114.
19. M. I. Skolnik, *Radar Handbook*, McGraw Hill Book Co., New York, 1970, pp. 17-53, 17-54.
20. Frank F. Kretschner, et al., "A Comparison of Noncoherent and Coherent MTI Improvement Factors," *IEEE Transactions on Aerospace and Electronic Systems*, vol. AES-19, no. 3, May 1983, pp. 398–404.
21. Skolnik, *Radar Systems*, p. 154.
22. David K. Barton, *Radars*, Volume 7: *CW and Doppler Radar*, Artech House, Dedham, Mass., 1978, p. 193.
23. Ibid., p. 198.
24. Skolnik, *Radar Handbook*, sections 4.1–4.4.
25. David K. Barton, *Radar System Analysis*, Artech House, Dedham, Mass., 1976, sections 7.1, 7.3, and 7.4.
26. William D. Stanley, et al., *Digital Signal Processing*, Reston Publishing Company, Reston, Va., 1984.
27. Skolnik, *Radar Systems*, sections 17.8 and 17.9.

28. Barton, *Radar System Analysis*, pp. 191–192.

29. Ibid., pp. 212–213.

30. M. B. Ringel, et al., ''F-16 Pulse Doppler Radar (AN/APG-66) Performance,'' *IEEE Transactions on Aerospace and Electronic Systems*, vol. AES-19, no. 1, January 1983, pp. 147–158.

31. Fredric J. Harris, ''On the Use of Windows for Harmonic Analysis with Discrete Fourier Transform,'' *Proceedings of the IEEE*, vol. 66, no. 1, January 1978, pp. 51–83.

32. R. E. Ziemer and J. A. Ziegler, ''MIT Improvement Factors for Weighted DFTs,'' *IEEE Transactions on Aerospace and Electronic Systems*, vol. AES-16, no. 3, May 1980, pp. 393–397.

15
PULSE COMPRESSION IN RADAR SYSTEMS

Marvin N. Cohen

15.1 INTRODUCTION

Pulse compression allows a radar system to transmit a pulse of relatively long duration and low peak power to attain the range resolution and detection performance of a short-pulse, high-peak power system. This is accomplished by coding the RF carrier to increase the bandwidth of the transmitted waveform and then compressing the received echo waveform.

The range resolution achievable with a given radar system is

$$\delta_r = \frac{c}{2B} \tag{15-1}$$

where

c = speed of light ($\approx 3 \times 10^8$ m/s)
B = bandwidth of the transmitted waveform

For a simple (noncoded) pulsed radar system, $B = 1/T$ where T is the transmitted pulse length. Thus

$$\delta_r = cT/2 \tag{15-2}$$

for a simple pulse system.

In a pulse compression system, the transmitted waveform is modulated, usually in phase or frequency, so that $B \gg 1/T$. Let $\tau = 1/B$. Then from Eq. (15-1)

$$\delta_r = c\tau/2 \tag{15-3}$$

and τ represents the effective pulse length of the system after pulse compression. Thus, a pulse compression radar can use a transmit pulse of duration T

and yet achieve a range resolution equivalent to that of a simple pulse system with a transmit pulse of duration τ. The ratio of the transmitted pulse length T to the system's effective (compressed) pulse length τ is termed the *pulse compression ratio* and is given by

$$CR = T/\tau \tag{15-4}$$

Average power on a single-pulse basis is given by Pt, where P is peak power and t is pulse duration. It follows that CR is the ratio of the average power transmitted by the pulse compression system to the average power transmitted by a simple pulse system, assuming that they both transmit the same peak power and achieve the same range resolution. Since $\tau = 1/B$,

$$CR = TB \tag{15-5}$$

that is, the compression ratio also equals the time–bandwidth product of the system. Pulse compression systems can, for many purposes, be characterized by their time–bandwidth products.[1-3]

Figure 15-1 illustrates the concept of pulse compression processing in a radar system. The dispersive delay line and the pulse compression filter are of primary interest in this discussion. The RF source generates a short pulse of duration τ, which then passes through a dispersive delay line. The output of the dispersive delay line is a pulse of duration T ($T \gg \tau$), but the bandwidth B of the signal

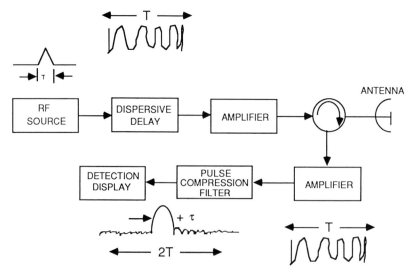

Figure 15-1. Pulse compression processing in a radar system.

is $1/\tau$—the bandwidth of the input pulse. This signal is then amplified and transmitted through the radar antenna. On receive, the signal is suitably processed and passed through the pulse compression filter. This filter often takes the form of the filter matched to the transmitted waveform, resulting in a compressed pulse of length $\tau = 1/B$. The compressed signal is then appropriately translated down in frequency, amplified, and displayed.

The compressed pulse is the result of a signal of duration T passing through its matched filter. The time extent of the response out of the matched filter is thus on the order of $2T$, not the compressed pulse length τ, as indicated in Figure 15-1. The responses outside of $|t| < \tau$ are termed *range sidelobes*. Since range sidelobes from a given range bin may appear as signals in adjacent range bins, they must be controlled in any pulse compression system. The first two of the following three measures are often used to quantify the level of these sidelobes. The third serves to quantify the loss in signal-to-noise ratio performance due to the use of a mismatched filter in the receiver. All three are usually given in decibels:

$$\text{Peak sidelobe level (PSL)} = 10 \log \frac{\text{maximum sidelobe power}}{\text{peak response}}$$

$$\text{Integrated sidelobe level (ISL)} = 10 \log \frac{\text{total power in the sidelobes}}{\text{peak response}} \qquad (15\text{-}6)$$

$$\text{Loss in processing gain (LPG)} = 10 \log \frac{\text{CR}}{\text{peak response}} .$$

The PSL is closely associated with the probability that a false alarm in a particular range bin is due to the presence of a target in a neighboring range bin. A low PSL is especially important in scenarios where a high density of targets of different cross sections are expected. The ISL, a measure of the energy distributed in the sidelobes, is also important in dense target scenarios, as well as when distributed clutter is present.

A receiver filter matched to the transmit waveform provides maximum signal-to-noise ratio, but one may employ mismatched pulse compression filters in the receiver to reduce ISL and PSL. The loss in signal-to-noise ratio due to mismatched as opposed to matched filtering is the LPG of the system. Various mismatched filtering techniques are discussed in Section 15.5.

Radar applications that require high-range resolution include object detection, object classification, terrain mapping, accurate ranging, and as an aid in any application in distributed clutter suppression. High-range resolution can be achieved either by transmitting a simple, short-duration pulse or by transmitting a lower-peak-power, coded pulse of greater duration and then compressing on receive.

A radar system that incorporates pulse compression processing rather than a simple pulse system to achieve high-range resolution provides the following potential advantages:

- Improved detection performance.
- Mutual interference reduction.
- Increased system operational flexibility.

On a single-pulse basis, the probability of detection at a fixed range increases with the signal-to-noise ratio, which in turn is proportional to the average power. As mentioned earlier, a pulse compression system can achieve higher average power than a short-pulse system when peak power and range resolution are equal by utilizing a longer transmit pulse. Thus, pulse compression allows the system designer to trade pulse length for peak power. This is important because of the realities and state of the art in component design. In general, it is not desirable to operate transmitters at high power for short durations; in fact, a pulse length of 10 ns is usually accepted as an appropriate lower bound for the minimum pulse duration achievable with today's transmitters. However, compressed pulses an order of magnitude shorter are obtainable with state-of-the-art pulse compression systems. Furthermore, the introduction of practical solid-state transmitters, which are characteristically low powered, has underscored the utility of reducing peak power requirements by means of pulse compression.

Mutual interference among radars can become severe in a dense, active EM environment. Pulse compression radars can be designed to reduce mutual interference by equipping each radar with a different modulation code and matched filter.

Finally, if the radar signal processor is designed so that various codes and compression ratios can be effected, a system using a transmitter operating at a fixed PRF, peak power, and duty cycle can still provide variable range resolution capability. This flexibility in range resolution may be exploited to incorporate target detection, acquisition, track, and identification modes in a single radar system.

Concomitant with these potential advantages are complications associated with relying on pulse compression as opposed to the transmission of a simple pulse of the requisite length and power to achieve the aims of a particular radar system. Among these complications are:

1. Increased system processing requirements due to requisite compression and sidelobe suppression processing.
2. The need for fine control of the waveform parameters that carry the code information.
3. Minimum range constraints.

Restrictions 1 and 2 can be translated into an increase in the radar/processor complexity and sophistication requirement—that is, system cost, size, and maintenance. Generally, pulsed monostatic radar systems cannot transmit and receive simultaneously. Thus, the minimum range of the system must be greater than one-half the range extent of the transmit pulse. Pulse compression systems therefore require longer minimum ranges than equivalent short-pulse systems. In fact, the ratio of the minimum ranges is given by the pulse compression ratio.

The precise nature and extent of the potential advantages and restrictions discussed above are heavily dependent on the pulse compression technique considered. Techniques based on frequency modulation and techniques based on phase modulation of the transmit waveform are well developed and have seen widespread application in modern radar systems. Polarization modulation techniques have been employed on an experimental basis. The following sections describe some of these techniques, how they may be implemented, and what their fundamental properties are.

15.2   FREQUENCY MODULATION TECHNIQUES

A radar's carrier frequency may be modulated to increase the bandwidth of the radar transmission and allow pulse compression on receive. Linear FM (LFM or chirp) is the oldest and most well-developed pulse compression technique, having first been proposed in the late 1940s.[4] Frequency stepping, which entails changing the frequency of the transmit signal discretely on a pulse-by-pulse or subpulse basis, is a more recent technique. Its development has been spurred by the flexibility of the technique and the recent advances in digital technology.[5]

LFM and stepped frequency pulse compression techniques are discussed in the following subsections. The waveforms are described and investigated, and various technologies and methods available for processing these waveforms are presented.

15.2.1   LINEAR FREQUENCY MODULATION

The LFM waveform consists of a rectangular transmit pulse of duration $T = t_2 - t_1$, as shown in Figure 15-2(a). The carrier frequency f is swept linearly (chirped) over the pulse length by an amount Δf, as depicted in Figure 15-2(b).

A pulse compression filter in the radar receiver is matched to the transmitted waveform so that the received signal experiences a frequency-dependent time delay. As shown in Figure 15-2(c), the higher-frequency components of the received signal experience correspondingly longer delay times than the lower-frequency components at the output of the dispersive delay line. The receiver is constructed so that the delay is proportional to the frequency where the proportionality factor is the negative of the slope (in time-frequency) of the trans-

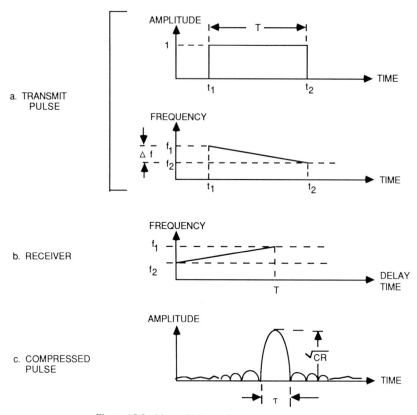

a. TRANSMIT
   PULSE

b. RECEIVER

c. COMPRESSED
   PULSE

Figure 15-2. Linear FM waveform and processing.

mitted waveform; that is, the receiver constitutes a filter matched to the transmitted waveform. The output signal from the pulse compression filter may be
characterized by an envelope having a higher amplitude and a narrower pulse
length than the transmitted envelope, as depicted in Figure 15-2(c). The discussion of the analytic form of the linear FM waveform below shows that the
compressed pulse length τ is given by

$$\tau = \frac{1}{\Delta f} \tag{15-7}$$

and thus, the pulse compression ratio CR is given by

$$CR = \frac{T}{\tau} = T\Delta f \tag{15-8}$$

Since Δf is the bandwidth B of this signal, the product $T\Delta f$, as noted earlier, is defined to be the system time–bandwidth product.

The range resolution of this pulse compression radar is given by

$$\delta_r = \frac{c}{2\Delta f} = (1/2)\ c\tau \tag{15-9}$$

For example, a pulse compression radar at K_u-band utilizing a 200-ns pulse could have a 500-MHz chirp bandwidth, resulting in a range resolution of $\delta_r = c/2\Delta f = 3.0 \times 10^8/(2 \times 500 \times 10^6) = 30$ cm. This resolution represents a hundredfold improvement over the resolution of the uncompressed 200-ns pulse from the same radar. Note that the compressed pulse length and, hence, the range resolution are independent of the transmitted pulse length; both are a function only of the frequency excursion Δf, as can be seen from Eqs. (15-7) and (15-9).

We now proceed to the equations that represent the general linear FM waveforms. Let the angular frequency of the carrier ω_0 be chirped over the pulse length T according to

$$\omega(t) = \omega_0 - \mu t \qquad |t| \leqslant T/2 \tag{15-10}$$

where the radian sweep rate μ is given by

$$\mu = \frac{2\pi\Delta f}{T} \tag{15-11}$$

If all target backscatter modulations and range losses are ignored, the received signal $s(t)$ has the form

$$s(t) = \begin{cases} \cos\ (\omega_0 t - (1/2)\ \mu t^2) & |t| \leqslant T/2 \\ 0 & |t| > T/2 \end{cases} \tag{15-12}$$

where the argument of the cosine function is the phase of the received signal— that is, the integral of the frequency function $\omega(t)$ of Eq. (15-10).

The matched-filter impulse response for the signal $s(t)$ is a time inversion of $s(t)$ (normalized in amplitude) given by

$$h(t) = \left(\frac{2\mu}{\pi}\right)^{1/2} \cos\left(\omega_0 t + \frac{1}{2}\ \mu t^2\right) \tag{15-13}$$

By translation to the frequency domain via the Fourier transform, simplifying,

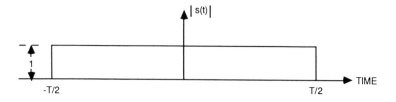

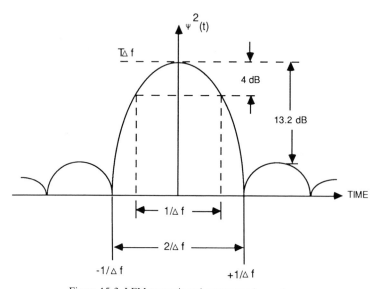

Figure 15-3 LFM transmit and compressed waveforms.

and passing back to the time domain, one can show[6] that the output $\psi(t)$ resulting from passing $s(t)$ through $h(t)$ is given by

$$\psi(t) = \left(\frac{\mu T^2}{2\pi}\right)^{1/2} \frac{\sin\ (\mu Tt/2)}{(\mu Tt/2)}\ \mathrm{Re}\left[\exp\ \mathrm{j}\left(\omega_0 t\ +\ \frac{1}{2}\ \mu t^2\ +\ \pi/4\right)\right]\quad (15\text{-}14)$$

The envelope of the output power pulse $\psi^2(t)$ is shown in Figure 15-3, along with the corresponding input pulse $s(t)$.

The null-to-null pulse length is the distance between the first zeroes of $\psi^2(t)$, and these occur when

$$\frac{\mu Tt}{2} = \pm \pi$$

so that

$$t = \pm\frac{2\pi}{\mu T} = \pm\frac{1}{\Delta f} \tag{15-15}$$

The null-to-null pulse length $\tau(nn)$ is therefore

$$\tau(nn) = \frac{2}{\Delta f} \tag{15-16}$$

By convention, the compressed pulse length is defined as the distance between the points where

$$\frac{\mu Tt}{2} = \pm\frac{\pi}{2} \tag{15-17}$$

that is, between the points

$$t = \pm\frac{1}{2\Delta f} \tag{15-18}$$

Therefore, the compressed pulse length τ is given by

$$\tau = \frac{1}{\Delta f} \tag{15-19}$$

which corresponds to the points 4 dB down from the peak of $\psi^2(t)$, as can be computed by solving for t in Eq. (15-17) and substituting into Eq. (15-14).

Recall that the compression ratio, CR, of the system is the ratio of the uncompressed to compressed pulse lengths:

$$CR = T/\tau \tag{15-20}$$

Combining this with Eqs. (15-11) and (15-19), it follows that

$$CR = T \cdot \Delta f = \frac{\mu T^2}{2\pi} \tag{15-21}$$

which relates the radian sweep rate μ and the total frequency chirp Δf to the pulse compression ratio of the system.

An important feature of linear FM that makes it the most suitable compres-

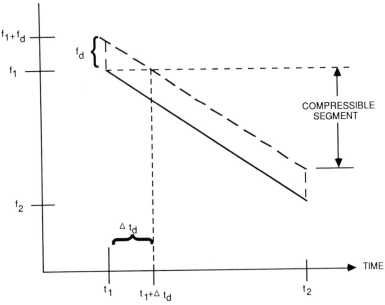

Figure 15-4. The range–Doppler coupling of linear FM.

sion code for many applications is its relative insensitivity to degradation in response to Doppler-shifted signals. Although a more general description of this insensitivity must be postponed until the introduction of the ambiguity diagram in Section 15.6, a discussion of the phenomenon is presented here, since linear FM itself can be derived as a second-order approximation to the ideal, completely Doppler-invariant waveform.[7]

Let the solid line in Figure 15-4 represent the time–frequency return from a stationary target. Now assume a moving target at the same range, but with a radial velocity v sufficient to impart a Doppler shift of f_d Hz on the return signal. Its return is represented by the dotted line parallel to, but shifted up from, the original in frequency. Since the compression filter is matched to the stationary return, only that portion of the shifted return falling in the frequency region between f_1 and f_2 will effectively compress. The effect of this Doppler-shifted return passing through the compression filter will thus be twofold: (1) a fractional loss in output power, given by $(\Delta t_d / T)$, due to the foreshortened compressible segment and (2) a delay in compression of

$$\Delta t_d = \frac{T}{\Delta f} f_d \qquad (15\text{-}22)$$

seconds, which corresponds to a decrease in apparent range of $\Delta t_d / \tau$ range bins.

For high pulse compression ratios, $T \gg \tau$. It follows that $(\Delta t_d/T) \ll (\Delta t_d/\tau)$. Thus, the moving target's return may compress almost fully (small $\Delta t_d/T$), yet there may be a significant shift in the target's apparent range (large $\Delta t_d/\tau$).

This range–velocity coupling may be highly desirable for applications in which detection is the primary goal, while it may present a problem for applications in which accurate range or velocity measurements are required.

Linear FM is often employed in long-range surveillance and track radars. The AN/FPS-85 and PAVE PAWS (AN/FPS-115) ground-based, very-high-frequency, phased-array radars are examples of such operational systems. A typical PAVE PAWS search pulse is 5 ms long and is linearly chirped 0.1 MHz in frequency, resulting in a range resolution of 150 m and a pulse compression ratio of 500:1. In the track mode, transmit pulse widths between 0.25 and 16 ms and chirp bandwidths of 1.0 MHz are employed.[8]

15.2.2 FREQUENCY STEPPING

As described in the previous section, linear FM uses continuous modulation of the carrier frequency to encode the transmit pulse for subsequent compression. When discrete modulation of the carrier frequency is used to encode the transmission, the technique is termed *frequency stepping*.

Figure 15-5 illustrates three different versions of frequency stepping. All have the same pulse compression characteristics at zero Doppler, but they may have very different characteristics when some Doppler shift is present. Generally, a frequency-stepped waveform consists of N subpulses, each at a different frequency. Consider the linear stepped-frequency pulse illustrated in Figure 15-5(a), where

τ_T = transmit subpulse length
τ_c = compressed pulse length
N = number of subpulses
f_k = frequency of the kth subpulse for $k = 1, \ldots, N$
$\Delta f_s = f_k - f_{k-1}$ = subpulse frequency step for $k = 1, \ldots, N$
$\Delta f = N\Delta f_s$ = total frequency excursion

Since $\Delta f_s > 1/\tau_T$ and $\Delta f_s < 1/\tau_T$ result in undesirable compressed waveform characteristics,[5] we assume that $\Delta f_s = 1/\tau_T$. This waveform may be thought of as an approximation of a linear FM waveform of duration T and total bandwidth $B = N \cdot \Delta f_s = N/\tau_T$. From Eq. (15-19), the compressed pulse length is therefore given by

$$\tau_c = \frac{1}{B} = \frac{\tau_T}{N} \tag{15-23}$$

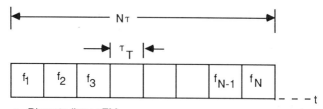

a. Discrete linear FM.

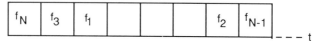

b. Scrambled frequency stepping.

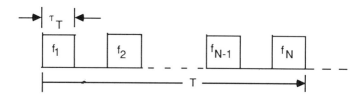

c. Interpulse frequency stepping.

Figure 15-5. Frequency-stepped waveforms. (By permission from Nathanson, ref. 5; © 1969 McGraw-Hill Book Co.)

From Eq. (15-20), it follows that the pulse compression ratio is

$$CR = \frac{T}{\tau_c} = \frac{N\tau_T}{\tau_T/N} = N^2 \qquad (15\text{-}24)$$

which is also the time–bandwidth product of the system.

A scrambled, frequency-coded pulse is depicted in Figure 15-5(b). We assume that the component frequencies of this pulse are the same as those of the linearly stepped pulse, but here they occur randomly in time. Since the total bandwidth is the same as in Figure 15-5(a) and range resolution is the reciprocal of bandwidth, it follows that $\tau = \tau_T/N$ and $CR = N^2$, just as before.

Essentially the same analysis applies to the multifrequency pulse train of Figure 15-5(c). Note that this last implementation is by definition an *interpulse*, rather than an *intrapulse*, technique.

Although all three of the frequency-stepped waveforms discussed above behave identically in response to returns from objects with zero radial velocity to the radar, their Doppler characteristics are quite different. Here, suffice it to say that the linearly stepped-frequency pulse has Doppler characteristics approximating those of linear FM, whereas the Doppler characteristics of the other two implementations depend on the order and spacing employed. See Skolnik[10] and Nathanson[5] for more detailed discussions of these waveforms' range-Doppler characteristics.

15.2.3 GENERATION AND PROCESSING OF FREQUENCY-CODED WAVEFORMS[2,10]

Linear FM waveforms are usually generated at a low power level and then amplified. The waveform may be generated either actively or passively and may be processed in the receiver using analog or digital techniques.

Figure 15-6 is a conceptual block diagram of a linear FM radar system that generates and processes the waveform passively. In the transmitter section, the FM waveform can be generated by driving a dispersive delay line with an impulse. The resulting linear FM waveform is amplified and transmitted through the antenna. The received signal is mixed down to IF and passed through a dispersive delay line that is the matched filter to the transmitted waveform. In many cases, the same dispersive device with input and output ports switched may be used by both the transmitter and receiver.

Dispersive delay lines may be EM or ultrasonic devices. EM implementations include lumped-constant circuits and tapered folded-tape meander lines.[11] Ultrasonic devices include bulk wave and surface acoustic wave (SAW) devices. Bulk wave devices include metallic (steel or aluminum) dispersive strips, quartz diffraction grating delay lines, and diffraction gratings on a strip.[12] SAW delay lines consist of a piezoelectric substrate such as a thin slice of quartz with input and output interdigital transducers arranged on the surface. Since the propagation takes place along the surface of the device, input/output taps can be placed wherever desired, allowing flexibility not present with bulk wave devices.

A typical SAW device for a linear FM pulse compression radar has a bandwidth of 500 MHz and an uncompressed pulse width of 0.46 μs.[13] Other designs

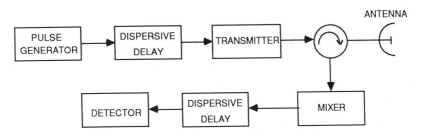

Figure 15-6. Conceptual block diagram for an LFM pulse compression radar system.

have achieved pulse widths of up to 150 μs and pulse compression ratios of up to 10,000:1.[14] Each method of dispersive delay has different characteristics and preferred regions of bandwidth and pulse width duration. See reference 10 (p. 155) for a more detailed discussion of these characteristics.

Linear FM waveforms may be generated using active circuits.[11] One common analog technique is to apply a programmed control voltage to a voltage-controlled oscillator (VCO) whose frequency varies with the applied voltage.

Interpulse stepped-frequency waveforms may be generated by driving a TWT with a frequency synthesizer which changes frequencies on a pulse-by-pulse basis. Since the receiver center frequency can be stepped pulse-by-pulse to follow the transmissions, the instantaneous receiver bandwidth can be kept relatively small. Pulse compression may be effected in the receiver by performing a digitally implemented FFT on the collected frequency samples. Extremely large time–bandwidth products can be achieved with this technique.

15.3   PHASE MODULATION TECHNIQUES

This section treats the coding of a transmitted radar pulse of duration T by dividing it into N subpulses, each of duration $\tau = T/N$, and coding these subpulses in terms of the phase of the carrier. The *binary phase (biphase) codes* can be represented simply by pluses and minuses, where a plus subpulse designation represents no phase shift and a minus subpulse designation represents a carrier phase shift of π radians. *Polyphase codes* are more complex and allow for any of M phase shifts on a subpulse basis, where M is called the *order of the code* and the possible phase states are

$$\phi_i = \frac{2\pi}{M} i \qquad i = 0, \ldots, M - 1 \qquad (15\text{-}25)$$

The process of coding a transmit pulse in phase and then using this modulation to effect pulse compression on receive is most easily exhibited by an example based on biphase coding of the pulse and compression through a tapped delay line matched filter. Figure 15-7 shows the transmission of a 13-element biphase code where the abscissa represents time, the ordinate represents amplitude, + represents no phase change in the carrier, and − represents a phase shift of π radians. The autocorrelation, or compressed output in response to a point target, may be found by passing this coded waveform through its matched filter, given as a tapped delay line in Figure 15-7(b). The autocorrelation is computed by passing the code through the filter from right to left, shifting one subpulse (τ seconds) for each computation. After each shift, the code and filter values that occupy the same positions are multiplied, and the results of all of these multiplications are added. This value is output, the code is shifted one more subpulse to the left, and the process is repeated. Thus, after 12 shifts of

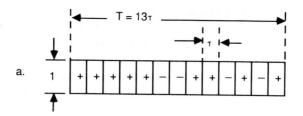

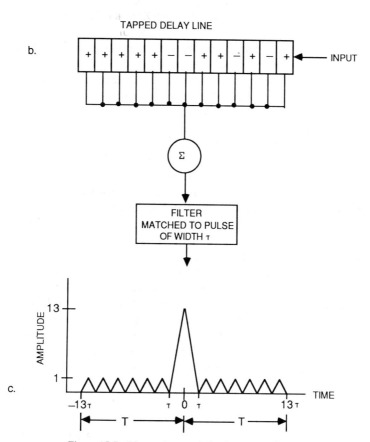

Figure 15-7. Binary phase-coded pulse compression.

the code through the filter, we obtain

code		+	+	+	+		+	−		−	+		+	−		+		−	+
× filter	+	+	+	+	+		−	−		+	+		−	+		−		+	
Σ		1	1	1	1		−1	1		−1	1		−1	−1		−1		−1	= 0,

where we take the product of two like signs to be 1 and the product of two different signs to be -1. Similarly, after 13 shifts

code	+	+	+	+	+	−	−	+	+	−	+	−	+	
× filter	+	+	+	+	+	−	−	+	+	−	+	−	+	
Σ	1	1	1	1	1	1	1	1	1	1	1	1		$= 13,$

which is the main peak of the compression function. Figure 15-7(c) gives a plot of all of the values so obtained. The system total time response for a point target echo is on the order of $2T$, where T was the transmit pulse duration; the compressed (3-dB) pulse length is on the order of τ, where τ was the subpulse length; and the effective amplitude of the peak response is 13 times that of the amplitude of the inserted coded pulse. The response between $t = -\tau$ and $t = \tau$ is the compressed signal. The responses for $|t| > \tau$ represent the range sidelobes generated by the system.

In general, given a transmit pulse of duration T which is coded in N subpulses each of duration τ, matched filtering will result in a compressed pulse that has an effective peak amplitude N times the input pulse and a resolving capability equivalent to that of a pulse of width τ. The pulse compression ratio is therefore

$$CR = N = \frac{T}{\tau} = B \cdot T \qquad (15\text{-}26)$$

where, as usual, $B = 1/\tau$.

In the remainder of this subsection, we discuss in some detail particularly useful binary phase codes such as the Barker codes, combined Barker codes, and pseudorandom codes. We also describe the Frank polyphase codes and the sidelobe-cancelling Welti and Golay codes.

15.3.1 Barker Codes

Barker codes are the binary phase codes with the property that the peak sidelobes of their autocorrelation functions are all less than or equal to $1/N$ in magnitude, where N is the code length and the output signal voltage (i.e., the voltage of the maximum output) is normalized to 1. The class of all known Barker codes, along with the ISL and PSL (see Section 15.1) of their matched systems, are given in Table 15-1. Extensive searches for Barker codes of length greater than 13 have been conducted by various researchers, but none have been found. Barker codes of length greater than 13 and less than several thousand do not exist.[15] In fact, those listed in Table 15-1 are thought to be the only ones possible.

The great attractions of the Barker codes are that (1) their sidelobe structures contain the minimum energy that is theoretically possible and (2) this energy is

Table 15-1. The Known Barker Codes.

Length of Code	Code Elements	PSL (dB)	ISL (dB)
1	+	—	—
2	+ −, + +	−6.0	−3.0
3	+ + −, + − +	−9.5	−6.5
4	+ + − +, + + + −	−12.0	−6.0
5	+ + + − +	−14.0	−8.0
7	+ + + − − + −	−16.9	−9.1
11	+ + + − − − + − − + −	−20.8	−10.8
13	+ + + + + − − + + − + − +	−22.3	−11.5

uniformly distributed among the sidelobes. Because of these properties, Barker codes are sometimes called *perfect codes*. The search for longer perfect codes (to achieve higher pulse compression ratios) having so far failed, it is natural to seek out longer *near-perfect* or ''good'' codes—codes whose sidelobe structures are nearly uniform and contain a minimal amount of energy. MacMullen has compiled a list of such codes.[16]

15.3.2 Combined Barker Codes

One scheme to generate codes longer than 13 bits is the method of forming combined Barker codes using the known Barker codes. For example, in order to devise a system with a 20:1 pulse compression ratio, one may use either the 5 × 4 or the 4 × 5 Barker code. The 5 × 4 Barker code consists of the 5-bit Barker code, each bit of which is the 4-bit Barker code. Thus, the 5 × 4 combined Barker code is the 20-bit code

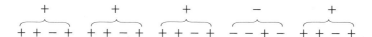

Combined codes consisting of any number of individual codes can be analogously defined. The filter associated with a combined code is a combination of the filters matched to the individual codes. The individual codes (and corresponding filters) are called the *subcodes* (or *subsystems* or *components*) of the full *code* (*system*). The matched filter for a combined code may be implemented directly as a tapped delay line whose impulse response is the time inverse of the code on as a combination of subcode-matched filters.

Figure 15-8 is an example of a combined matched filter for the 5 × 4 combined Barker code. The first stage of the filter (on the right) is simply the matched filter to the inner (4-bit) code. The second stage represents a filter matched to the 5-bit code, except that the active taps are spaced four taps apart. This filter is equivalent to the 20-bit tapped delay line matched filter (their impulse re-

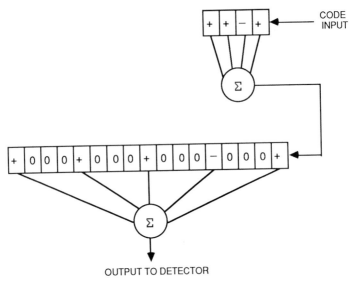

Figure 15-8. Combined matched filter for 5 × 4 combined Barker code.

sponses are identical); however, the number of active arithmetic elements (+ or − multiplies) in the combined filter is 9, the sum of the subcode lengths, and the number in the 20-bit filter is 20, the product of the subcode lengths. These results generalize to codes which are the combination of any number of subcodes.

There are various advantages to creating high-pulse compression ratio systems using combined Barker codes and filters:

- By combining the subcodes in various ways, one can generate numerous codes of various lengths appropriate for different modes in a single radar system (such as surveillance, track, and identification).
- As indicated above, the number of arithmetic elements required for processing is the sum of the subcode lengths when implementing a combined filter as opposed to the product obtained when implementing the matched filter as a single tapped delay line.
- The analytic procedure for deriving the tap weights of a length n ISL-optimized filter (as described in Section 15.5.2) requires the solution of a system of n linear equations in n unknowns. As n grows large, the solution becomes difficult. These sidelobe reduction techniques may be applied to a long combined code by utilizing a combination of filters optimized for each subcode.[12]
- The sidelobe characteristics of a combined system are easily and inexpensively studied, since knowledge of the sidelobe characteristics of its components is essentially all one needs to know. In particular, the PSL of a

system is approximately the PSL of its weakest component, the ISL is approximately the root sum square of the subsystems' ISLs, and the LPG is approximately the sum of the individual LPGs (the ISL and LPG show some sensitivity to order).[17]

15.3.3   Pseudorandom Codes

Another set of long binary phase codes that are relatively easy to generate, have good sidelobe properties, and can be changed algorithmically are the *pseudorandom sequences* (*PN codes*). The PN codes of primary interest are the ones of maximal length (*maximal-length binary shift register sequences, maximal-length sequences*, or simply *m-sequences*). Because of the properties mentioned above, as well as their spread spectrum characteristics, these sequences have proved popular for some radar and many communications applications.[18]

These sequences can be generated by initializing, in any nonzero state, a binary shift register with feedback connections such as the one depicted in Figure 15-9, clocking the system to circulate the bits, and picking off the appropriate output. The result of this process is a sequence of length $2^n - 1$, where n is the number of shift registers employed. In order for the output to be a maximal length and, therefore, a nonrepeating sequence, the feedback paths must correspond to the nonzero coefficients of an irreducible, primitive polynomial modulo 2 of degree n. An example of such a polynomial of degree five is $(1) + (0)x + (0)x^2 + (1)x^3 + (0)x^4 + (1)x^5$. The five-stage shift register that corresponds to this polynomial is given in Figure 15-9. Note that the constant "1" terms corresponds to the feedback from the adder to the first bit in the register. The "1" coefficients of the x^3 and x^5 terms correspond to the feedback paths from the third and fifth registers to the adder. In general the shift registers may be initialized in any nonzero state to generate an m-sequence. An initial state of 0, 1, 0, 0, 0 is shown.

Table 15-2 is adapted from Nathanson.[19] For each possible degree (number

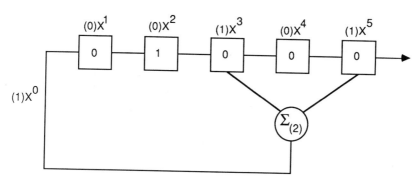

Figure 15-9. Maximal-length binary shift register with initial state.

Table 15-2. Maximum-Length Pseudorandom Codes and Their Properties.

Degree (number of stages) and length	Polynomial octal	Lowest peak sidelobe amplitude	Initial** conditions, decimal	Lowest rms sidelobe amplitude	Initial conditions, decimal
1 (1)	003*	0	1	0.0	1
2 (3)	007*	−1	1, 2	0.707	1, 2
3 (7)	013*	−1	6	0.707	6
4 (15)	023*	−3	1, 2, 6, 8 10, 11, 12	1.39	2, 8
5 (31)	045*	−4	5, 6, 26, 29 (9 conditions) 2, 16, 20, 26	1.89 1.74 1.96	6, 25 31 6
6 (63)	103*	−6	1, 3, 7, 10 26, 32, 45, 54 (9 conditions) (9 conditions)	2.62 2.81 2.38	32 35 7

7 (127)	203*	−9	1,54	4.03	109
	211*	−9	9	3.90	38
	235	−9	49	4.09	12
	247	−9	104	4.23	24,104
	253	−10	54	4.17	36
	277	−10	14, 20, 73	4.15	50
	313	−9	99	4.04	113
	357	−9	15, 50, 78, 90	4.18	122
8 (255)	435	−13	67	5.97	135
	453	−14	(20 conditions)	5.98	254
	455	−14	124, 190, 236	6.10	246
	515	−14	54	6.08	218
	537	−13	90	5.91	90
	543	−14	(10 conditions)	6.02	197
	607	−14	(6 conditions)	6.02	15
	717	−14	124, 249	5.92	156

*Only single Mod-two adder required.
**Mirror images not shown.
Source: Ref. 19, p. 465

of shift registers employed) between 1 and 8, the shift registers that yield m-sequences with the best peak and root mean square sidelobe levels are given. The octal numbers of column 2 can be translated into binary numbers to obtain the polynomial coefficients appropriate for defining the feedback connections. The binary representations of the decimal integers of column 4 represent the initial states of the shift register that yield the minimum peak sidelobe amplitudes catalogued in column 3. Column 5 gives the minimum root mean square sidelobe levels for the codes, and column 6 gives the decimal equivalents of the binary initial conditions necessary to achieve those levels. Note that shift registers for mirror images of the defined codes are not included in the table. Thus, the total number of these best codes is twice those explicitly defined.

By referring to the list of irreducible polynomials in the appendix of reference 20, one can construct the appropriate feedback connections to generate any m-sequence of degree 1 through degree 34 (output length $2^{34} - 1$).

For large $N = 2^n - 1$, the peak sidelobe out of the filter matched to such a code when the signal is normalized to 1 is approximately $\left(\dfrac{1}{N}\right)^{1/2}$ in voltage. The actual value varies with the particular sequence. For example, with $N = 127$, the PSL varies between -18 and -19.8 dB, as opposed to the -21 dB predicted by the above rule of thumb. As N increases, the rule-of-thumb approximation improves.

Most interesting, perhaps, is the fact that in response to a continuous, periodic flow of one of these codes through its matched filter, the output is a periodic peak response of N (in voltage) and a flat range sidelobe response of -1. That is, for a periodic code response, the pseudorandom codes are perfect.

The fact that the PSL is inversely proportional to the square root of the code length makes m-sequences appropriate codes for pulsed radar applications in which only a few closely spaced targets are expected in the field of view; however, the relatively high ISL of these output waveforms make them unsuitable for high target-density and extended clutter situations. Furthermore, either a full tapped delay line or a bank of shift register generators, one for each code bit, must be employed in the receiver to compress at all ranges. These considerations are some of the primary reasons that m-sequences have been more popular in communications and CW radar applications (see Chapter 13) than in pulsed radar applications.

15.3.4 Polyphase Codes

In general, a *polyphase code* employs the M possible phase states $\phi_k = \dfrac{2\pi}{M} k$, for $k = 0, \ldots M - 1$, to code a long, constant-amplitude pulse so that the output resulting from the matched system has particular, designed-for properties.

The class of *Frank polyphase codes* are discrete approximations to a linear FM waveform.[5,21] In general, for each integer M, there is a Frank code of length $N = M^2$ that uses the phase shifts $2\pi/M$, $2(2\pi/M)$, . . . , $(M - 1)(2\pi/M)$, 2π. The ratio of the peak signal to the peak sidelobe for the Frank code of length N approaches $\pi\sqrt{N}$ for large values of N (as opposed to $\sqrt{N}$ for pseudorandom codes); thus, the Frank codes provide an interesting alternative to pseudorandom codes when long codes are required for a radar application in which extended clutter or a high-density target environment is expected.

Furthermore, as would be expected since the Frank codes are discrete approximations to linear FM, autocorrelation functions of Frank codes degrade less rapidly than those of binary phase codes in the presence of Doppler shifts. Therefore, Frank codes should be considered in applications in which the Doppler sensitivity of binary codes presents a problem.

The Frank codes also exhibit the range–Doppler coupling inherent in linear FM waveforms; however, since Frank codes are discrete, the smooth degradation of the peak response experienced with the latter may appear as a loss of detection at certain intermediate velocities in the former—that is, blind speeds exist that the system engineer must be cognizant of and take into account. Codes similar to Frank codes but with improved Doppler tolerance have been investigated by Kretschmer and Lewis.[22]

The Welti and Golay codes are sidelobe-cancelling codes.[23,24] They are sets of pairs of codes with the property that the sum of the autocorrelations of the two codes in a pair sum to twice the voltage of a single autocorrelation at the peak and to zero elsewhere. The Welti codes form a large, general set of polyphase codes that have this property. The Golay codes form the subset of these sidelobe-cancelling codes that are binary. Figure 15-10 exhibits a pair of Golay codes, their autocorrelations, and the zero-sidelobe sum of their autocorrelations. As the figure shows, the key to the sidelobe-cancelling property of Golay code pairs is that the range sidelobes of one are equal in amplitude and opposite in sign to the range sidelobes of the other.

Although these codes may seem to represent the perfect solution to the sidelobe suppression problem, the difficulties involved in their implementation as well as their sensitivities to echo fluctuations and Doppler shift make them less than ideal.

15.3.5 Generation and Processing of Phase-Coded Waveforms

Phase-coded waveforms may be generated either actively or passively in the radar transmitter. Active generation is often accomplished by switching the RF signal between a $0°$ delay and a $180°$ delay line using low-power diode switches. Passive generation can be accomplished using SAW devices in much the same way that FM waveforms are generated.[25]

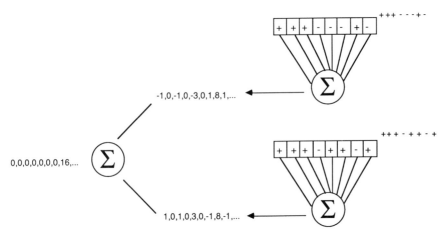

Figure 15-10. Golay (sidelobe-cancelling) code pair of length 8.

Phase-coded waveforms may be compressed using analog or digital techniques for all-range or single-range-bin compression. All-range compression techniques are employed in surveillance radar applications where the range to the target is not known a priori. Single-range-bin cross-correlation techniques are often employed in tracking radar applications where only a single known range bin is of interest.

Analog all-range compressors are often implemented as a tapped delay line, as depicted in Figure 15-11. Here we assume that the transmit subpulse width is τ and the 5-bit Barker code is transmitted. The individual delays may be implemented by using lumped constant delay lines, quartz delay lines, or a SAW device. Each implementation has particular characteristics and problems associated with its use, but analog all-range compressors are generally most appropriate for fairly low pulse compression ratio systems,[26] though advances in SAW technology have allowed compression of codes up to length 2047.[27]

Digital all-range pulse compression may be achieved using binary shift registers as the equivalent of digital tapped delay lines. An example of such an implementation is given in Figure 15-12. The received signal is split into I and Q channels at IF, beat down to video, sampled and A/D converted at the chipping rate, $B = 1/\tau$. Separate digital shift registers/correlators are implemented for each channel, and the sum of the squares of the outputs of each channel is formed and passed on to a video detector. Dual-channel implementation is necessary to avoid a 3-dB system loss.[28] Signal bandwidths up to 100 MHz[29] and extremely high pulse compression ratios can be achieved with digital techniques.[30] Note that the time domain processing indicated in Figure 15-12 may be replaced with frequency domain processing using a digital implementation of the FFT.

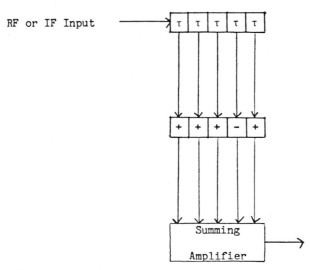

Figure 15-11. Analog all-range phase-coded pulse compression.

The processing required for single-range-bin pulse compression is much simpler than that required for all-range processing. In the former case, active cross-correlation processing as depicted in Figure 15-13 will suffice. In this implementation, the received signal is correlated with a copy of the transmitted signal that has been suitably delayed (to compress the range bin of interest). Like all-range compressors, single-range-bin compressors may be implemented using analog or digital devices. The primary applications of this technique are in communications and radar tracking modes.

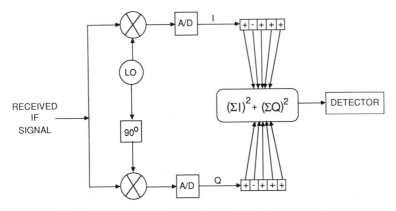

Figure 15-12. Digital all-range phase-coded pulse compression.

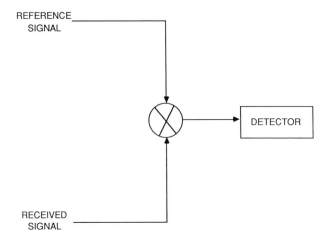

Figure 15-13. Single-range-bin phase compression.

15.4 POLARIZATION MODULATION TECHNIQUES

An EM wave is right circularly (RC) polarized when its vertical (V) E-field modulation is advanced 90° with respect to its horizontal (H) E-field modulation. The wave is said to be left circularly (LC) polarized when the V component lags the H component by 90°. The experimental intrapulse polarization agile radar (IPAR) system codes its transmit pulse in RC and LC polarization on a subpulse basis and uses this modulation to effect pulse compression on receive.[31]

Figure 15-14 is a block diagram of the IPAR system. The transmit waveform is generated by switching the relative phase between the H and V channels from +90° to −90° according to a prestored binary code at rates of up to 100 MHz. The signals are then amplified and transmitted through a dual polarized feed to yield a pulsed RF carrier that is also modulated in polarization on a subpulse basis. As in transmission, two channels are employed on receive. The signals are translated down in frequency and applied to a relative phase detector which extracts the received polarization-modulation code. The resulting signal is then compressed in a high-speed (100-MHz) correlator.

The concept is best illustrated by a simple example. Suppose a long pulse of length T coded in polarization is transmitted into the five subpulses RC, RC, RC, LC, and RC. Furthermore, suppose the transmitted energy is reflected by a dihedral. Recall that the energy returned from odd-bounce scatterers such as flat plates and trihedrals has the reverse circular polarization sense of the illuminating energy (e.g., LC becomes RC and RC becomes LC) and that even-bounce scatterers such as dihedrals return the same sense polarization (e.g., RC

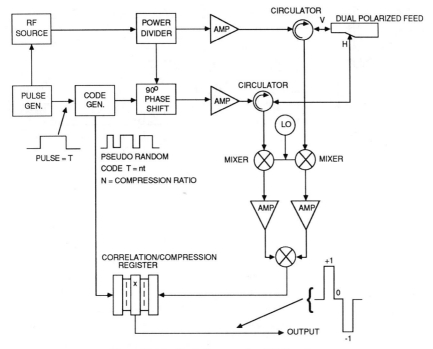

Figure 15-14. Block diagram of an IPAR system.

remains RC and LC remains LC).[32] Since we are assuming an even-bounce scatterer, the reflected waveform will be coded RC, RC, RC, LC, RC. When the signal is received and the relative phase ϕ between the two horizontal and vertical channels is detected on a subpulse basis, the signal $+90°$, $+90°$, $+90°$, $-90°$, $+90°$ is obtained. Taking the sine of ϕ, again on a subpulse basis, we obtain the 5-bit Barker code 1, 1, 1, -1, 1. Similarly, if the scatterer were odd-bounce, the sine of the relative phases would be -1, -1, -1, 1, -1. Both of these signals would compress perfectly (although with opposite signs) through a 5-bit Barker code matched filter.

The above example can be generalized to show that any binary code will give similar results; that is, any binary phase codes can be used equally well to perform pulse compression on a simple target utilizing polarization modulation.

Polarization coding is thus quite similar to biphase coding. For polarization coding, however, the uncoded reference signal is the V component of the trans-mitted waveform rather than a coherent local oscillator signal. Since the coding is contained in the relative phase between the H and V components of the signal, this waveform is insensitive to Doppler shifts and can be implemented with various RF carriers. This dependence on relative phase also causes the wave-form to be sensitive to target and clutter complexity which may be advantageous for certain applications, while, together with potential mutual interference of closely spaced scatterers, may be disadvantageous for other applications.

15.5 RANGE SIDELOBES AND WEIGHTING

As mentioned in Section 15.1, every method of pulse compression leads to the generation of range sidelobes. Since the sidelobes from any range bin may appear as targets in adjacent range bins, the suppression of sidelobes is critical in applications where high target densities, targets of varying reflectivity, or extended clutter are expected. For example, given two targets whose RCS, differ by 15 dB, the first sidelobes due to the larger target will cause a false alarm in a linear FM system designed to detect the smaller target unless sidelobe suppression is employed. (Recall that the first sidelobe is down approximately 13 dB in the matched-filter implementation of linear FM.)

In general, sidelobe suppression is achieved by tapering the matched-filter response by weighting the transmitted waveform, the matched filter, or both in either frequency or amplitude. The weighting that is employed should ideally be applied to both the transmitted waveform and the matched filter so as to retain the matched-filter characteristics and thus avoid a loss of signal-to-noise ratio (LPG); however, this is often impractical. Therefore, weighting is usually applied only to the matched filter, and the ensuing LPG is accepted as a necessary system loss.

The following subsections describe various mismatch-filter weighting functions used for sidelobe suppression. The discussion is divided into functions for FM waveforms and functions for discrete phase-coded waveforms.

15.5.1 Sidelobe Suppression for FM Waveforms

The same illumination functions used in antenna design to reduce spatial sidelobes can also be applied to the frequency domain to reduce the time sidelobes in pulse compression. A comparison of several types of spectral weighting functions for an LFM signal with a rectangular spectrum is shown in Table 15-3. The Dolph-Chebyshev weighting theoretically results in all sidelobes being equal; however, it is physically unrealizable.[9, p. 20-29] The Taylor weighting is a practical approximation to the Dolph-Chebyshev. The Taylor weighting with $\bar{n} = 6$ means the peaks of the first 5 sidelobes ($\bar{n} - 1$) are designed to be equal; the sidelobes then fall off at 6 dB per octave. Weighting the received-signal spectrum to lower the sidelobes increases the main lobe width and reduces the peak signal-to-noise ratio compared to unweighted pulse compression. These effects, catalogued in columns 3 and 4 of Table 15-3, are due to the filter not being matched to the received waveform. For example, reducing the sidelobes to a level of -42.8 dB with the Hamming weighting results in a loss in peak signal-to-noise ratio (LPG) of 1.34 dB and a broadening of resolution by 47%.

Table 15-3. Weighting Function Data.

Weighting function	PSL (dB)	Pulse widening	Mismatch loss (dB)	Far sidelobe fall-off rate
Dolph-Chebyshev	−40.0	1.35	—	1
Taylor, n̄ = 6	−40.0	1.41	−1.2	1/t
$k + (1 - k)\cos^n$				
Hamming (k=0.08, n=2)	−42.8	1.47	−1.34	1/t
Cosine-squared (k=0, n=2)	−32.2	1.62	−1.76	$1/t^3$
Cosine-cubed (k=0, n=3)	−39.1	1.87	−2.38	$1/t^4$
n = 1, k = 0.04	−23.0	1.31	−0.82	1/t
n = 2, k = 0.16	−34.0	1.41	−1.01	1/t
n = 3, k = 0.02	−40.8	1.79	−2.23	1/t

Source: Ref 1, p. 205

The nonlinear-FM, constant-amplitude waveform provides a compressed waveform with low time-sidelobes at the output of its matched filter without the LPG that is incurred with the linear-FM waveform and mismatched filter. The nonlinear variation of frequency with time is equivalent to amplitude-weighting the transmitted-signal spectrum in terms of sidelobe reduction, yet it allows maintaining the rectangular pulse shape that is desired for efficient transmitter operation.

15.5.2   Sidelobe Suppression for Phase Modulated Waveforms

Mismatched filtering for sidelobe suppression is often required with biphase-coded waveforms as well. Various weighting functions for this purpose have been studied and suggested.[5, p. 489] The technique discussed here is an optimal method for reducing the integrated sidelobe level of the code response.

Optimal integrated-sidelobe (ISL) suppression filters of any length can be derived analytically for any discrete phase-coded waveform.[17,33] Given a code of length M and a desired mismatched filter length N, one can derive the expression for the total energy in the sidelobe structure of the output waveform in terms of the unknown coefficients of the optimal filter. That the filter is constrained to yield minimum integrated-sidelobe levels can be interpreted as requiring all the partial derivatives of this expression with respect to the unknown coefficients to be zero. The result is a system of N linear equations in N unknowns whose solution is the set of optimal coefficients.

The filters so derived have the property of minimizing the ISL for a given filter mismatch length, as noted above. In addition, these filters drive the peak sidelobe level (PSL) down rather quickly as well.

Table 15-4 gives the performance of particular ISL-optimized filters for the six non-trivial Barker codes. These levels were computed assuming that the

Table 15-4. Optimal ISL Filter Performance.

Barker Code Length	Filter Length	PSL (dB)	ISL (dB)	LPG (dB)
3	11	−30.9	−21.9	1.2
4	18	−32.0	−23.0	1.7
5	19	−35.3	−25.3	0.6
7	31	−31.6	−20.6	1.4
11	39	−32.1	−18.9	0.9
13	35	−37.7	−24.9	0.2

filter weights and processor arithmetic were quantized to five bits. Note that significant sidelobe reduction is achieved with relatively little loss in processing gain (LPG). As an example, the 35 bit filter for the 13 bit code affords approximately 15 dB improvement in PSL and 13 dB improvement in ISL levels over the matched filter case (the matched filter characteristics are compiled in Table 15-1), while causing only a 0.2 dB LPG.

As a final comment, we note that the matrix inversion required to derive the requisite ISL-optimized filter coefficients becomes unmanageable for large compression ratios. However, if one forms a large code by combining Barker codes, then an effective sidelobe suppression filter can be formed by combining subcode ISL-optimized filters (see Section 15.3.2).

15.6 THE RADAR AMBIGUITY FUNCTION[2,10]

The autocorrelation function of the radar's transmitted waveform represents the time (range) response of the radar receiver's matched filter in the presence of a point target that has no radial velocity with respect to the radar. In general radar applications, there may often be undetermined radial velocity between the radar and the target. The mathematical functions called *time-frequency autocorrelation functions* or *ambiguity functions* were introduced by Woodward[34] to allow representation of the time response of a signal processor when a target has significant radial velocity. The ambiguity function, its basic properties, and specific function types are discussed below.

The output from a filter $v(t)$ at time T_R in response to a transmit waveform $u(t)$ that has been Doppler shifted f_d Hz is given by

$$X_{uv}(T_R, f_d) = \int_{-\infty}^{\infty} u(t)v^*(t + T_R) \exp(j2\pi f_d t) \, dt \qquad (15\text{-}27)$$

where

v^* = complex conjugate of v
$T_R = 0$ corresponds to the peak response when $f_d = 0$

If $u \equiv v$ (i.e., the matched-filter case) and $\Psi(T_R, f_d) = |X_{uv}(T_R, f_d)|^2$, it follows that

$$\Psi(T_R, f_d) = \left| \int_{-\infty}^{\infty} u(t)u^*(t + T_R) \exp(j2\pi f_d t)\, dt \right|^2 \qquad (15\text{-}28)$$

The function Ψ is called the *ambiguity function* of u, and its plot in time-frequency-amplitude space is the *ambiguity diagram* of u. The function $\Psi(T_R, f_d)$ has the following properties:[35]

1. The peak response of ψ occurs at $T_R = 0, f_d = 0$ and, in power, is given by $(2E)^2$, where $E = $ received energy.
2. The diagram is symmetric about $T_R = -f_d$.
3. When there is no Doppler, the ambiguity function is the autocorrelation function of the transmit waveform $u(t)$.
4. $|\Psi(0, f_d)|^2 = |\int u^2(t) \exp(j2\pi f_d t)\, dt|^2$; that is, the frequency profile at $T_R = 0$ is proportional to the spectrum of $u^2(t)$.
5. The volume under the entire ambiguity surface is independent of $u(t)$ and is equal to $(2E)^2$.

In terms of ramifications for radar waveform design, from property 5 above, one can deduce perhaps the single most important lesson to be learned from ambiguity function analysis: the total energy in the response over the entire frequency–range plane of a returned signal of fixed energy is likewise fixed; thus, attempts to reduce the energy content in any particular region must lead to increased energy content in other regions.

Transmit waveforms with different ambiguity diagrams are desirable for different applications. For example, if precise measurements of both range and Doppler of an echo source are required, the ideal waveform would have an ambiguity diagram consisting of a single peak at the origin sufficiently thin in both dimensions to achieve the desired resolutions. An idealized ambiguity function of this sort, generally called a *thumbtack ambiguity*, is depicted in Figure 15-15(a). Note that the maximum value of Ψ is $(2E)^2$ and that, since the total volume under Ψ must be $(2E)^2$, the thinner the peak (i.e., the more stringent the resolution requirements), the more the energy must be spread throughout the remainder of the response.

The result of requiring fine range and velocity resolution is increased processor requirements. Suppose a particular radar system utilizes a phase-coded pulse compression waveform that has a thumbtack-type ambiguity function. Let the pulse width be $T = 0.2$ ms and the modulation bandwidth be $B = 100$

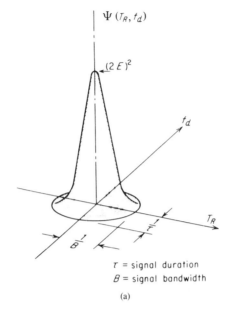

τ = signal duration
B = signal bandwidth

(a)

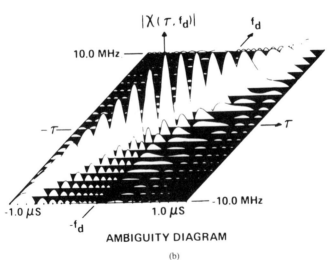

AMBIGUITY DIAGRAM

(b)

Figure 15-15. (a) Thumbtack ambiguity function. (b) An ambiguity diagram for a linear FM pulse of 1 μs duration and 10 MHz bandwidth. [(a) By permission, from Skolnik, ref. 2; © 1962 McGraw-Hill Book Co.; (b) By permission, from Brookner, ref. 10; © 1982 Artech House, Inc.]

MHz. Then the system's range resolution is given by

$$\delta_r = \left(\frac{1}{2}\right) c\tau = \frac{c}{2B} = \frac{c}{100 \times 10^6} \approx 1.5 \text{ m} \qquad (15\text{-}29)$$

and its Doppler resolution is given by

$$\delta_f = \frac{1}{T} = \frac{1}{0.2 \times 10^{-3}} = 5 \text{ kHz} \qquad (15\text{-}30)$$

Further, assume that the maximum target Doppler of interest falls between $+50$ and -50 kHz and that the required range coverage is 100 km. Then full range and Doppler coverage requires that 100 kHz/5 kHz = 20 Doppler channels must be implemented for each of the $100/1.5 \times 10^{-3} \approx 6.7 \times 10^4$ range cells of interest. Although a thumbtack ambiguity diagram is optimum for unambiguous target location in both range and velocity, the price of instrumenting the 20 Doppler channels per range gate may be excessive compared to the total radar cost and processor complexity.

For certain applications, it may in fact be desirable to accept some ambiguity in target location in range and Doppler to reduce processor complexity. An example of one such waveform is given in Figure 15-15(b). Here, the "knife edge" surface slanting across the range–Doppler plane represents the ambiguity diagram of a linear FM pulse of 1 μs duration and 10 MHz bandwidth. Note that the responses from targets separated by 0.1 μs in range and 1.0 MHz in Doppler will be almost identical, and thus they will be unresolvable; however, full range and velocity coverage can be achieved with relatively few Doppler channels per range bin.

Thus, there is no single ideal waveform for all situations, but particular waveforms are well suited for particular applications. With this in mind, we present a discussion of several commonly used waveforms and compare them in terms of their ambiguity functions. The waveforms selected for comparison are:

1. A simple 1-μs pulse.
2. A linear FM pulse with 10 MHz sweep and 1 μs duration.
3. A 13-bit Barker phase code pulse of 1 μs duration.

Figure 15-16 compares the ambiguity functions of each of these three waveforms by superimposing the 3-dB contours of the ambiguity functions for each waveform. Note that all three waveforms have the same Doppler resolution near $T_r = 0$ because all of them have the same basic time envelope—a 1-μs rectangular pulse. The Doppler resolution is equal to the reciprocal of this rectangular pulse envelope, which, in this example, is 1 MHz. The time or range resolution

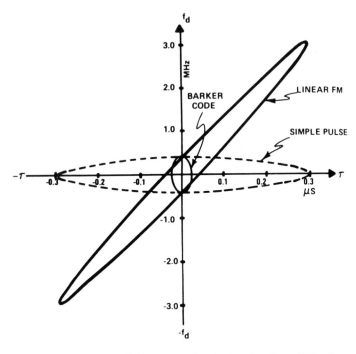

Figure 15-16. Ambiguity function 3-dB contours for simple pulse, linear FM pulse and 13-bit Barker code. (By permission, from Brookner, ref. 10; © 1982 Artech House, Inc.)

is directly proportional to the transmitted bandwidth. The 13-MHz bandwidth of the 13-bit Barker coded pulse provides the highest time resolution, approximately 0.192 μs. The 10- and 1-MHz bandwidths of the linear FM and simple pulse provide time resolutions of approximately 0.025 and 0.25 μs, respectively. The FM pulse experiences a Doppler–range cross-coupling, as evidenced by the skewing of its 3-dB contour. For the actual value of range or Doppler to be determined, the other would have to be known a priori. Thus, three quite different response functions can be produced with three waveforms having the same time duration and amplitude.

15.7   SUMMARY

In general applications, radar waveforms may be modulated in phase or frequency to increase the bandwidth of the transmitted pulse. This enhanced bandwidth may then be used by matched (or mismatched) filtering on receive to increase the range resolution of the radar system. This general technique, called *pulse compression*, is often used in modern radar systems to maintain a required range resolution while increasing the average power on a single-pulse basis. Each technique for effecting pulse compression has a unique set of characteristics depending on both theory and the state of the art of the technology involved. Waveform choice and design thus continue to pose challenging problems to the radar system engineer.

Implementation of pulse compression in a radar system requires a stable signal source for generating the transmitted carrier, a method for coding the transmitted carrier, and implementation of the appropriate signal processing in the receiver to compress backscattered radar signals. All of these factors involve increased radar system complexity, cost, and chance for failure.

The paragraphs below summarize some of the most salient properties of simple pulse, FM, and phase-coded waveforms with respect to their utility and use in radar systems.

Simple pulse radar systems are the least complex, least expensive, and easiest to maintain. Simple pulse waveforms are used in older-generation radars with magnetron transmitters, in radars where cost is a primary constraint, and in systems where range resolution and detection requirements can both be met using a simple pulse.

Of all pulse compression techniques, LFM is the oldest and most well developed. It is used to improve detection performance while maintaining range resolution. High time–bandwidth products and fine range resolution are possible. LFM is particularly useful for moving-target surveillance, since its sensitivity to Doppler shift is low. LFM results in a range–Doppler ambiguity function that does not allow for precise determination of either unless the other is extracted some other way.

Frequency stepping represents the discrete approach to FM, making use of the recent gains in digital technology. It can be used in much the same way as LFM, with the added feature that the order of the transmission frequencies can be easily varied to add flexibility and to shape the waveform's Doppler response. Interpulse stepping can achieve extremely fine range resolution and large time–bandwidth products.

Phase-coding techniques may also be employed to improve detection performance without affecting range resolution. Biphase coding of the carrier is the most well developed of these techniques. Extremely high pulse compression ratios are achievable, though resolution is limited compared with FM techniques because of instantaneous bandwidth limitations in current digital technology. The discrete nature of the coding makes these waveforms easy to generate and particularly flexible. They may be implemented with a bank of Doppler filters to give precise range and Doppler information, and can be used in conjunction with MTI to improve subclutter visibility.

15.8 REFERENCES

1. C. E. Cook and M. Bernfield, *Radar Signals*, Academic Press, New York, 1967.
2. M. I. Skolnik, *Introduction to Radar Systems*, 2nd ed., McGraw-Hill Book Co., New York, 1980.
3. A. W. Rihaczek, *Principles of High-Resolution Radar*, McGraw-Hill Book Co., New York, 1969.
4. J. R. Klauder, A. C. Price, S. Darlington, and W. J. Albersheim, "The Theory and Design of Chirp Radars," *Bell System Technical Journal*, vol. 39, no. 4, July 1960, pp. 745–808.
5. F. E. Nathanson, *Radar Design and Principles*, McGraw-Hill Book Co., New York, 1969.
6. C. E. Cook, "Pulse Compression, Key to More Efficient Radar Transmission," *Proceedings of the IRE*, vol. 48, no. 3, March 1960, pp. 310–316.
7. Skolnik, *Radar Systems*, p. 427.
8. D. R. Moss, "Updated EMC Analysis of the PAVE PAWS (AN/FPS-115) at Otis AFB," Final Report, Contract no. F-19628-78-C-006, Electromagnetic Compatibility Analysis Center, IIT Research Institute, Annapolis, Md., July 1978.
9. M. I. Skolnik (Ed.-In-Chief), *Radar Handbook*, McGraw-Hill Book Co., New York, 1969.
10. E. Brookner, *Radar Technology*, Artech House, Dedham, Mass., 1982.
11. Skolnik, *Radar Systems*, p. 424.
12. Brookner, *Radar Technology*, p. 138.
13. Skolnik, *Radar Systems*, p. 425.
14. Brookner, *Radar Technology*, p. 173.
15. R. Turyn, "On Barker Codes of Even Length," *Proceedings of the IEEE* (correspondence), vol. 51, September 1963, p. 1256.
16. MacMullen, "Radar Short Course Notes," Technology Service Corp., Washington, D.C., 1978.
17. M. N. Cohen, "Binary Phase-coded Pulse Compression," Internal Report No. 1293-R-0021, Norden Systems, Norwalk, Conn., 1979.
18. R. C. Dixon, *Spread Spectrum Systems*, John Wiley & Sons, New York, 1976.
19. Nathanson, *Radar Design*, p. 465.
20. E. W. Peterson and E. J. Weldon, Jr., *Error Correcting Codes*, 2nd ed., MIT Press, Cambridge, Mass., 1972, Appendix C.
21. R. L. Frank, "Polyphase Codes with Good Nonperiodic Correlation Properties," *IEEE Transactions on Information Theory*, vol. 9, January 1963, pp. 43–45.
22. F. F. Kretschmer, Jr., and B. L. Lewis, "Doppler Properties of Polyphase Coded Pulse Compression Waveforms," NRL Report 8635, Naval Research Laboratory, Washington D.C., September 1982.
23. G. R. Welti, "Quaternary Codes for Pulsed Radar," *IRE Transactions on Information Theory*, vol. 7, June 1960, pp. 82–87.
24. M. J. E. Golay, "Complementary Series," *IRE Transactions on Information Theory*, vol. 7, No. 2, April 1961, pp. 82–87.
25. Brookner, *Radar Technology*, p. 138.
26. Nathanson, *Radar Design*, p. 469.
27. Brookner, *Radar Design*, p. 174.
28. Nathanson, *Radar Design*, p. 472.
29. M. N. Cohen, E. S. Sjoberg, and E. E. Martin, "The Intrapulse Polarization Agile Radar," Proceedings Microwave Systems Applications and Technology, 1983 (MSAT-'83), March 1983, pp. 483–494.
30. Brookner, *Radar Technology*, p. 139.

31. M. N. Cohen and E. S. Sjoberg, "Intrapulse Polarization Agile Radar," *Proceedings of Radar-82*, Institute of Electrical Engineers Conference Publication No. 216, October 1982, pp. 7–11.

32. H. Jasik, *Antenna Engineering Handbook*, McGraw-Hill Book Co., New York, 1961.

33. M. H. Ackroyd and F. Ghani, "Optimum mismatched filters for sidelobe suppression," *IEEE Transactions on Aerospace and Electronics Systems*, vol. AES-9, March 1973, pp. 214–218.

34. P. M. Woodward, *Probability and Information Theory, with Applications to Radar*, McGraw-Hill Book Co., New York, 1953.

35. Skolnik, *Radar Systems*, p. 412.

16
SYNTHETIC APERTURE RADAR

Robert N. Trebits

16.1 OPERATIONAL CONSIDERATIONS

High angular resolution radar imagery can be obtained using conventional *non-coherent* radar systems by employing a large antenna or operating at a short wavelength. Airborne platforms, however, impose severe physical limitations on the size of the antenna which can be used and still maintain acceptable flight characteristics of the aircraft. The use of shorter wavelengths requires higher mechanical tolerances, is limited by generally lower power sources, and must accommodate higher atmospheric losses.

Consider an airborne radar system operating in the side-looking mode, in which the radar antenna is oriented normal to the aircraft flight path and boresighted at some depression angle ψ from the horizontal plane of the aircraft. Figure 16-1 depicts such a side-looking airborne radar (SLAR) scenario and the ground area illuminated by the radar. *Cross-range (along-track) resolution* is defined as the resolution in the direction of the flight path; *range resolution* is defined as the resolution normal to the flight path.

Assuming the pulse length–limited geometry of the antenna orientation and the area of a typical resolution cell, shown in Figure 16-2, note that return signals from two distinct targets are nonoverlapping in time if these targets are separated by a slant distance ΔR_s given by

$$\Delta R_s \geq \frac{c\tau}{2} \tag{16-1}$$

where τ is the duration of the transmitted pulse. We then can define the ground-range resolution δ_R to be the projection onto the ground of the slant range resolution ΔR_s:

$$\delta_R = \frac{c\tau}{2 \cos \psi} \tag{16-2}$$

where ψ = antenna depression angle, as discussed in Chapter 10

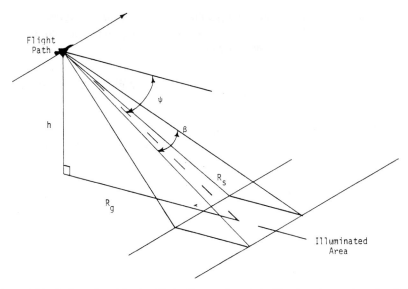

Figure 16-1. Radar oriented in the side-looking mode and the illuminated area along the flight path.

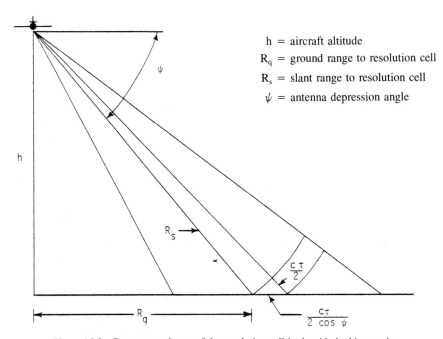

h = aircraft altitude

R_q = ground range to resolution cell

R_s = slant range to resolution cell

ψ = antenna depression angle

Figure 16-2. Geometry and area of the resolution cell in the side-looking mode.

If a pulse 1 ns in duration is transmitted at a depression angle for which $\cos \psi \approx 1$, then a 15-cm ground-range resolution is realizable. Obviously, an equivalent range resolution could also have been achieved using a long transmitted pulse and pulse compression on reception. The resulting range resolution would still be determined from Eq. (16-2), keeping in mind that τ now represents the compressed pulse duration.

If the half-power angular beamwidth of the antenna pattern is β radians, the corresponding cross-range resolution δ_x at a slant-range R_s is

$$\delta_x \approx R_s \beta \tag{16-3}$$

Define d_x to be the dimension of the antenna in the cross-range, or along-track, dimension. Assume operation of the radar at its diffraction limit at a wavelength λ so that the antenna beamwidth is given by

$$\beta \approx \frac{\lambda}{d_x} \tag{16-4}$$

Substitution of Eq. (16-4) into Eq. (16-3) defines the azimuth resolution:

$$\delta_x = \frac{\lambda R_s}{d_x} = \frac{\lambda R_g}{d_x \cos \psi} \tag{16-5}$$

where R_g is the ground range to the target cell.

Now consider a few implications of the algebraic form of Eq. (16-5). The cross-range resolution can be improved by either decreasing the wavelength of the radar signal or increasing the cross-range dimension of the antenna. These design alternatives are limited in their actual application by practical considerations. Attenuation of the radar waves tends to increase with higher frequencies (shorter wavelengths) due to increased scattering and absorption by water droplets and various molecules in the air. A maximum size constraint must necessarily be imposed on the antenna by availability of space on the aircraft and by considerations of the aerodynamic integrity of the aircraft. Furthermore, fabrication of a diffraction-limited radar antenna with $d_x \geq 10^3 \lambda$ is prohibitively costly and difficult.

For comparison purposes, assume that $d_x \approx 10^3 \lambda$, ψ is small, and the ground-range R_g is 10 km. Eq. (16-5) then predicts that an azimuth resolution of ~ 10 m is realizable for a conventional SLAR system.

It is apparent from these estimated values of achievable range and azimuth resolutions that there can be a difference of as much as two orders of magnitude between them. A primary motivation for the development of synthetic aperture radar systems is to make the range and cross-range resolutions commensurate.

16.2 CONCEPT OF SYNTHETIC APERTURE RADAR

Synthetic aperture radar (SAR) can provide significant improvement in cross-range resolution over that given by a real-aperture (SLAR) system. The SAR concept employs a coherent radar system and a single moving antenna to simulate the function of all the antennas in a real linear antenna array. This single antenna sequentially occupies the spatial positions of a synthesized array, as shown in Figure 16-3. All received signals, both amplitude and phase, are stored for each antenna position in the synthetic array. These stored data are then processed to re-create the image of the illuminated area seen by the radar.

The typical SAR configuration is side-looking, in which the radar antenna is oriented normal to the aircraft flight path and downward at some appropriate depression angle. Figure 16-4 illustrates this scenario and the ground area illuminated by the radar. Cross-range resolution is defined in the direction of the flight path and is also referred to in the literature as the along-track resolution.

Since a physically long linear array is the basis of comparison, it is instructive to recall some of the characteristics of such an array. In the usual array arrangement, several identical antennas are equally spaced along a straight line, and transmission and reception are accomplished through all of the antenna elements, as discussed in Chapter 6.

The radiation pattern for this linear array can be defined as the product of the radiation pattern of a single antenna element and an array factor. This array factor is a function of both the number of elements in the array and the spacing between adjacent elements. The array factor in general causes the lobes to be narrower than the antenna radiation pattern of a single element, so that the array factor tends to determine the gross radiation pattern of the entire array. The half-power beamwidth β, in radians, of the array will be

$$\beta = \lambda/L \qquad (16\text{-}6)$$

where L is the total length of the array. The cross-range resolution for this array

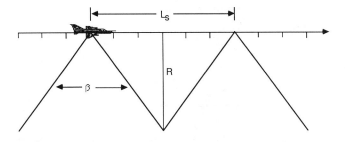

Figure 16-3. Synthetic array configuration.

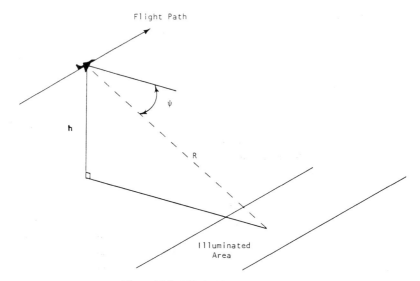

Figure 16-4. Side-looking SAR geometry.

at a slant range R_s is given by

$$\delta_x = \frac{\lambda R_s}{L} \tag{16-7}$$

The effective half-power angular beamwidth β_s of the synthetic array is given by

$$\beta_s = \frac{\lambda}{2L_s} \tag{16-8}$$

where L_s is the length of the synthetic aperture. The factor of 2 characterizes the synthetic array system and arises because the round-trip phase shifts determine the effective radiation pattern of the synthetic array, whereas the phase shifts for a physical array are provided only during reception.

With d_x as the horizontal aperture of the single antenna, the length of the synthetic aperture is given by

$$L_s = \frac{\lambda R_s}{d_x} \tag{16-9}$$

since L_s is the distance through which the aircraft flies while a target is within the beamwidth of the aircraft antenna. The cross-range resolution δ_x of the synthetic array is given by

$$\delta_x = \beta_s R_s \qquad (16\text{-}10)$$

or, by substituting Eqs. (16-8) and (16-9) into Eq. (16-10),

$$\delta_x = (\lambda/2L_s)R_s = (\lambda R_s/2)(d_x/\lambda R_s) = d_x/2. \qquad (16\text{-}11)$$

Note from Eq. (16-11) that δ_x is not range dependent, in contrast to the cross-range resolution of a physical linear array. Finer resolution can be obtained by making the single antenna still smaller, exactly the opposite of what one would do to increase the azimuth resolution of a single-antenna conventional SLAR.

The synthetic aperture, or array, may be either focused or unfocused. In the unfocused case, integration of the stored signals is accomplished with no effort made to shift the phases of signals differentially from various positions within the aperture. A result of this straightforward integration is a limitation of the aperture length due to these same phase shifts. The maximum usable aperture length is conventionally defined as that range where the distance from the target cell to the extremities of the array is $\lambda/8$ greater than the distance from the target cell to the center of the array. From the geometry of Figure 16-5, we see that this restriction results in the expressions

$$\left(R_s + \frac{\lambda}{8}\right)^2 = \frac{L_s^2}{4} + R_s^2 \qquad (16\text{-}12)$$

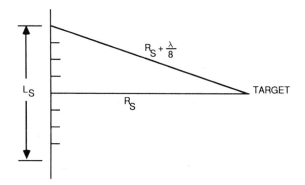

Figure 16-5. Maximum usable aperture in an unfocused synthetic array.

and

$$\lambda\left(R_s + \frac{\lambda}{16}\right) = L_s^2 \qquad (16\text{-}13)$$

Since $R_s \gg \lambda/16$, Eq. (16-13) becomes

$$L_s \approx (\lambda R_s)^{1/2} \qquad (16\text{-}14)$$

Replacing L_s in Eq. (16-8) by Eq. (16-14), we obtain

$$\beta_s \approx \tfrac{1}{2}(\lambda/R_s)^{1/2} \qquad (16\text{-}15)$$

Replacing β_s in Eq. (16-10) by Eq. (16-15), we obtain

$$\delta_x \approx \tfrac{1}{2}(\lambda R_s)^{1/2} \qquad (16\text{-}16)$$

In the focused case, an appropriate phase shift is added differentially to the signals from each position within the array so that the signals all travel the same electrical path length. The full potential of the $d_x/2$ cross-range resolution can then be realized, independent of the range of the target. Figure 16-6 shows a comparison of the cross-range resolutions obtainable from a conventional an-

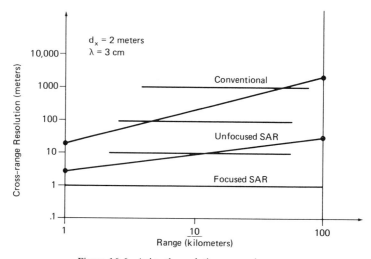

Figure 16-6. Azimuth resolution comparison.

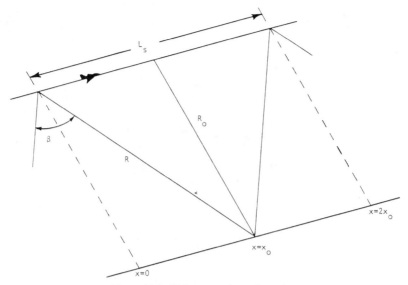

Figure 16-7. SAR geometric configuration.

tenna and from both a focused and an unfocused synthetic aperture, where the wavelength is 3 cm (X-band) and the antenna aperture is 2 m in the cross-range direction. Focusing is part of the SAR signal processing and is discussed further later in this chapter.

The salient features of the radar signal returned to the antenna can be determined by considering a point target located at position x_0 along the ground-track and at slant range $R_s = R_0$ from the aircraft, as shown in Figure 16-7. If a pulsed, sinusoidal signal of radian frequency ω_0 is transmitted, the return signal $s(t)$ will have the form

$$s(t) = A \cos \omega_0[t - 2R(t)/c] \tag{16-17}$$

where the coefficient A includes such factors as the RCS of the target, the inverse fourth power dependence with range, the antenna gain pattern, and so on. In the case where

$$|x(t) - x_0| \ll R_0 \tag{16-18}$$

the slant range $R(t)$ can be approximated by

$$R(t) \approx R_0 + \frac{[x(t) - x_0]^2}{2R_0} \tag{16-19}$$

The return signal $s(t)$ can then be expressed as

$$s(t) = A \cos \omega_0 \left[t - \frac{2R_0}{c} - \frac{(x(t) - x_0)^2}{R_0 c} \right] \qquad (16\text{-}20)$$

If V is the velocity of the aircraft, then $x(t) = Vt$ and

$$s(t) = A \cos \omega_0 \left[t - \frac{2R_0}{c} - \frac{(Vt - x_0)^2}{R_0 c} \right] \qquad (16\text{-}21)$$

The instantaneous radian frequency ω of the return signal is the time derivative of the phase:

$$\omega(t) = \omega_0 - \frac{2V\omega_0}{R_0 c} (Vt - x_0) \qquad (16\text{-}22)$$

The Doppler frequency shift f_d generated by the aircraft motion relative to the fixed ground target can be derived by substituting angular frequency $f = \omega/2\pi$ into Eq. (16-22):

$$f_d(t) = - \frac{2V}{R_0 \lambda} (Vt - x_0) \qquad (16\text{-}23)$$

At time $t = 0$ [$x(t) = 0$], the return signal frequency is Doppler shifted upward by an amount $(2Vx_0)/(R_0\lambda)$. The Doppler shift remains positive, but decreases linearly with the time as the aircraft closes on the target. When the target is directly abeam of the aircraft, $Vt = x_0$, and the return and transmitted signals have the same frequency; thus, the Doppler shift is zero. After the aircraft passes by the target, the Doppler shift becomes negative and decreases linearly with time until the target passes out of the antenna half-power beam pattern at $t = 2x_0/V$, ($x = 2x_0$) and $f_d = -2Vx_0/R_0\lambda$. This linear behavior of the Doppler frequency with time is illustrated in Figure 16-8.

The total Doppler bandwidth Δf_d is then

$$\Delta f_d = \frac{4Vx_0}{R_0\lambda} \qquad (16\text{-}24)$$

If T is the time period within which a target is illuminated by the real antenna beam, then this look time T, or aperture time, is

$$T = \frac{L_s}{V} = \frac{2x_0}{V} = \frac{R_0\lambda}{Vd_x} \qquad (16\text{-}25)$$

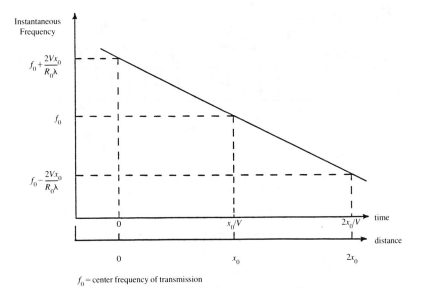

f_0 = center frequency of transmission

Figure 16-8. Linear Doppler shift.

and the total Doppler bandwidth is

$$\Delta f_{\mathrm{d}} = \frac{2V^2 T}{R_0 \lambda} = \frac{2V}{d_{\mathrm{x}}} \qquad (16\text{-}26)$$

Matched-filter processing will achieve an effective time resolution of $1/\Delta f_{\mathrm{d}}$. The equivalent spatial resolution δ_{x} along the flight path will be $V/\Delta f_{\mathrm{d}}$, or

$$\delta_{\mathrm{x}} = \frac{V}{\Delta f_{\mathrm{d}}} = \frac{d_{\mathrm{x}}}{2} \qquad (16\text{-}27)$$

which is the same as the azimuth resolution given by Eq. (16-11).

The time–bandwidth product $T\Delta f_{\mathrm{d}}$ represents the SAR processing gain of the matched-filter processing, or

$$T\Delta f_{\mathrm{d}} = \frac{2R_0 \lambda}{d_x^2} \qquad (16\text{-}28)$$

Alternatively, the SAR processing gain can be represented in terms of the radar pulse repetition frequency (PRF), which must be at least twice the Doppler bandwidth Δf_{d} to satisfy the Nyquist sampling criterion. Thus, the PRF is set

Table 16-1. Comparison of Linear Pulse Compression and SAR Characteristics.

Characteristic	Linear Pulse Compression	Synthetic Aperture Radar
Signal frequency history	$f = f_0 - \dfrac{\mu t}{2\pi}$ for $\lvert t \rvert < \tau/2$ μ = frequency slope	$f = f_0 - \dfrac{2V}{\lambda R_0}(Vt - x_0)$ for $t < T$
Signal bandwidth	$\Delta f = \dfrac{\mu t}{2\pi}$ FM excursion	$\Delta f_d = \dfrac{4Vx_0}{\lambda R_0} = \dfrac{2V}{d_x}$ Doppler bandwidth
Distance resolution	$\delta_R = \dfrac{c}{2\Delta f} = \dfrac{\pi c}{\mu t}$ Range resolution	$\delta_x = \dfrac{V}{\Delta f_d} = \dfrac{d_x}{2}$ Cross-range resolution
Processing gain (time–bandwidth product)	$\tau \Delta f = \dfrac{\mu \tau^2}{2\pi}$	$T\Delta f_d = \dfrac{2R_0 \lambda}{d_x^2}$

equal to $2\Delta f_d$ to determine the SAR processing gain:

$$T\Delta f_d = \frac{R_0 \lambda}{2Vd_x}(\text{PRF}). \tag{16-29}$$

Table 16-1 summarizes and compares the pertinent characteristics of linear pulse compression and SAR processing.[1] The two techniques have the same mathematical formulations, permitting identical analytical treatment, but the ranges of analogous parameters are significantly different. In a linear pulse compression implementation, an FM excursion of 150 MHz results in an effective range resolution of 1 m. For a SAR system with an antenna aperture in the cross-range direction of 2 m and a platform velocity of 150 m/s, the Doppler bandwidth will be only 150 Hz, which corresponds to an effective cross-range resolution of 1 m, commensurate with the (compressed) range resolution.

16.3   SAR DATA PROCESSING

16.3.1   General Considerations

SAR signal processing can be mathematically described as either a correlation or a filtering process which includes all the coherent returns stored during an aperture time. The SAR image reconstruction process thus requires a large data storage and dynamic range capability and the ability to process all these data within one interpulse period. In addition, focusing of the data is necessary for optimum cross-range resolution, and compensation for nonuniform airplane motion is certainly desirable. Two general classes of SAR data processing have been used for image reconstruction: one uses (analog) optical techniques and one uses (digital) electronic processing techniques.

16.3.2 OPTICAL PROCESSING

Historically, optical SAR processing was developed first. During the 1950s, photographic film was the only data storage medium having the necessary bandwidth and dynamic range characteristics. The high data processing speeds required the use of optical filtering, focusing, and masking techniques to re-create the terrain image stored on photographic film. As might be expected, such techniques have certain inconveniences. One must cope with chemical film development and handling, which is usually carried out in a ground-based vehicle rather than in the aircraft. The resulting time lag between data acquisition and image formation, while relatively unimportant for commercial terrain mapping, is unacceptable in a tactical application.

The bipolar video return signals, which are the radar end product of coherent heterodyning and phase detection, are used to intensity modulate the electron beam of a CRT, as shown in Figure 16-9. A DC bias level is added to this bipolar signal so that the modulating signal is always positive. When a signal is received from the nearest range in the illuminated target area, the electron beam of the CRT begins a vertical sweep. This sweep is synchronized so that it ends at the time when the return signal from the farthest range has been

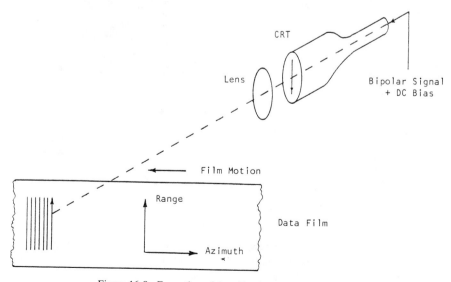

Figure 16-9. Formation of data film from a bipolar signal.

received. Therefore, the entire received video signal corresponding to one transmitted pulse is translated into a brightness-modulated vertical CRT trace which is then projected onto a photographic data film. The data film moves with a translational speed synchronized to the velocity of the aircraft so that the next pulse causes a vertical line to be projected onto the film adjacent to the last line. The intensity modulation of the CRT and the development of the film are both controlled to preserve a linearity relationship between the light amplitude transmissivity of the film and the bipolar radar video signal amplitude.

With appropriate scale factors, the horizontal position on the data film will correspond to the along-track position of the radar antenna, and the vertical position on the data film will correspond to a slant range from the antenna. If the PRF is high enough to meet sampling requirements, the vertical stripes on the data film may be considered a two-dimensional continuum.

The data film presents a recorded interference pattern not unlike that of an optical hologram created with coherent light. The azimuth history of a particular target cell depicted in Figure 16-10 corresponds to an arc on the film of width (in ground range) $\delta_R = c\tau/(2 \cos\psi)$, where τ is the effective pulse width (pulse may be chirped) and ψ is the antenna depression angle. The arc length L represents the synthetic aperture corresponding to the slant range R_s and is longer for targets at greater range [see Eq. (16-9)].

Figure 16-11 presents a simplified diagram of a typical coherent optical processor. A laser illuminates the data film will coherent visible light. A qualitative description of the ensuing procedure follows. The film, acting as a continuum of one-dimensional (along-track) Fresnel zone plates, diffracts the laser light and causes an optical reconstruction of the radar reflectivity distribution of the actual target field.

Antenna amplitude weighting can be accomplished by inserting an appropriately shaded transparency with uniform phase thickness next to the data film. Just as the target field lies tilted with respect to the aircraft orientation, so the optically created image is tilted with respect to the plane of the data film. A conical lens provides the proper phase shifts which must be added to focus the image for all ranges, erects the image, and moves it to infinity.

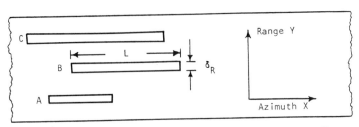

Figure 16-10. Azimuth histories on data film of targets at (A) near range, (B) medium range, and (C) far range.

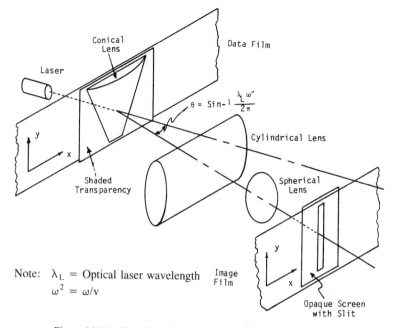

Figure 16-11. Simplified diagram of a coherent optical processor.

One notes that range resolution is present on the data film itself, that is, it is only necessary to preserve this range resolution throughout the optical correlation. The conical lens, curved only in the azimuth direction, does not destroy this range resolution. However, the locus of range focus is on the data film, while the locus of azimuth focus has been moved to infinity. Therefore, a cylindrical lens, curved in the range dimension, is also used to move the locus of range focus to infinity. A spherical lens then focuses both locus planes from infinity onto an image film which is moving synchronously with the data film. The image film must be exposed by means of a vertical slit in an opaque screen to remove parallax effects of imaged targets at different azimuth positions. Since an offset frequency (DC bias) was used to preserve Doppler sense, the slit will be located at an angle θ off the optic axis given by

$$\theta = \text{arc sin} \left[\frac{\lambda_L \omega'}{2\pi} \right] \tag{16-30}$$

where λ_L is the wavelength of the coherent illumination and $\omega' = \omega/V$. The image film then represents the optical reconstruction of a radar image as seen from the antenna on the aircraft. The processor is seen to act like the optical analog to the radar system.[2]

16.3.3 DIGITAL PROCESSING

The digital implementation of the SAR signal processor involves data storage, motion compensation, and correlation of the digitized radar video signal. The coherent radar data are processed in a manner utilizing amplitude and phase information, so that I and Q radar video signals are created within the radar receiver. The I and Q signals are digitized, filtered, and stored in processor memory for as long as the maximum SAR look time requires. The SAR signal correlator then performs motion compensation and correlation to create an output signal amplitude proportional to the re-created image density.

Figure 16-12 outlines the general flow of data processing, from left to right. The I and Q radar video signals are produced by mixing the coherent reference oscillator (COHO) output signal and a 90° phase-shifted version of it with the radar IF signal. Both I and Q video signals are then low-pass filtered to remove the upper sideband generated in the mixing process. The resulting I and Q signals then represent baseband SAR signals which contain all target backscatter amplitude and phase information.

Two A/D converters transform the I and Q video signals into the digital format required by the digital SAR signal processor. The dynamic range requirement for the A/D converters is determined by the dynamic range of the analog video signals themselves, plus the compression gain of the pulse compression processor. (In this discussion, the pulse compression process itself is omitted.) The A/D conversion rate must be commensurate with the bandwidth of the (compressed) radar pulse. Thus, a 50-ns radar pulse requires a receiver bandwidth of 20 MHz and a minimum A/D conversion rate capability for the I and Q channels of the same amount.

The first mathematical function performed within the SAR signal processor

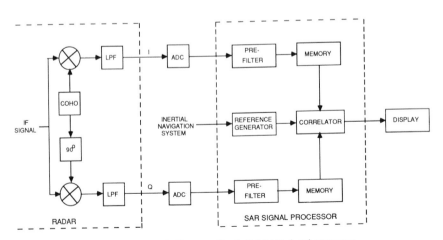

Figure 16-12. Block diagram of a digital SAR signal processor.

is that of prefiltering each channel. When the processed Doppler bandwidth is much less than the total terrain backscatter bandwidth, as seen by the real antenna, the radar PRF represents an oversampling condition. Therefore, the SAR signal processor data rate can be slowed down to match the processed Doppler bandwidth. The prefilter functions to decrease the data rate required of the remainder of the SAR signal processor.

The memory must be large enough to hold all of the digital words corresponding to all range bins over the range swath covered by the radar and to the maximum synthetic aperture length. Data come into each memory in a range bin by range bin fashion, corresponding to a single transmission. The SAR data correlator, however, will operate on virtually the entire content of each memory within a single interpulse period. When all the data received from one transmission are put into memory, the data received from the oldest transmission data are discarded from memory, so that the processor memory contains only those returns corresponding to terrain along the synthesized aperture.

Motion compensation information is provided to the SAR processor from the aircraft inertial navigation system via a reference generator. These signals compensate the radar video signal for changes in aircraft velocity due to airframe vibration, wind gusts, and aircraft maneuvers.

The output of the SAR signal processor is a signal representing terrain backscatter intensity versus range (time) on a pulse-by-pulse basis. If these data are laid side by side, a re-creation of the terrain image is displayed in near real time. Figure 16-13 describes the details of the correlator itself within the SAR

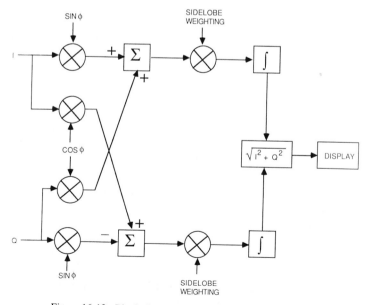

Figure 16-13. Block diagram of an SAR digital correlator.

signal processor. The mechanism of the correlator is that of a Fourier transform of the time domain I and Q radar signals. The digital functions shown in Figure 16-13 implement this transform. The video mixing involves four multiplications and two additions, using phase references (sine and cosine of ϕ_d) generated from the aircraft inertial navigation system, a linear phase term that steers the synthetic beam, and a quadratic phase term that focuses the synthetic aperture. The summers then pass the lower sideband only.

Weighting function signals multiply each summer's output to control the sidelobes of the synthetic antenna pattern. The weighted signals are then integrated using delay line integrators, for example, prior to amplitude detection. The signal intensity versus time (range) output signal is generated by the conventional square root of the sum of the squares of the I and Q signal components prior to storage or display to the operator.

16.4 SAR SYSTEM CONSIDERATIONS

In the actual implementation of a synthetic aperture radar and its associated processsor, several considerations affect the system design. These factors are discussed briefly, and means for overcoming or circumventing the limitations are suggested.

16.4.1 Phase Errors

The range-independent cross-range resolution

$$\delta_x = \frac{d_x}{2} \tag{16-31}$$

is not realizable in practice due to the presence of phase errors in the system. These phase errors can arise from two sources: atmospheric phenomena and random variations in the vector velocity and orientation of the aircraft. Researchers in high-resolution radar at the University of Michigan have determined that the phase errors induced by such aircraft gyrations are several orders of magnitude greater than those caused by variations in atmospheric propagation properties.[4]

The degradation of realizable azimuth resolution due to phase errors can be minimized in either of two ways. One technique is to reduce the magnitude of the phase errors by placing the antenna on an inertial platform. Such a platform may be stabilized through the use of gyros, for instance. In addition to using an inertial platform, changes in the vector velocity and orientation of the aircraft can be detected by accelerometers mounted on the plane. Proper phase corrections to the radar return signal can then be made to compensate for the deviation from a linear flight path. This phase correction is a function of the depression angle from the horizontal plane of the aircraft or, equally, a function of range. When these compensation corrections for aircraft motion are taken into account, the expression for the cross-range resolution is seen to depend on the standard deviation of radial acceleration, slant range R_s, aircraft velocity V, and wavelength λ:

$$\delta_x = \frac{R_s}{2V}\left[\frac{\lambda\Delta\alpha}{\pi}\right]^{1/2} \tag{16-32}$$

where $\Delta\alpha$ is the uncompensated rms deviation of radial acceleration.[4]

16.4.2 Range Limitations

A restriction on range is implied by Eq. (16-32) when a particular selection of aircraft velocity, transmission wavelength, and desired azimuth resolution is made. Additional constraints on range arise from considering range ambiguity and from satisfying the sampling theory criterion.

The PRF must be limited so that the return signal from a target at a selected maximum slant range R_{max} arrives at the antenna before the next pulse is transmitted. This range for a particular choice of PRF is called the *maximum unambiguous range*, as discussed in Chapter 1. The time between these successive pulses, $T_0 = 1/\text{PRF}$, has to satisfy this equation:

$$T_0 \geq \frac{2R_{max}}{c} \tag{16-33}$$

The resulting PRF will then be restricted by

$$\text{PRF} \leq \frac{c}{2R_{max}} \tag{16-34}$$

$$2\left[\frac{4Vx_0}{R_0\lambda}\right] = \frac{4V}{d_x} \leq \text{PRF} \tag{16-35}$$

since $2x_0 = R_0\beta = R_0\lambda/d_x$.

If we combine the restrictions on the PRF imposed by the selection of the maximum unambiguous range and by the sampling theorem, the allowable range of the PRF becomes bracketed by

$$\frac{4V}{d_x} \leq \text{PRF} \leq \frac{c}{2R_{max}} \tag{16-36}$$

For a PRF which satisfies this set of inequalities, the maximum range coverage will be such that

$$\frac{4V}{d_x} \leq \frac{c}{2R_{max}} \qquad (16\text{-}37)$$

When δ_x is substituted for $d_x/2$ in Eq. (16-37), we see that the maximum range coverage is limited to

$$R_{max} = \frac{d_x c}{8V} = \frac{\delta_x c}{4V} \qquad (16\text{-}38)$$

This is not a particularly stringent limitation for radar since c, the velocity of light, is very large compared to aircraft velocities. Figure 16-14 indicates the maximum unambiguous range corresponding to various aircraft velocities for a 2-m aperture antenna.

If one applied the same principles to a sonar signal, where the velocity of pressure wave propagation is ~ 1500 m/s, the range coverage as derived from Eq. (16-38) would be extremely restrictive.[5] For an aperture of 30 cm and a

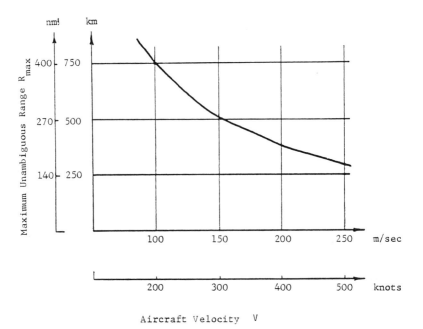

Figure 16-14. Maximum unambiguous range as a function of aircraft velocity for a 2-m aperture antenna.

ship velocity of 2 m/s, the resulting maximum range coverage is only

$$R_{max} = \frac{0.3 \times 1500}{8 \times 2} \approx 30 \text{ m}$$ (16-39)

16.4.3 Signal Strength Considerations

The power P_r received at the antenna is given by the familiar radar range equation (monostatic operation):

$$P_r = \frac{P_t G^2 \sigma \lambda^2}{(4\pi)^3 R^4}$$ (16-40)

where

P_t = transmitted power (peak)
G = gain of TR antenna
σ = RCS of illuminated target area

and the atmospheric attenuation effects discussed in Chapter 1 have been ignored. An expression will now be derived for the required transmitter power for a SAR system. Since SAR systems will likely include linear pulse compression (LPC) for improved range resolution, the processing gain associated with that process will also be included.

The processed signal-to-noise (S/N) ratio for a radar system employing LPC and SAR processing is given by[1]

$$\frac{S}{N} = (\tau\Delta f)(T\Delta f_d)\left[\frac{P_t G^2 \sigma \lambda^2}{(4\pi)^3 R^4 k T_0 B F_n}\right]$$ (16-41)

where the processing gain due to LPC is denoted by $\tau\Delta f$, the processing gain due to SAR is denoted by $T\Delta f_d$, and

k = Boltzmann's constant = 1.38×10^{-23} joules/degrees Kelvin
T_0 = absolute temperature in degrees Kelvin
B = receiver bandwidth
F_n = receiver noise figure

Since the time T for which a target is illuminated by the real antenna beam equals $R\lambda/Vd_x$, and the radar PRF must be at least twice the highest Doppler frequency, the processed signal-to-noise ratio may be expressed as

$$\frac{S}{N} = \frac{\tau\Delta f (PRF) P_t G^2 \sigma \lambda^3}{2(4\pi)^3 R^3 k t_0 B F_n d_x V}$$ (16-42)

The product $P_t\tau(\text{PRF})$ represents the average transmitted power P_{avg}. The receiver bandwidth B is usually chosen to match the FM bandwidth Δf. With these substitutions the S/N ratio expression becomes

$$\frac{S}{N} = \frac{P_{\text{avg}}G^2\sigma\lambda^3}{2(4\pi)^3R^3kT_0F_nd_xV} \qquad (16\text{-}43)$$

It is useful to express the antenna gain G in terms of an effective aperture A_e:

$$G = \frac{4\pi A_e}{\lambda^2} \qquad (16\text{-}44)$$

This effective aperture can then be expressed in terms of antenna dimensions. If it is assumed that the radar antenna is elliptical, with semiaxis dimensions d_x and d_y and an efficiency factor of 50%, then A_e is given by

$$A_e = \frac{\pi}{8} d_x d_y \qquad (16\text{-}45)$$

Substitution of these expressions yields

$$\frac{S}{N} = \frac{\pi P_{\text{avg}}d_x d_y^2\sigma}{512R^3kT_0F_nV\lambda} \qquad (16\text{-}46)$$

which represents the signal-to-noise ratio at the output of a radar receiver that uses both pulse compression and SAR processing.

The following observations can be made. The signal-to-noise ratio for a target of RCS σ is:

1. Proportional to the third power of the antenna "diameter."
2. Inversely proportional to the third power of the range.
3. Inversely proportional to the wavelength.
4. Inversely proportional to the aircraft platform velocity.

For distributed clutter, the RCS is described by the expression

$$\sigma = \sigma^0\delta_R\delta_x\sec\theta \qquad (16\text{-}47)$$

where

σ^0 = RCS per unit area
δ_R = range resolution
δ_x = cross-range resolution
θ = incidence angle of the radar beam from the horizontal

The corresponding expression for the signal-to-noise ratio for distributed clutter is then

$$\frac{S}{N} = \frac{\pi P_{avg} \sigma^0 \delta_R \sec\theta d_x^2 d_y^2}{1024 R^3 k T_0 F_n V \lambda} \tag{16-48}$$

The significant difference between the signal-to-noise ratio equations for a target and for distributed clutter is that the latter contains a fourth power dependency in antenna dimension. For example, calculate the target signal-to-noise ratio for the following set of radar, target, and geometric parameters:

$P_{avg} = 10$ W
$d_x = 2$ m
$d_y = 0.2$ m
$\sigma = 1$ m^2
$F_n = 2$
$V = 150$ m/s
$\lambda = 0.03$ m (X-band)
$R = 10$ km
$T = 290°$ K

For these parameter values, Eq. (16-46) gives a target signal-to-noise ratio of $1.36 \times 10^5 = 51.3$ dB. For a ground reflectivity of 0.01 (-20 dB), a 20° depression angle, and a 2-m range resolution, Eq. (16-48) yields a clutter-to-noise ratio of $2.90 \times 10^3 = 34.6$ dB.

16.4.4 Squint Mode Operation

In the usual high-resolution SLAR configuration, the radar antenna is pointed in a direction normal to the flight path. In a tactical application in which the detection and identification of hostile targets are of utmost importance, a more practical configuration would be one in which the radar antenna is pointed at some angle toward the direction of flight, as illustrated in Figure 16-15. This squint angle mode of operation allows a greater tactical advantage at the expense of sacrificing a portion of that along-track resolution obtainable with a side-looking mode of operation.[6]

The theoretical along-track resolution for a SAR in the squint angle mode is $\delta_x \sec\theta_s$, where δ_x is the cross-range resolution in a 90° side-looking mode and the squint angle θ_s is the angle between antenna boresight and the direction normal to the flight path. When random motion effects are taken into account,

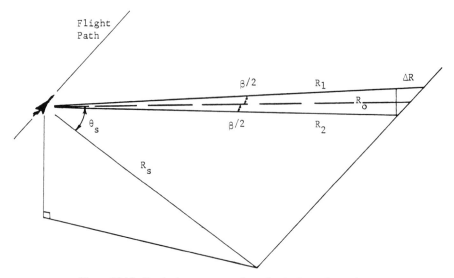

Figure 16-15. Synthetic aperture configuration in the squint mode.

the along-track resolution in the squint angle mode is

$$\delta_x = \left[\frac{R}{2V \cos \theta_s} \right] \left[\frac{\lambda \Delta \alpha}{\pi} \right]^{1/2} \tag{16-49}$$

where all parameters have the same meanings as before.

It can also be seen that the slant range to a target will vary as it passes through the beamwidth β of the antenna. This difference in range, ΔR, is given by

$$\Delta R = R_1 - R_2 \approx R_0 \beta \tan(\theta_s + \beta/2) \tag{16-50}$$

If the distance ΔR is greater than the range resolution, the target must be tracked in range. Range corrections applied to a given target are essentially correct for all targets if the antenna bandwidth β is small and the target range is large in comparison to the synthetic aperture L_s. The correction is made by varying the PRF so that the arrival times of returns from a given echo have the same interval between them as would have been the case in the side-looking mode:

$$\text{PRF}_{\text{corrected}} = \frac{\text{PRF}_{\text{original}}}{1 + \dfrac{2V}{c} \sin \theta_s} \tag{16-51}$$

As in the side-looking mode, the maximum unambiguous range is bounded by the PRF.

For the squint mode, the Doppler bandwidth is smaller than that for the side-looking case. Therefore, the PRF can be lower, and the unambiguous range can be increased. The platform radial velocities in the directions R_1 and R_2 are, respectively, given by

$$V_1 = V \sin(\theta_s + \beta/2)$$

and

$$V_2 = V \sin(\theta_s - \beta/2) \tag{16-52}$$

The Doppler bandwidth is given by

$$\Delta f_d = \frac{2}{\lambda}(\Delta V) = \frac{2}{\lambda}(V_1 - V_2) \tag{16-53}$$

or

$$\Delta f_d = \frac{4V}{\lambda} \cos \theta_s \sin\left(\frac{\beta}{2}\right) \tag{16-54}$$

For small β, Eq. (16-54) becomes

$$\Delta f_d = \frac{2V\beta}{\lambda} \cos \theta_s \tag{16-55}$$

This expression is seen to differ from Eq. (16-24) only by the cosine of the squint angle ($\beta = 2x_0/R_0$).

As an example, consider an antenna squint angle of $50°$. Relative to the side-looking geometry ($\theta_s = 0°$), Eq. (16-55) indicates that the Doppler bandwidth decreases by 36%, and Eq. (16-49) indicates that the cross-range resolution degrades by 56%.

The PRF would be

$$PRF = 2\Delta f_d = \frac{4V\beta}{\lambda} \cos \theta_s \tag{16-56}$$

and the maximum unambiguous range would be

$$R_{max} = \frac{c}{2PRF} = \frac{c\lambda}{8V\beta \cos\theta_s} \tag{16-57}$$

An interesting point is the lack of dependence of lateral slant range R_s on the squint angle:

$$R_s = R_{max}\cos\theta_s = \frac{c\lambda}{8V\beta}. \qquad (16\text{-}58)$$

This equation yields the same lateral slant range as the side-looking case. With appropriate variations of those system parameters that determine squint angle, scanning with the synthetic beam is possible. Such scanning would greatly enhance the versatility of this radar system and should increase the area that can be interrogated by an aircraft.

16.5 APPLICATIONS OF SAR TECHNOLOGY

16.5.1 Types of Applications

SAR applications can be generally classified as either large area or small area imaging. Large area imaging is a surveillance function and is generally performed by a dedicated radar system in either a side-looking or a squint-mode antenna configuration. Small area imaging is performed for navigation or weapon delivery purposes from either radar antenna configuration and, hence, has a tactical flavor. Surveillance SAR applications are often referred to as *strip mapping*, since the imaged terrain corresponds to a long strip whose width is a function of pulse rate and aircraft altitude. Tactical SAR applications are referred to as *spotlight mapping*, since illumination times are short compared to those for strip mapping. Spotlight mapping is often included with other radar modes such as air-to-air track, terrain avoidance, or terrain following in a multifunction radar. Note that Doppler beam sharpening is a special case of a spotlight-mapping SAR application.[7]

16.5.2 Surveillance

Terrain surveillance has been a significant beneficiary of the development of SAR technology. In 1970, the Goodyear Aerospace Corporation and the Aero Service Division of the Western Geophysical Company of America, a subsidiary of Litton Industries, made available a SAR system for commercial surveying.[8] Since the radar signals are little impacted by terrain cloud cover, SAR surveillance of rain forests has been accomplished over extensive areas of the Amazon Basin in Brazil, Venezuela, Columbia, Peru, and Bolivia. In fact, the entire country of Brazil has now been surveyed in this manner.

Figure 16-16 is an SAR image of a section of Venezuela obtained during the Aero Service/Goodyear mapping program. The area shown covers $1°$ of latitude and $1.5°$ of longitude. Note the relief features due to elevated terrain. The

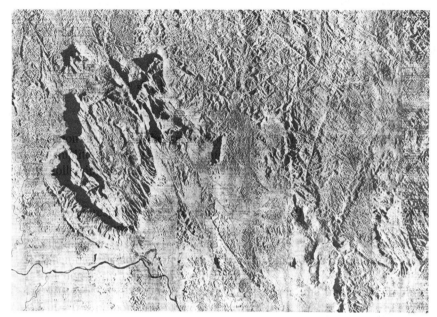

Figure 16-16. SAR image of an area in Venezuela made by Goodyear Aerospace. (Image Courtesy of Goodyear Aerospace) [8]

shadow effect is real and results from a lack of microwave illumination in the indicated areas.[8]

Other surveillance applications include the study of continental drift through the recognition of ground fault features and the study of basin drainage patterns as shown by lake and river radar signatures. Ground faults stand out well in the reconstructed SAR image because of the wide area coverage and the low illumination angle usually employed. Such fault location is vitally important in construction planning and earthquake research. Because water reflects little microwave energy at the look angles used, lakes, rivers, and other bodies of water appear dark in the reconstructed SAR image, permitting identification of large watershed sources for conservation planning and water resource management.

Figure 16-17 is a SAR image of the Trenton, Michigan, area obtained by the Environmental Research Institute of Michigan (ERIM). The resolution shown is 20 × 20 m, with 16 looks integrated. Note the dark area due to the low-reflectivity river and the distinct man-made patterns of roadways, bridges, and buildings.

In 1978 the National Aeronautics and Space Administration (NASA) launched its SEASAT-A satellite into an elliptical orbit around the earth. SEASAT sensors include a SAR system with a single-look resolution capability of 7 × 25 m from an average altitude of approximately 800 km. The SEASAT SAR has

Figure 16-17. SAR image of the area around Trenton, Michigan, made by the Environmental Research Institute of Michigan. (Image Courtesy of ERIM)

provided extensive data describing surface topographic features and ocean-related phenomena including ocean swell and wave features, Gulf Stream shear boundaries, sea ice characteristics and distribution, and the effects of hurricane-level winds on the sea surface.[9]

Figure 16-18 is a SAR image of the Newport Beach/Santa Ana region in Southern California obtained from the SEASAT-A SAR and processed by the Jet Propulsion Laboratory of the California Institute of Technology. The radar image is 32 × 36 km with a 25-m resolution and four looks integrated. Newport Beach is located in the lower center of the image, with the Huntington Beach pier to the left. Disneyland is a circular bright feature in the upper left part of the image, with the dark square parking lot directly below it. Figure 16-19 is a SEASAT-A SAR image of the sea around Nantucket Island.

16.5.3 INVERSE SAR

The theory of SAR data processing is based on the linear relationship between the target's along-track position and the Doppler-shifted signal reflected from the target. Thus, high-resolution target imagery can be obtained from the appropriate processing of the Doppler return history from a target. It is not necessary, however, that the radar motion be the generator of the target Doppler signature history. If a target itself is rotating and the radar is stationary, target returns containing Doppler-shifted information are still created. These return

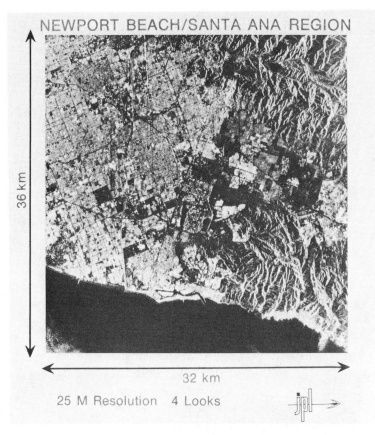

NEWPORT BEACH/SANTA ANA REGION

36 km

32 km

25 M Resolution   4 Looks

Figure 16-18. SAR image of the Newport Beach/Santa Ana region in Southern California made by the SEASAT-A system and processed by the Jet Propulsion Laboratory. (Image Courtesy of JPL.)

signals can be processed in a manner similar to that previously described to re-create a high cross-range resolution target image.[10] This technique is often re-ferred to as *inverse synthetic aperture radar (ISAR)*. ISAR techniques have been used to produce target images for objects on rotating platforms, for satellites rotating in orbit, and even for ships at sea which roll rhythmically on the sea surface.

Figure 16-20 depicts a visual image of the U.S. Belknap-class cruiser and an ISAR image (not collected simultaneously). The vertical streak is caused by the motion of a rotating radar antenna with respect to the overall motion of the ship.[11]

Another application of ISAR techniques is the mapping of planetary bodies

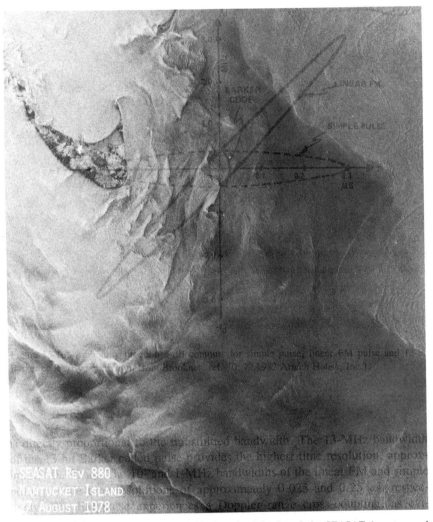

Figure 16-19. SAR image of the sea around Nantacket Island made by SEASAT-A system and processed by the Jet Propulsion Laboratory. (Image Courtesy of JPL)

from the earth.[2] Let the angular velocity vector be ω_r, as indicated in Figure 16-21, about an axis perpendicular to the line of sight. The radial velocity at a point on the planet a distance z from the body center will be $\omega_r z$. From Eq. (16-23), the magnitude of the resulting Doppler shift from this point will be

$$f_d = \frac{2\omega_r z}{\lambda} \qquad (16\text{-}59)$$

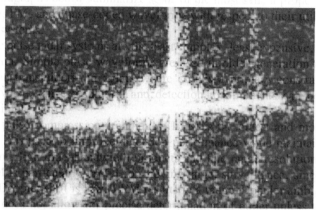

Figure 16-20. Visual and ISAR images of a U.S. Belknap-class cruiser (not recorded simultaneously). (Photograph and Image Courtesy of Defense Electronics)

If the return radar signals are Doppler processed in the manner previously described, the scattering centers on the planet will be resolved in the z, or cross-range, dimension. For coherent processing of the signals over time ΔT, the Doppler frequency resolution obtainable is $1/\Delta T$ Hz. Since the Doppler frequency shift corresponding to the radial velocity $\omega_r z$ is given by Eq. (16-59), a Doppler resolution of $\delta f_d = 1/\Delta T$ implies a cross-range resolution of

$$\delta_x \approx \frac{\lambda}{2\omega_r \Delta T} = \frac{\lambda}{2\Delta\theta} \tag{16-60}$$

where $\Delta\theta$ is the angular change in the position of the planet.

If, in addition, range resolution is provided by the radar system, a two-dimensional radar image of the planet can be determined. Using these and similar

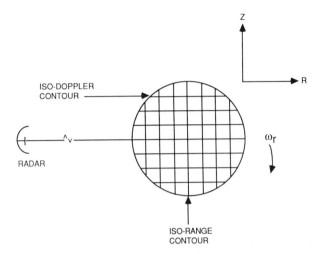

Figure 16-21. SAR in a planet-mapping configuration.

techniques, researchers have produced radar reflectivity maps of the planets Mercury, Venus, Mars, and Jupiter, in addition to our own moon. Correlation of radar scattering centers has been reasonably good with optically visible surface features for both Mars and the moon. In the case of perpetually shrouded Venus, there has even been an indication of cratering detected with high-resolution radars.[12]

Figure 16-22 is an SAR image of the planet Venus obtained at the Arecibo Observatory in Puerto Rico. The cross-range resolution varies between 10 and 20 km. Data in the empty band had not been processed when this image was published in 1979.[12]

16.5.4 Meteorological Applications

In its usual application, SAR data processing reconstructs the image of the illuminated area by coherent signal integration about the zero Doppler frequency. Target cross-range location is then defined with respect to zero radial velocity in relation to the aircraft. In radar meteorology, the atmospheric wind structure within storms can be deduced by the use of SAR imaging about nonzero Doppler frequencies. Energy reflected from falling hydrometeors moving in wind fields will be Doppler shifted with respect to the ground. SAR data processing can be performed which images only those target areas having specific radial velocities, at all ranges from the aircraft through the storm. Thus, three-dimensional storm dynamics can be mapped from aircraft or satellites with far greater cross-range resolution than could be obtained with conventional Doppler weather radars.[13]

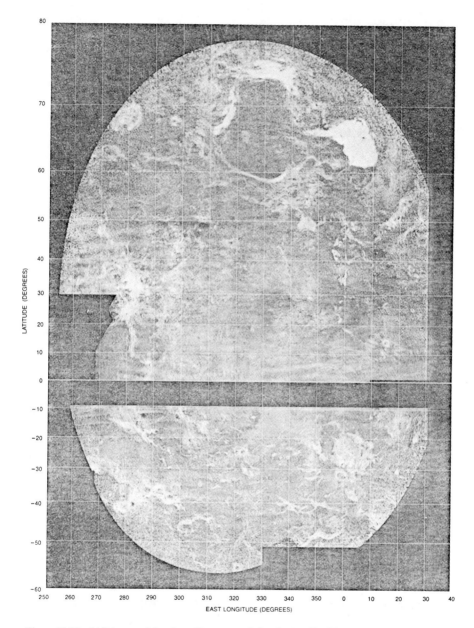

Figure 16-22. SAR image of the planet Venus recorded at the Arecibo Observatory in Puerto Rico. (Image Courtesy of Scientific American)

16.6 SUMMARY

The SAR system produces high cross-range resolution imagery by processing stored backscatter amplitude and phase information obtained by sequentially transmitting and receiving EM energy along a linear flight path, re-creating in time a long linear antenna array. The mathematical relationship on which the SAR concept is based is the equivalence between the cross-range target position and the instantaneous Doppler shift of the radar energy backscattered by the target.

The SAR concept requires a coherent radar implementation, a large volume of data storage, and a large amount of data processing. Early implementations relied on optical film for data storage and optical processing (coherent illumination, lenses, and transparencies) for re-creating the ground imagery seen through the "eye" of the radar. State-of-the-art implementations use computer memory for data storage and digital signal processing to re-create the image; this procedure is as close to real-time processing as is realizable in a SAR system.

Applications of SAR technology include geographic mapping, high-resolution surveillance, ground/sea feature identification, planetary mapping, and rotating object imagery. Platforms for which SAR technology can be implemented include aircraft, satellites, and missiles. Indeed, SAR image reconstruction can be performed in virtually any situation where there is a deterministic relationship between a target's spatial position or angular aspect and its instantaneous Doppler-shifted radar signature. Future trends in SAR technology will include accommodation of nonuniform platform motion, platform maneuvering, and bistatic transmitter/receiver geometries.

16.7 REFERENCES

16.7.1 Cited References

1. R. N. Trebits, "A Near-Millimeter Synthetic Aperture Radar Concept," Final Report for the U.S. Army Research Office, February 1979.
2. W. M. Brown and L. J. Porcello, "An Introduction to Synthetic Aperture Radar," *IEEE Spectrum*, vol. 6, no. 9, September 1969, pp. 52–62.
3. J. C. Kirk, Jr., "A Discussion of Digital Processing in Synthetic Aperture Radar," *IEEE Transactions on Aerospace and Electronic Systems*, vol. AES-11, no. 3, May 1975, pp. 326–337.
4. W. M. Brown, "Synthetic Aperture Radar," *IEEE Transactions on Aerospace and Electronic Systems*, vol. AES-3, No. 2, March 1967, pp. 217–229.
5. E. Rhodes, "Synthetic Apertures," in *Principles of Modern Radar*, Georgia Institute of Technology, Atlanta, September 1971.
6. R. C. Heimiller, "Theory and Evaluation of Gain Patterns of Synthetic Arrays," *IRE Transactions on Military Electronics*, vol. MIL-6, no. 2, April 1982, pp. 122–129.
7. E. Brookner, "Synthetic Aperture Radar Spotlight Mapper," *Radar Technology*, 1977, pp. 251–262.

8. B. Miller, "Side-Looking Radar Plays Key Role in Mapping," *Aviation Week & Space Technology*, July 17, 1972, pp. 44–46.

9. C. Elachi, "Radar Images of the Earth from Space," *Scientific American*, vol. 247, no. 6, December 1982, pp. 54–61.

10. W. M. Brown, ibid.

11. B. P. McCune and R. J. Drazovich, "Radar with Sight and Knowledge," *Defense Electronics*, vol., August 1983, pp. 80–96.

12. G. H. Pettengill, D. B. Campbell, and H. Masursky, "The Surface of Venus," *Scientific American*, vol. 243, no. 2, Aug. 1980, pp. 46–57.

13. J. I. Metcalf and W. A. Holm, "Meteorological Applications of Synthetic Aperture Radar," Final Report on Grant No. NSG-5220 for NASA, Georgia Institute of Technology, Atlanta, February 1979.

16.7.2 Additional References

1. L. J. Cutrona, E. N. Leith, L. J. Porcello, and W. E. Vivian, "On the Application of Coherent Optical Processing Techniques to Synthetic-Aperture Radar," *Proceedings of the IEEE*, vol. 54, no. 8, August 1966, pp. 1026–1032.

2. C. W. Sherwin, J. P. Ruina, and R. D. Rawcliffe, "Some Early Developments in Synthetic Aperture Radar Systems," *IRE Transactions on Military Electronics*, vol. MIL-6, April 1962, pp. 111–115.

3. H. L. McCord, "The Equivalence Among Three Approaches to Deriving Synthetic Array Patterns and Analyzing Processing Techniques," *IRE Transactions on Military Electronics*, vol. MIL-6, April 1962, pp. 116–121.

4. L. J. Cutrona and G. O. Hall, "A Comparison of Techniques for Achieving Fine Azimuth Resolution," *IRE Transactions on Military Electronics*, vol. MIL-6, April 1962, pp. 119–121.

5. L. J. Cutrona, W. E. Vivian, E. N. Leith, and G. O. Hall, "A High-Resolution Radar Combat-Surveillance System," *IRE Transactions on Military Electronics*, vol. MIL-5, April 1961, pp. 127–131.

6. Kendall Preston, Jr., *Coherent Optical Computers*, McGraw-Hill Book Co., New York, 1972.

7. J. V. Evans and Tor Hagfors, *Radar Astronomy*, McGraw-Hill Book Co., New York, 1968.

8. Robert O. Harger, *Synthetic Aperture Radar Systems*, Academic Press, New York, 1970.

9. M. I. Skolnik, *Radar Handbook*, McGraw-Hill Book Co., New York, 1970.

10. J. J. Kovaly, "Radar Techniques for Planetary Mapping with Orbiting Vehicles," *Annals of the New York Academy of Sciences*, vol. 187, January 1972, pp. 154–176.

11. K. Tomiyasu, "Tutorial Review of Synthetic-Aperture Radar (SAR) with Applications to Imaging of the Ocean Surface," *Proceedings of the IEEE*, vol. 66, no. 5, May 1978, pp. 563–583.

12. G. H. Pettengill, D. B. Campbell, and H. Masursky, "The Surface of Venus," *Scientific American*, vol. 243, no. 2, August 1980, pp. 46–57.

13. H. Jensen, L. C. Graham, L. J. Porcello, and E. N. Leith, "Side-Looking Airborne Radar," *Scientific American*, vol. 236, no. 10, October 1977, pp. 84–95.

14. Robert O. Harger, "An Optimum Design of Ambiguity Function, Antenna Pattern, and Signal for Sidelooking Radars," *IEEE Transactions on Military Electronics*, vol. MIL-9, nos. 3, 4, July–October 1965, pp. 264–278.

15. L. J. Cutrona, W. E. Vivian, E. N. Leith, and G. O. Hall, "A High-Resolution Radar Combat-Surveillance System," *IRE Transactions on Military Electronics*, vol. MIL-5, January 1961, pp. 127–131.

16. J. Holahon, "Synthetic Aperture Radar," *Space/Aeronautics*, vol. 40, November 1963, pp. 88–93.

17. P. M. Kelly et al. "Data Processing for Synthetic Aperture Radar," NAECON Conference, 1964, pp. 194–200, Dayton, Ohio, May 11–13, 1964.

18. J. H. Mims and J. L. Farrell, "Synthetic Aperture Imaging with Maneuvers," *IEEE Transactions on Aerospace and Electronic Systems*, vol. AES-8, no. 4, July 1972, pp. 410–418.

19. W. M. Brown et al., "Synthetic Aperture Processing with Limited Storage and Presumming," *IEEE Transactions on Aerospace and Electronics Systems*, vol. AES-9, no. 2, March 1973, pp. 166–176.

20. W. E. Kock, "A Holographic (Synthetic Aperture) Method for Increasing the Gain of Ground-to-Air Radars," *Proceedings of the IEEE*, vol. 59, March 1971, pp. 426–427.

21. R. H. MacPhie, "The Circular Synthetic Radar," *IEEE Transactions on Aerospace and Electronic Systems*, vol. AES-9, no. 4, July 1973, pp. 608–611.

22. L. J. Cutrona and G. O. Hall, "A Comparison of Techniques for Achieving Fine Azimuth Resolution," *IRE Transactions on Military Electronics*, vol. MIL-6, no. 2, April 1962, pp. 119–121.

23. A. W. Rihaczek, *Principles of High-Resolution Radar*, McGraw-Hill Book Co., New York, 1969.

24. C. Wu, "Electronic SAR Processors for Space Missions," Proceedings of the Synthetic Aperture Radar Technology Conference, March 1978, Sponsored by NASA and the Physical Science Laboratory of New Mexico State University.

25. V. C. Tyree, "Custom Large Scale Integrated Circuits for Spaceborne SAR Processors," Proceedings of the Synthetic Aperture Radar Technology Conference, March 1978, Sponsored by NASA and the Physical Science Laboratory of New Mexico State University.

26. E. N. Leith, "Photographic Film as an Element of a Coherent Optical System, Photographic Science and Engineering, vol. 6, March–April 1962, pp. 75–80.

27. E. N. Leith, "Optical Processing Techniques for Simultaneous Pulse Compression and Beam Sharpening," *IEEE Transactions on Aerospace and Electronic Systems*, vol. AES-4, no. 6, November 1968, pp. 879–885.

28. E. N. Leith and A. L. Ingalls, "Synthetic Antenna Data Processing by Wavefront Reconstruction," *Applied Optics*, vol. 7, no. 3, March 1968, pp. 539–544.

29. J. L. Farrell, J. H. Mims, and A. Sorrell, "Effects of Navigation Errors in Maneuvering SAR," *IEEE Transactions on Aerospace and Electronic Systems*, vol. AES-9, no. 5, September 1973, pp. 758–776.

30. J. C. Kirk, Jr., "A Discussion of Digital Processing in Synthetic Aperture Radar," *IEEE Transactions on Aerospace and Electronic Systems*, vol. AES-11, no. 3, May 1975, pp. 326–337.

31. R. W. Bayma and P. A. McInnes, "Aperture Size and Ambiguity Constraints for a Synthetic Aperture Radar," *Record of the IEEE 1975 International Radar Conference*, April 1975, pp. 499–504.

32. J. C. Kirk, Jr., "Digital Synthetic Aperture Radar Technology," *Record of the IEEE 1975 International Radar Conference*, April 1975, pp. 482–487.

33. R. K. Raney, "Synthetic Aperture Imaging Radar and Moving Targets," *IEEE Transactions on Aerospace and Electronic Systems*, vol. AES-7, no. 3, May 1971, pp. 499–505.

34. K. Tomiyasu, "Ocean-Wave Cross-Radial Image Error in Synthetic Aperture Radar Due to Radial Velocity," *Journal of Geophysical Research*, vol. 80, November 1975, p. 4555.

35. L. J. Porcello, "Turbulence-Induced Phase Errors in Synthetic-Aperture Radars," *IEEE Transactions on Aerospace and Electronic Systems*, vol. AES-6, no. 5, September 1970, pp. 636–644.

36. W. M. Brown and J. F. Riordan, "Resolution Limits with Propagation Phase Errors," *IEEE Transactions on Aerospace and Electronic Systems*, vol. AES-6, September 1970, pp. 657–662.

37. K. A. Graf and H. Guthart, "Velocity Effects in Synthetic Apertures," *IEEE Transactions on Antennas and Propagation*, vol. AP-17, September 1969, pp. 541–546.

38. W. E. Brown, Jr., C. Elachi, and T. W. Thompson, "Radar Imaging of Ocean Surface Patterns," *Journal of Geophysical Research*, vol. 81, May 1976, pp. 2657-2667.
39. W. L. Grantham, E. M. Bracalente, W. L. Jones, and J. W. Johnson, "The Seasat-A Satellite Scatterometer," *IEEE Journal of Oceanic Engineers*, vol. OE-2, April 1977, pp. 200-206.
40. R. O. Pilon and C. G. Purves, "Radar Imagery of Oil Slicks," *IEEE Transactions on Aerospace and Electronic Systems*, vol. AES-9, no. 5, September 1973, pp. 630-636.
41. R. Rawson, F. Smith, and R. Larson, "The ERIM Simultaneous X- and L-band Dual Polarization Radar," *Record of the IEEE 1975 International Radar Conference*, April 1975, pp. 505-510.
42. J. C. de Leon, "Synthetic Array Radar to Map the Surface of Venus," *Microwaves*, vol. 16, January 1977, pp. 12-14.
43. R. A. Kallas, "Synthetic Apertures Studied for Satellite Radar," *Microwave System News*, vol. 8, no. 9, September 1, 1978, pp. 56-67.
44. W. H. Gregory, "Grumman's Ecosystems Growing," *Aviation Week & Space Technology*, vol. 97, no. 8, August 21, 1972, pp. 48-55.
45. W. E. Kock, "Holography Can Help Radar Find New Performance Horizons," *Electronics*, vol. 43, no. 21, October 12, 1970, pp. 80-88.
46. R. O. Harger, "SAR Ocean Imaging Mechanisms," *Spaceborne Synthetic Aperture Radar for Oceanography*, Ed. R. C. Beal, P. S. DeLeonibus, and I. Katz, Johns Hopkins Oceanographic Studies No. 7, Johns Hopkins Press, 1981, pp. 41-52.
47. J. H. Jolley and C. Doston, "Synthetic Aperture Radar Improves Reconnaissance," *Defense Electronics*, vol. 13, no. 9, September 1981, pp. 111-118.
48. E. Brookner and B. R. Hunt, "Synthetic Aperture Radar Processing," *Trends and Perspectives in Signal Processing*," January 1982.
49. R. L. Withman, "A System for the Real Time Exploitation of Digital Synthetic Aperture Radar Data," IEEE 1981 National Aerospace and Electronic Conference, Dayton, Ohio, May 1981, pp. 18-22.
50. J. G. Mehlis, "Synthetic Aperture Radar Range—Azimuth Ambiguity Design and Constraints," IEEE International Radar Conference, Washington, D.C., Jet Propulsion Laboratory, 1980, pp. 143-152.
51. R. O. Harger, "The Side-Looking Radar Image of Time-variant Scences," *Radio Science*, vol. 15, no. 4, July-August 1980, pp. 749-756.
52. R. K. Raney, "SAR Response to Partially Coherent Phenomena," *IEEE Transactions on Antennas and Propagation*, vol. AP-28, no. 6, November, 1980, pp. 777-787.

PART 5
TRACKING RADAR TECHNIQUES
AND APPLICATIONS

- Range Tracking—Chapter 17
- Angle Tracking—Chapter 18
- Doppler Frequency Tracking—Chapter 19

After the most basic function of target detection, target tracking is a primary radar operation. PART 5 addresses three areas of radar target tracking: range tracking, angle tracking, and Doppler frequency tracking. In general, range tracking is a prerequisite to and inherent in both angle and Doppler frequency tracking systems. Chapter 17 defines and introduces range tracking and describes techniques for achieving range tracking in several radar applications.

Target angular position information is essential in such diverse radar applications as air-traffic control, antiaircraft artillery control, missile guidance, satellite tracking, etc. Chapter 18 presents and describes several radar angle tracking techniques including conical scan, sequential lobing monopulse, and track-while-scan. Applications, advantages, and disadvantages of each technique are discussed including performance and accuracy in a noise or jamming environment.

In PART 4, radar techniques were discussed that use the Doppler frequency shift due to motion to effect discrimination between moving and stationary targets. Chapter 19 presents concepts and techniques for Doppler frequency tracking for several radar types including CW, High PRF, Medium PRF, Low PRF, phased array and mechanical scan. Doppler frequency tracking not only provides clutter rejection that enables continuous range and angle track of a single desired target but also can significantly increase the signal-to-noise or signal-to-interference ratio due to longer integration time provided by narrow bandwidth filters in the tracking loops.

17
RANGE TRACKING

Joseph A. Bruder

17.1 INTRODUCTION

17.1.1 Overview

Range tracking is the continuous estimation of slant range from the radar to a specific target. It is accomplished by measuring the time required for each transmitted signal to travel from the radar to the target of interest and back to the radar. Recall from Chapter 1 that slant range can be determined from the equation

$$R = \frac{c}{2} T$$

where

$R =$ slant range
$c =$ velocity of light
and $T =$ time delay between transmission and reception of the reflected signal from the desired target

Also recall that the factor 2 appears in the denominator because the distance between the target and the radar must be traversed twice.

Barton summarizes six basic steps in range tracking:[1]

1. Generation of recognizable features in the transmitted wave (e.g., discrete pulses of the low-duty cycle).
2. Storage and delay of a reference signal (e.g., a rectangular gate or strobe marker matched to the signal width).
3. Measurement of the relative delay of reference and received signals (e.g., viewing of the two signals on an expanded A-scope).
4. Correction of the reference delay to obtain coincidence of maximum correlation between reference and received signals.
5. Measurement of reference delay.
6. Encoding and reading out of reference delay to indicate signal delay and, hence, target range.

17.1.2 History of Range Tracking

In the early uses of radar, range tracking was accomplished manually. The radar operator observed the desired target signal on the radar display (A-scope, J-scope, etc.) and attempted to maintain alignment of a reference marker with the target signal. The reference marker may have been a range cursor, an electronic pulse, or a pair of brackets that were displayed by deflection or intensity modulation on the display. The radar operator positioned the reference marker over the target signal and maintained the range track by manipulating manual controls such as hand wheels or cranks. Mechanically operated range cursor scales placed over the face of the display tube were also used as reference range indicators. Continuous range and range rate information was obtained by mechanically translating the position and movements of the hand crank or cursor into shaft rotations suitable for driving data output devices such as dials, potentiometers, or synchros. Through the use of expanded displays, a proficient operator could achieve tracking bandwidths approaching 1 Hz over a short period when the signal-to-noise ratio was high. But when the signal faded into the noise or became obscured because of interference, even the expert operator had difficulty maintaining an accurate range track. That problem was alleviated somewhat through the use of *aided range tracking*. In aided tracking, operation of the manual range marker control produced both range position and range rate commands for the reference marker; hence, under constant range rate conditions, the reference marker automatically tracked the target signal after the initial adjustments by the operator. Corrections were necessary only when the radial range rate of the target with respect to the radar changed. Aided tracking is employed in some modern tracking radars when they are operated in a manual range track or range search mode.

17.1.3 Range Tracker Applications

Range tracking is an important function of modern radars. Automatic determination of the target range requires the use of a range tracker. Target range and range rate information generated by the range tracker can also be applied to a display for observation by the operator.

One of the most important applications of range trackers is to provide target range and range rate information to weapon system computers to enable the direction of a missile or ballistic projectile to the target. In radars that automatically track the target in angle, most angle tracking circuits use range gating information provided by the range tracker to separate the tracked target from other potential targets. Instrumentation radars record target range data as well as angle data to provide a record of target position versus time. Radar altimeters include a range-tracking function to provide a readout of aircraft altitude above the terrain.

17.1.4   Requirements for Range Tracking

Three basic radar modes are required to establish range track: search, acquisition, and track. The search mode is normally required to discover the target's presence in the radar's field of view. A typical search scenario requires scanning the antenna in azimuth (and possibly elevation) and presenting the radar data from the sector of interest on a PPI or other radar display, such as that shown in Figure 17-1. When a target of interest is identified in the search mode, the radar is switched to the acquisition mode.

The target acquisition process can be either manual, semiautomatic, or completely automatic. For manual acquisition, the operator locates the desired target on the display, then stops the antenna scan and points the antenna along the target direction (searchlight). The operator then positions a range cursor (and possibly even a range rate control) on the target. After the antenna bearing and range cursors are manually positioned on the target, the operator can switch to the track mode.

For semiautomatic target acquisition, the operator uses a light pen or other means to designate (identify) the target for the radar's computer. The computer then slews the antenna and range reference to the target location.

For automatic target acquisition, a target selection philosophy must be implemented to enable the range tracker to select the target of interest from among many other reflectors. Radar altimeters, for example, include an automatic target acquisition function in that they acquire and track the return from the target that is closest to the aircraft.

When the target is acquired, the radar is switched either manually or automatically to the range track mode. The range tracker then continues to track the

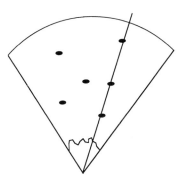

Figure 17-1. PPI display.

target as long as it remains in the radar field of view, unless it loses the target due to a low signal-to-noise ratio, clutter, or other target interference.

17.2 ANALOG RANGE TRACKING TECHNIQUES

17.2.1 General Characteristics

The reaction time requirements and sophistication of modern radar systems preclude reliance on manual range tracking; therefore, automatic tracking systems are employed in most modern radars.

The basic elements of an analog automatic range-tracking system are shown in Figure 17-2. Normal or raw video signals from the receiver are fed to a time discriminator, which produces an output signal that is a function of the difference in range between the range sampled by the range tracker and the actual range to the target of interest. The time discriminator generates a video signal which has an average value that is proportional to that range difference or, in other words, the range tracking error. In most range tracking systems, the error signal is then operated on by two integrators which provide radial range rate and slant range data as outputs. Slant range data are routed to the range delay generator, which provides range-gate enable pulses that are delayed from the zero-range synchronizing trigger by an amount determined by the slant range data. The range-gate enable pulses are fed to the time discriminator, and the tracking loop is closed. The magnitude of the error signal is a measure of the difference between the actual target range and the range at which the range gate is enabled. The sign of the error signal determines whether the slant range corresponding to the position of the range-gate enable pulse must be increased or decreased to minimize the range tracking error.

While there are a number of possible configurations for range tracking circuits, the split-gate tracker has gained the greatest prominence in analog range tracker implementations.

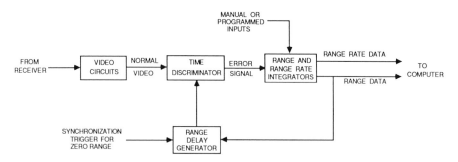

Figure 17-2. Basic elements of an automatic range tracking circuit.

17.2.2   Split-Gate Range Tracker

The most common configuration for range tracker implementations is that of a split-gate tracker (also referred to as an *early-gate/late-gate tracker*). A split-gate tracker is designed to track the centroid of the video corresponding to the desired target. The heart of the split-gate tracker is the time discriminator, and the basic configuration for a split-gate tracker is shown in Figure 17-3. The radar video signals from all of the targets in view of the radar antenna beam are presented as raw video (ungated video) to the input of the early gate, the late gate, and, in some configurations, the on-target gate. The early gate passes only the first half of the tracked target video signal. The late gate starts at the end of the early gate and passes only the second half of the target video. The on-target gate passes the complete (both halves) tracked target video. The output of the on-target video gate is not required in the range tracker itself; this gated target video is often used in many radar configurations for subsequent signal processing and display.

The first-half or early-gate video is integrated during the early-gate enable period. A typical early-gate integrator circuit is shown in Figure 17-4, and charging waveforms are shown in Figure 17-5. During the early-gate enable period, the field effect transistor (FET) acts as a short circuit and connects the gated video return to the resistor, capacitor (R_s, C) integrator. If the target video signal is a rectangular pulse, then the integrator (the capacitor is initially discharged) will charge up to output voltage V_0 by the end of the early-gate-enable period. The charging circuit time constant (τ) is generally several times greater than the time width of the early gate. Thus, the R_s, C charging circuit operates in a more linear region of the charging curve, and the voltage across the capacitor is approximately proportional to the target video energy during the early-gate enable period. At the end of the early gate, the FET is turned off rapidly so that the FET appears as an open circuit to the R_s, C integrator. Also, the output R_s, C integrator is normally isolated by the high input impedance of a voltage follower amplifier. The charge on the integrator capacitor and, thus, the integrator voltage are held until the dump gate transistor is enabled to short the capacitor to ground.

The late-gate integrator is identical to the early-gate integrator, except that the FET is closed (shorted) during the late-gate enable period. Thus, the output of the circuit is a voltage approximately proportional to the target video energy during the late-gate enable period.

The voltage outputs from the early-gate and late-gate integrators are then applied to a difference amplifier, as shown in Figure 17-3. If the target video energy during the early-gate enable period is equal to that during the late-gate enable period, the two integrators will have equal output voltages. Thus, the difference amplifier input voltages will be equal and, consequently, the output

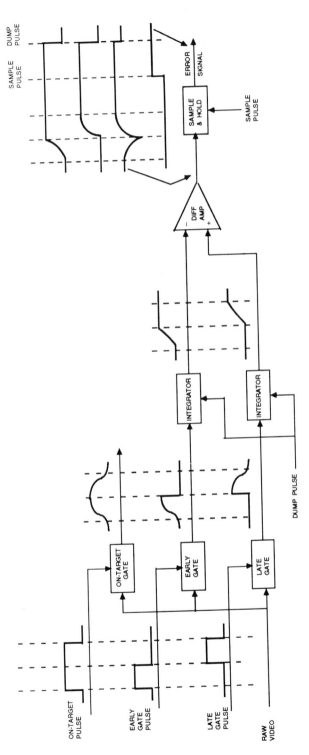

Figure 17-3. Split gate discriminator.

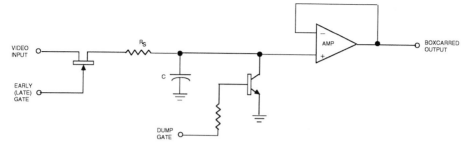

Figure 17-4. Typical boxcar circuit.

will be zero volts. This condition exists when the range tracker is correctly following the centroid of the target return. The difference amplifier output will be nonzero, however, if the partition between the early and late gates is not centered on the target video. For example, if the early-gate/late-gate partition is on the latter half of the target pulse, then more target video energy will be present during the early-gate period than during the late-gate period. Thus, the early-gate integrator output will be larger than the late-gate integrator output, and the difference amplifier output will be a negative voltage.

After the output of the difference amplifier has had sufficient time to settle, the difference amplifier output is sampled and held until the next target video pulse is ready to be sampled. After the difference amplifier output is sampled, a dump pulse is generated to initialize the early-gate and late-gate integrators

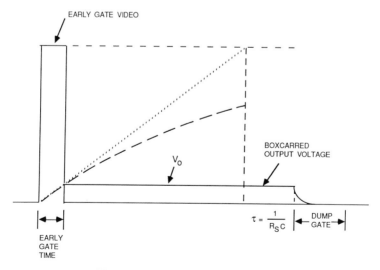

Figure 17-5. Boxcar charging waveforms.

in preparation for the next target video pulse. The sample-and-hold circuit output voltage is proportional to the distance of the early-gate/late-gate partition from the centroid of the target video, and the polarity of the voltage indicates whether the partition is before or after the centroid.

If the target video pulse were rectangular and the early-gate and late-gate times were each equal to the video pulse width, then the time discriminator output voltage would appear, as shown in Figure 17-6, as the target video pulse slews past those early and late gates. Thus, as long as the target range is within twice the pulse width of the early-gate and late-gate times, the time discriminator will generate a positive or negative error voltage.

The error voltage output of the time discriminator is applied to a servo amplifier, as shown in Figure 17-7. The outputs of the servo amplifier are voltages corresponding to range (R), range rate $(\dot{R})$, and possibly acceleration $(\ddot{R})$. The range voltage is coupled into a comparator circuit along with a linear ramp voltage that is proportional to the range. When the linear ramp voltage equals the range voltage, the voltage comparator generates a pulse that triggers the timing circuit. The trigger pulse normally occurs a fixed time interval before the target range and initiates fixed time-delay circuits that generate the early-gate, late-gate, on-target gate, sample pulse, and dump pulse. These gates and pulses occur at fixed times relative to the trigger pulse time.

A typical comparator circuit is shown in Figure 17-8. A ramp circuit generer ates a negative ramp that is linearly related to time. Resistors R1 and R2 form a summing network that adds the two input voltages. When the absolute value of the negative ramp voltage increases to a value that is equal to the range voltage, the summed input voltage into the comparator causes the voltage com

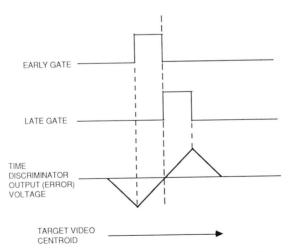

Figure 17-6. Time discriminator output versus target position.

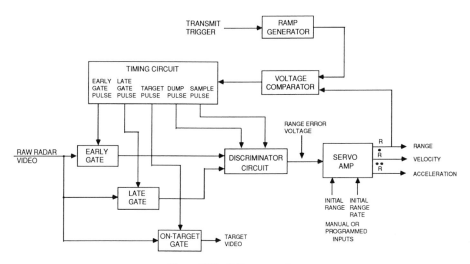

Figure 17-7. Split-gate tracker.

parator to change states rapidly. This transition then triggers the time-delay circuits.

The timing of the early gate and late gate relative to the trigger pulse is shown in Figure 17-9. The delay from the range trigger to the early-gate/late-gate partition is fixed and is normally set equal to the time between the pretrigger pulse and the transmit pulse time. If the target moves out in range, more of the video return energy will appear in the late-gate time than in the early-gate time, thus generating a range error voltage. This range error voltage will cause the range voltage to increase by an amount ΔV_R, causing the range trigger time to

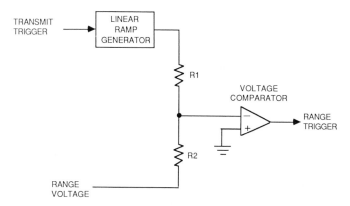

Figure 17-8. Range trigger circuit.

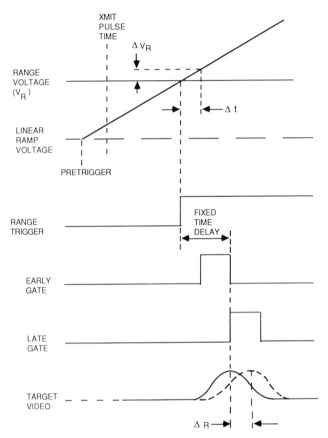

Figure 17-9. Split-gate tracker timing waveforms.

increase correspondingly by Δt. Thus, the range tracker is capable of automatically tracking a moving target.

17.2.3 Leading-Edge Range Tracker

For certain applications, it is more desirable to track the leading edge, rather than the centroid, of the target return. For example, a radar altimeter normally reads out the altitude associated with the range of the return from the nearest terrain, since this is normally the return from directly below the aircraft, and is not interested in returns from farther ranges, since these normally come from terrain to the side of the aircraft. Also, leading-edge trackers are sometimes used to reduce range scintillation which often occurs when target length in the range aspect exceeds the radar pulse width.

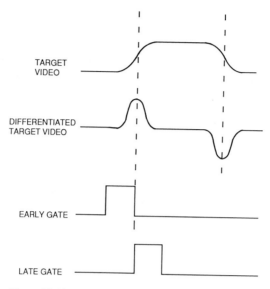

Figure 17-10. Leading-edge tracker, differentiating type.

One method for implementing a leading-edge tracker is to differentiate the target video with a capacitor-resistor circuit such as shown in Figure 17-10. The differentiated target video can then be tracked by a conventional split-gate tracker. The advantage of this implementation is that it allows the same split-gate tracker to be used as either a centroid tracker or a leading-edge tracker simply by switching out or in the differentiator.

Another method for implementing a leading-edge tracker is to use a video threshold and generate a fixed-width pulse whenever the target video voltage exceeds the threshold as shown in Figure 17-11. Here, careful attention must be paid to the setting of the threshold to prevent receiver noise from triggering the leading edge pulse and to prevent target amplitude from causing excessive variations the leading-edge track point. The resulting leading-edge pulse can then be used directly to shut off the range counter.

Figure 17-11. Leading-edge tracker (radar altimeter).

17.2.4 Trailing-Edge Range Tracker

There are also times when a trailing edge-tracker is required. For example, when tracking a missile, it is desired to track the missile and not the extended return from the missile exhaust plume. If the missile is traveling away from the radar, a trailing-edge tracker can be used to keep the range tracker on the missile rather than the plume.

17.3 DIGITAL RANGE TRACKING

17.3.1 Implementation Trend

Digital range trackers can be implemented using hybrid analog and digital circuitry (such as the split-gate digital range tracker) or all digital circuitry. With the continuing development of high-speed digital circuitry including very high speed integrated circuit (VHSIC) technology, the tendency will be toward implementing greater portions of the range trackers using digital circuitry.

17.3.2 Split-Gate Digital Range Trackers

For digital range trackers, digital circuits replace all or part of the analog circuitry. The fundamental advantages of the split-gate digital range tracker are that it eliminates the requirement for an ultralinear ramp generator and provides direct numerical readout of the target range[2].

An example of a split-gate digital range tracker is shown in Figure 17-12. In this configuration, the linear ramp generator is replaced by a high-speed digital counter, and the range voltage is replaced by a target range register. The digital

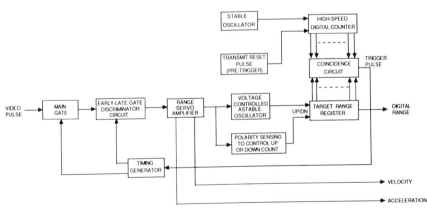

Figure 17-12. Digital split-gate tracker.

range tracker is initially reset to zero count by a pretrigger pulse at a fixed time before the transmit pulse. Immediately after the pretrigger pulse, the stable oscillator (clock) causes the digital counter to count up at the clock rate. A coincidence circuit compares the digital output of the counter with the output of the range register; when the counter reaches the number in the range register, the coincidence circuit generates a trigger pulse. This trigger pulse starts the generation of the timing pulses in much the same manner as in the analog split-gate tracker. The timing pulses then cause an early-gate/late-gate discriminator, such as that shown in Figure 17-3, to determine the range error. This range error voltage then drives the servo amplifier integrators to output the range rate, $\dot{R}$, and possibly the acceleration, $\ddot{R}$. The range rate output then controls the variable-frequency astable oscillator, which outputs clock pulses at a rate proportional to the absolute value of the range error input voltage. In addition, a comparator circuit senses the polarity of the range rate output and, in turn, controls whether the range register counts up or down at the rate of the output frequency from the voltage-controlled astable oscillator. Thus, if the target range is increasing, the range rate out of the servo amplifier will be of the proper polarity to cause the range register to count up at that range rate.

An alternative configuration of a range counter circuit is shown in Figure 17-13. In this circuit, the target range register receives the target range error and up/down command in the same manner as the previous circuit. The main difference is that the target range from the target range register is loaded into the high-speed counter just prior to the pretrigger pulse time. The output of the stable oscillator is normally gated off; at the pretrigger (or transmit) pulse time, the oscillator is gated to the down-count input of the high-speed counter. Thus, the counter starts counting from the target range down toward zero. When the counter reaches zero, a trigger pulse is generated which initiates the early-gate and late-gate timing pulses from the timing generator and shuts off the clock gate in the high-speed counter.

The frequency of the stable oscillator clock into the high-speed counter determines the range increment in the range register. The required clock frequency

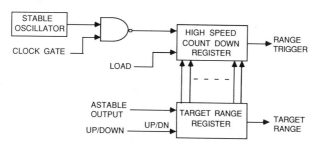

Figure 17-13. Alternative range counter circuit.

(C_f) is given by the formula

$$C_f = \frac{492}{R_I} \quad \text{MHz}$$

where R_I is the desired range increment expressed in feet. Thus, if the required range increment is 1 yard, then the frequency of the stable clock must be 164 MHz. The high-frequency counter and gating circuitry would have to be capable of handling clock rates in excess of 164 MHz.

If range resolution finer than that provided by the clock rate is required, then various interpolation techniques can be used. In many cases, range filters are required to smooth the output range display in order to reduce the range jitter; the same circuitry can also increase the range resolution. In a digital range filter such as that shown in Figure 17-14, the present range measurement along with the nine previous range measurements are stored in the R_n through R_{n-9} registers. The outputs of the registers are digitally added, and the output of the adder is then divided by 10. The output of the divider is then an average of the 10 range measurements, with potentially an additional decimal digit of resolution. The ability of this circuit to provide improved range resolution is critically dependent upon the timing relationship between the radar pulse timing and the stable clock frequency. Ideally, the PRF should be a subharmonic of the stable oscillator clock frequency, except that the phase of the stable clock should advance by 36° every PRF. In this manner, the range circuitry would scan the range interval and provide optimal timing for the averaging process. This can also be approximated by ensuring that the stable clock transitions are random with re-

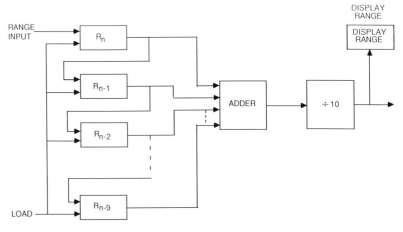

Figure 17-14. Digital range filter.

spect to the pretrigger time transitions. A third alternative would be to make the radar PRF an exact subharmonic of the stable clock and depend upon the randomness of the range measurement. This last method would not provide the optimum range resolution, however, but would at least avoid the effects of beat frequencies between the PRF and subharmonics of the clock frequency.

When using a range filter such as that shown in Figure 17-14, the user must be aware of the lag introduced in the indicated range of the target. For high-velocity targets, this lag can be significant, since the indicated range is actually that of the average of the current range plus that of the previous nine pulses.

The operation of the split-gate range tracker is similar to that of a digital range tracker, as shown in Figure 17-15. The timing diagram shown is that for a count-down counter (shown in Figure 17-13). The count stored in the range register is transferred to the high-speed counter during range load time. At the pretrigger pulse time, the high-frequency counter starts counting down. When it reaches zero, the counter issues a radar trigger which then initiates the early-

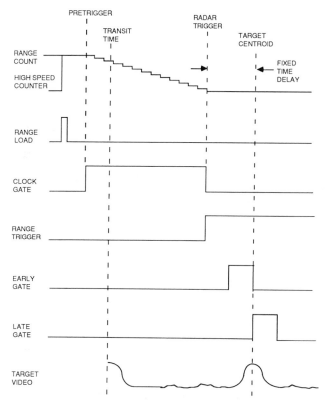

Figure 17-15. Digital range tracker timing.

gate, late-gate, and other required timing pulses. If the target has moved out in range since the last radar PRF, more energy will be present in the late gate, thus causing a range error signal to develop. This range error signal will increase the range count in the range register, effectively moving out the range trigger and, hence, the early-gate/late-gate partition. In this manner, the range tracker will keep adjusting the timing to keep the early-gate/late-gate partition centered on the target video, and the range stored in the range register will be that of the target range.

17.3.3 Modern Digital Range Tracking Techniques

The range trackers described previously were either full analog or partial digital range trackers with hard-wired logic. In the past, these techniques were required because the response times required were too fast for existing logic and microprocessor circuits. With the development of VHSIC and high-speed 16-bit and 32-bit microprocessors, however, many of the previous restrictions should disappear. The future should see continuing development of full digital processor range trackers.

A full digital processor range tracker converts the analog radar video signal to a digital format and subsequently uses a combination of integrated circuit, large scale integration techniques, and microprocessor logic. Larger-scale range trackers utilize mini-computers or full-size computers for their tracking functions. The advantages of using full digital processing are numerous. Analog and partial digital range trackers are typically hard-wired with fixed characteristics such as gate widths, tracker response, and discriminator characteristics. Microprocessor-controlled range trackers provide the capability to adapt to the target characteristics and, thus, provide greater flexibility under varying range tracking conditions. Also, hard-wired range trackers can only track a single target at a time. Digital range trackers, on the other hand, through modularization and computer control, can potentially track multiple targets simultaneously. Also, digital range trackers are more suited to complex range tracking tasks such as track-while-scan.

Digital range trackers can be implemented in a variety of configurations, depending upon the application and the complexity of the task. One possible configuration for a full digital range tracker is shown in Figure 17-16. The raw radar video signal is converted directly into digital format by a very high speed A/D converter. The conversion rate of the A/D converter has to be less than that of the stable oscillator period and should at minimum provide several samples across the video pulse width. The digitized radar video is input to a high-speed digital register, and the video data is clocked into the register by the pulses from the window gate, as shown in Figure 17-17. The digitized video data entered in register 1 is that present during the window period, which the range tracker keeps centered on the target position.

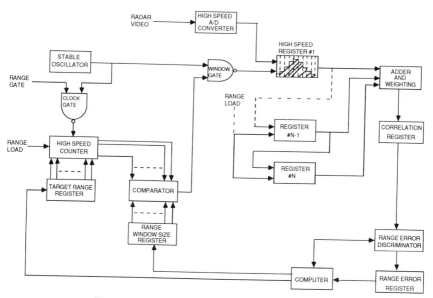

Figure 17-16. Block diagram of a digital range tracker.

The sample digitized video in high-speed register 1 is correlated at each range interval in the range window with the sampled data recorded during corresponding range intervals for the previous $n-1$ pulses. For each range interval, the digitized number in that cell of all of the registers is added with an appropriate weighting factor and stored in a corresponding cell in the correlation register. Once the correlation process is completed, the range error discriminator in conjunction with the computer determines the position of the video centroid with respect to the center of the range window. The output of the range error discriminator is a range error which is subsequently stored in the range error register. The computer then processes the range error data and determines the number by which the range stored in the target range register should be increased or decreased. The updated target range is entered into the high-speed counter at range load time, which occurs just prior to the next transmit pulse. The range load pulse also causes the contents of high-speed register 1 to be transferred to register 2, which is a standard-speed register. In a similar manner, data in register 2 are transferred to register 3, and correspondingly data in register $n-1$ are transferred to register n. At the transmit pulse time, the clock gate pulse starts counting down the high-speed counter. When the count in the range register reaches that present in the range window register, the comparator on the high-speed counter output gates goes high. The window gate then causes pulses from the stable oscillator to clock the new digitized target data into high-speed register 1. It continues to gate through clock pulses until the end of the range window period.

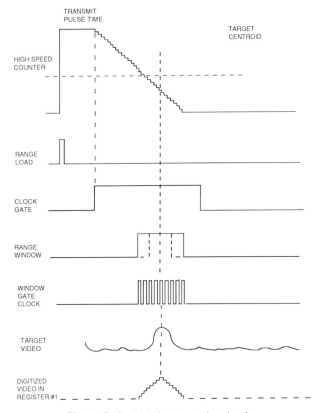

Figure 17-17. Digital range tracker signals.

The computer controls the width of the range window, which can be changed depending on the pulse width or target characteristics. The computer can also be used to control other range tracker aspects such as correlation weighting, range discriminator characteristics, and range data smoothing.

17.4 SERVO LOOP ANALYSIS

The servo loop in a range tracker reacts to the range error from the time discriminator and changes the tracker reference range to keep the reference centered on the actual range of the target. Detailed descriptions of servo loops and analyses of servo loop stability criteria are available in several texts.[3,4] A brief description is included here, however, to highlight the effects of the different types of range servos on the radar tracking performance.

As Figure 17-7 shows, the range servo loop includes the discriminator, the servo amplifier, and the range comparator circuits. The servo amplifier, depending upon the type of servo, provides the tracker and external circuits with

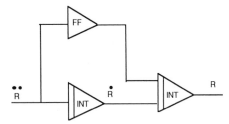

Figure 17-18. Type II servo loop.

the target range, range rate, and possibly acceleration. Initial target range and sometimes target range rate are input to the servo amplifier during the acquisition mode to initialize the target parameters in the range tracker. The initial range and range rate commands can be entered either by an operator or by a computer. When switched from the acquisition mode to the track mode, the servo tracks the designated target.

There are basically three types of range servo loops, appropriately designated type I, II, and III. A type I servo consists of a single range integrator whose output voltage is proportional to the range. The output range stays constant as long as the error voltage input is zero, and a constant-velocity target requires a constant error voltage input. Thus, the input error voltage is proportional to the target velocity. For a constant-velocity target, the output range voltage lags the actual target range.

A type II servo, such as that shown in Figure 17-18, includes a second integrator and, normally, a feed-forward amplifier for loop stability. The error voltage into the first integrator is proportional to the target acceleration; the input to the second integrator is proportional to the target range rate. A type II servo will follow a constant velocity target with respect to the radar with zero lag. In contrast, an accelerating target or a constant-velocity target with a changing aspect angle with respect to the radar will be tracked with a lag proportional to the acceleration component.

A type III servo, potentially, has the ability to track even an accelerating target with zero lag. A type III servo, however, has critical stability requirements, and the tracker design must prevent range noise from being interpreted as an acceleration component.

The servo loop bandwidth affects the performance of the range tracker. Reducing the loop bandwidth will reduce the effects of range tracking noise, but, at the same time, will increase the lag components in the tracker. A tight (small) loop bandwidth will minimize target break-locks that often occur when the target passes isolated clumps of clutter or other targets. Since acquisition is more difficult when the tracker loop bandwidth is small, it is generally necessary to increase the bandwidth until the target is acquired during the acquisition phase.

Normally, the acceleration component is limited by the aircraft or target dy-

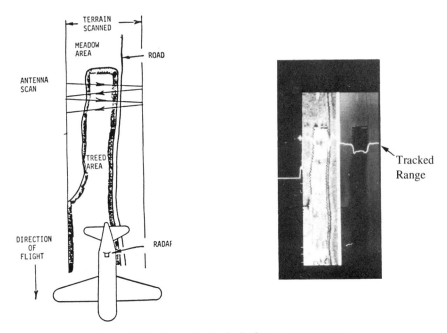

Figure 17-19. Range tracker for backscatter measurements.

namics. In certain cases, however, the acceleration component can be virtually unlimited. In the application shown in Figure 17-19, the mission was to map the RCS of the terrain swath overflown by the aircraft. The purpose of the range tracker was to keep the video samplers on the centroid of the returns from the terrain. When the radar antenna scanned laterally across a tree line, however, the range to the return could decrease virtually instantaneously by up to several hundred feet for low radar antenna depression angles. The recorded terrain map for this case is displayed on the right of the figure, along with the range trace adjacent to the radar map. The range trace shows the range slewing approximately 300 ft, with a 1-ms response time at the edges of the tree line.

17.5 RANGE TRACKING ACCURACY

Two terms that are often confused in radar are range resolution and range accuracy. *Range resolution* is the ability of a radar to resolve (or separate) two targets closely spaced in range, both of which are within the antenna beamwidth. In general, for a pulsed single frequency radar, it is necessary for two equal amplitude targets to separated by at least the pulse length in range ($\tau_0/2c$, τ_0 is the pulse duration) before a determination can be made that two targets are present. For targets of unequal amplitudes, an even greater separation may be required.

Range accuracy is the ability to determine within a given range error increment (E_r) the actual range from the radar to the target. For a target with high signal to noise ratio, the E_r term can be a very small fraction of a radar pulse width. Range tracking accuracy depends upon a number of factors[5] including radar systematic errors, target dependent errors, and propagations effects. Propagation effects for most radars have little effect on the radar's range accuracy. Target induced range errors include range scintillation from extended range targets, and range jitter from targets in multipath situations.

The prime consideration for a radar system designer is to minimize the radar induced range errors, which can be categorized[5] as range biases and noise. Range bias is produced by the inaccurate determination of the target delay time relative to zero range, which can be caused by a number of factors including drift in the time discriminator, inaccurate zero range setting, and instabilities in the range clock frequency. For accurate range measurement systems, a digital range counter clocked from a highly stable frequency source is normally employed to minimize range error bias.

Range noise can be caused by several sources including thermal noise, PRF jitter, and other instabilities in the radar system. In a well designed radar system, PRF jitter and other systematic induced noise sources can be reduced so that the primary source of range noise is that due to thermal noise. According to Barton,[1] the equation for the standard devation of range error (σ_{r_1}) is given by

$$(\sigma_{r_1})_{opt} = \frac{1}{2B\sqrt{S/N}} \qquad \text{for } (B\tau_0 \geq 1)$$

where B is the receiver bandwidth.

This result assumes that the approximate range to the target is known, so that the measurements include the target signal in the entire measurement period along with the noise.

The above equation determines the range noise on a single pulse basis. When the range is determined by observing the target range over several pulses, then the range noise (given by the standard deviation of the range, σ_r) is reduced by the factor $1/\sqrt{n}$, where n is the number of pulses observed. For this case, Barton[1] gives the range noise as

$$\sigma_r = \frac{1}{2B\sqrt{f_r t_o}\,(S/N)}$$

where

f_r is the radar PRF, and
t_o is the observation time.

17.6 TRACK-WHILE-SCAN (TWS)

In many cases, a radar must continue the search process while tracking a target. These track-while-scan (TWS) radars include aircraft surveillance radars, multifunction military airborne radars, and phased array radars. In many cases, the TWS radar is capable of tracking multiple targets simultaneously. Depending upon the configuration, the radar can either provide full hemispheric coverage or cover a limited angular segment.

A TWS radar must track the target on a scan-to-scan basis or, in the case of a phased array radar, on a look-to-look basis. Because of the complexity of the TWS process and the need to store both present and past target positions and velocities for multiple targets, digital computers are generally required to provide TWS processing.

17.6.1 Range Track for Digital TWS

Computer TWS processing requires procedures or algorithms for associating current radar target observations with established target tracks, initiating new target tracks, and computing the information for displays or for other system inputs. Hovanessian[6] has outlined a procedure for a general TWS system. The general sequence for TWS processing is shown in Figure 17-20.

Current observations of target positions are inherently performed in polar coordinates (range, azimuth angle, and elevation). However, rectilinear coordinates are more convenient for computer processing of aircraft or other target tracks. For a typical application, the target position is normally stored as the northerly component, the easterly component, and the altitude or vertical component. In order to convert the radar measurements to the rectilinear coordinate system, the measured range to the target must be multiplied by $\vec{N}$, $\vec{E}$, and $\vec{V}$, which are the unit direction cosines for the northerly, easterly, and vertical components, respectively. The resulting target components represent the target position expressed in rectangular coordinates with the radar at the origin.

After the coordinate transformation has been performed, the observed target position must be correlated with the established target tracks stored in the computer. If the target position is near the predicted target position for one of the previously established tracks and the difference between the observed and predicted positions is within the preset error bound, then a positive correlation is obtained. If the observed target does not correlate with any of the existing tracks, then a new track is established for the target. If the observed target correlates with two or more of the established tracks, then an established procedure such as that described by Hovanessian[6] must be followed by the computer in assigning the observation to a particular established track.

After the observed targets are associated with established tracks or new tracks are initiated, estimated target positions must be computed for each target, along with predictions of the target positions for the next radar scan. The current

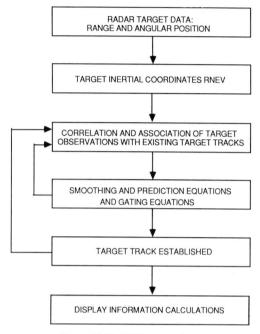

Figure 17-20. TWS processing.

estimated target positions are computed by digital filtering of the current observed target position, along with a weighted estimate of previous target observations associated with that target track. The predicted target positions for each track are then computed based upon the current target position estimate, the time between scans, and the velocity components along each of the directional cosines. The new predicted target positions are then used in the correlation process for each target observation on the next radar scan.

For newly established target tracks, if Doppler information is available from the radar, the computer can determine the radial velocity of the aircraft with respect to the radar; however, it cannot determine the lateral velocity with respect to the radar based on a single observation. Thus, for a new target track, the predicted target position includes a larger zone of uncertainty than that for a previously established track. The zone of uncertainty becomes even greater if Doppler information is not available from the radar.

The present target position can be calculated by the equations given by Hovanessian[6]:

$$R_N = R\vec{N}$$

$$R_E = R\vec{E}$$

$$R_V = R\vec{V}$$

where R_N, R_E, and R_V are the target positions in the respective rectilinear co-ordinates, R is the target range, and $\vec{N}$, $\vec{E}$, and $\vec{V}$ are the unit directional cosines.

The target velocity components in the three rectilinear directions can be given by the terms $\dot{R}_N$, $\dot{R}_E$, and $\dot{R}_V$. The target velocity can then be computed using the following equation[6]:

$$V_T = [(\dot{R}_N)^2 + (\dot{R}_E)^2 + (\dot{R}_V)^2]^{1/2}$$

where V_T is the target velocity. With the knowledge of the present target position and the data stored on previous positions of the target (expressed in rectilinear coordinates), it is possible to estimate, using appropriate smoothing equations, the target position, velocity, and acceleration components. Using this information, the computer can predict the target location, $R_{\Delta t}$, on the next radar scan Δt seconds later using the following equation[6]:

$$R_{\Delta t} = R_{now} + \Delta t \dot{R}_{now} + [(\Delta t)^2/2]\ddot{R}_{now}$$

17.6.2 TWS Application

Today's air traffic controllers monitoring sectors near large metropolitan areas rely primarily on computer synthesized radar displays in which the computers include built-in TWS. Using predicted target locations, the air-traffic control computer uses this information to present an air-traffic controller with an electronic cursor on his display to enable him to keep track of target positions between scans and to provide the necessary predicted target location for correlation with observed targets on the next radar scan. In addition to displaying the information to an air-traffic controller, the predicted target trajectory can be used to alert the controller to possible collision courses of two aircraft.

A typical air-traffic control display such as that shown in Figure 17-21 takes full advantage of the computer's TWS capabilities. All commercial aircraft operating in air-traffic control zones are equipped with beacon transponders that transmit information on the aircraft's identity, altitude, and other features such as aircraft type. This information is associated with the stored target position in the computer and is also displayed via alphanumerics on a computer-synthesized radar display. The computer prints a symbol (such as a square) at the predicted target position for the next radar scan along with the aircraft alphanumerics. When the radar scans by the target, the target return blip is painted on the screen at the position determined by the radar which should correspond to the predicted position. Following the target blip paint, the target symbol advances to the position predicted for the next scan. This enhances the operator's ability to track assigned aircraft easily. In addition, the computer can dis-

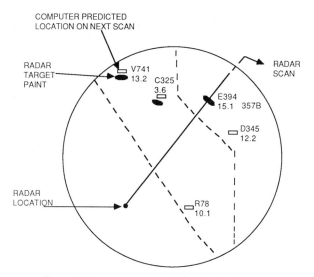

COMPUTER PREDICTED
LOCATION ON NEXT SCAN

RADAR
TARGET
PAINT

V741
13.2   C325
         3.6

RADAR
SCAN

E394
15.1   357B

D345
12.2

RADAR
LOCATION

R78
10.1

Figure 17-21.   Computer-synthesized radar display.

play a simulated radar map of important geographic features, aircraft glide paths, and other information useful to the controller.

17.7   SUMMARY

Automatic range tracking systems can perform precise range tracking of targets under circumstances that would tax even the most experienced radar operators. In this chapter, the trackers have been categorized into two basic types: analog and digital.

Analog range trackers employ hard-wired linear-type circuit elements to implement the tracking function. Normally, they are employed with tracking antenna systems that keep the boresight directed on the target with automatic angle tracking or with manual positioning of the antenna to keep the target within the beam. This type of tracker is typically easier to implement than the digital tracker and can provide good tracking of targets with low signal-to-noise ratios or fluctuating cross sections. Thus, analog trackers are most commonly used when only one target at a time requires tracking. Manual or semiautomatic acquisition of targets is generally used for analog trackers, and depending upon the bandwidth of the servo system, critical matching of target position and velocity by the operator may be required.

Digital range trackers are generally more complex than analog trackers, and therefore can be called upon to perform more complex tasks. With the advent of high-speed digital microprocessors and VHSIC circuits, the use of digital trackers will be greatly facilitated, resulting in trackers that are more adaptable and controllable by software. Complex tracking functions such as TWS and multiple-target tracking are well suited for digital tracker implementations, though at the present time large mainframe computers are used for more complex tracking requirements. Acquisition is generally not as critical for a digital tracking system, since it can adapt to target acquisition needs more readily than an analog tracker. In addition, a digital range tracker is more suitable for automatic as well as manual acquisition because of the adaptability of its acquisition algorithms. Further, digital trackers are also more suitable when digital processing techniques are to be used to enhance target detection, clutter rejection, and other target discrimination techniques.

17.8 REFERENCES

1. D. K. Barton, *Radar System Analysis*, Prentice-Hall, Englewood Cliffs, N.J., 1964, chap. II.
2. M. I. Skolnik, *Radar Handbook*, McGraw-Hill Book, New York, 1970, chap. 21.
3. W. M. Humphrey, *Introduction to Servomechanism Systems Design*, Prentice Hall, Inc., New Jersey, 1973.
4. F. M. Gardner, *Phaselock Techniques*, Wiley & Sons, New York, 1979.
5. R. S. Berkowitz, *Modern Radar*, John Wiley & Sons, New York, 1965, chap. 7.
6. S. A. Hovanessian, *Radar Detection and Tracking Systems*, Artech House, Dedham, Mass., 1978, chap. 9.

18
ANGLE TRACKING

George W. Ewell and Neal T. Alexander

18.1 OVERVIEW

One important function of many radar systems is the accurate determination of target position. Target position information is essential in such diverse applications as air-traffic control, antiaircraft artillery direction, missile guidance, and satellite tracking. Perhaps the simplest example of a tracking system is an operator with a grease pencil marking target positions on a plan position indicator. On a more complex level, target azimuth and elevation may be determined by using conventional scanning antennas and appropriate scanning formats. When accurate target position and motion information is desired, however, special radars called *tracking radars* are often used.[1-3] Some of the more common tracking radar systems use conical scan, sequential lobing, or phase or amplitude monopulse techniques to sense target position. Each of these techniques involves the use of information obtained from offset antennas or antenna beams to develop signals related to the angular errors between the target position and the boresight axis of the tracking antenna.

For applications where several targets must be tracked simultaneously, or where surveillance and tracking must be maintained over an extended area, special systems called *track-while-scan (TWS)* systems have been developed.

18.2 POINT-TRACKING RADAR SYSTEMS

18.2.1 Sequential Lobing Systems

One of the earliest tracking radar techniques employed was sequential lobing. Recent advances in switched-feed lenses and phased-array antennas have revived interest in sequential lobing techniques.

A sequential lobing tracking radar sequentially generates two overlapping beams which have a small angular separation. These beams may be generated by using a single reflector antenna having two slightly separated feeds, by switching between the adjacent feeds of an electronically scanned lens, or by changing the phasing of the various elements of a phased-array antenna. Operation of a sequential lobing radar system may be described by referring to Figure 18-1. The signal received from the target when the antenna beam is in

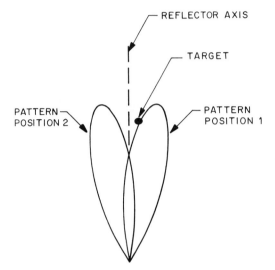

Figure 18-1. Antenna beam switching in a sequential lobing system.

position 1 is larger than that received when the beam is in position 2; the difference between these signals indicates the direction of the error and provides a certain measure of the magnitude of this angular error. If both azimuth and elevation data are desired, additional beams are required.

18.2.2 Conical Scan Systems

Conical scan tracking is conceptually similar to sequential lobing. A typical conical scan pulsed radar system may be represented by the block diagram in Figure 18-2. The beam axis is offset from, and rotates about, the axis of the reflector (this motion is called *nutation*). The signal returned from an off-axis target is amplitude modulated at the conical scan rate, and the amplitude and phase of this modulation indicate the magnitude and direction of the angular error. The received signal is processed to provide error signals that cause the antenna to point toward the desired target.

A conical scan may be generated by using a rotating offset horn or dipole feed, by rotating a tilted Cassegrain subreflector, or by movement of the main or primary reflector. Some of these techniques are illustrated in Figure 18-3.

The key to the operation of a conical scan system is that the amplitude of the signal received from a target that is offset from the axis of rotation bears a definite relationship to the angular offset of the target, while the phase of the envelope of the signal is related to the direction in which the target is offset. The use of position reference information from the antenna beam to detect scan

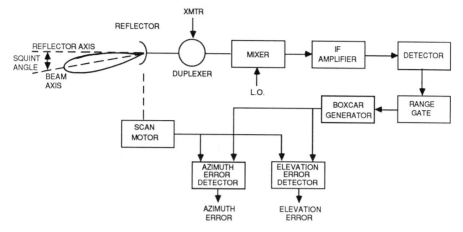

Figure 18-2. Simplified block diagram of a conical scan tracking radar.

modulation synchronously facilitates the generation of error signals in the two principal angular coordinates.[4]

Since a conical scan system utilizes amplitude changes to sense position, amplitude fluctuations at or near the conical scan frequency will adversely affect the operation of the conical scanning radar system. Tracking systems which obtain their direction from a single received pulse—monopulse systems—were developed to circumvent this problem.

18.2.3 Monopulse Systems

The amplitude comparison monopulse radar employs information obtained simultaneously from two overlapping beams, as shown in Figure 18-4. From these two patterns, the sum and difference patterns are generated. That is, the difference pattern is formed by subtracting the received voltage in pattern 1 from that of pattern 2, and the sum pattern is formed by adding the received voltages in patterns 1 and 2.

A block diagram of a simplified monopulse radar for tracking in one angular coordinate is shown in Figure 18-5. In this simplified system, the signal is transmitted through both horns by means of a hybrid junction, and the two received signals are fed to a "magic tee" or similar junction that forms the sum (Σ) and difference (Δ) signals, which are then amplified in two identical channels. As can be seen from Figure 18-4(b), the difference signal indicates the magnitude of the error, but the error direction in ambiguous. The ambiguity is resolved by comparing the phase of the difference signal with the phase of the sum signal. In Figure 18-4, if the target is to the left of the reflector axis, the

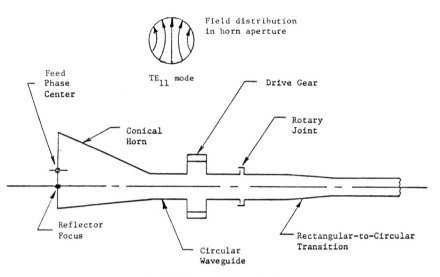

(a) Nutating conical scan feed.

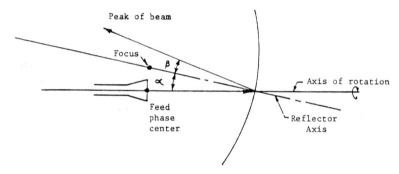

(b) Conical scan antenna employing a tilted reflector.

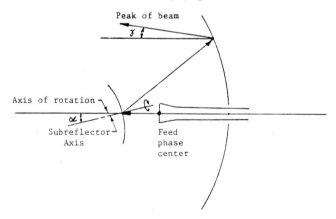

(c) Conical scan antenna (Cassegrain) employing a tilted subreflector.

Figure 18-3. Methods of generating a conical scanning antenna beam.

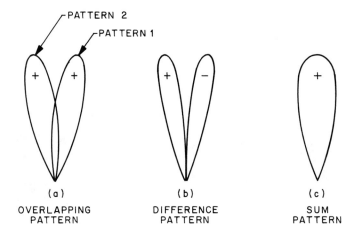

Figure 18-4. Generation of sum and difference signals in a monopulse system.

difference signal is in phase with the sum signal. If the target is to the right of the reflector axis, the difference signal is 180° out of phase with the sum signal. The detector in Figure 18-5 uses the Σ and Δ signals to generate an error signal which may also be used to align the antenna with the target. While this approach applies strictly to CW systems, appropriate range gating of the Σ and Δ signals at the detector output permits its use in a pulsed system.

There are a number of ways in which such a monopulse antenna may be realized. Perhaps the simplest approach is to use an arrangement of four horns and appropriate hybrid junctions, as shown in Figure 18-6, to illuminate a reflector and generate the required offset beams. In monopulse feed design, it is important to consider not only the usual antenna design parameters of gain, beamwidth, and sidelobe levels, but also a factor called *error slope* (proportional to Δ/Σ), which is a measure of the inherent tracking capability of the

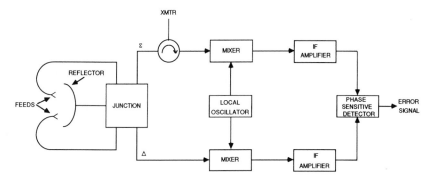

Figure 18-5. Simplified block diagram of a monopulse tracking radar system.

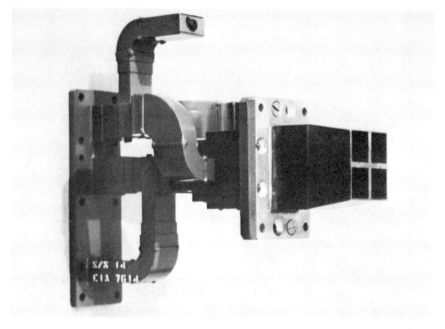

Figure 18-6. Example of four-horn feed.

antenna. Attempts to optimize both the sum and difference patterns simultaneously with respect to these parameters by using a small number of horns often present an impossible problem. For this reason, feed designs using larger numbers of feed horns[5] and higher-order modes within each horn[6] have been developed. Table 18-1 lists examples of practical monopulse feeds for both single- and dual-plane implementations. Figure 18-7 shows examples of the logical extension of the simple four-horn feed to a five-horn configuration which phys-

Table 18-1. Practical Monopulse Feeds.

Single-plane multihorn
Single-plane multimode
Dual-plane multihorn
  Four-horn
  Two-horn dual-mode
  Two-horn multimode
  Twelve-horn
  Four-horn triple-mode
  Five-horn
Single-horn multimode

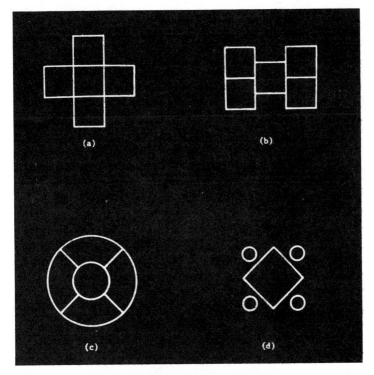

Figure 18-7. Five-horn feed configurations.

ically separates the sum and difference antenna channels, and Figure 18-8 shows an example of a two-horn dual-mode feed which exhibits increased gain and improved antenna patterns resulting from the removal of the center septum of the four-horn feed of Figure 18-6. Schematic diagrams of two complex but very efficient monopulse feeds are shown in Figure 18-9. A practical example of the implementation of the four-horn triple-mode feed of Figure 18-9(b) is shown in the center of Figure 18-10; the corrugated plastic devices in the horn aperture are phase-correction lenses.

Cassegrain reflector systems which use a main reflector and a subreflector, as shown in Figure 18-3(c), have been developed to avoid the feed blockage associated with a front-fed reflector and to allow the feed structure to be placed behind the main reflector. This is particularly useful in the case of monopulse systems because of the complexity of the feed system. Figure 18-11 shows a "twistreflector" antenna developed to reduce aperture blockage of the subreflector in Cassegrain-type reflectors. This particular Cassegrain design employs a flat subreflector. The subreflector of a twistreflector antenna consists of an

Figure 18-8. A two-horn dual-mode feed.

array of wires that is transparent to one polarization but reflects the orthogonal polarization. Incident radiation passes through the subreflector to the main re-flector, which uses a wire-dielectric-reflector sandwich to rotate the reflector polarization. This signal is then reflected by the subreflector to the feed horn.

In addition to straightforward approaches using front-fed and subreflector-focused antennas, other approaches to realize monopulse antennas using planar arrays, focusing lenses, and optically fed phased arrays have been used. Some of these approches are illustrated in Figure 18-12. An example of a planar slot array antenna usable for monopulse systems is shown in Figure 18-13; a mono-pulse beam-forming waveguide network can be implemented in reduced-height waveguide on the back of the array face. The front face of a space-fed phased array employing the feed of Figure 18-10 is shown in Figure 18-14; the array employs some 5000 elements.

Phase monopulse which measures the relative phase between returns in two parallel beams generated by spatially displaced apertures is becoming popular with planar array antennas but not reflector systems.

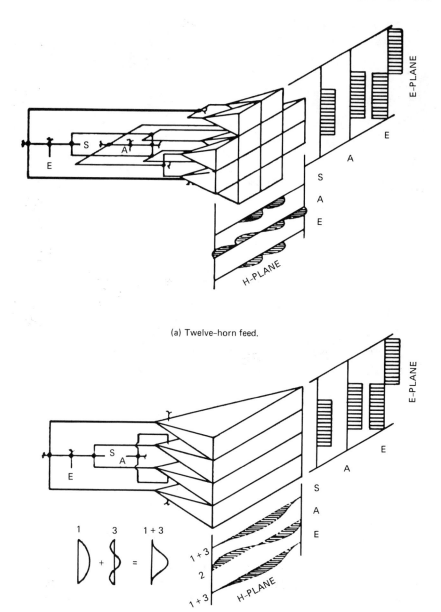

(a) Twelve-horn feed.

(b) Four-horn triple-mode feed.

Figure 18-9. Twelve-horn and four-horn triple-mode feeds.

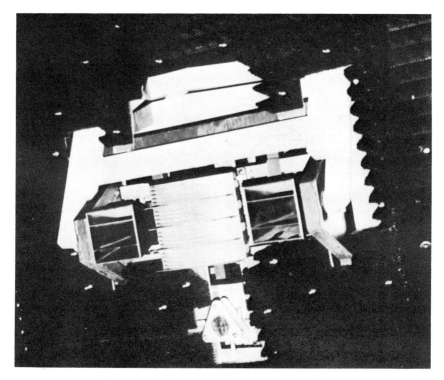

Figure 18-10. Example of a four-horn triple-mode feed.

18.2.4 Hybrid Systems

There are also systems which combine features of monopulse and conical scan systems. One such system, shown in Figure 18-15, uses a mechanically rotating resolver or an electronic switch to generate a conical scan or sequential lobing signal from a monopulse antenna.[7] This is accomplished in the case of electronic switching by sequentially adding to the sum signal the azimuth error, the elevation error, the negative of the azimuth error, and the negative of the elevation error. Thus, a scan-on-receive-only signal is formed from the monopulse signals. Such a system does not require carefully balanced receiver channels or a mechanical conical scan feed, can switch at a high rate, and denies scan rate information to a potential enemy. If a resolver is used, a conical-scan-on-receive-only (pseudo conical scan) signal is formed.

Another similar but improved technique is scan with compensation,[8,9] which uses a conical scan system that rotates *two* beams, as shown in Figure 18-16(a). Since two simultaneous samples are now available, this system can suppress the effects of off-axis jamming signals and target amplitude fluctuations on sys-

Figure 18-11. A Cassegrain twistreflector monopulse antenna.

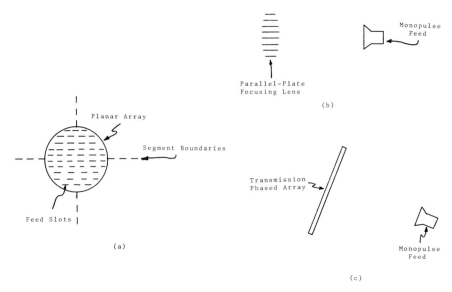

Figure 18-12. Other methods of obtaining monopulse patterns: (a) planar arrays, (b) focusing lenses, and (c) optically-fed phased arrays.

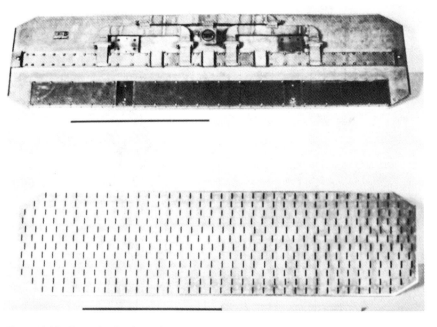

Figure 18-13. Example of a planar slot-array antenna of a type suitable for monopulse applications.

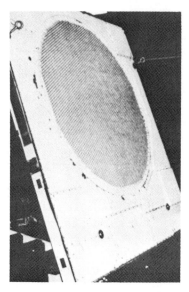

Figure 18-14. Example of a space-fed array monopulse antenna.

tem tracking performance, but at the expense of a lower signal-to-noise ratio relative to a full monopulse system.

A scan-with-compensation system may also be realized using a monopulse antenna and a resolver or electronic switch to add and subtract the difference signals sequentially to and from the sum signal in the two channels, as shown in Figure 18-16(b). This implementation is also called *two-channel monopulse*. A resolver which could be used in such a system is shown in Figure 18-17. Such a device introduces two orthogonal modes corresponding to the two difference signals into circular waveguide and extracts a signal using a rotating loop or slit. This signal, which is the multiplexed difference signal, is then added to and subtracted from the sum signal to form the two signals for processing in the scan-with-compensation system.

18.3 TWS RADAR SYSTEMS

In a number of applications, it is desirable to survey an area while continually tracking a target or to track several targets simultaneously; these applications often use TWS systems. TWS systems may be implemented in a number of different configurations. As noted before, perhaps the simplest TWS system is a human operator that uses a grease pencil to plot target position on a plan position indicator or a B-type display can be employed to determine the target track. In general, however, TWS systems are more complex and use automatic

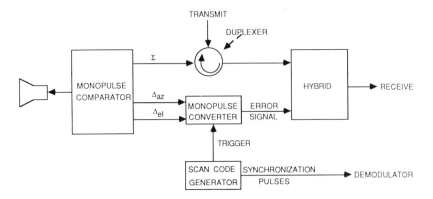

a. Using a monopulse converter.

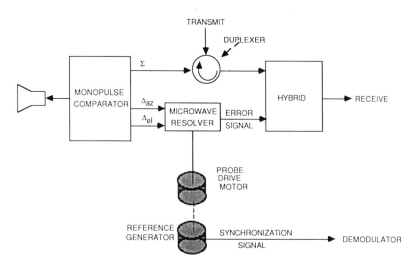

b. Using a microwave resolver.

Figure 18-15. Block diagrams of a "single-channel" monopulse system.

tracking techniques to determine the target position. A simplified block diagram
of one such system is given in Figure 18-18. In this system, the output of a
scanning antenna is used to map received power on a storage device, such as a
charge storage tube; on succeeding scans, the return is compared to the previously stored information on a cell-by-cell basis. Moving targets are indicated
by scan-to-scan changes in the clutter map. Alternatively, this system can em-

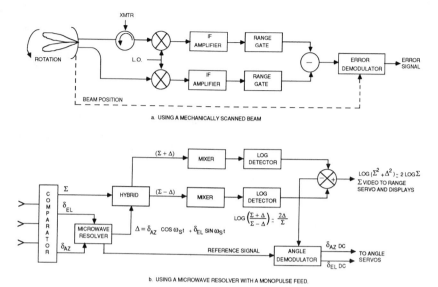

a. USING A MECHANICALLY SCANNED BEAM

b. USING A MICROWAVE RESOLVER WITH A MONOPULSE FEED.

Figure 18-16. Simplified block diagrams of two possible scan-with-compensation systems.

ploy an MTI cancellation filter to select the moving targets; however, targets moving tangentially can not be detected.

More complex and more accurate systems often use both range tracking and angle tracking with a rapidly scanning antenna beam; a simplified block diagram of one such system is shown in Figure 18-19. In practice, an operational system involves tracking in two angular coordinates rather than one, thus in-

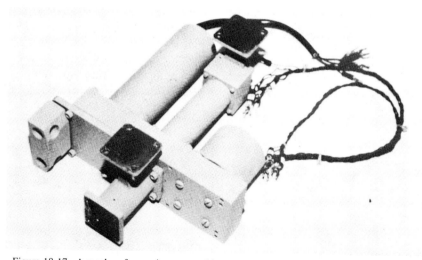

Figure 18-17. A resolver for use in a scan-with-compensation system using a monopulse feed.

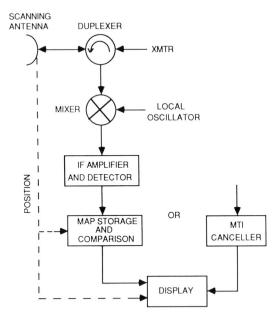

Figure 18-18. Simplified block diagram of a TWS system employing map storage or MTI cancellation.

creasing system complexity. Range and azimuth tracking can be performed automatically by conventional techniques. Split-gate range trackers can be used for range tracking, and similar techniques can be used for azimuth tracking.

Azimuth tracking accuracy can be increased by using separate, rather than contiguous, azimuth gates, as shown in Figure 18-20. This system tends to drive the scan gate positions so that the returns in the two azimuth gates are equal; if the returns in the two azimuth gates are not equal, the system determines that the center between the gates is displaced from the true target azimuth and makes a correction. When the center between the gates is properly aligned

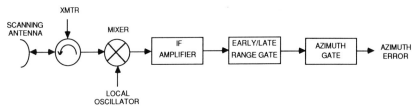

Figure 18-19. Simplified block diagram of a TWS system employing range and azimuth gating of target returns.

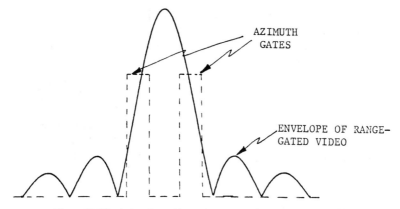

Figure 18-20. Azimuth tracking using noncontiguous azimuth gates.

with the target position, the energy in the gates is equal and the error signal is zero; similar concepts are used in split-gate range trackers.

Analysis of the performance of split-gate angle trackers indicates that an adequate signal-to-noise ratio, symmetry of the antenna beam, careful control of the sidelobe levels, and a smooth tapering of the main lobe are necessary for good performance.

Figure 18-21. A TWS system using either the organ-pipe scanner or the two geodesic lens antennas for tracking targets.

A system using TWS techniques is shown in Figure 18-21. The system uses either an organ-pipe, pencil-beam dish scanner for elevation scan or orthogonal, fan-beam, lens-type scanning antennas to track in the two orthogonal 45° planes.

Many TWS systems, particularly, beam-agile phased arrays, use predictor-corrector algorithms, commonly known as *GH*, *GHK*, or *alpha-beta filters*.[10] Such algorithms use the previous track history to predict where target should be during the next radar look interval, to sense the difference between the predicted and measured positions, and to incorporate a correction into the algorithm. Such techniques are typically used in connection with digital signal-processing techniques that require a high-speed digital computer. The phased-array system of Figure 18-14 is an example of such a system.

18.4 ERROR SOURCES

Target tracking inaccuracies are a primary concern of the tracking radar system designer. Several factors contribute to these errors, including target amplitude noise, radar servo noise, radar receiver noise, variations in apparent target center reported by the radar, multipath reflections from the earth's surface, and electronic countermeasures (ECM).

Servo noise is due to inaccuracies in the drive train, position sensors, and feedback circuits of the antenna mount and is essentially independent of the range to the target. Receiver noise is defined here as the effect of thermal noise within the receiver. Tracking errors due to receiver noise vary inversely with the signal-to-noise ratio; therefore, they increase with the range. Angle noise, also called *glint*, is the tracking error caused by variations in the apparent position of a complex target, and its effects are most severe at short range. Amplitude noise is the effect of variations in the amplitude of the received signal and primarily affects sequential lobing and conical scan radar systems which perform angle measurements over a period of time. The importance of several of the error components in tracking radars as a function of range is indicated in Figure 18-22.

18.4.1 Receiver Noise

The presence of receiver noise limits the maximum tracking range attainable with a tracking radar system. Barton[11] has analyzed the errors due to the presence of noise in the receiver output. He derived an expression for monopulse tracking error given by (for S/N ≥ 6 dB):

$$\sigma_\theta = \frac{\theta_3}{k_m[2B\tau(S/N)]^{1/2}} \tag{18-1}$$

and, if the loop is closed through a servo system, the receiver thermal noise

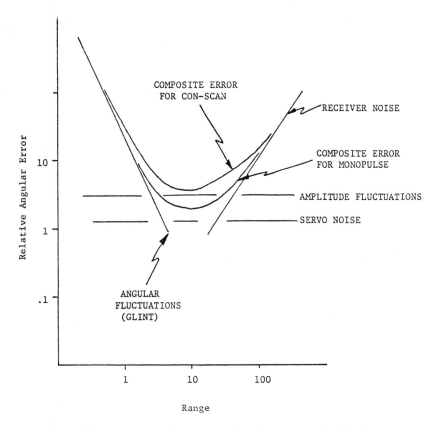

Figure 18-22. Importance of components of tracking error as a function of range.

error becomes

$$\sigma_t = \frac{\theta_3}{k_m[B\tau(S/N)(f_r/\beta_n)]^{1/2}} \tag{18-2}$$

In a conical scan system, the error for the closed-loop system is

$$\sigma_t = \frac{1.4\theta_3}{k_s[B\tau(S/N)(f_r/\beta_n)]^{1/2}} \tag{18-3}$$

and for a TWS system

$$\sigma_\theta = \frac{0.5\theta_3}{[n(S/N)]^{1/2}} \tag{18-4}$$

In the above expressions,

σ_θ = rms angle measurement error, open loop
σ_t = rms angle pointing error, closed loop
θ_3 = one-way antenna 3-dB beamwidth
k_m = monopulse error slope
k_s = conical scan error slope
S/N = on-axis signal-to-noise ratio
B = equivalent noise bandwidth
τ = pulse width
f_r = pulse repetition frequency
β_n = servo noise bandwidth
n = number of pulses integrated

For a signal-to-noise ratio smaller than approximately 6 dB, or for fluctuating targets, more complex expressions must be employed.

18.4.2 Amplitude Scintillation

In conical scan or sequential lobing systems, variations in target amplitude (scintillation) over the measurement interval produce additional errors given by[11]

$$\sigma_s = \frac{\theta[W(f_s)\beta_n]^{1/2}}{k_s} \tag{18-5}$$

where

σ_s = angular rms error due to amplitude scintillation
$W(f_s)$ = power spectral density of amplitude scintillation power near f_s, expressed as (fractional modulation)^2/hertz

Unfortunately, little information concerning the shape of $W(f)$ is available for many targets of interest. This error component, which is independent of the target range, limits the accuracy of most sequential systems.

18.4.3 Target Angular Noise (Glint)

Many radar targets of practical interest are complex targets consisting of many scatterers. The apparent position of a complex target is a function of how the contributions from the scatterers on the target recombine at the radar antenna. Variations in the apparent target position (also called *glint, angular scintillation*, or *phase return error*) can significantly reduce tracking accuracy at short ranges and result in complete loss of the target track.

In the case of a single-point target, the apparent location of the target is the center of the spherical wavefronts reradiated by the target. If the target is complex, the wavefronts are no longer spherical and the problem is considerably more complicated. The analysis for complex targets can be simplified by assuming that the target is far enough from the observer so that a uniform plane wave is incident upon the antenna aperture.

Stratton[12] or Paris and Hurd[13] show that the following expressions are equivalent for uniform plane waves:

- The direction of the Poynting vector S (direction of power flow).
- The gradient of the phase of the electric field ($\nabla\Phi$).
- Normal to the phase front.

All of these expressions are used in the literature to indicate the position of the target as sensed by a tracking radar.[14-16] However, they are seldom used directly because of the complexity of the required calculations. A development that simplifies the computations for a target consisting of a number of isotropic point scatterers is given by Peters and Weimer.[17]

A two-dimensional target consisting of an array of point scatterers is shown in Figure 18-23. The apparent location of this array of isotropic scatterers, as sensed by a distant radar, is

$$X = \text{Re} \sum_{i=1}^{m} \frac{A_i X_i \exp j\theta_i}{A_i \exp j\theta_i} \tag{18-6}$$

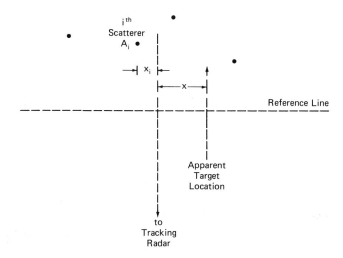

Figure 18-23. Hypothetical target consisting of an array of point scatterers.

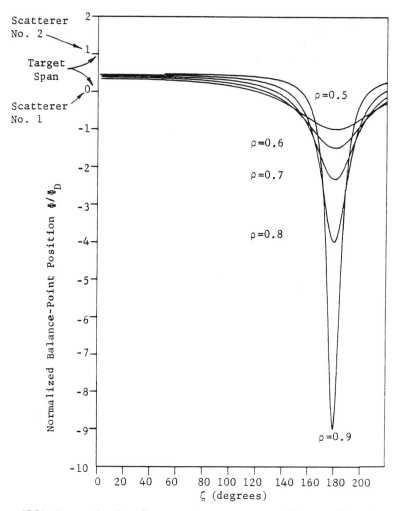

Figure 18-24. Apparent location of a two-scatterer target, as sensed by a tracking radar, as a function of the relative phase between the returns for various values of relative scatterer strength (ρ).

where

  A_i = amplitude of the electric field radiated from the ith scatterer
  X_i = coordinate of the ith scatterer on the reference line
  θ_i = phase of the electric field reradiated from the ith scatterer referenced to the reference line
  Re = take the real part of

A number of important characteristics of glint associated with complex tar-

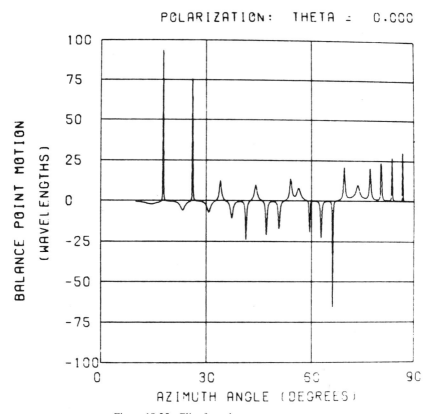

Figure 18-25. Glint for a three-scatterer target.

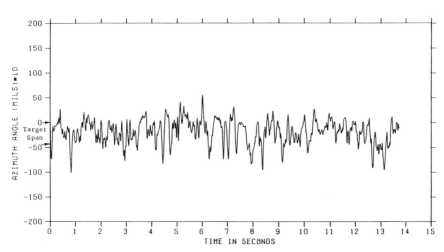

Figure 18-26. Angular errors for a C-45 aircraft.

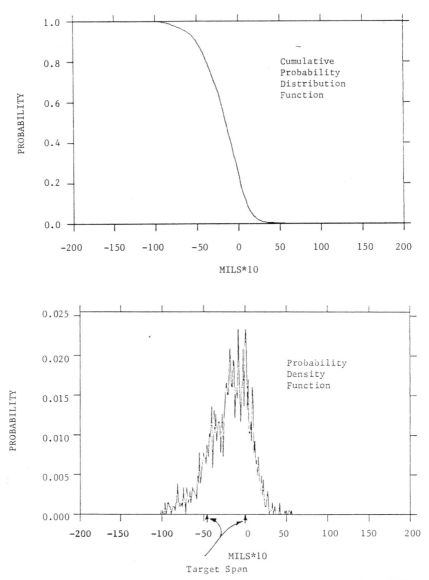

Figure 18-27. Probability densities and distributions for the flight shown in Figure 18-26.

gets are highlighted by considering the behavior of a complex target consisting of two scatterers. For such a two-scatterer model, the indicated position is a function of the relative amplitudes of the returns from the two scatterers and the relative phase between these returns, as shown in Figure 18-24. Note that the apparent target location may lie outside the physical target span and that maximum glint errors correspond to destructive interferences, that is, to nulls

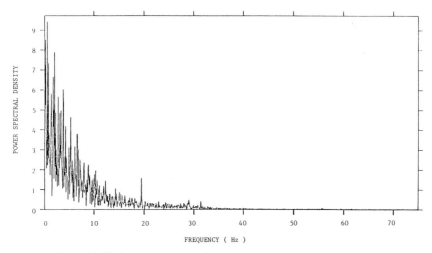

Figure 18-28. Power spectral density for the flight shown in Figure 18-26.

in signal amplitude. A similar plot for a three-scatterer target (a cylinder with a length of 9λ and a diameter of 1.5λ) is shown in Figure 18-25.

The tracking errors associated with real targets have the appearance of a random process. Figure 18-26 shows the apparent angular error for a particular flight of a C-45 aircraft, and Figure 18-27 shows the probability density functions and distributions for the same run. Note that there is a significant probability that the apparent target location lies completely outside the physical limits of the target. The power spectral density computed for this run is given in Figure 18-28. Note that low-frequency components contribute heavily to angular errors. Since these components are difficult to reduce by filtering or integration, while still retaining a good tracking response, integration is a relatively ineffective method for reducing glint effects on a tracker.

Attempts to model the return from complex targets analytically have met with varying amounts of success. The assumption of a large number of independent scatterers leads to a statistical glint model,[18] but there are significant differences between the behavior of such models and real targets.[19] Scattering-center theory has been applied to justify deterministic glint models,[20] but much work remains to be done in this area.

Empirical data show that the rms angular glint error associated with aircraft tracking is usually between 10 and 25% of the length of the aircraft normal to the radar.

18.4.4 Multipath Errors

When a low-altitude target is illuminated by a radar system, or for higher-angle situations involving appreciable antenna sidelobes, energy can enter the track-

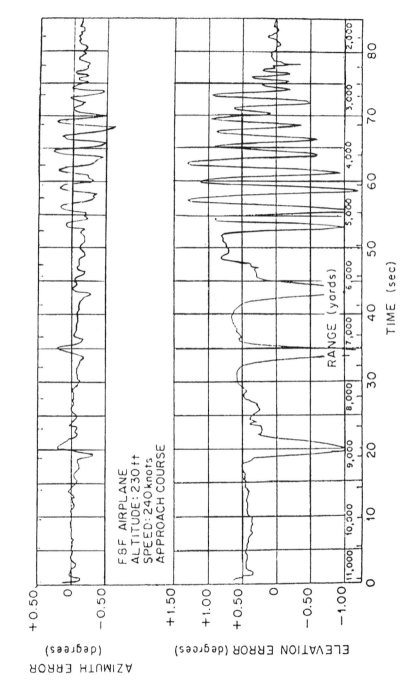

Figure 18-29. Azimuth and elevation tracking errors for a low-flying aircraft target.

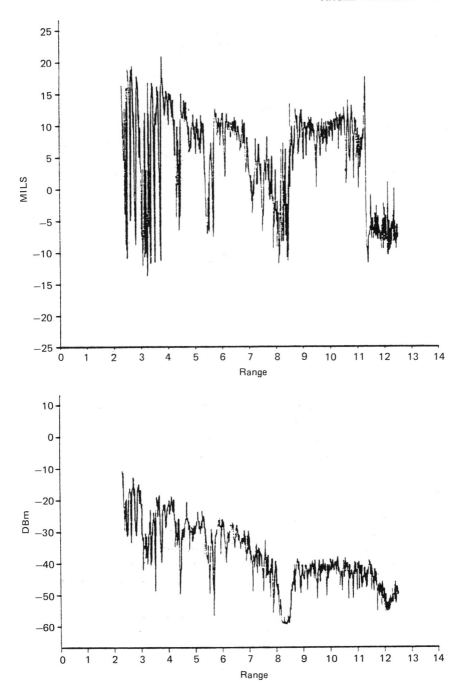

Figure 18-30. Elevation angle error and amplitude variations for an aircraft flying 100 ft above the ocean surface.

ing antenna by two separate paths: a direct path from the target and an indirect path involving energy reflected from the surface of the earth. The effect of such surface reflections is that the antenna "sees" both the actual target and an image target.[21,22] These two targets produce signals which combine in the antenna to produce errors similar to those for a two-scatterer target. Figure 18-29 shows azimuth and elevation angular errors for an aircraft flying at an altitude of 230 ft. The effect of ground reflections on elevation tracking is evident. There is essentially no multipath effect on azimuth tracking accuracy; however, some crosstalk from the elevation channel or a slight glint effect at close ranges is present. Figure 18-30 shows the angular error and variations in received power from another aircraft flying at an altitude of 100 ft above the ocean surface. Barton[23] has analyzed the effects of multipath returns and has formulated an expression for rms tracking error given by

$$\sigma_m = \frac{\rho\theta_3}{[8G_{se}(\text{peak})]^{1/2}} \tag{18-7}$$

where

σ_m = rms angle multipath error
θ_3 = one-way 3-dB antenna beamwidth
ρ = reflection coefficient
$G_{se}(\text{peak})$ = ratio of the antenna gains for target and multipath signals

Attempts to reduce multipath effects on radar tracking accuracy include the use of frequency agility, polarization agility, high-resolution antennas, clutter fences, and complex indicated angle processing techniques. An excellent survey of these techniques has been published by Barton.[21]

18.4.5.   Angle Tracking ECM/ECCM

Operation of tracking systems in an environment where ECMs (which may be active or passive) are employed presents a relatively difficult problem. Chaff consists of very small passive reflectors or absorbers suspended in the atmosphere and may consist of metal foils, metal-coated dielectrics, or even aerosols. Chaff usage is intended to screen existing targets or generate false targets,

Table 18-2.   Spot Noise Characteristics.

Narrowband noise-like energy
Produces high power density
Jammer scans and tracks radar operating frequency
Typically jams only a single radar

Table 18-3.   Barrage Noise Characteristics.

Noise-like energy radiated over a wide bandwidth
Greater than 100 MHz up to an octave
Jamming effectiveness dependent on power density (watts/megahertz)
May be amplitude modulated

and chaff is typically employed in conjunction with aircraft tactics and active ECM such as velocity gate pull-off (VGPO). Chaff usually has a limited effective reflecting bandwidth and also moves slowly at the prevailing wind velocity. The effective frequency bandwidth of a single chaff dipole is approximately $\pm 5\%$, but various dipole sizes are often intermixed to increase the effective chaff bandwidth. Wide-range frequency diversity is one electronic counter-countermeasure (ECCM) technique which can be relatively effective against certain types of chaff. Since there is often an appreciable difference between the velocity of chaff and the velocity of the desired targets, MTI techniques are effective in reducing the effects of chaff in many applications.

Active ECM techniques which are often employed against tracking systems include spot, barrage, or swept noise jamming, and angle deception techniques. Range gate pull-off (RGPO) is almost always used in conjunction with angle deception ECM. The general characteristics of spot and barrage noise jamming are listed in Tables 18-2 and 18-3, and several angle deception techniques are given in Table 18-4. The effects of various noise-jamming techniques can be reduced by incorporating receivers of wide dynamic range, carefully controlling antenna sidelobe levels, incorporating sidelobe blanking or cancelling circuits, and providing for frequency-diverse operation. Deception techniques are somewhat more difficult to defeat and often necessitate the use of monopulse or scan-with-compensation systems to assure reliable target tracking.

Table 18-4.   Angle Deception ECM Techniques.

Type	Victim Systems	ECCM
Inverse gain		
Inverse scan	Active TWS/conical scan	COSRO/SORO*
Con-scan noise	COSRO	SWC†
Angle gate pull-off	TWS	SORO
Angle gate ECM	TWS/SORO	Variable scan rate
Synchronized	Monopulse	Passive angle track
Blinking	Conical scan	
Cross-eye	Monopulse	Leading-edge
	Conical scan	tracking

*COSRO = conical-scan-on-receive-only.
SORO = scan-on-receive-only.
†SWC = scan-with-compensation.

Table 18-5. Cross-Polarization Jamming.

Senses polarization of the victim radar
Retransmits cross-polarized signal (noise or pulse)
Angle errors introduced via the radar's cross-polarized pattern
Effective against (active and passive) monopulse, conical scan, and sequential lobing

The possibility of cross-polarization jamming (see Table 18-5) can significantly influence tracking antenna design. For paraboloidal reflector antennas, there is often an appreciable cross-polarized component, and it has been observed that there is cross-coupling of the cross-polarized error components; that is, cross-polarized energy entering the antenna from directions off the reflector axis results in tracking errors in the other coordinate. Such cross-polarized errors can introduce substantial tracking error and may cause complete loss of track under some conditions. The cross-polarized component is associated with the surface curvature of the reflector and is greater for small values of the ratio of focal length to reflector diameter (f/d). For example, for a particular 37-wavelength-diameter antenna, the cross-polarized component was at a level of approximately -16 dB relative to the copolarized component, for an f/d ratio of 0.25. This value decreased to -28 dB for an f/d ratio of 0.60.[24] Of course, the polarization behavior of the antenna needs to be controlled well out into the sidelobe region, necessitating both a low antenna sidelobe level and a small cross-polarized component. Methods of reducing the effects of cross-polarization jamming include employing gridded subreflectors (as in Figure 18-11) and array antennas (Figures 18-13 and 18-14) which can be designed to have very low cross-polarized lobes.

Operation of TWS systems in an ECM environment reveals a particular advantage of a TWS system relative to single-target trackers. Conventional deception techniques may readily degrade the performance of automatic tracking loops; however, the ability of TWS systems to revert easily to manual operator-aided tracking makes such systems relatively difficult to jam by deception. However, the jamming effectiveness of other ECM techniques such as barrage, spot, and swept noise jamming or chaff tends to be relatively unaffected by a TWS implementation. Therefore, conventional ECCM techniques such as frequency diversity, narrow beamwidths, and careful control of sidelobe and backlobe radiation are all important for proper operation of TWS in such an environment. The use of SORO techniques which attempt to deny scan rate information to the jammer will provide another advantage for these systems.

18.5 REFERENCES

1. D. K. Barton, *Radar System Analysis*, Prentice-Hall, Englewood Cliffs, N.J., 1964, chaps. 9 and 10.

2. M. I. Skolnik, *Introduction to Radar Systems*, McGraw-Hill Book Co., New York, 1962, chap. 5.
3. M. I. Skolnik, *Radar Handbook*, McGraw-Hill Book Co., New York, 1970, chap. 21.
4. J. B. Damonte and D. J. Stoddars, "An Analysis of Conical Scan Antennas for Tracking," *IRE National Convention Record*, vol. 4, pt. 1, 1956, pp. 39–47.
5. P. W. Hannan, "Optimum Feeds for All Three Modes of a Monopulse Antenna I: Theory" and "II: Practice," *IRE Transactions on Antenna and Propagation*, vol. 9, September 1961, pp. 444–454, 454–461.
6. D. D. Howard, "A New Broadband, Efficient, and Versatile Single-Horn Feed for the Monopulse Tracking Radar," NRL Report 5559, National Research Laboratory, Washington, D.C., 1960.
7. M. I. Skolnik (Ed.), *Radar Handbook*, McGraw-Hill Book Co., New York, 1970, pp. 21–30.
8. H. Sakamoto and P. Z. Peebles, "Conopulse Radar," *IEEE Transactions on Aerospace and Electronic Systems*, vol. AES-14, no. 1, January 1978, pp. 199–208.
9. P. A. Bakut et al., *Questions of the Statistical Theory of Radar*, Defense Technical Information Center, AD 6457775, 1966.
10. T. R. Benedict and G. W. Bordner, "Synthesis of an Optimal Set of Radar Track-While-Scan Smoothing Equations," *IRE Transactions on Automatic Control*, vol. 6, July 1962, pp. 27–32.
11. Barton, *Radar System Analysis*, pp. 275–290.
12. J. A. Stratton, *Electromagnetic Theory*, McGraw-Hill Book Co., New York, 1941, Chap. V.
13. D. T. Paris, and F. K. Hurd, *Basic Electromagnetic Theory*, McGraw-Hill Book Co., New York, 1969, chap. 7.
14. A. S. Locke, *Guidance*, D. Van Nostrand Co., Princeton, N.J., 1955, pp. 440–442.
15. James B. Angell, *Errors in Angle Radar Systems Caused by Complex Targets*, MIT Research Laboratory of Electronics, October 1951.
16. Dean D. Howard, "Radar Target Angular Scintillation in Tracking and Guidance Systems Based on Echo Signal Phase Front Distortion," *Proceedings of the NEC*, vol. 15, 1959, pp. 840–849.
17. L. Peters and F. C. Weimer, "Tracking Radars for Complex Targets," *Proceedings of the IEE (GB)*, vol. 110, no. 12, December 1963, pp. 2149–2162.
18. R. H. Delano, "A Theory of Target Glint or Angular Scintillation in Radar Tracking," *Proceedings of the IRE*, vol. 41, December 1953, pp. 1778–1785.
19. J. T. Coleman and F. W. Someroski, "A Survey and Analysis of Radar Glint as Applied to Air Defense Missile System Design," Final Report, Contract No. DA-01-021-AMC-14693, U.S. Army Missile Command, August 1967.
20. A. I. Gaheen, A. A. Orentas, and C. A. McGraw, "A Deterministic Glint Model and Its Applications," *14th Annual Tri-Service Radar Symposium Record*, vol. 1, October 1968, pp. 297–336.
21. D. K. Barton, "Low-Angle Radar Tracking," *Proceedings of the IEEE*, vol. 62, June 1974, pp. 687–704.
22. D. K. Barton, "Multipath Error in Elevation-Scanning Radars," *Twentieth Annual Tri-Service Radar Symposium Record*, 1974, pp. 151–158.
23. Barton, *Radar System Analysis*, pp. 327–331.
24. Skolnik, *Radar Handbook*, pp. 21–50.

19
DOPPLER FREQUENCY TRACKING

Guy V. Morris

19.1 INTRODUCTION

Continuous wave (CW), pulsed Doppler, and moving target indicator (MTI) radars, described in Chapters 13, 14, and 15, respectively, use the Doppler effect to discriminate between moving and stationary objects. These radars provide a method of detecting targets in the presence of ground clutter. Doppler frequency tracking provides clutter rejection that enables continuous range and angle tracking of a single moving target. The single-target tracking mode, in turn, provides the highest data rate and potentially the most accurate measurement of current target position. Although clutter rejection is the most common application, the tracking bandwidths can be narrowed considerably from those used for detection, providing an improvement in the signal-to-noise or signal-to-interference ratio. Since velocity or, more precisely, Doppler frequency is measured directly, predictions of future target position can be made more accurately.

When single-target tracking, that is, the antenna is pointed at the target continuously, the target accelerations between sample times may be small enough so that adequate tracking accuracy is provided by simple tracking loops. However, at short ranges, the accelerations may necessitate higher-order prediction methods. Also, when tracking multiple targets, especially with a mechanically scanned antenna, the time between samples is much greater. In these cases, Kalman filters can greatly improve tracking accuracy. Kalman filters permit higher-order modeling of the expected target motion and adaptively weight the current observation and track history. The α-β tracker is widely used and may be viewed as a special case of the Kalman filter, having nonadaptive weights. Table 19-1 summarizes the types of radars for which Doppler tracking will be discussed.

The topics covered in this chapter include:

1. A brief review of the Doppler effect.
2. An explanation of the fundamentals of Doppler frequency tracking, first using a stationary CW radar example and then extending it to high pulse repetition frequency (PRF) pulsed Doppler radar.

Table 19-1. Typical Doppler Tracking Methods.

Tracking Mode	Radar Type	Sampling Rate		Doppler Tracking Bandwidth	Typical Tracking Methods
Single target (continuous sampling)	CW	IF bandwidth	(highest)	Narrowest	Phase-lock loop
	High PRF	PRF			
	Low PRF	PRF			α-β; simplified Kalman filter
Multiple target (intermittent sampling)	Phased array	PRF burst; frequent revisit			
	Mechanical scan	PRF burst; less frequent revisit	(lowest)	Broadest	Multistate Kalman filter

3. Additional considerations applicable to low- and medium-PRF radars that also include range tracking loops.
4. Adaptations of the tracking methods that are used when the radar is not stationary, such as aircraft and missile radars.

Throughout this chapter, we will use the terms *measurement* (the current Doppler frequency, as indicated by the frequency measurement method of the system, such as a Doppler filter bandwidth); *filtering*, also commonly called *smoothing* (combining a number of measurements to form an estimate of the current Doppler that is presumably more accurate than a single measurement); and *prediction* (estimating the position, velocity, or acceleration at some time in the future).

19.2 DOPPLER EFFECT

The round trip path length between the radar and a target can be expressed in terms of the number of wavelengths as:

$$\text{Number of wavelengths} = \frac{2R}{\lambda_0}$$

where

λ_0 = carrier wavelength
R = range to the target in consistent units

The phase of the received echo relative to the transmission is:

$$\phi = 2\pi\left(\frac{2R}{\lambda_0}\right) \quad \text{radians.}$$

If the target is moving at a constant radial velocity and if a CW radar is used, then the Doppler frequency relative to the transmission that is observed at the radar is:

$$f_d = \frac{1}{2\pi}\frac{d\phi}{dt} = -\frac{2}{\lambda_0}\left(\frac{dR}{dt}\right) = -\frac{2V}{\lambda_0}$$

where f_d is in hertz and V is the target radial velocity. The negative sign merely indicates that when the range to the target is decreasing, the received echo is at a higher frequency than the transmission.

Similarly, in a pulsed radar, a phase change is observed from pulse to pulse:

$$\Delta\phi = 2\pi \left(\frac{2\Delta R}{\lambda_0}\right) \quad \text{radians}$$

where ΔR is the change in range between pulses. If $\Delta\phi < 2\pi$, then the radar can determine the Doppler frequency unambiguously from a series of measurements. If $\Delta\phi$ is an integral multiple of 2π, then the target appears not to be moving and cannot be distinguished from stationary objects; the velocities which result in this condition are referred to as *blind speeds*. If $\Delta\phi > 2\pi$ but is not an integral multiple, the Doppler effect can still be used, but the measured frequency will not be an accurate measure of target velocity. The true target Doppler will be the measured value plus some integer times the PRF; this condition is referred to as a *velocity ambiguity*. Resolution of blind speeds and velocity ambiguities is discussed further in Section 19.4.

19.2.1   Frequency Measurement Accuracy

The error in measuring a single isolated frequency in the presence of noise is:

$$\delta f = \frac{2}{\pi\tau_0\sqrt{S/N}} = \frac{2\Delta f}{\pi\sqrt{S/N}}$$

where

δf = rms frequency error
Δf = spectral width
S/N = signal-to-noise ratio
τ_0 = observation time

This result is directly applicable to CW radar or to a single pulse if τ is taken to be the pulse width. Reference 1 contains a compendium of formulas derived for various cases, such as trapezoidal pulses. The usual situation in the frequency measurement section of pulsed Doppler trackers is that some number of samples, N_p, of the return are measured and analyzed by an FFT or other filter. The above equation is then interpreted by using $\tau_0 = N_p/\text{PRF}$ or $\Delta f = \text{PRF}/N_p$. If a number of independent frequency measurements, N_0, are averaged and if the Doppler frequency is constant, then the rms error will be further reduced by $\sqrt{N_0}$. An analysis of various sources of error is presented in Reference 2.

19.3 CW RADAR

A simplified block diagram of a CW tracking radar is shown in Figure 19-1. The signal-to-interference benefits of Doppler tracking must be applied to the angle error channel as well. Also, when range must be measured, a secondary modulation is applied to the transmit signal. Each of these functions must be accommodated by the Doppler tracker.

Figure 19-2(a) depicts the transmit and receive spectra for an infinite observation time and in the absence of ranging modulation. For a finite observation time, τ, the line spectra is replaced with one having bandwidths of $1/\tau$, as shown in Figure 19-2(b). Target accelerations and practical equipment considerations limit the upper bounds of τ. Equipment limitations may include transmitter stability and minimum achievable Doppler filter bandwidths. Target acceleration and the relative motions of the parts of a complex target broaden the observed spectra.

A Doppler tracking loop is shown in Figure 19-3. For the moment, it will be described as if it were an analog mechanization. The primary tracking function is indicated by the bold signal flow path. A narrowband tracking filter is synthesized by down converting with a voltage-controlled oscillator (VCO) so that the desired signal falls within a narrow bandpass filter. This is the first in a series of successive steps of bandwidth reduction. The signal-to-receiver noise improvement is the ratio of the IF to Doppler filter bandwidths. Frequently, however, the most important benefit is the attenuation of large unwanted signals such as ground clutter, transmitter leakage, or other interference. The center frequency of the bandpass filter is chosen so that the ability to discriminate between opening and closing velocities, that is, negative and positive Doppler frequencies, is retained.

The elements of phase-sensitive detector through VCO represent a phase-locked loop. In the quiescent state, the VCO frequency is $f_D + f_2$ and represents a smoothed estimate of the target Doppler plus a constant offset f_2. The mea-

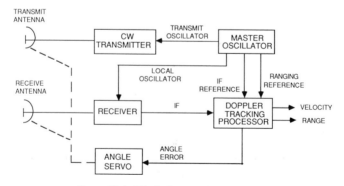

Figure 19-1. Block diagram of a CW radar.

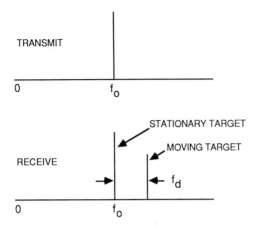

a. Infinite observation time.

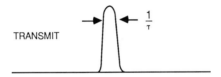

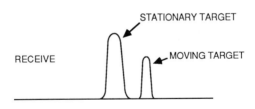

b. Finite observation time.

Figure 19-2. Spectra for CW radar.

surement accuracy is established by the bandpass filter and phase-sensitive detector bandwidths. The low-pass filter and integrator perform the filtering function. Figure 19-3 depicts a *single-integrator loop*, commonly called a *type 1
tracking loop*. This system tracks a constant-velocity target without bias but has
a fixed bias error when tracking a target with constant acceleration. *Double-
integrator* or *type 2 loops* are often used to track a constantly accelerating target.
The integrators also provide the prediction function. For example, if the signal

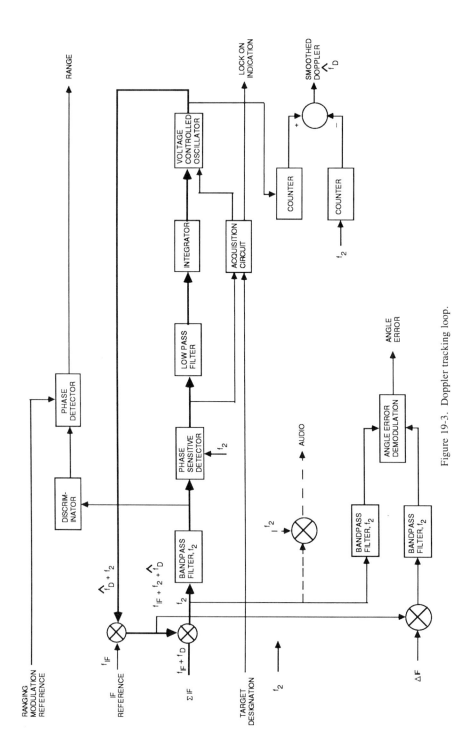

Figure 19-3. Doppler tracking loop.

is lost to a single-integrator loop, the loop predicts that the target velocity is constant. Similarly, for a double integrator, the loop holds the previous acceleration. For most CW radars, these simple tracking loops are adequate because the high data rate ensures small changes in measured velocity between samples. More sophisticated filtering and prediction methods will be discussed later.

Several other functions are often included with Doppler frequency tracking. The acquisition circuits sweep the VCO about the designated frequency until a target is detected above the threshold. Matched bandpass filters are used in each of the monopulse angle tracking channels.

Ranging.

After target tracking has been established, some systems add a ranging modulation to the transmitted signal. Sinusoidal FM is a common choice. The modulation frequency is chosen so that the range can be determined unambiguously by measuring the phase difference between the transmitted and received modulation waveforms. The target Doppler has spectral sidebands at integer multiples of the modulating frequency. The target Doppler may be translated to a convenient IF, and one of the sidebands can be detected for range measurements.

Audio Output.

Complex targets with moving parts or vibration modulate the target Doppler and produce spectral sidebands. These same mechanical movements can produce sound waves in air. If the target Doppler is translated to zero frequency, the resulting sidebands within the audio region can be input to the operator headset. Frequently, there is a close resemblance between the actual sound and that produced by the radar, and the operator is able to detect the type of target. Spectral analysis and pattern recognition processing can be used to perform the function automatically. Spectral sidebands are not always helpful and must be accounted for in the design. They can cause false lock-ons and other spurious effects.

19.3.1   Application To High-PRF Pulsed Doppler

Typical high-PRF systems operate at transmitter duty factors of 0.3 to 0.5 and with a PRF sufficiently high that the target Doppler is unambiguous. A single antenna is duplexed between transmit and receive. A bandpass filter is placed in the receiver IF to eliminate all frequencies except those in a region around the central spectral line, as shown in Figure 19-4. The filter output is essentially CW, and tracking is accomplished as in Figure 19-3.

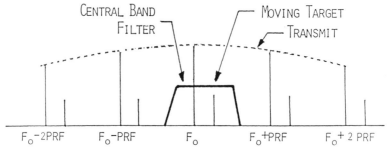

Figure 19-4. High-PRF spectra.

The radar is blind to targets at ranges such that the echo arrives when the receiver is gated off. This characteristic is called *eclipsing*. The time during which a target is eclipsed is inversely proportional to the closing rate. One solution is to switch sequentially between a set of slightly different PRFs chosen to ensure that the target is uneclipsed on at least one of the PRFs throughout the range of interest. Another method is to measure the signal strength in the Doppler tracker and initiate a PRF change when approaching eclipse.

19.3.2 Digital Implementations

The tracking implementation of Figure 19-3 can be implemented in digital form. Generally, the IF frequency representing zero Doppler is translated into zero frequency, eliminating one intermediate down conversion, as shown in Figure 19-5. The sequence of in-phase (I) and quadrature (Q) sample pairs preserves the ability to discern between opening and closing Doppler frequencies. The A/D conversion rate must be greater than or equal to the Nyquist criterion of one sample pair every $1/B$ seconds, where B is the IF bandwidth. The digital equivalents of the function of Figure 19-5 can be synthesized. For example, an integrator is an add and accumulate operation. Translating the spectrum of the time waveform by ω_s is achieved by rotating each of the complex sample pairs by sampled values of $(\cos \omega_s t + j \sin \omega_s t)$.

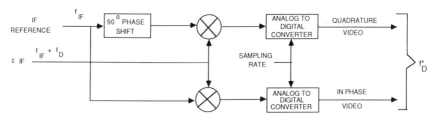

Figure 19-5. Translation to zero IF.

The flexibility of digital processing often permits the same hardware to be used in several radar modes, such as surveillance and track. For some applications, an FFT spectral analysis which synthesizes a contiguous bank of narrowband filters covering the entire Doppler region might be used for initial detection and for frequency measurement during tracking. A particular spectral line is selected for tracking. Somewhat different concepts of filtering and prediction are necessary and will be discussed later. The FFT simultaneously provides information on ranging and target-induced modulation sidebands. In practice, the optimum bandwidths necessary to accomplish tracking and sideband demodulation may be sufficiently different that separate FFTs with different sample sizes may be required.

19.4 RANGE-GATED PULSED DOPPLER RADARS

In a range-gated pulsed Doppler radar, the Doppler tracking loop and the range tracking loop are interrelated. The Doppler tracking loop operates upon video sampled at the range indicated by the range tracker. Range tracking can be accomplished by several methods, but an early/late gate approach will be used for illustration. Before the signals in the early and late gates are compared to derive the range tracking error, however, each is passed through a Doppler tracking filter to reject returns from stationary objects at the same range.

Figure 19-6 depicts the I video at the input to the A/D converter of Figure 19-5 overlaid for nine consecutive transmitted pulses. The round trip distance to the target changes by an RF wavelength during this interval. The Q channel response appears to be similar, except that its maximum occurs at the time when

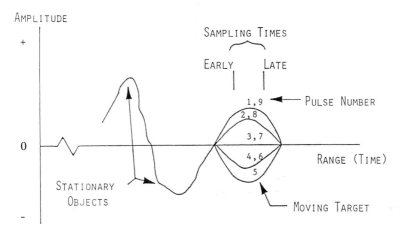

Figure 19-6. Pulsed Doppler radar time waveform.

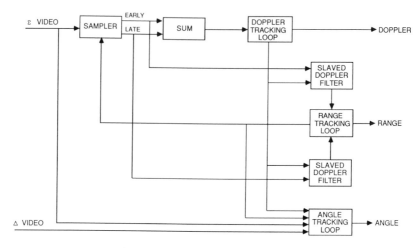

Figure 19-7. Range-gated Doppler Tracking.

the I channel response is zero. The amplitude is sampled at a fixed range at each PRF interval. This set of samples is frequently referred to as *range-gated video*. The range-gated video from each range of interest is filtered to select the desired signal. FFTs, clutter cancellers, or Doppler bandpass filters may be used.

Figure 19-7 is a simplified block diagram of range-gated Doppler tracking. The details of the circuits needed to initiate simultaneously tracking in range, Doppler, azimuth, and elevation have been omitted. Frequently, a different set of bandwidths and gains are used during acquisition until a stable track is achieved. For many military applications, the transition from search to track is a nontrivial problem. Pointing the antenna at an unfriendly target signals the intent to initiate tracking. The unfriendly target may attempt to prevent lock-on by maneuvers or electronic means. Doppler tracking may be performed on the early or late gate alone or on the sum. The sum provides some smoothing of range-timing changes.

19.4.1 Velocity Ambiguities

As discussed earlier, the choice of transmitter frequency and PRF often results in ambiguity of the measured Doppler frequency. In a range/Doppler tracking radar, unambiguous radial velocity is also available from the range tracker as dR/dt and is usually sufficient to resolve the ambiguity. The ambiguity can also be resolved by varying the PRF and observing the shift in measured Doppler.

19.4.2 Blind Speeds

One method of preventing the tracked target acceleration from resulting in a blind speed is to permit the Doppler tracking loop to adjust the PRF. This, of course, does not help when the true radial target velocity approaches zero.

19.4.3 Kalman and α-β Trackers

The benefits of Kalman filtering have been known for some time.[3] Implementation of the Kalman filter can impose heavy computational demands. The α-β tracker,[4] which is now recognized as a simplified subset of the Kalman filter, has seen wide application because of its simplicity. Also, a body of theory for suboptimal filters, each with simplifications appropriate for a certain class of applications, has been developed to reduce the computational load.[5,6] Two technological factors have created a growth in Kalman filter applications during the 1980s. First, low-cost, high-speed digital computing capability has made Kalman filters practical for more applications. Second, an increase in electronically scanned antennas has resulted in more multiple-target tracking systems. The application of Kalman filtering to velocity tracking will be described by first considering a system with only velocity measurements and employing an α-β tracker. Then the concept will be extended to Kalman filtering and finally to an application in which both range and velocity are measured independently.

The α-β equations when applied to velocity tracking become:
Smoothing

$$\hat{\dot{R}}_N = \hat{\dot{R}}_{PN} + \alpha(\dot{R}_N - \hat{\dot{R}}_{PN}) \tag{19-1}$$

$$\hat{\ddot{R}}_N = \hat{\ddot{R}}_{PN} + \beta \frac{(\dot{R}_N - \hat{\dot{R}}_{PN})}{T} \tag{19-2}$$

Prediction

$$\hat{\dot{R}}_{P(N+1)} = \hat{\dot{R}}_N + \hat{\ddot{R}}_N T \tag{19-3}$$

$$\hat{\ddot{R}}_{P(N+1)} = \hat{\ddot{R}}_N \tag{19-4}$$

where

$\hat{\dot{R}}_N$ = smoothed estimate of current velocity
$\hat{\ddot{R}}_N$ = smoothed estimate of current acceleration
$\dot{R}_N$ = measured velocity
T = time between samples

$\hat{R}_{P(N+1)}$ = predicted velocity T seconds later
α, β = predetermined smoothing constants
$\hat{R}_{PN}$ = predicted velocity at the time of measurement

Eqs. (19-1) and (19-2) are used to compute the smoothed estimates of the current velocity and acceleration. These values are computed immediately after the current measurement of velocity, $\dot{R}_N$, and are frequently part of the radar output data stream. The physical interpretation of Eq. (19-1) is straightforward. The quantity $(\dot{R}_N - \hat{R}_N)$ is the difference between the measured and predicted velocities. A portion of that difference determined by α is added to the prediction to form the smoothed estimate. This same difference when divided by T represents an acceleration difference that is weighted by β in an analogous manner. The performance of the tracker depends on the choice of α and β, but the choices are not independent. For $\alpha = \beta = 0$, the current estimate is ignored. For $\alpha = \beta = 1$, the current estimate is simply the current measurement, and no smoothing is provided.

The filter reduces the variance of measurement noise at steady state (constant acceleration). The variance reduction ratio (VRR) is given by:

$$\text{VRR} = \frac{2\alpha^2 + 2\beta + \alpha\beta}{\alpha(4 - 2\alpha - \beta)}.$$

Clearly $(4 - 2\alpha - \beta) > 0$ is required for stability. The selection of α and β is application dependent and usually represents a compromise between smoothing (variance reduction) and transient response. Benedict and Bordner's criterion,[4] when applied to velocity tracking, minimizes the total velocity output error (steady state variance plus transient error due to a step change in acceleration) through the choice of $\beta = \alpha^2/(2 - \alpha)$.

The computational sequence for the α-β tracker is to (1) use the current measurement to update smoothed velocity and acceleration [Eqs. (19-1) and (19-2)] and (2) predict the velocity and acceleration at a time T seconds later [Eqs. (19-3) and (19-4)]. The value of T may range from a few milliseconds for a continuous tracking or agile beam radar to several seconds for a mechanically scanned antenna radar.

The α-β tracker described above is capable of tracking a target having a constant acceleration with a mean error of zero. Eq. (19-4), although it appears trivial, clearly states that the system model predicts constant acceleration. The constants α and β are usually predetermined, although multiple sets and some criteria such as range, operator selection, and so on to select the desired set could be used.

A heuristic simplified view of the Kalman filter is one in which (1) the optimum weighting coefficients (somewhat analogous to α and β) are dynamically computed each update cycle and (2) a more precise model of the target dynamics can be used. The benefits of this additional complexity include an improvement in tracking accuracy, a running measure of the accuracy with which target coordinates are being estimated, and a method for handling measurements of variable accuracy, nonuniform sample rate, or missing samples.

Kalman filtering is usually described by matrix notation, used below to provide a transition between the α-β tracker and more sophisticated models.

Prediction Eqs. (19-3) and (19-4) may be rewritten as

$$\begin{bmatrix} \hat{R}_{P(N+1)} \\ \hat{\dot{R}}_{P(N+1)} \end{bmatrix} = \begin{bmatrix} 1 & T \\ 0 & 1 \end{bmatrix} \begin{bmatrix} \hat{R}_N \\ \hat{\dot{R}}_N \end{bmatrix}$$

or, more compactly, as

$$\hat{R}_{P(N+1)} = \Phi \, \hat{R}_N \tag{19-5}$$

where

$$\hat{R}_N = \begin{bmatrix} \hat{R}_N \\ \hat{\dot{R}}_N \end{bmatrix} = \text{current (smoothed) state vector}$$

$$\hat{R}_{P(N+1)} = \begin{bmatrix} \hat{R}_{P(N+1)} \\ \hat{\dot{R}}_{P(N+1)} \end{bmatrix} = \text{predicted state vector}$$

$$\Phi = \begin{bmatrix} 1 & T \\ 0 & 1 \end{bmatrix} = \text{transition matrix}$$

The transition matrix contains the assumed system model, that is, the calculations used to predict the state at the next sampling time. Selection of a system model is a compromise between computation complexity and tracking accuracy. Approximations in the model and target maneuvers not in accordance with the model introduce tracking errors. In a more general Kalman formulation, an additional term, a modeling noise vector, is added to the right side of Eq. (19-5).

Smoothing Eqs. (19-1) and (19-2) may be rewritten as

$$
\begin{bmatrix} \hat{R}_N \\ \hat{\dot{R}}_N \end{bmatrix} = \begin{bmatrix} \hat{R}_{PN} \\ \ddot{R}_{PN} \end{bmatrix} + \begin{bmatrix} \alpha & 0 \\ \dfrac{\beta}{T} & 0 \end{bmatrix} \left(\begin{bmatrix} 1 & 0 \\ 0 & 0 \end{bmatrix} \begin{bmatrix} \dot{R}_N \\ \ddot{R}_N \end{bmatrix} - \begin{bmatrix} 1 & 0 \\ 0 & 0 \end{bmatrix} \begin{bmatrix} \hat{R}_{PN} \\ \hat{R}_{PN} \end{bmatrix} \right)
$$

or, more compactly, as

$$
\hat{R}_N = \hat{R}_{PN} + B_N (M \, \dot{R}_N - M \, \hat{R}_{PN}) \tag{19-6}
$$

where

$$
B_N = \begin{bmatrix} \alpha & 0 \\ \dfrac{\beta}{T} & 0 \end{bmatrix} = \text{weighting coefficients matrix}
$$

$$
\dot{R}_N = \begin{bmatrix} \dot{R}_N \\ \ddot{R}_N \end{bmatrix} = \text{measurement vector}
$$

$$
M = \begin{bmatrix} 1 & 0 \\ 0 & 0 \end{bmatrix} = \text{measurement matrix}
$$

and the other terms are as previously defined.

For notational completeness, the measurement vector contains all of the system states. The product $Y_N = M \, \dot{R}_N$ specifies which quantities are actually measured at each sample time (only $\dot{R}$ in this example). The Kalman formulation adds on additional term, the observation error vector N_N, to the formulation of Y_N.

The power (and the computational complexity) of the Kalman filter results from the dynamic computation of the weighting coefficients matrix, B_N. The filter computes the weights such that if the error in the measurement, Y, increases relative to the error of the smoothed estimate, then the coefficients will be reduced to weight the measurement less heavily. The weighting coefficient matrix is computed using the matrix equation:

$$
B_N = C_{PN} \, M^T (E_N + M \, C_{PN} \, M^T)^{-1} \tag{19-7}
$$

where

$$C_{PN} = \begin{bmatrix} \text{mean square velocity error} & \begin{array}{l}\text{cross-correlation between} \\ \text{velocity and acceleration} \\ \text{errors}\end{array} \\ \begin{array}{l}\text{cross-correlation between} \\ \text{velocity and acceleration} \\ \text{errors}\end{array} & \text{mean square acceleration error} \end{bmatrix}$$

= predicted covariance matrix

superscript T = transposed matrix

superscript -1 = inverse matrix

E_N = covariance of measurement error

The Kalman filter has an additional prediction and an additional correction or smoothing equation. These provide a running measure of tracking accuracy. Correction

$$C_{N^+} = [I - B_N M] \, C_{PN} \tag{19-8}$$

Prediction

$$C_{P(N+1)} = \Phi \, C_{N+1} \Phi^T \tag{19-9}$$

The computation sequence is summarized in Figure 19-8.

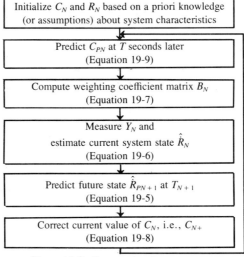

Figure 19-8. Computational sequence.

Kalman filters permit, and are most often applied to, higher-order system models. The previous discussion used a simple model. Only one quantity, velocity, was measured directly, and the model predicts constant acceleration. One example of a higher-order model is a range-gated pulsed Doppler radar which measures both range and Doppler.[7] Another example is an airborne radar in which measurement of own-ship motion is used to improve the prediction and tracking accuracy discussed in a subsequent paragraph.

19.5 APPLICATION TO AIRBORNE AND MOBILE RADAR

When the radar is in motion, objects which are stationary with respect to the ground are now moving with respect to the radar. Figure 19-9(a) depicts the resulting spectrum for a CW radar. For a high-PRF radar, this spectral shape is replicated around each spectral line at $f_0 \pm n\text{PRF}$.

In a range-gated radar, the range gate containing the target to be tracked will also contain mainlobe clutter if the target is at low altitude. The range gating eliminates the returns from short ranges through the antenna sidelobes. The mainlobe clutter peaks appear at multiples of the PRF. However, the clutter Doppler will generally be ambiguous and cannot be visualized as merely being shifted from the nearest PRF line.

(A) CW OR CENTRAL LINE OF HIGH PRF RADAR

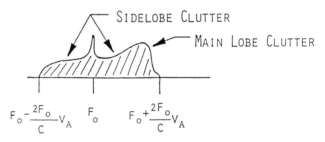

(B) RANGE GATED PULSED DOPPLER MAIN LOBE CLUTTER

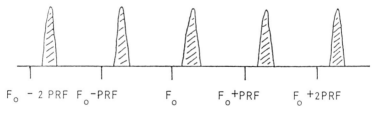

Figure 19-9. Received clutter spectra when radar is moving.

The nominal center of the mainlobe clutter video spectrum is given by

$$f_{MLC} = \left[\frac{2V_A}{\lambda} \cos \theta \right] \text{ modulo PRF}$$

where

V_A = aircraft velocity
λ = transmitted wavelength
θ = angle between the velocity vector and the antenna boresight

Airborne velocity tracking loops often include a loop to track mainlobe clutter and translate the spectrum so that mainlobe clutter falls at zero frequency. The loop used to track the desired target should contain provisions to maintain the track when the target passes through the mainlobe clutter. This occurs not only as it does in all Doppler radars when the target radial velocity passes through zero, but also at other times due to the ambiguities in both clutter and target Doppler.

The spectral width of the mainlobe clutter is given by

$$\Delta f_{MLC} = \left[\frac{2V_A}{\lambda} \sin \theta \, \Delta \theta \right]$$

where $\Delta\theta$ is the antenna beamwidth. For a mobile radar, the antenna size is frequently restricted. The clutter spreading can occupy the total spectral region between PRF lines and result in no clutter-free region for detection or tracking. The antenna beamwidth is proportional to λ/L, where L is the physical antenna length. Thus, the clutter spectral width is independent of the transmitter frequency. Therefore, the only options for providing a clutter-free region are increasing the antenna size, restricting the antenna angle, slowing the aircraft, or increasing the PRF. These trade-offs have resulted in the incorporation of medium-PRF modes in some airborne radars. The medium-PRF mode is characterized by ambiguities in both range and Doppler.

19.5.1 Compensation For Own Vehicle Motion

Consider the geometry of Figure 19-10. Even when the radar and the target being tracked are moving with constant velocity vectors, the relative geometry between the two produces a nonconstant velocity. In general, the acceleration, rate of change of acceleration, and so on are also not constant. In short range encounters, the accelerations can be several g's (g is the acceleration due to gravity). The simplest of the prediction methods discussed earlier, the single

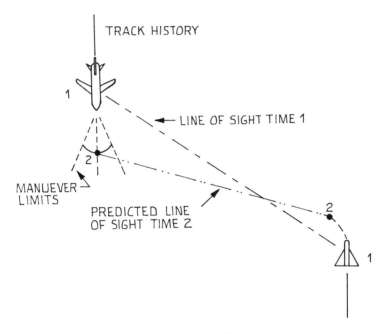

Figure 19-10. Airborne tracking geometry.

integrator, predicts zero acceleration between updates. It depends upon frequent updates and a Doppler measurement bandwidth wide enough to accommodate the shift in Doppler between updates to maintain the track. The more sophisticated tracking methods such as Kalman filtration can model the engagement geometry and make a prediction which includes geometric effects. The models generally chosen predict that both the target and the radar continue to follow the trajectory established by the measurement history.

If either the target or the radar initiates a maneuver, this constitutes a transient input to the tracking loop. If the radar and target are at positions 1 at one update time, when at the next update time they could be anywhere on arc 2 representing the maneuver limits of the aircraft. The target maneuvers must be measured and tracked by the radar. However, it is not necessary to widen the Doppler measurement bandwidth to accommodate the part of the Doppler frequency shift that is due to the radar motion. An inertial measurement of own-vehicle acceleration in the direction of the target can be used to aid the prediction function. The measured acceleration can be used as an additional input to the integrator that controls the VCO in the tracking logs of Figure 19-3 or as an additional direct measurement in a Kalman filter tracking mechanization. An accelerometer may be mounted on the antenna to provide the input directly in the proper coordinate system, or input may be supplied by the aircraft navigation system.

19.6 REFERENCES

1. M. I. Skolnik, *Radar Handbook*, McGraw Hill, Book Co., New York, 1970, chap. 4.

2. D. K. Barton, *Radar Systems Analysis*, chap. 12, Prentice-Hall, Englewood Cliffs, N.J., 1964.

3. R. E. Kalman, "New Results in Linear Filtering and Prediction Theory," *Transactions of the ASME*, vol. 83D, March 1961, pp. 95–108.

4. R. T. Benedict and G. W. Bordner, "Synthesis of an Optimal Set of Radar Track-While-Scan Smoothing Equations," *IRE Transactions on Automated Control*, vol. AC-7, July 1962, pp. 27–32.

5. F. A. Faruqi and R. C. Davis, "Kalman Filter Design for Target Tracking," *IEEE Transactions on Aerospace and Electronic Systems*, vol. AES-16, no. 4, July 1980, pp. 500–508, including "Corrections to . . ." vol. AES-16, no. 5, pp. 740.

6. R. J. Fitzgerald, "Simple Tracking Filters: Closed-Form Solution" *IEEE Transactions on Aerospace and Electronic Systems*, vol. AES-17, no. 6, November 1981, pp. 781–785.

7. R. J. Fitzgerald, "Simple Tracking Filters: Position and Velocity Measurement," *IEEE Transactions on Aerospace and Electronic Systems*, vol. AES-18, no. 5, September 1982, pp. 531–537.

PART 6
TARGET DISCRIMINATION AND RECOGNITION

- Polarimetric Fundamentals and Techniques—Chapter 20
- Target Recognition Considerations—Chapter 21

Radar researchers have long believed that through detection and utilization of all the target information present in an electromagnetic signal scattered from an unknown target, significant information about the nature (i.e., point scatterer versus area scatterer) and identity (i.e., tank versus truck) of that target could be derived. However, this has proved a difficult and elusive goal, especially for targets not displaying Doppler signatures, stationary targets. On the other hand, the detection and identification of stationary targets could be extremely important from an operational military standpoint. Accordingly, in recent years, a considerable amount of attention has been focussed on this problem by the radar research and development community. The next two chapters represent one of the first attempts at summarizing some of the basic principles of stationary target discrimination and recognition.

In Chapter 20, polarization fundamentals and techniques to achieve stationary target discrimination and recognition based on full utilization of the information contained in the polarization scattering matrix (PSM) are presented. An introduction to polarization scattering matrix theory and a brief tutorial on elementary polarization theory precedes consideration of techniques for employing the PSM in discrimination and recognition of stationary targets.

A more general treatment of the problem of target recognition based on radar-derived signatures is presented in Chapter 21. The overall goal of this chapter is threefold: (1) to introduce some of the basic issues associated with the problem of target recognition, (2) to discuss and compare various target recognition techniques, and (3) to provide examples of applications of target recognition techniques, thereby pointing out the limitations (and unrealized potential) associated with this technology.

20
POLARIMETRIC FUNDAMENTALS AND TECHNIQUES

William A. Holm

20.1 INTRODUCTION

The requirement of a radar system (1) to detect a target of interest, (2) to discriminate between that target and target-like clutter, and (3) to recognize that the target belongs to a target class of interest to the radar operator is technically severe, especially if the target is stationary. Target detection is accomplished by using well-known noise and clutter thresholding techniques such as constant-false-alarm-rate (CFAR) detection. In relatively benign clutter scenarios, high (>0.9) probabilities of detection with low (<0.00001) probabilities of false alarms are readily obtainable using these techniques. After detecting a target, the radar attempts to discriminate between *potential* targets of interest and strong, target-like clutter returns that were passed as targets in the detection phase. If the target is moving, discrimination can be accomplished by using one or more of the well-known spectral (Doppler) discrimination techniques, such as MTI techniques. These techniques enable target–clutter discrimination for clutter-to-target signal ratios of 30 to 40 dB (and higher). The discrimination of stationary targets from clutter is a much more difficult task.

In the recognition phase, the radar attempts to determine whether a target belongs to any of the target classes of interest. (These three phases of target identification—detection, discrimination, and recognition—are described in more detail in Chapter 21.) If the target is moving, then spectral processing techniques can be employed in an attempt at recognition. The recognition of stationary targets, however, is a more difficult task. Stationary target discrimination and recognition is addressed in this and the next chapter.

All of the backscatter signature characteristics of the target and clutter must be exploited to achieve stationary target discrimination or recognition. The more detailed these signatures are, the more likely it is that some signature characteristics will allow discrimination/recognition. In recent years, radar designers have expended considerable effort in an attempt to design transmit waveforms and corresponding matched-filter receivers to achieve target discrimination/recognition. Every waveform parameter (amplitude, frequency, phase) and the vector nature of EM waves, polarization, has been exercised singly and in combination (e.g., polarization techniques combined with frequency agility) to attain this goal. In this chapter, polarization fundamentals and techniques to achieve stationary target discrimination and recognition are discussed. This discussion is preceded by a brief review of elementary polarization theory followed by an introduction to polarization scattering matrix theory.

20.2 ELEMENTARY POLARIZATION THEORY

EM radiation is a vector quantity, that is, it possesses polarization. Consider a monochromatic plane EM wave propagating along the $+z$-axis of a right-handed coordinate system. The electric field, $\vec{E}$, of this wave may be represented as

$$\vec{E} = E_{0x} \cos(\omega t - kz + \alpha_x)\hat{x} + E_{0y} \cos(\omega t - kz + \alpha_y)\hat{y} \qquad (20\text{-}1)$$

where

$$\hat{x}, \hat{y} = \text{unit vectors along the } +x\text{-axis (horizontal) and } +y\text{-axis (vertical), respectively,}$$
$$\omega = 2\pi f \, (f \text{ is frequency}),$$
$$k = 2\pi/\lambda \, (\lambda \text{ is wavelength}),$$
$$\alpha_x \text{ and } \alpha_y = \text{absolute phases,}$$
$$t = \text{time.}$$

Eq. (20-1) can be written as

$$\vec{E} = \text{Re}[e^{i(\theta + \alpha_x)}(E_{0x}\hat{x} + E_{0y}e^{i\delta}\hat{y})] \qquad (20\text{-}2)$$

where

$$i = \sqrt{-1},$$
$$\theta = \omega t - kz,$$
$$\text{Re} = \text{``take the real part of'',}$$
$$\delta = \alpha_y - \alpha_x \text{ is the relative phase difference between the two components.}$$

Dropping the "Re" and time harmonic factor, $e^{i\theta}$, and remembering that they are implicitly understood, $\vec{E}$ can be expressed in matrix notation as follows:

$$\vec{E} = e^{i\alpha_x} \begin{bmatrix} E_{0x} \\ E_{0y}e^{i\delta} \end{bmatrix} \qquad (20\text{-}3)$$

where

$$\hat{x} = \begin{pmatrix} 1 \\ 0 \end{pmatrix} \quad \text{and} \quad \hat{y} = \begin{pmatrix} 0 \\ 1 \end{pmatrix} \tag{20-4}$$

The amplitudes E_{0x}, E_{0y} and the relative phase δ determine the polarization of the wave.

In general, the tip of the $\vec{E}$-vector traces out an ellipse in the x-y plane (fixed z) as time evolves (see Figure 20-1). The major axis of the ellipse is rotated by an angle ϕ with respect to the x-axis. For a nonrotated ellipse, $\delta = \pi/2$. Thus, Eq. (20-3) may be written as

$$\vec{E} = e^{i\alpha} \begin{bmatrix} \cos\phi & -\sin\phi \\ \sin\phi & \cos\phi \end{bmatrix} \begin{bmatrix} E_{0x'} \\ iE_{0y'} \end{bmatrix} \tag{20-5}$$

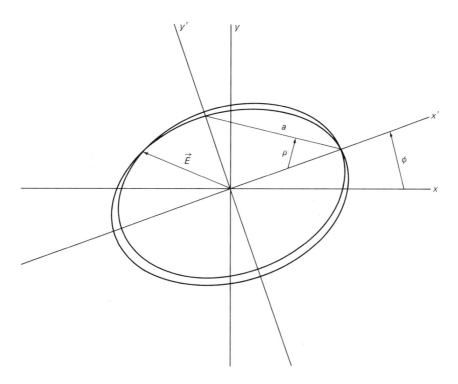

Figure 20-1. Elliptical polarization.

where $E_{0x'}$ $E_{0y'}$ are components with respect to the rotated primed coordinate system and $-90° \leq \phi \leq 90°$. From Figure 20-1, $E_{0x'}$ and $E_{0y'}$ can be expressed in terms of a and ρ as

$$E_{0x'} = a \cos\rho \qquad (20\text{-}6)$$
$$E_{0y'} = a \sin\rho$$

where $-45° \leq \rho \leq 45°$. Thus, Eq. (20-5) becomes (dropping the absolute phase factor)

$$\vec{E} = a \begin{bmatrix} \cos\phi & -\sin\phi \\ \sin\phi & \cos\phi \end{bmatrix} \begin{bmatrix} \cos\rho \\ i \sin\rho \end{bmatrix} \qquad (20\text{-}7)$$

which can be rewritten as

$$\vec{E} = a e^{\phi J} e^{\rho K} \begin{pmatrix} 1 \\ 0 \end{pmatrix} \qquad (20\text{-}8)$$

where

$$J = \begin{bmatrix} 0 & -1 \\ 1 & 0 \end{bmatrix} \quad \text{and} \quad K = \begin{bmatrix} 0 & i \\ i & 0 \end{bmatrix}. \qquad (20\text{-}9)$$

Thus, any arbitrary polarization can be written as two rotation operators, $e^{\phi J}$ and $e^{\rho K}$, operating on the horizontal polarization. The first rotation $e^{\rho K}$ is a complex rotation which gives the proper ellipticity to the polarization. The second rotation, $e^{\phi J}$, is a real rotation which gives the proper orientation to the elliptical polarization (proper orientation of the major axis with respect to the horizontal). Any polarization state can thus be represented as a point on a unit sphere, called the *Poincaré sphere*,[1] generated by these two successive rotations from the point representing horizontal polarization (see Figure 20-2).

As mentioned above, ρ can range from $-45°$ to $45°$. The sign of ρ gives the "sense" of the polarization, that is, right-hand or left-hand elliptically polarized ($\rho = \pm 45°$ is circularly polarized; $\rho = 0°$ is linearly polarized). As might be expected, there is no consensus on what sign convention to adopt. The optics convention[2] is as follows: "Right-hand elliptical polarization is that polarization in which the $\vec{E}$-field rotates in a *clockwise* direction at a fixed point in space as seen by an observer *toward* whom the wave is propagating." Adoption of this convention makes positive ρ correspond to right-hand elliptical polarization.

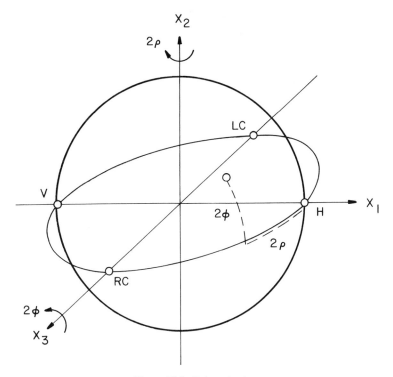

Figure 20-2. Poincaré sphere.

The IEEE has adopted a sign convention that is opposite to that of the optics convention.[3] Adoption of the IEEE convention makes positive ρ correspond to left-hand elliptical polarization.

A final word of caution is in order. Specification of the sign convention adopted does not completely specify whether an arbitrary polarization given by Eqs. (20-3), (20-5), (20-7), or (20-8) is right-hand or left-hand elliptically polarized. This is due to the fact that the sign of the phase, θ, in the time harmonic factor is arbitrary, that is, some choose to write the time harmonic factor as $e^{-i\theta}$. Reversal of the sign of θ changes the polarization sense. This is a problem only when an $\vec{E}$-field is written in the form given by Eqs. (20-3), (20-5), (20-7), or (20-8), since the time harmonic factor, $e^{i\theta}$, is not explicitly given.

The two most common basis sets used in polarization theory are the linear (H, V) basis and the circular (R, L) basis. Circular polarization is given by $\rho = \pm 45°$; that is, from Eq. (20-7),

$$\vec{E}_{\pm} = e^{\mp i\phi}\frac{a}{\sqrt{2}}\begin{pmatrix} 1 \\ \pm i \end{pmatrix}. \tag{20-10}$$

Table 20-1. Right-Hand and Left-Hand Circular Polarization Conventions.

$\vec{E}_+ = \dfrac{1}{\sqrt{2}}\begin{bmatrix}1\\i\end{bmatrix}$		Convention	
		Optics	IEEE
Time–Harmonic Factor	$e^{i\theta}$	RHC	LHC
	$e^{-i\theta}$	LHC	RHC

$\vec{E}_- = \dfrac{1}{\sqrt{2}}\begin{bmatrix}1\\-i\end{bmatrix}$		Convention	
		Optics	IEEE
Time–Harmonic Factor	$e^{i\theta}$	LHC	RHC
	$e^{-i\theta}$	RHC	LHC

Notes:
$\theta = \omega t - kz$.
LHC = right-hand circular.
RHC = left-hand circular.

To construct a circular basis set from $\vec{E}_\pm$, one must (1) normalize $\vec{E}_\pm$, that is, $|\vec{E}_+ \cdot \vec{E}_+^*| = |\vec{E}_- \cdot \vec{E}_-^*| = 1$ (where * denotes complex conjugation), which makes $a = 1$; (2) choose a "base direction," that is, choose a value for ϕ, and (3) choose a convention for right-hand and left-hand circular polarization. Normally, the $+x$ (horizontal) axis is chosen as the base direction, that is, $\phi = 0$. Therefore,

$$\vec{E}_\pm = \frac{1}{\sqrt{2}}\begin{pmatrix}1\\\pm i\end{pmatrix}. \tag{20-11}$$

Finally, whether $\vec{E}_+$ is right-hand or left-hand circular polarization depends on the sign convention chosen (optics or IEEE) and whether the implicit time harmonic factor is $e^{i\theta}$ or $e^{-i\theta}$, where $\theta = \omega t - kz$. These choices are summarized in Table 20-1.

20.3 POLARIZATION SCATTERING MATRIX THEORY

There are two general approaches to the theoretical development of radar scattering theory. One approach is that of Sinclair and Jones and is based on the description of transmitted and received backscattered EM energy in terms of complex voltage measurements.[4] The other approach is that of Mueller and Stokes and is based on the description of transmitted and received backscattered EM energy in terms of real power measurements.[5]

In the case of time–harmonic targets for which the return signal is fully polarized, the Jones calculus and the Mueller calculus lead to the same final result. The scattering operators of both approaches may be expressed in terms of a common set of five physically significant parameters. The Mueller calculus, however, naturally accommodates time-varying targets for which the return signal is partially polarized. The time-averaged scattering operator in the Mueller calculus for partially polarized signals depends on nine parameters.

Both the Jones and Mueller calculi are used here to develop the theory of radar polarization scattering. The theory presented and the notation used in the remainder of this section closely parallel the treatment on radar polarization theory given by J. R. Huynen.[6,7]

When a target is illuminated by a plane wave transmitted from an antenna with a transmit polarization $\vec{a}$, the backscattered complex voltage, V, collected at the terminals of the receiver antenna with receive polarization, $\vec{b}$, may be expressed as

$$V = S\vec{a} \cdot \vec{b} \qquad (20\text{-}12)$$

where S is the 2×2 complex scattering operator, or matrix, that characterizes the target's scattering properties. (The scalar product of two complex vectors

$$\vec{r} = \begin{bmatrix} r_1 \\ r_2 \end{bmatrix} \text{ and } \vec{s} = \begin{bmatrix} s_1 \\ s_2 \end{bmatrix}$$

is defined here as $\vec{r} \cdot \vec{s} = r_1 s_1 + r_2 s_2$).

The units of the scattering matrix are determined when the units of $\vec{a}$ and $\vec{b}$ in Eq. (20-12) are defined. The units of S vary from user to user, and various systems of units have been employed. One of the more common systems employed is the following: let

$$\vec{a} = a_0 \hat{a}$$
$$\vec{b} = b_0 \hat{b} \qquad (20\text{-}13)$$

where $\hat{a}$ and $\hat{b}$ are the normalized, unitless polarization vectors that represent the transmit and receive antenna polarization, respectively, and a_0 and b_0 are scalars that are defined such that the scattering matrix can now be defined by the equation

$$\sigma_{ab} = |S\hat{a} \cdot \hat{b}|^2 \tag{20-14}$$

where σ_{ab} is the RCS that appears in the radar range equation and is measured in square meters. (The subscripts a and b appear on the symbol for RCS, σ, as a remainder that the RCS does indeed depend on the polarizations of the transmit and receive antennas.) The product $a_0 b_0$ is now measured in volts/meter and includes all radar–target range dependencies. The scattering matrix is measured in meters.

The vectors $\hat{a}$ and $\hat{b}$ and the scattering matrix S are expressable in terms of a two-dimensional complex basis, usually linear:

$$\hat{x} = \begin{pmatrix} 1 \\ 0 \end{pmatrix}; \quad \hat{y} = \begin{pmatrix} 0 \\ 1 \end{pmatrix} \tag{20-15}$$

or the circular basis given by Eq. (20-11). If a linear basis is used, then S is given by

$$S = \begin{bmatrix} S_{HH} & S_{VH} \\ S_{HV} & S_{VV} \end{bmatrix} \tag{20-16}$$

where, for example, S_{HV} is the complex target backscatter amplitude received with a vertically polarized antenna when a horizontally polarized wave was transmitted. Only the case where the scattering matrix is symmetric, i.e., $S_{VH} = S_{HV}$, will be considered further. This case is applicable to the monostatic radar problem in the absence of magnetic, e.g., Faraday rotation, effects.

An eigenvalue analysis of the scattering matrix gives physical insight into the scattering process. The characteristic eigenvalue problem for the scattering matrix for the monostatic radar case is given by

$$S\vec{a} = s\vec{a}* \tag{20-17}$$

where * denotes complex conjugation. [The complex conjugate appears in Eq. (20-17) because the relevant eigenvalue problem here is to find the transmit polarization that will be proportional to the optimum receive polarization of the transmitting antenna when scattered from the target.] There are two solutions (eigenvectors) to Eq. (20-17), $\vec{a}_1$ and $\vec{a}_2$, with corresponding eigenvalues s_1 and s_2. The orthonormal eigenvector solutions to Eq. (20-17) can be written in the polarization vector notation given by Eq. (20-8) (with $a = 1$, $\phi = \psi$, and $\rho = \tau$) as

$$\vec{a}_1 = e^{\psi J} e^{\tau K} \hat{x}$$
$$\vec{a}_2 = e^{(\psi + \pi/2)J} e^{-\tau K} \hat{x}.$$

(20-18)

That is, the arbitrary polarization vector, $\vec{E}$, given by Eq. (20-8) (with $a = 1$) becomes the eigenvector solution, $\vec{a}_1$, to Eq. (20-17) when the parameters ϕ and ρ take on the specific values ψ and τ, respectively. In addition, if ψ and ρ in $\vec{a}_1$ are then replaced by $\psi + \pi/2$ and $-\tau$, respectively, then the orthogonal eigenvector, $\vec{a}_2$, results.

As indicated above, the parameter τ determines the ellipticity of the polarization eigenvector $\vec{a}_1$ ($0°$ for linear, $\pm45°$ for circular) and has the range $-45° \leq \tau \leq 45°$. It is also a measure of target symmetry with respect to right and left circular polarization ($0°$ for symmetric; $\pm45°$ for totally nonsymmetric). The parameter ψ determines the angle that the major axis of the elliptical polarization eigenvector $\vec{a}_1$ makes with respect to $\hat{x}$ and has the range $-90° \leq \psi \leq 90°$. It is also a measure of the orientation of the target.

The associated eigenvalues can be expressed as

$$s_1 = m e^{i2\nu}$$
$$s_2 = m \tan^2\gamma e^{-i2\nu}$$

(20-19)

where the parameter ν is related to the number of bounces of the reflected signal and has the range $-45° \leq \nu \leq 45°$. The parameter γ is the target polarizability angle and has the range $0° \leq \gamma \leq 45°$. It is a measure of the target's ability to polarize incident unpolarized radiation ($0°$ for full polarizability; $45°$ for no polarizability). Finally, the parameter m provides an overall measure of target size or RCS. It is the amplitude of the maximum return from the target, that is, the amplitude of the optimum polarization that maximizes the RCS of the target.

The relative scattering matrix (equivalent to the scattering matrix to within a phase factor) can thus be expressed in terms of the five physically relevant parameters (m, ψ, τ, ν, γ) as

$$S = U* \begin{bmatrix} s_1 & 0 \\ 0 & s_2 \end{bmatrix} U^+$$

(20-20)

where + denotes the Hermitian adjoint, that is, the complex conjugate and

Table 20-2. Five Physically Relevant Parameters of the Scattering Matrix S.

Parameter	Physical Significance	Range
m^2	Maximum RCS of target (obtained when optimum polarizations are used)	$(0 < m < \infty)$
ψ	Orientation angle	$(-90° \leq \psi \leq 90°)$
τ	Symmetry angle	$(-45° \leq \tau \leq 45°)$
ν	Odd/even bounce angle	$(-45° \leq \nu \leq 45°)$
γ	Polarizability angle	$(0° \leq \gamma \leq 45°)$

transpose, and U is the unitary matrix given by

$$U = [\vec{a}_1, \vec{a}_2]. \tag{20-21}$$

A summary of the five physically relevant parameters, their physical significance, and their ranges are given in Table 20-2. Shown in Figures 20-3 through 20-6 are some simple targets and their corresponding values for ψ, τ, ν, and γ. Finally, the scattering matrices (in the H, V basis) and their parameter values for these simple targets are shown in Table 20-3.

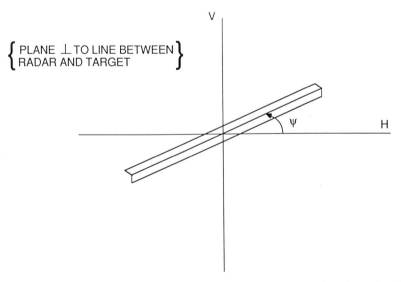

Figure 20-3. The parameter ψ for a simple target. ψ is a measure of target orientation angle ($0°$, horizontal; $\pm 90°$, vertical).

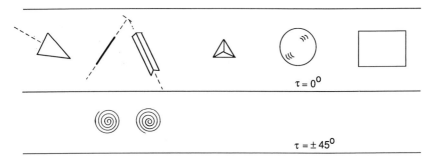

Figure 20-4. The parameter of τ for some simple targets. τ is a measure of target symmetry ($0°$, symmetric; $\pm 45°$, nonsymmetric).

The four angular parameters (ψ, τ, ν, γ) also specify two polarization states, called *null polarizations*, of the scattering matrix, S, such that when S operates on these null polarizations the orthogonal polarizations result. Therefore, the measured backscattered RCS given by Eq. (20-14) for a target represented by S is zero (a null) when the transmitted and received antenna polarizations are one of these null polarizations. The null polarizations and, thus, the normalized ($m = 1$) relative scattering matrix can also be represented geometrically (by

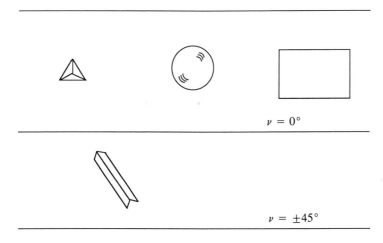

Figure 20-5. The parameter ν for some simple targets. ν is a measure of target even/odd bounce ($0°$ odd bounce; $\pm 45°$, even bounce).

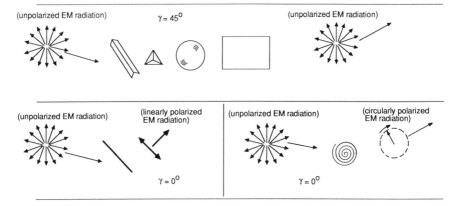

Figure 20-6. The parameter γ for some single targets. γ is a measure of target target polarizability (0°, full polarizability; 45°, no polarizability).

two points) on the Poincaré sphere. To see how the null polarizations are obtained from ψ, τ, ν, and γ and how they are represented on the Poincaré sphere, first consider the scattering matrix as given by Eq. (20-20). The scattering matrix can also be written as

$$S = U^*(\psi, \tau, \nu) \begin{bmatrix} m & 0 \\ 0 & m \tan^2\gamma \end{bmatrix} U^+(\psi, \tau, \nu)$$

where now

$$U(\psi, \tau, \nu) = e^{\psi J} e^{\tau k} e^{\nu L} \qquad (20\text{-}22)$$

Table 20-3. Scattering Matrices and Parameters for Some Simple Targets.

Target	Scattering Matrix	ψ	τ	ν	γ
Flatplate, Trihedral, Sphere	$\begin{bmatrix} 1 & 0 \\ 0 & 1 \end{bmatrix}$	Arbitrary	0°	0°	45°
Diplane at $\psi = \psi_a$	$\begin{bmatrix} \cos 2\psi_a & \sin 2\,\psi_a \\ \sin 2\psi_a & -\cos 2\,\psi_a \end{bmatrix}$	ψ_a	0°	±45°	45°
Helix	$\dfrac{1}{2}\begin{bmatrix} 1 & \pm i \\ \pm i & -1 \end{bmatrix}$	Arbitrary	±45°	Arbitrary	0°

and

$$L = \begin{bmatrix} -i & 0 \\ 0 & i \end{bmatrix}.$$

(20-23)

Thus, the scattering matrix is obtained from the matrix

$$\begin{bmatrix} m & 0 \\ 0 & m \tan^2\gamma \end{bmatrix}$$

by three successive unitary transformations (rotations): first a ν rotation, then a τ rotation, and finally a ψ rotation.

Now consider two line segments (as shown in Figure 20-7) drawn in the x_1-x_3 plane from the center of the Poincaré sphere to the surface such that the x_1-axis bisects the angle between them. Let the angle between the two segments be equal to 4γ, and let the points where they intersect the sphere be labeled N_1 and N_2. Consider a third line segment drawn from the center of the sphere to the point on the sphere that represents horizontal polarization. Let all three line segments be rigidly connected, producing a "fork." Now let the three sequential rotations (ν, τ, ψ) mentioned above to obtain S be three rotations about the

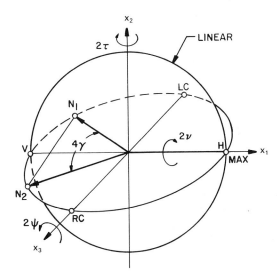

Figure 20-7. Polarization fork.

x_1, x_2, and x_3 axes by the amounts 2ν, 2τ, and 2ψ, respectively. These rotations will rotate the fork inside the Poincaré sphere. After the fork has undergone these three rotations, the intersection of the handle of the fork with the sphere represents the eigenvector $\vec{a}_1$ of S. The polarizations $\vec{n}_1$ and $\vec{n}_2$ associated with the points N_1 and N_2 are the null polarizations of S. A more complete theory of null polarization is presented on pages 82–86 of reference 6. Null polarization techniques have been shown to be useful in the suppression of unwanted clutter returns, especially from the sea.[8]

The Mueller/Stokes approach to scattering matrix theory is now discussed. The two-dimensional, complex polarization vector $\vec{E}$, which represents an EM wave in the Jones approach to scattering matrix theory, is given by Eq. (20-8):

$$\vec{E} = ae^{\phi J}e^{\rho k}\hat{x}. \tag{20-24}$$

For constant values of ϕ and ρ, this expression represents a fully polarized EM wave. A fully polarized EM wave is one whose polarization state is constant in time. A signal whose polarization state varies in time in a random fashion is said to be partially polarized. These types of signals are of most interest in polarimetric radar technology.

When dealing with a random, time-varying polarization vector, that is, a partially polarized wave, it is natural to ask how the components of that vector vary in time. One way to answer this question is to form correlations of the components of the vector. For example, in the linear basis expressed by Eq. (20-15), the polarization vector in Eq. (20-24) can be written as

$$\vec{E} = \begin{bmatrix} E_x \\ E_y \end{bmatrix} \tag{20-25}$$

The correlations of the components of this vector are given by $\langle E_x^* E_x \rangle$, $\langle E_y^* E_y \rangle$, $\langle E_x^* E_y \rangle$, where the brackets $\langle \ \rangle$ denote a time or an ensemble average. The polarization vector in the Stokes/Mueller approach to scattering matrix theory is called the *Stokes vector*, and its components are constructed from these correlations. Therefore, the Mueller calculus naturally accommodates time-varying targets for which the return signal is partially polarized. Another, equivalent approach, is the covariance or density matrix approach, in which a three by three Hermitian matrix is constructed from correlations of the vector components (or functions of the vector components) above. The interested reader is referred to Reference 9 and 10 for further details.

The Stokes vector of an EM wave represented by the two-dimensional, complex polarization vector, $\vec{a}$, given by Eq. (20-24) or Eq. (20-25) is defined as

$$\vec{g} = \begin{bmatrix} g_0 \\ g_1 \\ g_2 \\ g_3 \end{bmatrix} = \begin{bmatrix} \langle \vec{E} \cdot \vec{E}* \rangle \\ \langle iJ\vec{E} \cdot \vec{E}* \rangle \\ \langle iL\vec{E} \cdot \vec{E}* \rangle \\ \langle -iK\vec{E} \cdot \vec{E}* \rangle \end{bmatrix} . \tag{20-26}$$

[The Stokes vector in optics is more commonly written with the second component given last, that is, as

$$\vec{g} = \begin{bmatrix} g_0 \\ g_2 \\ g_3 \\ g_1 \end{bmatrix} . \tag{20-27}$$

Also, some authors choose to replace g_1 with $-g_1$. The convention adopted by Huynen[6] and given by Eq. (20-26) is adopted here.] In terms of the linear components E_x and E_y, the Stokes vector is given by

$$\vec{g} = \begin{bmatrix} \langle |E_x|^2 + |E_y|^2 \rangle \\ \langle 2Im(E_x^*E_y) \rangle \\ \langle |E_x|^2 - |E_y|^2 \rangle \\ \langle 2Re(E_x^*E_y) \rangle \end{bmatrix} . \tag{20-28}$$

In terms of the a, ϕ, and ρ parameters, the Stokes vector is given by

$$\vec{g} = \begin{bmatrix} \langle a^2 \rangle \\ \langle a^2\sin2\rho \rangle \\ \langle a^2\cos2\rho \, \cos2\phi \rangle \\ \langle a^2\cos2\rho \, \sin2\phi \rangle \end{bmatrix} . \tag{20-29}$$

Of course for fully polarized waves, the time-averaged brackets in Eqs. (20-26), (20-28) and (20-29) can be omitted. For fully polarized waves, notice that the following relationship among the components of the Stokes vector holds:

$$g_0^2 = g_1^2 + g_2^2 + g_3^2 \quad \text{(fully polarized)}. \tag{20-30}$$

For partially polarized waves, the following inequality holds:

$$g_0^2 > g_1^2 + g_2^2 + g_3^2 \quad \text{(partially polarized).} \tag{20-31}$$

And for unpolarized waves, $g_1 = g_2 = g_3 = 0$.

If $\vec{g}$ is the Stokes vector corresponding to the transmit antenna polarization, $\vec{a}$, and $\vec{h}$ is the Stokes vector corresponding to the receive antenna polarization, $\vec{b}$, then the backscattered power received by the radar is given by

$$P \sim \sigma_{ab} = |S\vec{a} \cdot \vec{b}|^2 \equiv M\vec{g} \cdot \vec{h}. \tag{20-32}$$

The operator M is represented by a 4×4 real matrix called the *Stokes reflection* or *Mueller matrix*, and from Eq. (20-32) can be expressed in terms of the five parameters $(m, \psi, \tau, \nu, \gamma)$ by

$$M = \begin{bmatrix} A_0 + B_0 & F & C_\psi & H_\psi \\ F & -A_0 + B_0 & G_\psi & D_\psi \\ C_\psi & G_\psi & A_0 + B_\psi & E_\psi \\ H_\psi & D_\psi & E_\psi & A_0 - B_\psi \end{bmatrix} \tag{20-33}$$

where

$$H_\psi = C \sin 2\psi$$
$$C_\psi = C \cos 2\psi$$
$$G_\psi = G \cos 2\psi - D \sin 2\psi$$
$$D_\psi = G \sin 2\psi + D \cos 2\psi$$
$$E_\psi = E \cos 4\psi + B \sin 4\psi$$
$$B_\psi = -E \sin 4\psi + B \cos 4\psi$$
$$A_0 = Q f \cos^2 2\tau$$
$$B_0 = Q(1 + \cos^2 2\gamma - f \cos^2 2\tau)$$
$$B = Q[1 + \cos^2 2\gamma - f(1 + \sin^2 2\tau)]$$
$$C = 2Q \cos 2\gamma \cos 2\tau$$
$$D = Q \sin^2 2\gamma \sin 4\nu \cos 2\tau$$
$$E = -Q \sin^2 2\gamma \sin 4\nu \sin 2\tau$$
$$F = 2Q \cos 2\gamma \sin 2\tau$$
$$G = Q f \sin 4\tau$$
$$Q = m^2/(8 \cos^4 \gamma)$$
$$f = 1 - \sin^2 2\gamma \sin^2 2\nu$$

For time–harmonic (non-time-varying) targets, the nine parameters $(A_0, B_0, B_\psi, C_\psi, D_\psi, E_\psi, F, G_\psi, H_\psi)$ in M are not all independent, but are related by the following four auxiliary equations:

$$2A_0(B_0 + B_\psi) = C_\psi^2 + D_\psi^2$$

$$2A_0(B_0 - B_\psi) = G_\psi^2 + H_\psi^2$$

$$2A_0E_\psi = C_\psi H_\psi - D_\psi G_\psi$$

$$2A_0F = C_\psi G_\psi + D_\psi H_\psi. \qquad (20\text{-}35)$$

For time-varying targets, however, these nine parameters become independent for the time-averaged Mueller matrix:

$$R \equiv \langle M(t) \rangle. \qquad (20\text{-}36)$$

Huynen[6] has shown that if one writes the time-averaged parameters $\langle B_0 \rangle$, $\langle B_\psi \rangle$, $\langle E_\psi \rangle$, and $\langle F \rangle$ as

$$\langle B_0 \rangle = B_0^T + B_0^N$$

$$\langle B_\psi \rangle = B_\psi^T + B^N$$

$$\langle E_\psi \rangle = E_\psi^T + E^N$$

$$\langle F \rangle = F^T + F^N \qquad (20\text{-}37)$$

where B_0^T, B_ψ^T, E_ψ^T, and F^T satisfy Eq. (20-35), then one can uniquely form the following decomposition:

$$R = M_0 + N \qquad (20\text{-}38)$$

where

$$N = \begin{bmatrix} B_0^N & F^N & 0 & 0 \\ F^N & B_0^N & 0 & 0 \\ 0 & 0 & B^N & E^N \\ 0 & 0 & E^N & -B^N \end{bmatrix}. \qquad (20\text{-}39)$$

The M_0 matrix is specified by the usual five (m, ψ, τ, ν, γ) parameters.

The physical significance of the decomposition of the R matrix given by Eq. (20-38) is that any time-varying (fluctuating) target can be represented by a mean stationary (non-time-varying) target (M_0) plus a noise residue part (N)

which gives an indication of how that time-varying target varies from its mean stationary representation. The time dependence of the target can be due to either the motion (translational or vibrational) of the target or the motion of the radar sensor itself.

The power of the decomposition theorem lies in its ability to separate motional perturbations (which are almost always present) from the target signature. Thus, like targets that are observed in different states of motion can, it is hoped, be recognized as like targets. In addition, the amount of motion variance attributed to a given target may provide some insight into the nature of that target. For example, rigid (man-made) targets may be separable from nonrigid (natural clutter) targets by examining the characteristics of their respective N matrices.

A few final comments are in order. The Huynen decomposition of the average Mueller matrix is not the only decomposition possible. Indeed, it is possible to extract a mean stationary target from an average Mueller matrix in an infinite number of different ways. The Huynen decomposition, however, certainly has some physical merit and has been shown to be successful in separating targets from clutter. Finally, other decompositions are currently being investigated using the covariance or density matrix approach mentioned above. Using this approach, it is possible to uniquely decompose the density matrix representing the time-varying target (from which the return signal is partially polarized) into three density matrix components: a mean target component from which the return is fully polarized, a component representing a totally unpolarized return, and a component representing a partially polarized return. This "natural" decomposition is currently being investigated for its target/clutter discrimination potential.

20.4   VECTOR DISCRIMINANTS

There are, in general, three classes of target–clutter discrimination algorithms: (1) scalar, (2) vector, and (3) matrix. Scalar discriminants do not use any aspect of polarization, while vector and matrix discriminants are both polarimetric. With vector discriminants, some aspect of the vector nature of EM radiation, that is, polarization, is used in an attempt to discriminate between the target and the clutter, but no attempt is made to separate the effects that both the radar (antenna) and the target have on the polarimetric properties of the measured backscattered signal. Most polarimetric discriminants in existence today are of this type.

Matrix discriminants are based on backscattered data in which the polarimetric effects of the radar (antenna) on the data have been eliminated, leaving only target-induced polarization effects. In other words, matrix discriminants are based on measurement of the polarization scattering matrix. The benefits of measuring the scattering matrix for not only discrimination but also recognition are discussed in the next section.

Vector discriminants are usually based on some simple polarization-dependent scattering property that differs between a target of interest and clutter. For example, probably the simplest vector discriminant used today is the one that discriminates between a complex target (one with many scatterers) and rain clutter.[11]

Since the shapes of raindrops can usually be well approximated by spheroids, the backscatter from rain of an incident circularly polarized wave will have the opposite sense of circular polarization, that is, it will be cross-polarized. The backscatter from the complex target, however, will have both copolarized and cross-polarized components. Therefore, if the radar transmits a given circular polarization but receives only the copolarized component, then the rain clutter backscatter will have been effectively eliminated, albeit at the expense of some of the target backscatter signal.

A class of vector discriminants that have demonstrated some success against ground and sea clutter are *depolarization* vector discriminants. In the implementation of these discriminants, a signal with a given polarization, (e.g., vertical) is transmitted, and both the copolarized (vertical) and cross-polarized (horizontal) components are received. The larger the signal in the cross-polarized channel, the more the target has depolarized the incident wave. If the clutter is more complex than the target, then this depolarization should be greater for a clutter return than for a target return. A discriminant of this class known as the *polarization ratio discriminant* is formed by taking the ratio of the copolarized channel to the cross-polarized channel, for example, $|S_{VV}|/|S_{VH}|$.[12] This ratio should be larger for a target return that for a clutter return. This discriminant performs better with a frequency-agile waveform which enhances the depolarization effects in the clutter return.

Another class of vector discriminants that are closely related to the depolarization vector discriminants are the *odd/even bounce* vector discriminants. Again, these discriminants have demonstrated some success with relatively simple targets in relatively complex clutter. The idea behind these discriminants is that some targets consist of either predominantly odd- or even-bounce scatterers, while some clutter types consist essentially of an equal number of odd- and even-bounce scatterers. Therefore, if a circularly polarized signal is transmitted, then the return signal from these target will either be predominantly cross-polarized or copolarized, whereas the return signal from the clutter will contain equal amounts of cross- and copolarization.

A discriminant of this class known as *pseudocoherent detection* is implemented by transmitting a circular polarization and receiving the H and V polarizations.[13] If the receive signal is circularly polarized, as it is hoped a target signal would be, the relative phase between H and V. (i.e., the polarimetric phase) will be $\pm 90°$ (the sign depends on the sense of the circular polarization). If, on the other hand, the receive signal contains equal amounts of both senses of circular polarization, then the polarimetric phase can assume any value; it is

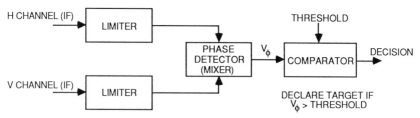

Figure 20-8. Pseudocoherent detection discrimination implementation. (From ref. 11).

hoped that for a clutter return, this phase will be uniformly distributed from $-90°$ to $+90°$. Thus, by thresholding the polarimetric phase, discrimination is effected. This discriminant also performs better with a frequency-agile waveform. An implementation of pseudocoherent detection is shown in Figure 20-8.

Many other vector discriminants exist, but they all have one thing in common: They all require the implementing radar to be in a specific polarization transmit/receive (TR) configuration. Therefore, in general, a radar configured to support one vector discriminant may not be configured to support another. This can be a severe limitation, since no one vector discriminant exploits all the target/clutter polarization information that is available to the radar sensor. Therefore, multiple vector discriminants may be necessary to achieve the discrimination performance level desired. This is especially true in dynamic target/clutter scenarios. These limitations of vector algorithms are overcome with the implementation of matrix algorithms.

20.5 MATRIX DISCRIMINANTS

Matrix dicriminants are potentially the most powerful polarimetric discriminants, since they can use all of the polarimetric information available in the scattering matrix to effect discrimination. This information includes all of the target geometric information, including maximum RCS, target orientation, target symmetry, odd/even bounce characteristics, target polarizability, and target complexity. In addition, this backscatter information is independent on the radar's transmit and receive antenna polarizations, effectively decoupling radar-induced polarization effects from the target data. (Radar system imperfections such as leakage between polarization channels will still corrupt the data.)

Once the scattering matrices have been measured in a data collection program to develop matrix discriminants, the same data base can also be used to develop vector discriminants (assuming all other algorithm requirements, such as frequency agility, have been satisfied). These vector discriminants can be formed and tested by selecting the appropriate transmit/receive antenna polarizations, calculating the mesured RCS from Eq. (20-14), and using this RCS to form the vector discriminant.

Several approaches have been taken in an attempt to achieve discrimination using matrix discriminants. One approach mentioned in Section 20.3 uses null polarizations.[8] The measured target RCS represented by the scattering matrix S is zero (a null) when the TR antenna polarization is a null polarization of S. Therefore, if a null polarization could be found for a given clutter type, simply transmitting and receiving this polarization would make the clutter invisible to the radar. Unfortunately, the radar–clutter interaction process is usually nonstationary, that is, the clutter return is partially polarized. Therefore, the null polarizations for a given clutter type are time dependent. Still, if during the observation time the null polarizations of the clutter remain fairly well localized on the Poincaré sphere, that is, do not vary greatly, then an optimum antenna transmit and receive antenna polarization can be found to maximize the target-to-clutter ratio.[14] Another approach suggested is to change the transmit and receive antenna polarizations adaptively in real time to maximize target-to-clutter ratio.[15]

Still another approach to target–clutter discrimination is to use parameters extracted from the scattering (Mueller) matrix as features in a multidimensional feature space and apply pattern recognition algorithms to these features to separate the target and clutter signatures.[16] This approach is also applicable for achieving target recognition and is discussed in the next section.

20.6   APPROACH TO STATIONARY TARGET RECOGNITION

Radar target recognition is the ability to determine whether a radar-detected target belongs to a target class of interest to the radar operator. The degree of difficulty in achieving target recognition depends on the definition of the target class of interest. For example, if the target class of interest is ''all blue chevrolets,'' then recognition with a radar sensor is virtually impossible. On the other hand, if the target class of interest is all targets other than clutter targets, then recognition reduces to discrimination. Most recognition requirements fall between these two extremes, for example, track vehicles versus wheel vehicles.

Whatever the target class of interest is, however, stationary target recognition is much more difficult than moving target recognition since the spectral techniques used in moving target recognition are not applicable to stationary targets. Stationary target recognition is currently being investigated by radar researchers, and the technology is still very immature. Some promising results have been obtained using high-resolution techniques, polarimetric techniques, and combinations of these techniques. Successful stationary target recognition using only a radar sensor will ultimately require the processing of all information contained in the radar backscatter. This means that a full-polarimetric, wide-bandwidth (high-resolution) radar waveform will be required. A full-polarimetric waveform is one from which the polarimetric scattering matrix can be extracted. In the remainder of this section, an approach to stationary target recognition using a full-polarimetric waveform is discussed.

Unless one target-viewing aspect angle, or range of aspect angles, can be guaranteed in a given operational scenario, a matrix recognition algorithm must enable the recognition of targets at all radar viewing aspect angles. Therefore, any approach to stationary target recognition using matrix algorithms should include the determination, or measurement, of scattering (Mueller) matrices at all aspect angles of all targets to be recognized. The aspect angle increment required (between measurements) will depend on the ratio of the average distance between the scatterers of the given target and the radar's transmit frequency; the larger this ratio, the smaller the aspect angle increment.

Parameters, such as the five physically relevant parameters $(m, \psi, \tau, \nu, \gamma)$ discussed in Section 20.3 can now be extracted from each of the target Mueller matrices and examined for any trends that will enable recognition. These parameters can be used to construct a target feature set containing, say, K features. A multidimensional feature space based on these K features can now be constructed. For low-resolution data, that is, data in which the individual targets are smaller than the radar's resolution cell, this feature space can have up to K dimensions. It is hoped that the targets will be separable in this space, that is, the data representing the different targets will be "clustered" in different parts of this feature space. As has been suggested,[17] standard, well-known statistical pattern recognition (PR) techniques[18] can then be employed to achieve the optimum recognition possible based on the amount of target separation present. These PR techniques essentially determine the decision surfaces that separate the clusters. Once these decision surfaces have been determined, the original "training" target data can be discarded, since the specification of these surfaces is the only information retained in an operational discrimination algorithm. A simple two-dimensional feature space containing data from three target classes is shown in Figure 20-9.

For high-range resolution, low-cross-range resolution data – that is data in which each target is contained in N range cells but is smaller in cross-range than the radar's cross-range resolution – a feature space of up to $K \times N$ dimensions can be constructed. Again, standard PR techniques can be employed in this feature space. For high-resolution data (N target range cells and P target cross-range cells), a feature space of up to $K \times N \times P$ dimensions can be constructed and PR techniques employed. High-resolution data also allow for the formation of target images and the application of image-processing techniques. Not only amplitude imaging (as is done with SAR data) but also other forms of imaging such as symmetry imaging, orientation imaging, or composite imaging are also possible.

Significant overlap between target clusters is possible in the multidimensional feature space approaches described above. This results in poor recognition per-

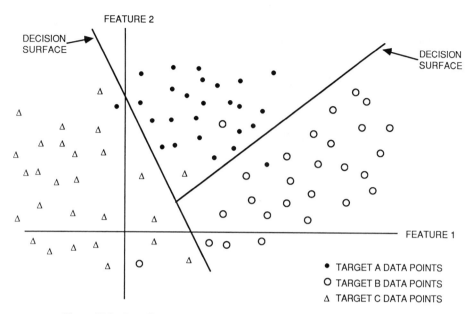

Figure 20-9. Two-dimensional feature space with data from three target classes.

formance. Significant cluster overlap is especially common for low-resolution data of highly complex targets. As mentioned above, this is caused by motional perturbations that corrupt the target signature. For example, even small changes in the viewing aspect angle of a complex target can result in severe backscatter fluctuation, producing a nonlocalized target cluster in feature space.

Feature space overlap of target data clusters, especially those clusters obtained from low-resolution data, can be reduced by selecting features from averaged Mueller (R) matrices. Target R matrices can be obtained by averaging over specified increments of aspect angle. The R matrices obtained for the target and clutter can then be decomposed using Huynen's decomposition theorem (see Section 20.3) and the five relevant target parameters (m, ψ, τ, ν, γ) extracted from the M_0 matrices. Features based on these parameters and the N-matrix parameters can then be selected.

20.7   SUMMARY

Discrimination and recognition of stationary targets by a radar is a difficult task. All of the information contained in the radar waveform will ultimately have to be utilized to accomplish this task. In this chapter, polarimetric fundamentals are discussed, and polarimetric techniques that enhance stationary target discrimination and recognition are described. By using polarimetry, the radar engineer is taking advantage of the geometric, or vector, nature of EM radiation. Not surprisingly, then, radar polarimetry provides target geometric information (e.g., orientation, symmetry) not otherwise available, and it is this information that can be used to help effect stationary target discrimination and classification.

Radar polarimetry has already been shown to be effective in stationary target discrimination. Most discriminants in use today are vector discriminants, although some matrix discriminants have been formulated and tested. Stationary target recognition using matrix algorithms is a relatively new field, and more research is needed if practical, robust algorithms are to be produced.

Perhaps the most promising new thrust toward solving the stationary target recognition problem has been in the area of combining full-polarimetric techniques with high-resolution recognition techniques. This ''ultimate'' radar waveform should give the definitive answer to the question, ''To what extent can targets be recognized using a radar sensor?''

20.8 REFERENCES

1. H. Poincaré, *Théorie Mathematique de la Lumière*, Vol. 2. G. Carré, Paris, 1892.
2. D. N. Lapedes (Ed.), *McGraw-Hill Dictionary of Scientific and Technical Terms*, 2nd ed. McGraw-Hill Book Co., New York, 1978, p. 1373.
3. Frank Jay (Ed.), *IEEE Standard Dictionary of Electric and Electrical Terms*, 2nd ed. Wiley-Interscience, New York, 1977, p. 601.
4. G. Sinclair, ''The Transmission and Reception of Elliptically Polarized Waves,'' *Proceedings of the IRE*, vol. 38, Feb. 1950, pp. 148–151, R. C. Jones; *Journal of The Optical Society of America*, vol. 46, 1956, pp. 126.
5. G. G. Stokes, Transactions of The Cambridge Philosophical Society *Trans Camb. Phil. Soc.*, 1852, pp. 399.
6. J. R. Huynen, ''Phenomenological Theory of Radar Targets,'' Ph.D. dissertation, Drukkerij Bronder-Offset N. V., Rotterdam, 1970.
7. J. R. Huynen, ''Phenomenological Theory of Radar Targets,'' in *Electromagnetic Scattering*, ed. P. L. E. Uslenghi, Academic Press, New York, 1978, p. 653.
8. S. Weisbrod and L. A. Morgan, ''RCS Matrix Studies of Sea Clutter,'' Teledyne-Micronetics Report No. R2-79, N00019-77-C-0494, AD B036684 January 1979.
9. Huynen, J. R. ''Towards a Theory of Perceptions for Radar Targets with Applications to The Analysis of Their Date Base Structures and Presentations,'' *Inverse Methods in Electromagnetic Scattering*, W.-M. Boerner, Editor, D. Reidel, Dordrecht, Holland, 1925, pp. 797–822.
10. Claude, S. R., ''Target Decomposition Theories in Radar Scattering,'' Electronics Letters, 21, pp. 22–24, 1985.
11. M. I. Skolnik, *Introduction to Radar Systems*, McGraw-Hill Book Co., New York, 1980, p. 504.
12. R. D. Hayes and J. L. Eaves, ''Study of Polarization Techniques for Target Enhancement,'' AF 33(615)-2523, AD-316270, Report A-871, Georgia Institute of Technology, Atlanta, August 1966.
13. J. D. Echard, et al., ''Discrimination Between Targets and Clutter by Radar,'' DAAG29-78-C-0044, Report A2230, Georgia Institute of Technology, Atlanta, December 1981.

14. G. A. Ioannidis, "Optimum Antenna Polarizations for Target Discrimination in Clutter," *IEEE Transactions on Antennas and Propagation,* vol. AP-27(3), May 1979, pp. 357–363.

15. A. J. Poelman, "The Applicability of Controllable Antenna Polarizations to Radar Systems," *Tijdschrift van het Nederlands Electronica en Radiogenootschap*, vol. 42, no. 2, 1979, pp. 93–106.

16. W. A. Holm, "Polarization Scattering Matrix Approach to Stationary Target/Clutter Discrimination," *Colloque International sur le Radar*, May 1984, pp. 461–465.

17. S. R. Cloude, "Polarimetric Techniques in Radar Signal Processing," *Microwave Journal*, vol. 26, no. 7, July 1983, pp. 119–127.

18. N. F. Ezquerra, "Radar Target Recognition: A Survey," *Colloque International sur le Radar*, May, 1984, pp. 281–285.

21
TARGET RECOGNITION
CONSIDERATIONS

N. F. Ezquerra

21.1 OVERVIEW

This chapter discusses fundamental considerations in radar target recognition. The overall goal of the chapter is threefold: (1) to introduce some of the basic issues associated with the problem of target recognition, (2) to discuss and compare various target recognition techniques, and (3) to provide examples of applications of target recognition techniques, pointing out the limitations (and unrealized potential) of this technology. An extensive (but not exhaustive) list of references provides sources for further information on different aspects of the problem.

21.2 INTRODUCTION

Target recognition has recently received a great deal of attention, and as a result, considerable literature exists. Rather than presenting a survey[1] of the entire field (which would require an extensive study far beyond the scope of this chapter), we treat the topic by highlighting specific techniques and representative applications, thus introducing an interesting facet of radar systems.

Prior to a technical discussion of target recognition methodology, it is appropriate to address some fundamental questions, including: Why is target recognition important? What does the term *target recognition* mean? How, in very general and qualitative terms, might the problem of recognizing targets by radar be approached? Preliminary answers to these questions will serve as a starting point for subsequent discussions.

21.2.1 Motivation

There are several reasons for introducing the concept of target recognition as part of a radar system. First, it is desirable not only to detect an object in the field of illumination of the radar beam but also to know something about the object beyond its mere presence. (This consideration applies, of course, to sensors in general, not only to radar.) In other words, basic target information such as location, velocity, and total cross section, which is the type of information usually derived by radar sensing, frequently may not provide sufficient target data. This demand for more target information (or more sophisticated ways of processing it) attests to the evolution of radar: It may no longer suffice simply to display the data on a scope or to track a target; the identity of the target itself may be required in some applications.

Second, the progress that has occurred in the field of signal processing has influenced the direction of research in radar data processing. This is evidenced in the availability of fast, light weight, and relatively inexpensive hardware which facilitates numerous radar tasks, both in terms of operating the radar system and in processing large amounts of data. Examples of such progress include the advent of very-large-scale integration (VLSI) and very-high-speed-integrated-circuit (VHSIC) technology, which may be capable of improving the computational capacity by two orders of magnitude.[2] An indirect consequence of this progress has been an increased interest in implementing target recognition techniques, since a wide variety of algorithmic procedures can be easily implemented with present signal-processing technology.

Third, certain characteristics of radar motivate the use of this sensor in target recognition applications. Although the choice of a sensing device depends largely on the specific recognition problem to be solved, radar sensors provide some significant advantages. For instance, environmental factors affect the propagation of radar wavelengths less than optical wavelengths.[3] Factors that contribute to image degradation include smoke, clouds, and adverse weather conditions; there may also be changes in the perspective of the sun or in target illumination that can greatly affect the quality of optical imagery. In addition to these propagation characteristics, the amount of information obtained at radar wavelengths may be more easily managed than the typically larger amount of information that is obtained with optical devices. These attributes thus make radar an attractive means of sensing targets for recognition applications.

Fourth, and perhaps most importantly, an increasingly large number of areas of application have placed greater emphasis on target recognition methodology. Among the various disciplines and research areas that employ target recognition techniques are geophysics and meteorology (e.g., weather radar), terrain and scene matching, and military applications (e.g., recognition of aircraft, buried mines, ground vehicles, or ships).[4-8] Hence, the ability of a radar system to identify a target automatically is a common goal in many fields of research. This wide-ranging applicability has further increased the importance of developing radar target recognition technology.

21.2.2 Definitions and Semantics

Difficulties due to definitions and semantics arise in the field of target recognition. Words such as *identification*, *recognition*, *classification*, and *discrimi-*

nation, among others, are used in different ways by different researchers. In an attempt to minimize confusion, we define the following terms as they are used in this chapter.

The *target* is the object that is illuminated by the radar beam and that interacts with the EM wave to produce a backscattered signal. Furthermore, the target is the object (or class of objects) of interest in the specific application. Hence, the phrase *target recognition* implies that only a specific, predetermined set of targets is recognized.

The term *recognition* can be defined with the aid of Figure 21-1, which depicts the flow of radar information as a sequence of three steps. The goal of this sequence is to address the question, What is the target that has been sensed by the radar? Since the overall goal of this process is to identify the target, the sequence shown in Figure 21-1 may be called the *target identification process*. This process includes separating noise and clutter interference from potential target returns, separating targets of interest from other targets, and assigning unknown targets to specific categories.

The first step in the target identification sequence is detection, which determines all radar resolution cells that contain potential targets. The detector should detect all true targets while minimizing the probability of false alarms due to background clutter, interference, and receiver noise. *Clutter* usually refers to objects that cause spurious radar returns or unwanted detections, such as grass, trees, and ocean waves. The probability of false alarm per resolution cell is typically 10^{-5} or less.

The second step in the target identification sequence is discrimination. The discriminator further reduces the number of false alarms by exploiting *target discriminants*, that is, differences between the scattering characteristics of targets of interest and other scatterers. Target discriminants can be derived by using techniques such as frequency agility, high-range resolution, polarimetric techniques, or any combination of these. The discriminator does not generally distinguish between different types of targets. Rather, it separates potential target returns from unwanted detections, providing target-like information as output. The topics of detection and noise are treated in Chapter 9 and thus are not discussed further in this chapter.

The third step in the target identification sequence is recognition. The recognizer processes the target-like information provided by the discriminator and makes a decision regarding the identity of the target. The decision-making steps are generally based on methodologies of such disciplines as pattern recognition, artificial intelligence, inverse scattering, and information and signal-processing

Figure 21-1. Target identification sequence.

Table 12-1. Glossary of Terms.

Identification	The entire signal-processing sequence, from the detection of a potential target to the automatic determination of the target's identity.
Detection	The initial step of separating noise and clutter-like noise from potential target returns. One detection method is the CFAR technique.
Discrimination	The further separation of clutter and target-like clutter from potential target returns. Examples of discrimination methods are those based on frequency agility, high resolution, or polarimetric processing and characterized by relatively simple algorithmic sequences.
Recognition	The automated decision-making process whereby features are extracted from the information provided by the previous stages of detection and discrimination, and a decision is made regarding the identity of the target.

theories. The recognition process may be viewed as having a threefold purpose: (1) to extract useful information (generally called *features*) from the radar return, (2) to separate targets of interest from all objects illuminated by the radar, and (3) to assign the target of interest to one of several categories or classes (i.e., classification).

Table 21-1 summarizes the definitions of the different stages of the target identification process. The remainder of this chapter deals with the recognition process itself and the three aspects of recognition outlined in the previous paragraph. The discussion includes examples of various applications, operational considerations, and present trends and future directions in target recognition.

21.2.3 Preliminary Remarks

Three basic considerations are emphasized at the outset. The first is that a general, theoretical methodology does not yet exist for selecting and implementing radar target recognition techniques. The reason for this is unclear, and two opposing schools of thought offer possible explanations. According to one, the theoretical formalism associated with target recognition is not sufficiently mature or advanced, and thus a set of rules does not yet exist. According to the other, the subject is such that no set of rules can cover all applications under all possible conditions. In either case, there is no clearly defined sequence which can be followed to determine an optimal set of target recognition techniques.

The second basic consideration is that, once a set of appropriate target recognition techniques has been selected, the actual development, implementation, testing, and evaluation of these techniques are both time-consuming and iterative processes. For instance, an initial set of candidate target features usually does not provide optimal recognition performance; several iterations are usually necessary to select an adequate set of features.

Table 21-2. Journals and Conferences Dealing With Target Recognition Topics.

IEEE Transactions on Pattern Analysis and Machine Intelligence
IEEE Transactions on Acoustics, Speech, and Signal Processing
IEEE Transactions on Computers
IEEE Transactions on Systems, Man and Cybernetics
IEEE Transactions on Engineering in Medicine and Biology
IEEE Transactions on Information Theory
IEEE Transactions on Control
IEEE Proceedings on Pattern Recognition
IEEE Proceedings on Information and Control
IEEE Proceedings on Artificial Intelligence
Defense Electronics
Communications of the ACM
Machine Intelligence
International Joint Conference on Pattern Recognition
International Conference on Acoustics, Speech and Signal Processing
International Joint Conference on Artificial Intelligence
IEEE Computer Society Conference on Pattern Recognition and Image Processing
NATO Advanced Study Institute on Pattern Recognition Theory and Applications
IEEE-IEE International Radar Conference (sponsored by the IEEE in the United States and the
    IEE in the United Kingdom)
Tri-Service Radar Symposium
Non-Cooperative Target Recognition Conference

The third consideration has to do with the target recognition literature published over the past several years. An extensive amount of information covers numerous areas of application. As a representative fraction of the available literature, Table 21-2 provides the names of trade journals (most of which are available internationally), as well as the names of various conferences, symposia, and workshops. In addition, references 9 through 26 include numerous texts and handbooks that provide self-contained tutorial treatments of the subject. In the subsequent discussions, selected topics from some of these and other references are discussed in greater detail.

21.3 FUNDAMENTALS OF TARGET RECOGNITION

In its simplest conceptualization, the recognition process may be viewed as consisting of three major steps: (1) transmitting a set of signals to probe the target or targets, (2) extracting useful information from the backscattered signals, and (3) processing the extracted information in an automated, decision-making sequence. In a sense, the objective is to optimize these three interrelated aspects of the recognition process. The approaches followed to meet this objective generally fall into one of three broad categories, as illustrated in Figure 21-2. The emphasis is usually placed on (1) specific mathematical methods or

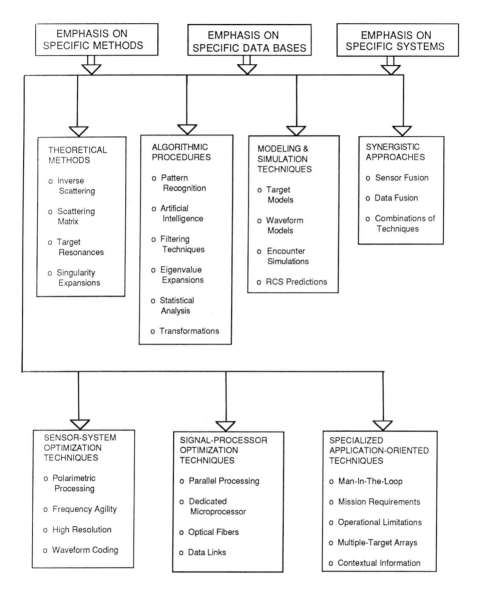

Figure 21-2. Approaches to target recognition.

algorithms (e.g., pattern recognition or artificial intelligence techniques), (2) specific data bases (e.g., SAR imagery or range-amplitude profiles), or (3) specific radar systems (e.g., concentrating on application-dependent issues such as man-in-the-loop interactions). Although this division appears arbitrary, it actually emerges rather naturally from the literature over the past several years.[1]

Placing the emphasis on any one of these approaches means that the resultant target recognizer is dependent on a specific method, data base, or sensor system. Consequently, it is difficult to assess the performance of most recognition systems in terms of overall usefulness and general applicability. However, due to practical considerations (such as the limitations imposed by specific missions, particular radar systems, or the availability of funds and resources), research on target recognition has traditionally been limited to a finite (and usually small) number of issues. As illustrated in Figure 21-2, the approaches include (1) theoretical methods, (2) modeling and simulation techniques, (3) algorithmic and artificial intelligence procedures, (4) synergistic approaches (fusion of multiple sensors or data types), (5) sensor system parameter optimization techniques, (6) signal processor optimization techniques, (7) procedures designed to address specific application-related issues (e.g., man-in-the-loop responses, limited response-time requirements, and possible exposure to multiple-target arrays), and (8) some combinations of these techniques.

A significant amount of effort has been devoted to developing, implementing, and testing these approaches. A mathematically rigorous and technically detailed treatment of each of these approaches is necessary to address fully the topic of target recognition—an ambitious undertaking that falls well outside the scope of this chapter. The following discussions are thus limited to some of the more important aspects of various methodologies, particularly techniques and applications that can be illustrated through simple examples.

21.3.1   Target Recognition Methodologies

Target recognition methodologies can be illustrated by considering the problem of recognizing ground vehicles using a pulsed radar system, as represented schematically in Figure 21-3. Such a problem is extremely difficult to solve, and a rigorously complete and operationally satisfactory solution is yet to emerge (the same may be said of recognition problems that deal with other sensors, such as infrared image recognition). The target recognition problem is difficult to solve because numerous factors have to be taken into consideration in an operational situation, including the following:

1. Factors related to the radar system itself
   a. System stability and noise.
   b. System parameters such as resolution, polarization, frequency, beam-width, and bandwidth.
   c. Signal-processing speed and memory capabilities.

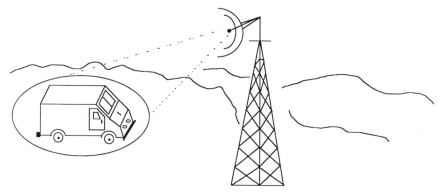

Figure 21-3. Schematic representation of a stationary-vehicle recognition problem.

2. Target-related considerations
   a. Changes that may occur in target appearance.
   b. Degree of similarity or dissimilarity between the target of interest and other targets.
   c. Number of targets that may be present.
   d. Target motion characteristics.
3. Environmental factors
   a. Atmospheric or propagation effects.
   b. Amount of background clutter interference that may be present.
   c. Variations in relative orientation between targets and radar system.
   d. Other possible sources of induced signal interference.
   e. Operational or mission-related limitations.

These factors contribute to the overall complexity of radar target recognition, which can therefore be viewed as a multifaceted problem. This point is illustrated in Figure 21-4, which shows three distinct yet interrelated aspects of target recognition: (1) the sensor system (i.e., the radar), (2) the scattering system (i.e., the target), and (3) the information-processing system (i.e., the recognition method). The underlying message of Figure 21-4 becomes clear: The target, sensor, and recognition systems are interrelated aspects of the overall problem; hence, the recognition method cannot be easily separated from the application it is intended to serve. This is the principal reason for the lack of a general methodology for recognizing targets. It is also the main reason that any such methodology must account for numerous interrelated operational considerations. As a starting point, a suitable approach would have to draw on such disciplines as EM scattering theory, as well as information- and signal-processing methods.

The vehicle recognition example illustrated in Figure 21-3 can be simplified by making several assumptions. For the radar systems, assume that (1) the range resolution is less than the extent of the target (such that the target may occupy several range bins, thereby producing a target profile or *signature*), (2) the an-

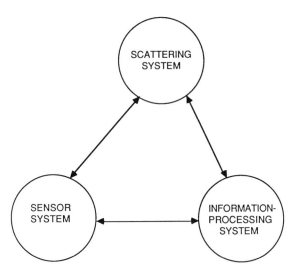

Figure 21-4. Three basic elements of scattering interactions.

tenna is dual-polarized, (3) the receiver has dual channels for processing orthogonal polarizations (e.g., horizontal and vertical) components of backscattered signals, (4) the levels of signal interference due to noise and clutter are
assumed to be negligible, (5) the relative orientation between the radar and the
target is assumed to be fixed (i.e., the target is always illuminated at the same
aspect angle), and (6) the other system characteristics are left unspecified to
allow some degree of generality.

For the recognition technique, we assume that (1) the overall goal is to distinguish between two classes of targets—a tank and a truck (without further
specifying a particular type for either vehicle)—and (2) the recognition technique is optimal. As pointed out earlier, target recognition approaches can be
assigned to three broad categories (depending on whether the emphasis is placed
on the sensor system, a specific data base, or a particular set of decision-making
techniques). For the recognition approach, we assume that the type of received
signal is given a priori: The target information consists of the total RCS for the
target and a target range profile associated with each polarization channel. In
other words, three sets of numbers are provided as raw target data (the RCS
value and two target signatures). An algorithmic approach is followed rather
than, for instance, artificial intelligence or purely theoretical methods. Admit-
tedly, this set of conditions and assumptions represents a highly simplified scenario, but the example illustrates the types of methods that can be implemented
in radar target recognition.

Based on the assumptions stated above, the problem of recognizing ground
vehicles can be reformulated in terms of specific questions: What type and
amount of information can be extracted from the received signals to represent

- Preprocessing = Feature Extraction
- Preclassification = Alien-Target Separation
- Classification = Assignation of Input to One Class

Figure 21-5. Stages in a recognition process.

optimally the targets of interest (i.e., tanks and trucks)? How can these targets of interest be distinguished from any other targets? How can tanks be separated from trucks? These questions provide some insight into the type of recognition process that is required.

Figure 21-5 shows the recognition process as a sequence of three steps: (1) preprocessing, (2) preclassification, and (3) classification. The preprocessing stage extracts useful information from the received signal; this function is more accurately called *feature extraction*. The preclassification stage uses the extracted features to determine whether the target in question is one of the targets of interest or an alien type of target (alien targets are extraneous to the application, that is, they are not targets of interest or are simply unrecognizable targets). If the target is one of interest, then the features are used in the classification stage to decide to which class the target should be assigned. Note that the simple linear sequence illustrated in Figure 21-5 is not the only possible process that can be implemented. For example, the features extracted for classification and preclassification may be different. The subsequent discussions illustrate how the tasks of preprocessing, preclassification, and classification can be performed in an algorithmic fashion.

21.3.2   Feature Extraction, Ranking, Selection, and Evaluation

The reason for extracting target features from the received signal is threefold: (1) to optimize performance, (2) to reduce the amount of information to be processed, and (3) to ensure robustness or invariance of the recognition system. The first consideration, optimization of performance, is a common goal of most recognition systems, where performance is usually interpreted as recognition accuracy or the probability of making correct decisions.

The second consideration, dealing with the amount of available information and sometimes referred to as *dimensionality reduction*, implies that reducing the amount of information to be processed also reduces the demands placed on the signal processor (in terms of processing time as well as processor memory). Not only will the processor generally be faster and less complex if the amount of data is initially reduced, but also subsequent steps in the decision-making sequence will benefit from such a reduction in data flow.

The third criterion, robustness, is also an important requirement of the selected features. The extracted target features should be invariant, or unchanging, with respect to changes in the operational scenario. In other words, the recognition algorithm should ideally yield a level of performance in the field that is consistent with the performance obtained during its development. This consistency in performance is commonly called *robustness* or *algorithm generalization*. In practice, there is little guarantee that the conditions in which the algorithm is to operate will remain unchanged (e.g., there is usually no control over target motion or environmental conditions). Nevertheless, an attempt should be made to ensure that at least some (expected) changes in the target or its environment will not alter recognition performance. Hence, a certain degree of robustness is viewed as a basic requirement to be fulfilled by the selected features.

Of the three criteria mentioned thus far (performance, low dimensionality, and robustness), the concept of robustness has come to be regarded as a very important consideration in the development of recognition systems.

Unfortunately, these three criteria cannot be regarded separately, since they are inextricably connected. Usually a trade-off must be made between optimizing performance, ensuring robustness, and reducing processor requirements. For instance, several different types of target features may have to be extracted to increase the amount of information significantly before the desired level of performance is attained.

The approaches to feature extraction may be subdivided into heuristics-based and deterministic methods. The former includes the selection of features based on intuitive arguments, whereas the latter includes well-defined analytical methods that usually transform the original target data by means of mathematical expansions or operations. Examples of intuitively-derived features are total RCS and the extent of the target signature in a range–amplitude profile (resembling the target's EM length). Examples of deterministic feature extraction techniques include Fourier transformations and eigenvalue expansions such as the Karhunen-Loeve expansion.[9]

A set of target features may actually be composed of different types of features. For instance, a composite set may consist of nine features: five Fourier coefficients [i.e., fast Fourier transform (FFT) terms], three Karhunen-Loeve (KL) expansion terms, and the measured RCS value. These features correspond to spectral characteristics (FFT terms), large eigenvalues (KL terms), and a physical measurement (RCS), thereby constituting a composite set of features with different physical interpretations.

In syntactic pattern recognition, target representation is achieved without the concept of features. These syntactic approaches rely on the development of a set of grammatical rules (as in a language), and target primitives (rather than features) are used in conjunction with the rules of syntax that define the language.

Feature selection procedures can be illustrated by considering a heuristics-based approach and the example of ground vehicle recognition of Figure 21-3. The radar information obtained from a single look of the truck is contained in the total RCS value and the horizontal (H) and vertical (V) polarization channel range profiles. These hypothetical target signatures are plotted in Figure 21-6 in terms of amplitude as a function of time (we assume that the returns are properly normalized so that the units can be ignored for the present discussion). For the sake of simplicity, let the features be selected intuitively by claiming

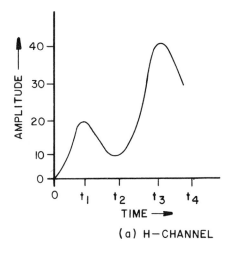

(a) H−CHANNEL

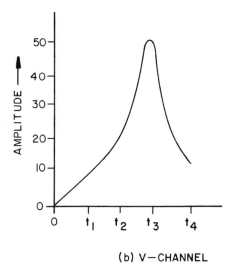

(b) V−CHANNEL

Figure 21-6. Hypothetical target returns for horizontal (H) and vertical (V) polarization channels. The units are left intentionally undefined, but it is assumed that the returns have been properly normalized.

that, based on our experience and a priori knowledge of the problem, we know that the backscattering characteristics of tanks and trucks are such that the most important target features are the total RCS and the dominant amplitude in each of the range profiles of Figure 21-6. Hence, the important features to be extracted from the target returns are (1) the largest amplitude in the H channel (with a value of 40), (2) the largest amplitude in the V channel (with a value of 50), and (3) the RCS value (let RCS = 30 in a properly normalized system of units). These three features can be denoted as $f_1 = 40$, $f_2 = 50$, and $f_3 = 30$. These features also define what is called a *feature vector*, denoted by $\bar{f}$; $\bar{f} = (f_1, f_2, f_3)$ or $\bar{f} = (H, V, RCS) = (40, 50, 30)$.

This feature vector $\bar{f}$ is illustrated in Figure 21-7, which shows how a collection of features defines the vector. As before, the letters H, V, and RCS are used to denote the dominant amplitudes in the horizontal and vertical polarization channels and the radar cross section, respectively. Each feature can be regarded as a variable in a three-dimensional feature space (H, V, RCS), where the three-dimensional feature vector can be represented by one point (the endpoint of the feature vector). In this sense, any n-dimensional vector can be reduced to one point in feature space, as illustrated in Figure 21-7 for the case of $n = 3$.

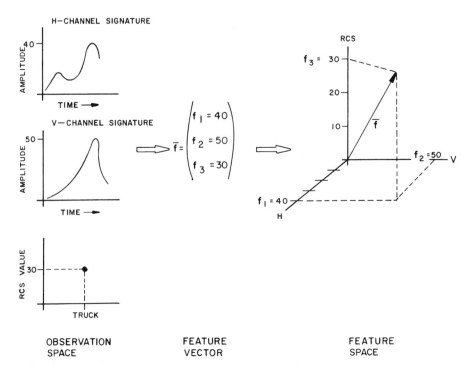

Figure 21-7. Feature vector construction from the hypothetical dominant amplitudes in H and V channels and total RCS value for the truck. These values are plotted in (H, V, RCS) feature space.

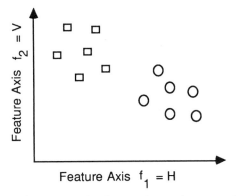

Figure 21-8. Hypothetical feature space showing two classes of targets: tanks (circles) and trucks (squares). The two features f_1 and f_2 correspond to dominant amplitudes in the horizontal (H) and vertical (V) polarization channels.

A collection of feature vectors (e.g., several looks at the truck) forms a group of points in feature space. Such a group of points is called a *cluster*. In the approach commonly taken in target recognition problems, the goal is to distinguish between clusters of different classes of targets. For instance, a tank may produce a particular cluster in feature space, whereas a truck may produce a different cluster. The difference between clusters in feature space may be in how the cluster points are distributed or perhaps in where the cluster is located. Clustering is illustrated in Figure 21-8, where a two-dimensional hypothetical feature (H, V) space shows the feature vectors corresponding to tanks (circles) and trucks (squares) clustering in separate regions of feature space. This cluster separation shows that the target features H and V, extracted from the radar returns, constitute an appropriate choice of features (the third dimension, RCS, was suppressed to simplify the figure).

Although the previous example illustrates the concepts of intuitively derived features, feature vectors, and feature spaces, the simplicity of the example is far from representative of real-world situations for several reasons. First, the appropriate features are rarely selected with such ease. Instead, the typical feature selection process is performed experimentally and iteratively, usually by comparing the results of different feature extraction methods, ranking the features according to different criteria (e.g., performance, dimensionality, and robustness), and determining whether the overall effectiveness requirements have been met. In addition, the feature dimensionality is normally higher than two or three; for feature vectors of higher dimensionality, computational considerations increase in importance (and the characteristics of the data cannot be easily visualized, since the corresponding clusters in feature space cannot be graphically represented). Furthermore, operational requirements generally place very strict demands on target features. For instance, in the example chosen, a rotation of the target about the radar line of sight would have yielded different target signatures in each polarization channel, such that the simple features H and V would probably not be as effective. Nevertheless, the importance of feature extraction cannot be overemphasized. The proper selection of features enhances recognition performance and simplifies the tasks of preclassification (i.e., alien-target separation) and classification (i.e., assignment of input to one particular target class).

Among heuristics-based approaches, orientation-independent descriptors of target geometries (e.g., size, shape, or other visibly distinguishable target characteristics) have been used to represent targets as point ensembles, and target-scattering characteristics such as amplitude ratios in high range-resolution profiles have been implemented in air-to-ground target recognition applications.[27,28] Similar methods have been used to define features in terms of the location and relative strength of scattering centers derived from RCS data, and targets have been modeled in terms of the minimum number of scatterers required to produce the expected RCS values.[29,30]

With respect to formal, deterministic approaches to feature extraction, the trend seems to be to apply eigenvalue or principal-component analysis such as the KL expansion or other transformations of the data that preserve some sort of symmetry relationship.[31] These include, for example, shift- and scale-invariant transformations such as the Fourier-Mellin transform and scale- and rotaton-invariant transformations such as the algebraic moment functions.[32,33] An interesting application of feature extraction techniques to two-dimensional radar imagery using algebraic invariant moments (as features) involves a comparison of optical and radar images.[5,12] The goal of this application was to select features that were invariant with respect to rotations and scale variations of the images, as shown in Figure 21-9. For a set of seven functions of algebraic invariants, the numerical values of the selected features did not vary appreciably when the images were rotated by 2° and 45°; the features were also invariant with respect to reduction to half-size and mirror imaging of the scenes.[5] These invariant moments are useful in cases where invariance with respect to such transformations is desired.

In general, these transformations of the data represent well-defined analytical tools for reducing dimensionality and ensuring a certain degree of robustness or invariance. However, they generally lack the natural appeal of intuitively derived features and may also render the extracted features difficult to interpret in a physical sense.

In addition to heuristics-based and deterministic approaches, syntactic methods and sensor-specific signal-generation techniques are used to represent target information. Syntactic pattern recognition, as previously mentioned, relies on using radar returns to construct sentences with which to describe targets of interest. The sentences corresponding to detected targets are then parsed or analyzed to infer the target identity according to a predetermined set of syntactical and grammatical rules.[34]

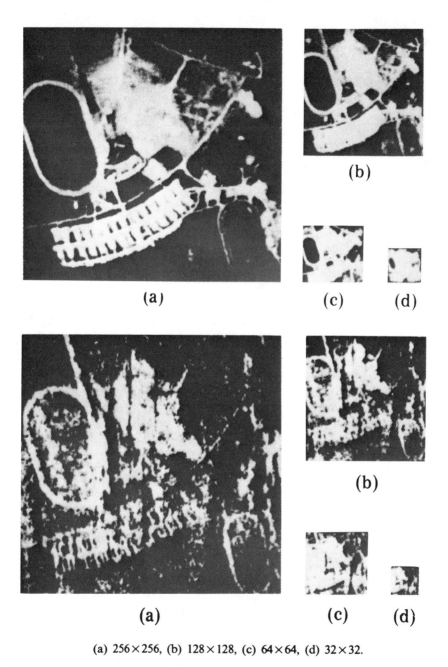

(a) 256×256, (b) 128×128, (c) 64×64, (d) 32×32.

Figure 21-9. Comparison of optical (top), radar (bottom), and noise-corrupted images of a stadium and parking lot. (By permission, from Wong, et al., ref. 5; © 1978 Academic Press, Inc.; and from Hall, ref. 12; © 1979 Academic Press, Inc.)

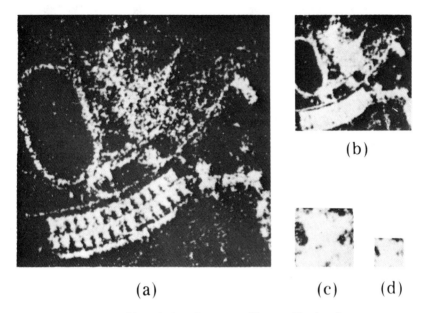

(a)   (c)   (d)

Figure 21-9. Noise-corrupted images *(Continued)*

Specific sensor-system designs or configurations provide various methods for generating radar signals and for extracting features from target returns. For example, a distinction between different targets may be observed through signal fluctuation differences,[35] that is, the fluctuations in amplitude and phase shift induced in the behavior of coherent signals. These signal fluctuation differences may also be used in conjunction with statistical decision tests for target classification.[36] Through the use of autoregressive time series analysis and linear prediction filtering, a sequential approach has been proposed[37] which is based on illuminating the target with consecutive pulses until a predetermined error level is reached. The aforementioned techniques represent some of the more common approaches used to derive target information and to extract target features. Table 21-3 provides a brief summary of these techniques and an overview of the methods used in feature extraction.

In a sense, the selection of target features is the most difficult and most important task in developing radar target recognition techniques. Further research in the general area of feature extraction and robustness to provide a better understanding of the interrelationships that may exist between feature extraction methods seems to be warranted.[38] At present, there seems to be a growing interest in heuristics-based methods, which is primarily due to the emergence of artificial intelligence techniques in target recognition.[39]

**Table 21-3. Examples of Feature Extraction Techniques
and Target Representation Methods.**

Intuitively derived features
   Target size and shape descriptors
   Descriptors of scattering centers or relationships between major scatterers
   Descriptors of measured target signatures

Heuristics-based rules
   Priority-driven decision rules
   Scenario-dependent target representations
   Multiple-target array descriptions

Discriminant and principal-component analysis
   Linear and higher-order functions of target data
   Eigenvalue expansions
   Statistical correlations and covariance analysis
   Linear regression

Transformations
   Moment invariants
   Fourier transformations
   Simplex transformations
   Other symmetry-preserving operations

System-dependent signal generation and processing
   Filtering techniques
   Waveform-coding techniques
   Features derived from specific sensor-system parameters or configura-
      tions

Theoretical methods
   Target expansions in terms of poles or resonances
   Scattering matrix parameters

21.3.3 Preclassification: Separation of Targets of Interest From Other Targets

Separating targets of interest from other radar targets is important in those applications where there is no control over the types of targets that may be encountered by the radar. In such instances, extraneous or alien targets must be separated from targets that are of interest. Otherwise, the recognition system will be forced to assign any alien target to one of the target categories of interest, thereby causing a misclassification. The task of separating alien targets from targets of interest is called *preclassification*.

Preclassification differs from other decision-making steps in a recognition system in one important aspect: The preclassifier makes a decision regarding the identity of certain targets without any a priori knowledge of them. More specifically, the preclassifier determines whether a given target is recognizable (i.e., is it a target for which prior information is available?) or unrecognizable (i.e., is it an alien target, for which prior information is not available?). This is the salient characteristic of a preclassification scheme: the ability to recognize an alien target with limited or no information about it.

Among the methods that can be used in preclassification are clustering analysis techniques.[40] Cluster analysis attempts to sort data into subsets such that each subset contains only targets that are similar to each other. There are numerous ways of constructing clustering algorithms, but most methods use a similarity measure or function to describe the degree of similarity between targets in any given subset. This similarity measure may vary, taking the form of a distance function, a correlation measure, or even a determinant.

Clustering techniques can be illustrated by considering once again the ground vehicle recognition problem of Figure 21-3. Using the idea of selecting intuitively derived features introduced in Section 21.3.2, let trucks and tanks be represented in terms of the dominant amplitudes in the H and V receive polarization channels (deleting RCS as a third feature to simplify matters). The resulting two-dimensional feature vector (H, V) thus represents a single look at each target, and numerous looks can therefore be plotted in a two-dimensional feature space, as shown in Figure 21-8. A simple clustering algorithm for this idealized case might proceed as follows:

1. Find the mean and standard deviation for each target class of Figure 21-8 along the H and V axes. Hypothetical means and standard deviations are shown in Figure 21-10.
2. Any given target of unknown origin is of interest if its corresponding feature vector (H′, V′) falls within the boundaries defined by two standard deviations from either of the means of the two classes. Otherwise, the target is labeled as an alien.

As illustrated in Figure 21-10, a clustering technique helps (1) to find the location of clusters (specified by the means in the simple two-step procedure outlined above) and (2) to determine the dispersion of the clusters or provide a measure of how they are distributed (represented by the standard deviations in this example).

The decision of whether a target falls into a particular cluster can be made by measuring, for example, the Euclidean distance between the feature vector corresponding to the unknown target (H′, V′) and the mean of the cluster center. In other words, if the mean of the cluster for the tank is the ordered pair (H_0, V_0) and the unknown target's feature vector is (H′, V′), then the Euclidean distance D between these two is given by

$$D = [(H' - H_0)^2 + (V' - V_0)^2]^{1/2}. \tag{21-1}$$

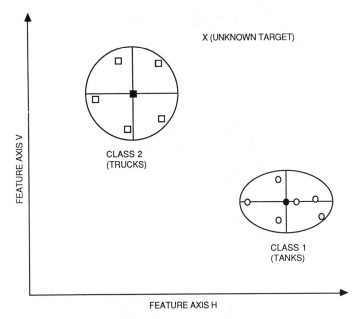

Figure 21-10. Feature space showing the feature vectors corresponding to tanks (circles) and trucks (squares). The filled-in circle (square) represents the mean for the cluster associated with the tank (truck). The dashed curves represent the two-standard deviation boundary for each cluster along the H and V axes. A target of unknown origin (x) is compared to its distance in feature space to each cluster mean.

Eq. (21-1) is, as expected, the expression for the distance between two points in a two-dimensional space. The distance D is subsequently compared to some threshold value, which in this example was taken to be the equivalent of two standard deviations. This simple example illustrates the use of the Euclidean distance as a measure of the similarity between targets within the context of a clustering approach.

The example presented above is, of course, highly idealized. First, the features H and V were assumed to have been conveniently selected to maximize the similarity among samples of the same target class. In reality, rarely are only two features used, and the clustering for each target class is seldom distributed so compactly and disjointedly from other classes. Second, one must usually deal with an indefinite number of alien targets, and it is therefore difficult to determine how to select the clusters of targets of interest properly. Third, it is very difficult to test a clustering scheme extensively, since representative data corresponding to alien targets are not normally available. In spite of these shortcomings, however, there are few ways to accommodate alien targets other than measuring the degree of dissimilarity with respect to targets of interest.

Table 21-4. Examples of Techniques Used in Separating Targets of Interest from Other Targets.

Hierarchical clustering methods
  Linkage methods
  Centroid methods
  Optimization techniques

Nonhierarchical clustering methods
  K-means method
  MacQueen's modified K-means method

Similarity measures
  Euclidean distance
  Weighted Euclidean distance
  Mahalanobis distance

Correlation measures

Table 21-4 lists some of the more common approaches to the problem of separating targets of interest from other targets. The table includes hierarchical techniques, in which the clusters are formed according to a set of rules that allow various degrees of similarity to be assigned to the data.[41] Various non-hierarchical techniques, such as the well-known K-means algorithm, sort through the target data with no predetermined ordering sequences.[40] In general, clustering techniques and similarity measures should be viewed as an aid in grouping the target data in order to find interrelationships between these groupings. Clustering techniques, however, may not offer much insight into feature invariance or overall robustness and should be viewed as constituents of a target recognition process rather than as candidates for fully autonomous target recognizers.

21.3.4   Target Classification

The process of classification involves the assignment of an input target signature to one of several classes. In practice, the classification algorithm is generally fast and the number of classes is finite. On the other hand, the actual development of this algorithm is a time-consuming and iterative exercise. If enough attention is paid to feature extraction, however, the task of developing a classifier is considerably simplified.[20]

An example of a classifier was given in Section 21.3.2, where a clustering procedure was discussed in the context of alien-target separation. The procedure illustrated in Figure 21-10 serves to distinguish not only between targets of interest and other targets, but also between the two classes of interest. This can

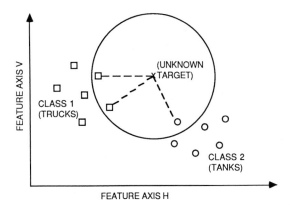

Figure 21-11. Illustration of the NN technique. A search is made for the (three) nearest neighbors of the unknown target X in (H, V) feature space, and the target is assigned to the category containing the largest number of neighbors. Thus, the target is assigned to Class 1, since there are two squares and only one circle closest to the target.

be accomplished by extending the algorithm to decide that targets that fall within one of the target clusters belong to the class (i.e., either tank or truck in this example) corresponding to that cluster. Some of the methods listed in Table 21-4 may also serve as classifiers.

Another example of a classification method is the nearest-neighbor (NN) technique.[20] This classifier is based on a nonparametric approach closely related to the Parzen window PDF estimator;[16] it is appealing primarily because it is a conceptually simple scheme. Considering Figure 21-11, assume that the classification problem is, as before, to distinguish between tanks and trucks, and that (based on heuristic arguments) the selected target features consist of the dominant amplitudes in each of the H and V receive channels. Hence, Figure 21-11 shows several target signatures for the two targets of interest based on these two-dimensional feature vectors (H, V). The target of unknown origin, whose location in feature space is marked with an X in Figure 21-11, is the location at which a search is begun for a specific number n of nearest neighbors. The figure illustrates the case for $n = 3$, that is, a search is made for the three neighbors that are nearest to the unknown target X. The mathematical expression for distance D can be Euclidean distance, as in Eq. (21-2):

$$D_{xi} = [(H_x - H_i)^2 + (V_x - V_i)^2]^{1/2} \qquad (21-2)$$

where

H_x = value of dominant amplitude in the H channel
V_x = value of dominant amplitude in the V channel

and the subscripts x and i correspond to the unknown target and to signature number i of either the truck or tank classes (i ranges from 1 to as many prototype signatures as are stored in memory). The algorithm for $n = 3$ may then proceed as follows:

1. Sort through all of the stored prototype signatures, for both tank and truck classes, until three signatures are found whose values of D [from Eq. (21-2)] are the smallest.
2. Assign the unknown target to the class containing the largest number of prototype signatures out of the three signatures found in step 1.

By inspecting Figure 21-11, we see that the unknown target X is closest to two signatures from Class 1 (trucks) and one signature from Class 2. Hence, the target in question is assigned to Class 1, which means that the target is classified as a truck. Note that if the unknown target X is actually neither a tank nor a truck, this technique would result in a misclassification by forcing the target to one of the only two available classes; this can be avoided by using a preclassification method discussed in the previous section.

The NN technique has numerous advantages and disadvantages. One appealing characteristic is that the cluster distributions in feature space are left undisturbed, since the clusters are faithfully stored in memory through their representative prototypes. Another advantage is the fact that the decision logic is simple (i.e., the unknown is assigned to the class containing the majority of nearest neighbors). On the other hand, sorting through the stored prototypes may be time-consuming, and the necessity of storing prototypes may place strict demands on memory capacity (especially in cases where there are several prototypes and high feature dimensionality). However, if there seems to be a relatively high degree of overlap of different class clusters in feature space (e.g., if the squares and circles of Figure 21-11 were scattered randomly in feature space), then an NN technique may be appropriate.

Some of the demands that the NN method places on processing time and storage capacity can be alleviated by modifying the algorithm such that only a predetermined number of prototype vectors are stored for each class. These prototypes are generated by computing localized means and variances of cluster centers which are representative of the cluster distribution of the target data. For instance, rather than storing all of the prototypes for both classes shown in Figure 21-11, simply store two prototypes: the mean for each class.

The NN approach to classification was selected for discussion because it is one of the simplest techniques to illustrate graphically. There are other techniques that can be implemented as classifiers. A significant portion of the research in target recognition has been devoted to developing, implementing, and testing algorithmic procedures based on PR techniques.[42] Statistical[43] as well as syntactic[44] pattern recognition methods are developed somewhat independently of the sensor system or the data type, since the objective of the PR methods is to separate the different target classes by means of boundary surfaces or by invoking the proper sentence-parsing techniques[21] (corresponding to statistical or syntactic methods, respectively). Other data-independent methods include clustering techniques[45] to find measures of similarity (or dissimilarity) between different target classes, as discussed in Section 21.3.2.

Statistical PR methods including linear, piecewise linear, and quadratic discriminant functions[9] of the features have also been used in ground-target classification applications.[46] Note that clustering and PR methods such as these provide well-defined analytical expressions that can be easily implemented in a decision-making sequence. If target features are selected judiciously, however, the likelihood that different algorithms will yield vastly different results may decrease. Hence, the major differences among PR classification algorithms (assuming that an optimal set of features is selected) will probably not lie in their respective error rates. Rather, the main differences will probably be in their respective speeds, processor memory requirements, and overall architectural complexity.

The interest in classification algorithms has extended beyond the PR methods mentioned thus far. Various attempts have been made to link the type of collected data directly to the algorithmic procedure. These data-dependent procedures include (1) algorithms that convert analog waveforms to binary sequences from which words are selected such that the target is described by the frequency of occurrence of binary words,[47] (2) algorithms that are based on digital spatial frequency filtering of curves of target return as a function of frequency,[48] and (3) algorithms that are based on periodically AM targets wherein the received signal is a function of target type, range, velocity, orientation, and noise content.[49]

Theoretical approaches to target classification have also been extensively investigated over the past several years. One of the most common techniques involves the application of inverse scattering methods.[50] In these methods, information concerning the scattering mechanism (i.e., the target) is obtained from integral expressions that are functions of the transmitted and received signals (this is in contrast to the direct-scattering problem, in which information about both the transmitted signal and the scatterer is known, and the received signal can thus be directly calculated). Inverse scattering methods may be used to help define features for specific sets of sensor systems and targets.[51] The integral expressions are usually difficult to solve, however, and most studies tend to concentrate on simple targets.* In addition to inverse scattering tech-

---

*This is mainly due to the difficulty of extracting the salient target characteristics embedded in integral functions of the inverse scattering solution. See, for instance reference 50.

niques, a phenomenological approach[52] based on the polarization scattering matrix has emerged in recent years as a practical method for describing complex targets within a theoretical framework, with applications in discriminating targets from background clutter interference.[53,54] Chapter 20 further discusses polarization scattering matrix considerations.

Theoretical approaches may be useful in deriving relationships between the scattering mechanism and the observed radar data, but they may prove difficult to validate experimentally. This drawback is mainly due to the difficulties that arise in attempting to generalize a theory to incorporate different complex targets or different radar system parameters. In some cases, the lack of appropriate data bases has also made it difficult to verify theoretical approaches.[54]

Recent advances in artificial intelligence (AI) have helped to define promising directions in target recognition methods.[39,55,56] In contrast to rigid algorithmic sequences (such as those of PR methods) and mathematically rigorous approaches (such as those used in theoretical models), AI techniques rely on heuristic-based procedures. For example, the rules of logic in AI systems may turn out to be rules of thumb that a human expert might use in decision-making.[57] Other useful AI techniques include (1) the representation of knowledge or information such that the stored pieces of information are linked by sets of rules or relationships and (2) methods for controlling or managing different functions of a recognition system through a prioritization system. One of the important characteristics of some AI systems is their ability to adapt to or learn from different operational situations. This degree of flexibility can be of great importance in recognition problems that involve the analysis of contextual information or the fusion of multiple sensors.[58]

Table 21-5 lists some of the methods and techniques that are widely used in developing classification algorithms and target recognition systems. When considering these approaches to target recognition, note that (1) the appropriate method is largely application dependent, (2) the actual development of a target recognizer is usually a time-consuming and iterative procedure, and (3) the recognition system should exhibit a certain degree of robustness in operational situations.

21.3.5 Training and Testing of Recognition Systems

One important consideration regarding the development and implementation of target recognition systems is that a general methodology for developing a target recognizer does not yet exist. Most target recognition problems, however, are generally addressed in two stages: (1) the training phase and (2) the recognition phase. This is schematically illustrated in Figure 21-12. The first phase is devoted to the development and construction of the recognition system and is, therefore, viewed as a non-real-time developmental process. The second phase

Table 21-5. Examples of Methods and Techniques Used in Recognition Systems.

Pattern recognition techniques
  Syntactic
  Statistical
  Structural

AI methods
  Heuristics-based systems
  Image understanding
  Knowledge-based and expert systems

Synergistic approaches
  Multiple sensor fusion
  Multiple data base fusion
  Fusion of multiple approaches

Theoretical techniques
  Inverse scattering methods
  Polarization scattering matrix approach
  Algorithms referenced to target models

Other approaches
  Semiautomatic (man-in-the-loop) systems
  Exploitation of specific sensor system parameters

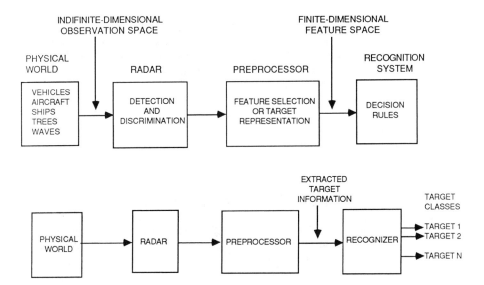

Figure 21-12. Training and recognition phases.

represents the actual implementation of the developed system, whether in a testing mode or an operational mode; the second phase thus represents the real-time, automated, decision-making portion of the recognition process.

In general, the training phase involves the selection and development of techniques for extracting target features (preprocessing) and making decisions. Examples of the methods used in preprocessing and decision making were illustrated in the simple ground vehicle recognition problem discussed in previous sections (e.g., intuitively derived features, clustering techniques, and NN classification). Approaches such as those previously discussed and summarized in Tables 21-3, 21-4, and 21-5 are further illustrations of techniques used in the development and construction of a target recognizer. After the basic structure of the recognition system has been designed, target data are used to determine the appropriate decision rules. For instance, the weighting coefficients of discriminant functions in statistical PR are obtained by using a portion of the available target data (the remaining portion of the data may be retained for testing). This procedure is commonly referred to as *training*, since the algorithmic procedure is trained to recognize the data corresponding to targets of interest.

After the basic algorithmic structure is defined in the training phase, the recognition system is then tested (or, if appropriate, is implemented in an operational situation). The recognition system should ideally be designed to be implemented in a few simple steps such that recognition can be executed quickly to exploit fully the automated character of the algorithmic sequence. At this point, issues such as speed, accuracy, robustness, and overall effectiveness can be further addressed.

In practice, the training and recognition (or testing) phases are repeated numerous times until the level of performance of the recognition system is determined and deemed appropriate. Different iterations may be required, for instance, to implement different feature extraction methods or to introduce different testing data to the recognizer. In such cases, the use of modeling and simulation techniques may prove useful, since simulated data can be used to represent different sensor system parameters or different target–sensor encounters. At present, increasing emphasis is being placed on field testing of recognition systems to determine system performance more precisely in realistic operational environments.[58]

21.4 CURRENT TRENDS AND FUTURE DIRECTIONS

The foregoing discussions outlined some approaches generally used for target recognition. Present trends in these approaches may be used to assess the state of the art and provide a basis for projecting future research directions in this field. One overall trend seems to be the emphasis on approaches which tend to draw from the methodologies of pattern recognition and AI. These approaches offer the advantage that numerous well-developed techniques can be applied to

**Table 21-6. Overview of Topics of Current
Interest in Target Recognition.**

Sensor system considerations
  MMW componentry
  Miniaturized signal processors
  Optimization of sensor system parameters
  Suppression of noise and clutter interference

Information-processing considerations
  AI methods
  Invariance of target features
  Polarimetric techniques
  Development of theoretical models

Operational considerations
  Overall robustness of target recognizer
  Fusion of multiple sensors
  Man-in-the-loop interactions and human factors
  Mission requirements
  Presence of multiple targets and changes in target appearance

a variety of problems. These techniques, however, generally provide no insight into the underlying physical processes that govern the scattering interactions and, therefore, do not address the fundamental problems associated with the interrelationships between the sensor system, scattering system, and information processing system. This realization has led to increased interest in considering synergistic approaches as well as real-time recognition dynamics.[58,59]

In addition, the VLSI and VHSIC technologies being developed may improve computational capability by as much as two orders of magnitude.[2] Other factors which have also surfaced as important considerations for recognition systems include the effects of variations in noise level on performance[48] and the limits on performance of millimeter wave (MMW) radar imaging systems due to spatial resolution and image quality. Table 21-6 provides an overview of current trends in radar target recognition. The list of topics of interest in this field is changing rapidly as new ideas and technologies are developed in hardware, software, and operations.

21.5 SUMMARY

The recognition of targets by radar is a subject of great interest, as shown by the number of research efforts, conferences, and publications devoted to this subject. The interest in target recognition may be due to several factors, including (1) the desire to know more about a detected target than its mere presence, (2) the applicability of target recognition techniques to numerous research areas, and (3) the degree of technological maturity that has been achieved in radar sensing, evolving from a predominantly detection-only aid to include advanced systems capable of highly sophisticated tasks. The degree of technological maturity has been achieved mainly by recent advances in radar componentry, the availability of fast and efficient signal-processing hardware, and the development of automated decision-making procedures.

In spite of the efforts devoted to radar target recognition, a methodology does not yet exist for developing fully automatic recognition systems that are both operationally robust and can be implemented in any type of application. The lack of such a general methodology is largely due to the fact that recognition systems are predominantly application dependent. In addition, the interrelationship between the sensor system, scattering system, and information-processing system pose multifaceted problems that warrant further investigation.

This chapter discussed some of the most widely used approaches to target recognition. The approaches generally fall into one of three broad categories, depending on whether the emphasis is placed on specific data bases, decision-making methods, or radar systems. Examples of these approaches are (1) theoretical methods, (2) algorithmic procedures, (3) synergistic approaches, (4) modeling and simulation techniques, (5) sensor system optimization techniques, and (6) specialized application-oriented methods. Some of these approaches were illustrated by a simple example to show how features may be derived intuitively. These features were subsequently used to separate targets of interest from other targets and then implemented in a simple classification scheme to assign the targets of interest to one of two possible categories. In conjunction with this simple example, numerous references were provided for those interested in obtaining further information.

Topics of current interest in target recognition include such issues as (1) overall robustness of the recognition system, (2) implementation of AI techniques, (3) utilization of high-speed, miniaturized signal-processing technology, (4) fusion of multiple sensors, and (5) use of a human operator for making some of the recognition decisions. These and other topics of interest have considerably expanded the research horizons in target recognition, offering new and challenging problems in this important aspect of radar sensing.

21.6 REFERENCES

1. N. F. Ezquerra, "Radar Target Recognition: A Survey," *Proceedings of the International Conference on Radar*, May 1984, pp. 281–285.
2. R. P. Shenoy, "Trends in Radar System Design in the Eighties and Beyond," *Journal Institute of Engineering of India*, ET 1, vol. 62, August 1981, pp. 12–17.
3. F. LeChevalier, "Radar Target and Aspect Angle Identification," *Proceedings of the Fourth International Joint Conference on Pattern Recognition*, Kyoto, Japan, IEEE Publication N78CH1331-8C, New York, pp. 398–400 (1979).
4. R. H. Blackmer, Jr., R. O. Duda, and R. Reboh, "Application of Pattern Recognition Tech-

niques to Digitized Weather Radar Data,'' Final Report on Contract 1-36072, SRI Project 1287, Stanford Research Institute, Menlo Park, Calif., 1973.

5. R. Y. Wong et al., ''Scene Matching with Invariant Moments,'' *Computer Graphics and Image Process.*, vol. 8, 1978, pp. 317–320.

6. J. D. Echard, J. A. Scheer, E. O. Rausch, W. H. Licata, J. R. Moore, and J. A. Nestor, ''Radar Detection, Discrimination, and Classification of Buried, Non-Metallic Mines,'' Final Report GIT/EES Project A-1828, Georgia Institute of Technology, Atlanta, February 1978.

7. J. D. Echard, E. E. Martin, D. L. Odom, H. G. Cox, ''Discrimination Between Targets and Clutter by Radar,'' Final Technical Report, Georgia Institute of Technology, Atlanta, December 1981.

8. N. F. Ezquerra, ''Stationary, Tactical Target Classification,'' paper delivered at the 1979 Non-Cooperative Target Recognition Conference, San Diego, October 1979.

9. J. T. Tou and R. C. Gonzalez, *Pattern Recognition Principles*, Addison-Wesley Publishing Co., Reading, Mass., 1974.

10. S. Watanabe, *Methodologies of Pattern Recognition*, Academic Press, New York, 1969.

11. K. S. Fu, *Digital Pattern Recognition*, Springer-Verlag Publishing Co., Berlin, 1980.

12. E. L. Hall, *Computer Image Processing and Recognition*, Academic Press, New York, 1979.

13. S. Watanabe, *Knowing and Guessing*, John Wiley & Sons, New York, 1969.

14. C. Chen, *Statistical Pattern Recognition*, Spartan Books, Boston, 1973.

15. L. Uhr, *Pattern Recognition, Learning, and Thought*, Prentice-Hall, Englewood Cliffs, N.J., 1973.

16. E. Patrick, *Fundamentals of Pattern Recognition*, Prentice-Hall, Englewood Cliffs, N.J., 1972.

17. J. K. Aggarwal, R. O. Duda, and A. Rosenfeld, *Computer Methods in Image Analysis*, IEEE Press, New York, 1977.

18. A. K. Agrawala, *Machine Recognition of Patterns*, IEEE Press, New York, 1977.

19. H. C. Andrews, *Introduction to Mathematical Techniques in Pattern Recognition*, John Wiley & Sons, New York, 1972.

20. R. O. Duda and P. E. Hart, *Pattern Classification and Scene Analysis*, John Wiley & Sons, New York, 1973.

21. K. S. Fu, *Syntactic Methods in Pattern Recognition*, Academic Press, New York, 1974.

22. K. Fukunaga, *Introduction to Statistical Pattern Recognition*, Academic Press, New York, 1972.

23. J. M. Mendal and K. S. Fu, *Adaptive Learning and Pattern Recognition Systems*, Academic Press, New York, 1972.

24. M. Minsky and S. Papert, *Perceptions*, MIT Press, Cambridge, Mass., 1969.

25. N. Nilsson, *Learning Machines*, McGraw-Hill Book Co., New York, 1965.

26. R. C. Gonzalez and M. G. Thomason, *Syntactic Pattern Recognition*, Addison-Wesley Publishing Co., Reading, Mass., 1978.

27. B. V. Dasarathy et al., ''Recognition of Targets characterized by Point Ensembles,'' *Southeastcon 81 Region 3 Conference Proceedings*, vol. 1, April 1981, pp. 655–659.

28. N. F. Ezquerra, ''Application of Pattern Recognition Techniques to Radar Signature Classification,'' Fourth International Conference on Pattern Recognition Proceedings, vol. 1, Dec 1980, pp. 350–353.

29. M. P. Hurst et al., ''An Approach to Target Identification,'' *Proceedings of the Antenna Applications Symposium*, vol. 1, September 1982, pp. 356–370.

30. J. Saillard et al., ''An Original Method for the Analysis of the Power Diffraction Pattern of a Radar Target for Shape Recognition Purposes,'' *Annales des Telecommunications*, vol. 36, May–June 1981, pp. 359–368.

31. H. E. Hunter, ''Application of ADAPT to Quick Look Classification of Composite Radar Signatures,'' Final Report No. 74-4, Adapt Service Corp., Reading, Mass., 1974.

32. L. H. Johnson, "The Shift- and Scale-Invariant Fourier-Mellin Transform for Radar Applications," Report AD-A093690 Lexington Labs, Inc., Cambridge, Mass. 1980.

33. G. A. Ioannidis et al., "Aircraft Target View Determination for Method of Moments Identification," Report AD-A103483, Hughes Aircraft Co., El Segundo, CA, 1981.

34. F. LeChevalier et al., "An Approach to Recognition Criteria of Radar or Sonar Targets," *Proceedings of the 1984 International Conference on Radar*, May 1984, pp. 269–274.

35. K. Von Schlachta, "A Contribution to Radar Target Classification," *Proceedings of the 1977 International Radar Conference*, vol. 1, 1977, pp. 135–139.

36. K. Von Schlachta, "Signal Properties and Target Models in the View of Target Classification at a Surveillance Radar," *Proceedings of the Advanced Study Institute*, Reidel Publishing Co., Goslar, West Germany, pp. 55–64, 1975.

37. C. W. Therrien, "A Sequential Approach to Target Discrimination," *IEEE Transactions on Aerospace and Electrical Systems*, vol. AES-14, May 1978, pp. 433–440.

38. L. Heng-Cheng et al., "Optimal Frequencies for Aircraft Classification," Report AD-A065697, Ohio State University, Columbus, 1979.

39. B. P. McCune et al., "Radar with Sight and Knowledge," *Defense Electronics*, vol. 2, August 1983, pp. 55–60.

40. M. R. Anderberg, *Cluster Analysis for Applications*, Academic Press, New York, 1973.

41. N. F. Ezquerra and L. L. Harkness, "Pattern Recognition Applications in Radar Data Processing," *Proceedings of the 1981 IEEE International Conference on Cybernetics and Society*, vol. 1, October 1981, pp. 24–28.

42. L. F. Pau, "Aerospace Applications of Pattern Recognition," *L'Aeronautique et l'Astronautique*, no. 42, March 1973, pp. 63–78.

43. N. F. Ezquerra et al., "Application of Pattern Recognition Techniques to the Processing of Radar Signals," *1982 IEEE International Conference on Acoustics, Speech and Signal Processing Proceedings*, vol. 2, June 1982, pp. 347–350.

44. F. Le Chevalier et al., "Syntactic Tomographic Synthesis Applied to Radar Classification," Congress Sur la Reconnaisance des Formes et Intelligence Artificelle, Tolouse, 1979.

45. R. L. Whitman et al., "Automatic Clustering of SAR Targets," *Proceedings of the National Aerospace and Electronics Conference*, vol. 2, May 1980, pp. 717–724.

46. N. F. Ezquerra, "Radar Target Recognition," *Proceedings of the 1982 IEEE International Radar Conference*, October 1982, pp. 262–265.

47. D. G. Kulchak, "Target Classification by Time Domain Analysis of Radar Signatures," Report AD-A055221, Air Force Institute of Technology, Wright-Patterson AFB, Ohio, 1977.

48. W. B. Goggins, "Identification of Radar Targets by Pattern Recognition," Report AD-764713, Air Force Institute of Technology, Wright-Patterson AFB, Ohio, 1973.

49. C. V. Stewart, "Identification of Periodically Amplitude Modulated Targets," Ph.D. thesis, Air Force Institute of Technology, Wright-Patterson AFB, Ohio, 1978, Report Microfilm Order #7818599.

50. Special Issue on Inverse Methods in Electromagnetics, *IEEE Transactions on Antennas and Propagation*, vol. AP-29, no. 2, March 1981.

51. J. W. Sammon et al., "Time-Domain Analysis for Inverse Scattering." Report #PAR-72-28-VOL-1, Pattern Analysis and Recognition Corp., Rome, NY, 1972. Also see M. Morag, "Radar Target Imaging by Time-Domain Inverse Scattering," Report AD-A102660/8, Naval Postgraduate School Master's Thesis, 1981.

52. J. R. Huynen, "*Phenomenological Theory of Radar Targets*," Ph.D. dissertation, Drukkerij Bronder-Offset N.V., Rotterdam, 1970.

53. S. R. Cloude, "Polarimetric Techniques in Radar Signal Processing," *Microwave Journal*, vol. 26, no. 7, March 1983, pp. 119–125.

54. W. A. Holm, "Polarization Scattering Matrix Approach to Stationary Target/Clutter Discrimination," *Proceedings of the International Conference on Radar*, May 1984, pp. 461–465.

55. E. M. Drogin, "AI Issues for Real Time Systems," *Defense Electronics*, August 1983, pp. 57–62.
56. R. G. Naedel, "Intelligent Associative Memory (IAM): An Overview," *Defense Science*, vol. 2, August 1983, pp. 14–20.
57. P. Winston, *Artificial Intelligence*, Addison-Wesley Publishing Company, Reading, 1980.
58. R. Johnson, "Automatic Target Recognition Fuse Sensors and Artificial Intelligence," *Defense Electronics*, vol. 1, April 1984, pp. 106–115.
59. B. V. Dasarathy, "Target Recognition System Design: Computational Structures with Response Time Characteristics Dictated by the Operational Environment," Report AB-B041709/7, M. and S. Computing, Inc., Huntsville 1979.

PART 7
RADAR ECCM

- Considerations and Techniques—Chapter 22

The last part of this book also deals with a subject not normally covered in most radar texts devoted to principles and fundamentals. Unfortunately, modern electronic warfare (EW) involving sophisticated electronic countermeasures (ECM) and electronic support measures (ESM) is causing radar vulnerability considerations, or radar electronic counter countermeasures (ECCM), to become of overriding importance in most modern operational engagements. The recent limited-scale military encounters in the Falkland Islands and over the Golan Heights are excellent examples of both the limitations of radar when proper attention is not given to ECCM and the immense advantage a properly trained and equipped force can now achieve in an engagement dominated by electronic warfare.

Chapter 22 introduces the complete spectrum of electronic warfare including the accepted definitions of ECM, ESM, and ECCM as the major elements of electronic warfare. However, the emphasis of the chapter is on the vulnerability of radar to EW actions and techniques to reduce a radar's vulnerability, ECCM. Each major element of the radar, from the antenna to the operator, is considered from an ECCM point-of-view. Finally, a matrix is presented illustrating possible EW techniques against radar and the specific ECCM that is most effective in reducing the effectiveness of that particular technique. Unfortunately, much of the specific information about the area of electronic warfare is sensitive militarily and is normally classified making this topic difficult to cover in any detail in the open literature. The discussion in Chapter 22 suffers from that limitation.

22
RADAR ECCM CONSIDERATIONS AND TECHNIQUES

Edward K. Reedy

22.1 INTRODUCTION AND OVERVIEW

From the very beginning of radar, attempts have been made to disrupt its use through various forms of electronic and nonelectronic countermeasures and associated techniques. These countermeasures include active jamming, or the attempt to introduce extraneous electronic signals into the radar receiver and processor, passive techniques such as chaff, decoys, and so on; intercept equipment and techniques such as direction finding (DF), radar warning receivers, and electronic intelligence (ELINT) receivers; and radar homing missiles or anti-radiation missiles (ARM). In addition, target evasive actions, maneuvers, and flight plans can be developed as countermeasures against radar.

Techniques included in the radar or as part of the radar's general operational philosophy primarily to counter these countermeasures are appropriately designated *radar counter-countermeasures*. Even though not all of the techniques are electronic, the general term *electronic counter-countermeasures (ECCM)* is normally used to refer to the collection of both passive or nonelectronic and electronic techniques used to counter or reduce the effectiveness of radar countermeasures used by the enemy.

Radar ECCM is a very broad subject. Several books are available which deal with radar ECCM either exclusively[1,2] or in conjunction with discussions of electronic warfare and countermeasures.[3-6] The subject is much too large to cover in detail in this chapter. Only a generalized summary or overview of radar counter-countermeasures is presented. This treatment deals with nomenclature, definitions, and semantics rather than with specific technical descriptions and equipment details.

22.2 ELECTRONIC WARFARE DEFINITIONS

Before proceeding further, it is important to define certain terms which relate to radar ECCM and more generally to electronic warfare. Figure 22-1 gives the U.S. Department of Defense's accepted definitions for electronic warfare and associated components.[7] As noted, electronic warfare includes all actions re-

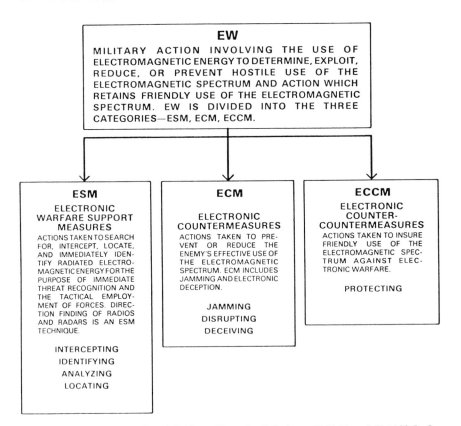

Figure 22-1. Electronic warfare definitions. (From the U.S. Army Field Manual FM 100-5, *Operations*).

quired to prevent hostile use of the EM spectrum and retain friendly use. Subelements include electronic support measures (ESM), generally passive electronic eavesdropping and location techniques, electronic countermeasures (ECM), active approaches to prevent or reduce the use of the EM spectrum by the enemy, and ECCM, actions taken to retain friendly use of the spectrum.

The remainder of this chapter deals almost exclusively with radar ECCM. Over the years, numerous ECCM techniques have been developed. Table 22-1 lists more than 150 types of radar ECCM.[8] Many of them will be discussed in this chapter.

22.3 GENERAL ECCM CONSIDERATIONS

Several basic considerations or objectives dictate radar ECCM strategy. The primary objective is always to negate the effects of the enemy's ECM on the

Table 22-1. An ECCM Lexicon.

Acceleration Limitation
Angle Sector Blanking
Angular Resolution
Audio Limiter
Aural Detection
Autocorrelation Signal Processing
Automatic Cancellation of Ex-
  tended Targets (ACET)
Automatic Threshold Variation
  (ATV)
Automatic Tuner (SNIFFER)
Automatic Video Noise Leveling
  (AVNL)
Back-Bias Receiver
Baseline-Break (on A-Scope)
Bistatic Radar

Broad-Band Receiver
Coded Waveform Modulation

Coherent Long-Pulse Discrimina-
  tion
Compressive IF Amplifier
Constant False Alarm Rate (CFAR)
  Cross-Gated CFAR

  Dispersion Fix (CFAR)
  IF Dicke-Fix CFAR (Dicke-Fix)
  MTI CFAR
  Unipolar Video CFAR
  Video Dicke-Fix CFAR (Dicke-
    Fix)
  Zero Crossings CFAR
Contiguous Filter-Limiter
Cross Correlation Signal Processing
CW Jamming Canceller
Detector Back Bias (DBB) (Same as
  Detector Balanced Bias)
Dicke-Fix

  Clark Dicke-Fix (Cascaded
    Dicke-Fix

  Coherent MTI Dicke-Fix
  Craft Receiver
  Dicke Log Fix
  IF Canceller MTI Dicke-Fix
  IF Dicke-Fix CFAR (Zero Cros-
    sings Dicke-Fix CFAR)

Gated FAGC
Instantaneous AGC
Manual Gain Control
Pulse Gain Control
Sensitivity-Time Control
Guard-Band Blanker
High PRF Tracking
High Resolution Radar
IF Diversity

IF Limiter
Image Suppressor
Instantaneous Frequency Correlator
  (IFC-CRAFT)
Integration
  AM Video Delay Line
    Integration
  Coherent IF Integration
  Coherent (IF) Integration
    (Moving Target

  Coherent (IF) Integration (Station-
    ary Target)
  Display Integration

  FM Delay Line Integration
  Noncoherent (Video) Integration
  Pulse Integration
  Video Delay-Line Integration
Inter-Pulse Coding (PPM)
Jamming Cancellation Receiver
Jittered PRF
Kirbar Fix
Least Voltage Coincidence Detector
Linear Intra-Pulse FM (CHIRP)
Lin-Log IF
Lin-Log Receiver
Lobe-on-Receiver Only LORO, Also
  (SORO)

Log Fix (Also, Log FTC)
Logarithmic Receiver
Logical ECCM Processing

Main Lobe Cancellation (MLC)
  Monopulse MLC
  Polarization MLC
Manually Aided Tracking
Manual Rate-Aided Tracking
Matched Filtering
Monopinch
Monopulse Tracker
MTI

  Area MTI (Velocity Filter)
  Cascaded Feedback Canceller
    (MTI)
  Clutter Gating (MTI)
  Coherent MTI
  Noncoherent MTI

  Pulse Doppler
  Pseudocoherent MTI

  Single-Delay Line (MTIC
    Canceller)
  Re-Entrant Data Processor
  Three-Pulse Canceller
  Two-PulseCanceller (Single De-
    lay Line MTI Cancellation

Multifrequency Radar
Multisimul Antenna
Phased Array Radar
Polarization Diversity
Polarization Selector
Post Canceller Log FTC
PRF Discrimination
Pulse Burst Mode
Pulse Coding and Correlation
Pulse Compression, Stretching
  (CHIRP)
Pulse Edge Tracking

Pulse Interference Elimination
  (PIE)
Pulse Shape Discrimination
Pulse-to-Pulse Frequency Shift
  (RAINBOW)
Pulse Width Discrimination (PWD)
Pulse Length Discrimination (PLD)
Random-Pulse Blanker
Random Pulse Discrimination
  (RPD)

Table 22-1. *Continued*

Instananeous Frequecy Dicke-Fix	Range/Angle Rate Memory
Noncoherent MTI Dicke-Fix	Range Gating
Video Dicke-Fix CFAR	Range Rate Memory
Diplexing	Scan-Rate Amplitude Modulation
Doppler-Range Rate Comparison	Short Pulse Radar
Double Threshold Detection	Side-Lobe Blanker
Electronic Implementation of Base-line-Break Technique	Side-Lobe Canceller
Fast Manual Frequency Shift	Side-Lobe Reduction
	Side-Lobe Suppression (SLS)
	Side-Lobe Suppression by Absorbing Material
Fast Time Constant (FTC)	
Fine Frequency	Staggered PRF
Frequency Agility	Transmitter Power
Frequency Diversity	Two Pulse Autocorrelation
Frequency Preselection (Narrow Bandwidth)	Variable Bandwidth Receiver
	Variable PRF
Frequency Shift	Variable Scan Rate
Gain Control	Velocity Tracker
Automatic Gain Control (AGC)	Video Correlator
Dual Gated AGC	Wide-Bandwidth Radar
Fast AGC	Zero Crossing Counter

Source: Reference 8.

radar. However, as previously indicated, *counter-countermeasures* is a generic term which includes *any thing or any action* resulting in the degradation of enemy ECM activities. It is certainly not limited to electronic techniques or approaches, but can include tactics, deployment, operational doctrines, and so on.

Although it is sometimes not obvious, radar ECCM is equivalent in a hostile EM environment to considerations of EM compatibility which involve techniques and approaches associated with reduction of the susceptibility of electronic equipment to interference—either man-made or natural. Another consideration sometimes overlooked is that natural ECM (clouds, inclement weather, ground returns, and other clutter) requires what can be thought of as ECCM. In this case, ECCM takes the form of clutter rejection processing such as MTI or CFAR processing.

General Robert S. Dixon, Tactical Air Command Commander, during testimony before Congress on the Air Force AWACS airplane stated a basic ECM/ECCM principle:[9] "The effective range of all radars can be reduced by jamming. This is not an argument against radars, it is recognition of a limitation." This statement can be paraphrased as follows: "Any radar can be jammed."

If one side is willing to expend the resources, any radar can be jammed. However, the objective of radar ECCM is to force the enemy to expend as much

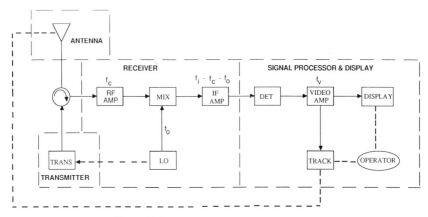

Figure 22-2. Block diagram of a typical radar.

of its limited resources as possible to jam or reduce the effectiveness of our radar systems.

Figure 22-2 is a block diagram of a typical radar. Components of the radar where ECCM can be implemented include the antenna, transmitter, receiver, and signal processor/display. These areas are identified in the figure.

Although not a part of the radar hardware, operational tactics and deployment philosophy represent another area for potential inclusion of ECCM features.

22.4 ANALYTICAL MEASURES OF JAMMING EFFECTIVENESS

Prior to discussing the areas of a radar system where ECCM features can be incorporated, several expressions relating radar basic operating parameters, jammer parameters, and the operating scenario are developed to provide analytical measures of the effectiveness of active ECM—primarily noise jamming.

Several analytical measures of jamming levels have been developed. The target reflected power received by the radar is given by

$$S = \frac{P_R G_{RT}^2 \lambda^2 \sigma (\text{SPG})}{(4\pi)^3 R_T^4} \qquad \text{watts} \qquad (22\text{-}1)$$

where

P_R = transmitted power (watts)
G_{RT} = radar antenna gain in direction of target
λ = wavelength (meters)
σ = target cross section (square meters)
R_T = range to target (meters)
SPG = radar signal-processing gain

Signal-processing gain is normally attributable to signal integration, either coherent or noncoherent.

For broadband noise jamming, the jammer power received in the bandpass of the radar is given by*:

$$J = \frac{P_J G_J G_{RJ} \lambda^2}{(4\pi)^2 R_J^2} \left(\frac{B_R}{B_J}\right) \qquad \text{watts} \qquad (22\text{-}2)$$

where

P_J = jammer power (watts)
G_J = jammer antenna gain in direction of radar
G_{RJ} = radar antenna gain in direction of jammer (may be sidelobe gain)
R_J = range to jammer (meters)
B_R = receiver bandwidth of radar (megahertz)
B_J = jammer noise bandwidth (megahertz)

Combining Eqs. (22-1) and (22-2) gives an expression for a commonly used measure of jammer effectiveness, the J-to-S ratio (J/S):

$$J/S = \left(\frac{4\pi}{\sigma(\text{SPG})}\right)\left(\frac{P_J}{P_R}\right)\left(\frac{G_J G_{RJ}}{G_{RT}^2}\right)\left(\frac{B_R}{B_J}\right)\left(\frac{R_T^4}{R_J^2}\right) \qquad (22\text{-}3)$$

The screening range, R_{TS}, is defined as the target range at which J/S = 1 or 0 dB. From Eq. (22-3), then,

$$R_{TS}^4 = \left(\frac{\sigma(\text{SPG})R_J^2}{4\pi}\right)\left(\frac{P_R}{P_J}\right)\left(\frac{G_{RT}^2}{G_J G_{RJ}}\right)\left(\frac{B_J}{B_R}\right) \qquad (22\text{-}4)$$

Eqs. (22-1) to (22-4) provide the basic analytical relationships for predicting a radar's performance in a jamming environment, for relating one radar's theoretical performance to another's and for evaluating the relative effectiveness of various ECCM techniques. Examples illustrating the use of these expressions are given later in this chapter.

22.5 ANTENNA-RELATED ECCM

The first and probably the most important area of the radar to be considered for incorporation of ECCM is the antenna. Since the antenna represents the transducer between the radar and the environment in which the radar must work, it

---

*If $B_R > B_J$, then assume that $B_R/B_J = 1$.

is the first line of defense against undesirable extraneous signals such as jamming.

22.5.1 Space Discrimination

By providing directivity to the transmitted and received EM energy, the antenna also allows space discrimination to be employed as an ECCM approach. Techniques for generating radar space discrimination include low sidelobes, sidelobe cancellers, sidelobe blanking, beamwidth control, and antenna coverage and scan control.

Consider a situation where a relatively isolated standoff jammer is employed against a radar. (A fixed ELINT or DF receiver could also be postulated.) By maintaining low sidelobes, thus preventing jamming energy from entering the radar receiver through the sidelobes, and either blanking or turning off the receiver while the radar is scanning across the azimuth sector containing the jammer or reducing the scan sector covered to prevent the radar from looking at the jammer, radar target detection effectiveness can be essentially maintained in all sectors, except where the jammer is centered. Such coverage control can be made to respond both adaptively and automatically to eliminate single spatially separated jammers and prevent detection of the radar radiation in defined regions by ELINT and DF receivers.

Smaller antenna beamwidths, or correspondingly higher antenna gain, can be employed to spotlight a target and burn through jamming. Multiple antenna beams can also be used to increase antenna directivity or gain and allow deletion of the beam containing a jammer .

Certain deception jammers depend on anticipation of the beam scan or on knowledge or measurement of the antenna scan rate. Random electronic scanning effectively prevents these deception jammers from synchronizing to the antenna scan rate, thus defeating this type of jammer.

Although they add complexity, cost, and possibly weight to the antenna, beamwidth, coverage, and scan control are valuable and worthwhile ECCM features of all radars.

The term in Eq. (22-3) which incorporates the antenna space discrimination ECCM, D_A, is

$$D_A = \left(\frac{G_J G_{RJ}}{G_{RT}^2} \right) \tag{22-5}$$

where

G_J = jammer antenna gain
G_{RJ} = radar antenna gain in direction of jammer
G_{RT} = radar antenna gain in direction of target

Normally, G_{RJ} represents antenna sidelobe gain and G_{RT} represents antenna mainlobe gain, since normal radar operation usually results in the antenna's mainlobe illuminating the target and the jamming signals entering the radar through the antenna sidelobes, as in the case of standoff jamming (SOJ) operations. A self-screening jammer (SSJ) attempts to inject a jamming signal through the antenna mainlobe by employing a jammer on board the target. In this case, G_{RJ} represents the antenna mainlobe gain, that is, $G_{RJ} = G_{RT}$, and

$$D_A = \left(\frac{G_J}{G_{RT}}\right) \tag{22-6}$$

If, by good antenna design or the addition of certain auxiliary circuits, the antenna sidelobes can be reduced to such a level that jamming energy is effective only when injected into the radar's mainlobe, then the SOJ will be effective only in preventing detection in the small sector (the width of the radar's azimuth beamwidth) centered on the jammer. Thus, low inherent antenna sidelobes or sidelobe control is extremely important and restricts jamming and detection to the mainlobe.

However, the level of sidelobes required to suppress jamming through the sidelobes is difficult to achieve in any practical antenna design. Dax[10] has estimated that sidelobes 60 dB or more below the peak mainlobe gain are required to make airborne jamming of a long-range, ground-based air defense search radar virtually ineffective. Dax also estimated that 40-dB sidelobes might be sufficient in some situations to counter an SOJ effectively. In other words, $G_{RJ}/G_{RT} \leqslant -40$ dB is required to prevent reduction of the radar's detection capability due to jamming injected into the radar receiver through the radar sidelobes.

Two additional techniques to prevent jamming from entering the radar's sidelobes are so-called sidelobe cancelling and sidelobe blanking. Both require the use of an auxiliary antenna or antennas. A simplified diagram of the basic elements of a sidelobe canceller is given in Figure 22-3. By adaptively controlling the phase and amplitude of the auxiliary channel signal and combining this signal with the main channel signal, a null in the composite antenna pattern response can be produced in the direction of a jammer. Through continuous adaptive control of the total antenna pattern, the null can be made to track the jammer. Although apparently simple in Figure 22-3, the adaptive sidelobe canceller system is relatively complex and requires one canceller loop per jammer. Approximately 20 to 25 dB of additional cancellation is possible with this technique.

Conceptually, the sidelobe blanker is less complex than the canceller. Figure 22-4 illustrates the basic principles of a sidelobe blanker. Note that an omnidirectional auxiliary antenna having a gain perhaps 3 to 4 dB larger than that

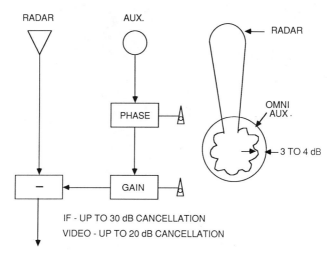

Figure 22-3. Elements of an adaptive sidelobe canceller.

of the sidelobes of the main antenna is required. When the signal in the auxiliary channel is compared with the main channel signal and found to be larger, the signal in the main channel must have been received through the sidelobes. In this case, the gate is opened, preventing the jamming signal from passing into the radar receiver and being displayed. The blanker is effective only for low duty cycle pulse or swept frequency jamming. High duty cycle and noise jamming effectively blanks the main channel most of the time, rendering the radar ineffective.

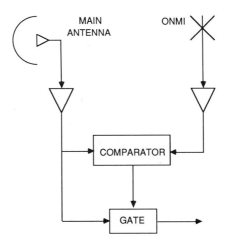

Figure 22-4. Simple sidelobe blanker.

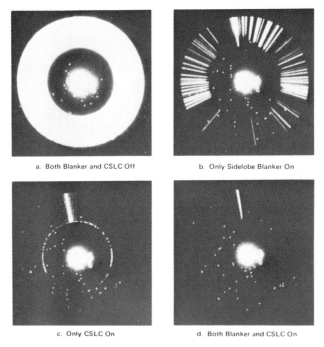

a. Both Blanker and CSLC Off   b. Only Sidelobe Blanker On

c. Only CSLC On   d. Both Blanker and CSLC On

Figure 22-5. Sidelobe blanker and canceller performance. (By permission, from Johnson, et al., ref. 11; © 1978 Horison House)

As an example of the effectiveness of sidelobe cancellers and blankers, consider a scenario where a delayed high duty cycle jammer is attempting to obscure targets from the radar's maximum range into a fixed range, as shown in Figure 22-5.* Both the coherent sidelobe canceller (CSLC) and the blanker are turned off in Figure 22-5(a), allowing the full jamming signal to pass through the radar receiver to the display. Obviously, the jammer is injecting considerable energy through the antenna sidelobes, since the PPI display is essentially ''white'' or saturated over the entire 360° of coverage. Figure 22-5(b) shows that the sidelobe blanker is only partially effective in countering the high duty cycle jamming. With only the sidelobe canceller on in Figure 22-5(c), most of the sidelobe jamming has been eliminated. The remaining noise is mainlobe jamming and a ring caused by the finite amount of time required for the phase and amplitude adjustments to occur during each pulse interval. This is the canceller ''lock-up'' time, during which the radar is susceptible to low duty cycle jamming. The sidelobe blanker eliminates this low duty cycle signal when it is also turned on, as shown in Figure 22-5(d). During these tests, the radar's MTI was not operating, which probably resulted in the high ground clutter at the short ranges.

---

*This example is taken from reference 11.

POLARIZATION JAMMING LEVEL REDUCTION (dB)

RADAR / JAMMER	POLARIZATION			
	HORIZ.	VERT.	LEFT CIRC.	RIGHT CIRC.
HORIZ.	0	∞*	3	3
VERT.	∞*	0	3	3
LEFT CIRC.	3	3	0	∞*
RIGHT CIRC.	3	3	∞*	0

POLARIZATION (left vertical axis label for RADAR rows)

* PRACTICAL LIMITS APPROXIMATELY 20 dB.

Figure 22-6. Jamming level reduction through polarization control.

A very general conclusion from this discussion is that very low antenna side-lobes are probably the single most effective means for reducing the effectiveness of SOJ.

22.5.2 Polarization

Figure 22-6 shows that some level of jammer suppression can be achieved through polarization control or mismatching of the radar and jammer polarizations. For example, if the radar is vertically polarized and the jammer horizontally polarized, an infinite amount of suppression can be theoretically achieved. Polarization isolation limitations in the antenna, however, restrict practical levels of suppression to about 20 dB; other theoretical polarization suppression levels are shown in Figure 22-6.

22.6 TRANSMITTER-RELATED ECCM

22.6.1 Increased Effective Radiated Power

As indicated in Eq. (22-3), the J/S is directly proportional to the ratio D_P of the jammer power and the transmitter power, or

$$D_P = \left(\frac{P_J}{P_R}\right) \tag{22-7}$$

Obviously, from Eq. (22-7), one brute-force approach to defeating active ECM or noise jamming is simply to increase the radar's transmitter power and thereby its effective radiated power (ERP $= P_R\, G_A$)* to the maximum amount com-

*G_A = antenna gain; normally, $G_A = G_{RT}$.

mensurate with the power available, antenna aperture size limitations, and power-handling limitations in the front-end components of the radar. This technique, when coupled with "spotlighting" the radar antenna on the target, can result in a significant increase in the radar's detection range by, in effect, burning through the noise jamming or simply overpowering the jammer.

However, in addition to being relatively inefficient in radar power management, such an approach is not effective against nonnoise jammers such as chaff, decoys, repeaters, spoofers, and so on. Also, high ERP has the added disadvantage of making the radar more vulnerable to detection and location by ESM equipment and ARMs. In fact, some radars incorporate power management to reduce the transmitted power when less power is required for target detection to decrease the probability of an ESM receiver's detecting the radar.

The preceding discussion, coupled with the considerations of antenna sidelobe levels presented in Section 22.5.1, supports an obvious conclusion: A radar designed to maximize search and track performance in a heavy jamming environment should have the highest transmitted power possible, minimum sidelobe levels, and maximum antenna gain.[12]

22.6.2 Transmitted Frequency and Waveform Coding

Another basic ECCM principle is that the use of complex, variable, and dissimilar transmitted signals places the maximum burden on the ECM. Implementation of this basic principle takes place in the radar transmitter and associated components such as the exciter and modulator.

Frequency agility usually refers to the radar's ability to change the transmitted frequency on a pulse-to-pulse or batch-to-batch basis, while *frequency diversity* normally refers to a much larger frequency change on a longer time scale. For example, the radar might change frequency by a few megahertz on a pulse-by-pulse basis and perhaps 100 to 500 MHz by retuning or adjusting the transmitter. The objective of frequency diversity is to force the jammer to spread its energy over as wide a frequency band as possible, increasing the B_J term in Eq. (22-3). Unfortunately, the transmitter frequency can rarely be changed by more than approximately 10% of the center frequency, and most broadband barrage or noise jammers anticipate such capability.

Some advantage can be gained by including the capability to examine the jammer signal, find "holes" or nulls in its transmitted spectrum, and select the radar frequency with the lowest level of jamming. This approach is particularly useful against pulsed ECM, spot noise, and nonuniform barrage noise; its effectiveness depends primarily on the extent of the radar agile bandwidth and the acquisition speed and frequency tracking of an "intelligent" jammer.

Frequency agility and diversity techniques represent a form of spread-spectrum ECCM in which the information-carrying signal is spread over as wide a frequency (or space, or time) region as possible to reduce detectability and make jamming more difficult.

Waveform coding includes PRF jitter, stagger, and coding, interpulse coding, and perhaps, shaping of the transmitted radar pulse. All of these techniques make deception jamming or spoofing of the radar difficult, since the enemy should not know or anticipate the fine structure of the transmitted waveform.

Intrapulse coding to achieve pulse compression may be particularly effective in improving target detection capability by radiation of a larger amount of average radar power without exceeding peak power limitations within the radar and by improving range resolution (larger bandwidth), which in turn reduces chaff returns and resolves targets to a higher degree. Since the peak power is lower, a pulse compression radar is also less detectable by ESM equipment than a conventional pulsed radar.

22.7 RECEIVER/SIGNAL PROCESSING-RELATED ECCM

22.7.1 Signal-Processing Techniques

Modern signal-processing techniques such as MTI and CFAR circuits provide effective radar ECCM capability, especially against chaff and various forms of noise jamming.[13] MTI techniques provide the radar with the potential for detecting and tracking aircraft through chaff clouds in much the same manner as it provides for the reduction of clutter returns from wind-blown rain, clouds, and ground clutter. CFAR circuits reduce the effects of jamming by increasing the radar's detection threshold in the presence of jamming. However, the sensitivity of the radar is reduced at the same time.

As an example of the use of Eq. (22-4) for radar screening range, and as an illustration of the effect of signal processing on a radar's performance in a jamming environment, consider the following scenario. An airborne SOJ is screening a penetrating aircraft by jamming through a ground-based search radar's sidelobes.

Radar parameters of interest

P_R = 200 kW
G_{RT} = 25 dBi (main beam)
G_{RJ} = 0 dBi (average sidelobes)
B_R = 10 MHz
SPG = 5 dB (integration gain)

Target cross section

σ = 5 dBsm (3.16 m^2)

Jammer parameters of interest

P_J = 200 W
G_J = 10 dBi
B_J = 100 MHz
R_J = 100 km

From Eq. (22-4), the calculated screening range R_{TS} for this scenario is 29.9 km. Thus, this jammer would maintain a J/S > 1 at target ranges in excess of 29.9 km. However, if the radar could achieve a 12-dB increase in the SPG (by integrating more pulses or perhaps using coherent integration), the R_{TS} could be increased to 59.8 km, double the previous value. Signal processing which results in a smaller radar bandwidth, B_R, also improves the radar's performance in a jamming environment, as can be seen by inspection of Eq. (22-4). Thus, signal processing in the form of integration gain and bandwidth reduction is an effective radar ECCM technique.

22.7.2 Gain Control/Antisaturation

Receiver saturation results in the virtual elimination of information about the target [consider an overdriven radar display which "whites out," as shown in Figure 22-5(a)]. One result of jamming is the possible saturation of the radar receiver and display. Depending on the intended application of the radar, special processing circuits may be included in the radar receiver to reduce the effect of clutter. These include fast time constant (FTC) receivers, wide dynamic range (i.e., log and lin-log receivers), and CFAR. Since these processing circuits prevent clutter and noise saturation and receiver false alarms, they are also effective in reducing the effects of noise jamming.

However, CFAR may have a negative side effect. In the presence of severe jamming, it may automatically increase the radar's detection threshold to prevent excessive false alarms. Since the radar display continues to appear normal, the operator may not detect that the radar is being jammed. Special displays are sometimes used to warn an operator that the radar is being jammed and indicate an approximate bearing to the jammer.

Other gain control and antisaturation circuits have been developed primarily to combat jamming. However, the human operator remains one of the most effective counter-countermeasures. A well-trained, alert operator can recognize the presence of jamming and take steps to reduce its effectiveness by using such simple techniques as manual receiver or display gain control.

Although a detailed description of the Dicke Fix circuit is beyond the scope of this chapter, it is another specific receiver antisaturation circuit used primarily to combat swept jamming. The interested reader is referred to Boyd[4] for a detailed discussion.

22.7.3 Signal Discrimination

Pulse width and PRF discrimination are effective countermeasures against pulsed jammers that try to confuse a radar and overload the signal processor. Figure 22-7 illustrates the operation of a pulse width discrimination circuit. The circuit, in effect, measures the width of each received pulse. If the received pulse is not

PULSE WIDTH DISCRIMINATION (PWD)

- DISCRIMINATES AGAINST PULSES WHOSE PW's DIFFER
  SIGNIFICANTLY FROM NORMAL RADAR ECHO
- CAN BE USED TO PERFORM AUTOMATIC GATING OF CLUTTER
  OR EXTENDED RETURNS

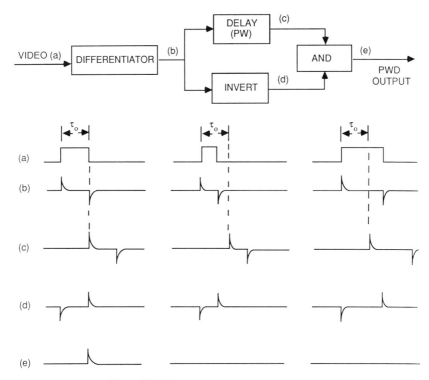

Figure 22-7. Pulse width discrimination circuit.

of approximately the same width as the transmitted pulse, it is rejected and not
passed to the signal processor or display.

22.8 OPERATIONAL/DEPLOYMENT TECHNIQUES FOR ECCM

To this point in the chapter, only electronic ECCM techniques have been considered. However, radar operational philosophy and deployment tactics may
also have a significant effect on the radar's vulnerability to ECM, especially
ESM and ARM. For example, emission control (EMCON) or on/off scheduling

REGION OF CONTINUOUS SURVEILLANCE

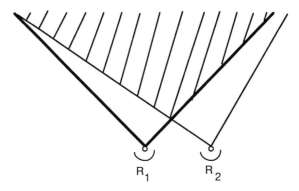

R_1 R_2

OPERATION:
  R_1 ON; R_2 OFF
  R_1 OFF AND RELOCATING; R_2 ON
  R_1 ON; R_2 OFF AND RELOCATING
  ETC.

ADVANTAGE:
  COMBINES EMISSION CONTROL WITH MOBILITY

DISADVANTAGE:
  TWO RADARS REQUIRED

Figure 22-8. Radar blinking ECCM.

of the radar's operation to include only those times when surveillance is required can reduce the probability of the radar location being found by DF equipment or radar homing and warning receivers.

Radar blinking (using multiple radars with scheduled on/off times under coordinated, centralized control), as shown in Figure 22-8, can confuse an ARM seeker and guidance or a DF receiver. The cost is high, however, since two radars are required. Radar decoys may also be employed to confuse DF receivers and ARMs. Decoys can also operate in conjunction with the radar in a blinking mode.

Proper site selection for ground-based radars can provide a degree of natural signal masking to prevent, for example, detection by ground-based ESM equipment. A high degree of mobility allows "radiate and run" operations which are designed to prevent the radar from being engaged by DF location techniques and "radar-bursting" weapons.

Table 22-2. ECM Versus ECCM

	Generic ECM	Typical ECM	ECCM
SEARCH RADARS	A. Active Denial	Jamming CW	Frequency diversity, frequency agility, fast time constant, gain controls, pulse width discrimination, bistatic radar, triangulation, spread spectrum, low sidelobe antenna, passive correlation
		Long pulse	All of the above plus integration
		Spot noise	Frequency diversity, frequency agility, integration, Dicke-Fix, CFAR, bistatic radar
		Barrage noise	Dicke-Fix, CFAR, log FTC, automatic video noise level, local oscillator off, higher ERP, longer time on target
		Medium and fast sweep jamming	Dicke-Fix, CFAR, variable IF bandwidth, integration, pulse compression, frequency agility, guard band blanking
		Impulse jamming	Dicke-Fix, CFAR
		Short pulse jamming	Integration, pulse compression
		Sidelobe repeaters	PRF jitter and integration, pulse compression, sidelobe blanking, bistatic radar
	B. Passive Denial	Chaff	MTI, Doppler radar, improved resolution (narrow beamwidth, short pulse, pulse compression) log FTC, gain controls, low scan rate on search
		Radar absorbing materials	Higher ERP, longer time on target
	C. Active Deception	Repeaters	Complex radar waveform, spread spectrum technique, frequency agility, PRF jitter and integration, sidelobe blanking
		False target generator	PRF jitter, frequency agility, lobe sidelobe antenna, diplexing
TRACKING RADARS	A. Active Denial	Jamming (various)	Frequency agility, frequency diversity (diplexing), higher ERP, pulse compression, angle track on jam
	B. Passive Denial	Chaff	MTI in angle track, Doppler radar, improved resolution (narrow bandwidth, short pulse, pulse compression)
		Radar absorbing materials	Higher ERP, frequency diversity
	C. Active Deception	Repeaters	Complex radar waveform, spread spectrum techniques, frequency agility, PRF jitter and integration
		Range and velocity gate stealers	PRF jitter, leading edge tracking, acceleration limiting, Doppler/range rate checking
		Inverse gain modulation (scan modulation jamming)	Monopulse, lobe on receive only, variable conical scan frequency, CONOPULSE
		Angle gate stealers (TWS radars)	Lobe on receive only
	D. Passive Deception	Decoys	Target/decoy trajectory characteristics, multistatic radars, bistatic operations
		Chaff	Higher resolution in range and angle, speedgate loop, MTI in angle tracking, leading/trailing edge range tracking
		Forward chaff	ECCM guard gates/coast

(By permission, from Johnston, ref. 17; © 1978 Horison House)

22.9 SUMMARY

Radar counter-countermeasures are included in the radar to reduce radar vulnerability to all types of ECM. In general, ECCM works to avoid the jammer frequency, reduce the jammer signal reaching the receiver, prevent receiver saturation, decrease the J/S ratio in a noise environment, discriminate between true and false targets, and provide a CFAR. To accomplish these objectives, ECCM must be considered during the radar system synthesis and design phase and designed into the radar, not added as an afterthought.

One of the areas of future emphasis in the development and evaluation of radar ECCM techniques and applications is sure to involve the exploitation of the millimeter wavelength (MMW) frequency spectrum (30-300 MHz) and its inherent advantages from an ECCM standpoint[14]. Indeed, one author has recently termed millimeter wave radar as "the new ECM/ECCM frontier".[15,16] In particular, MMW radar has unique capabilities to provide narrow beamwidths and high antenna gain, extremely covert operation by proper selection of the operating frequency, the opportunity to incorporate widely differing operating frequencies in dual mode, dual band sensors, and small size and high mobility. A more extensive development of MMW radar's inherent ECCM properties is given in References 15 and 16.

As this chapter indicates, specific radar ECCM is normally required to counter specific ECM. Table 22-2 illustrates this principle by identifying ECCM techniques which are effective against specific ECM types.[17]

22.10 REFERENCES

1. S. L. Johnston, *Radar Electronic Counter-Countermeasures*, Artech House, Dedham, Mass., 1979.
2. M. V. Maksimov et al., *Radar Anti-Jamming Techniques*, Artech House, Dedham, Mass., 1979.
3. L. B. Van Brunt, *Applied ECM*, EW Engineering, Inc., Dunn Loring, Va., 1978.
4. J. A. Boyd et al., *Electronic Countermeasures*, Peninsula Publishing, Los Altos, Calif., 1978.
5. H. F. Eustace, *The International Countermeasures Handbook, 1979–1980*, EW Communications, Palo Alto, Calif., 1980.
6. P. Tsipouras, et al., "ECM Technique Generation," *Microwave Journal*, vol. 27, no. 9, September 1984, pp. 38–74.
7. U.S. Army Field Manual, FM 100-5, *Operations*, Headquarters, Department of the Army, July 1976.
8. S. L. Johnston, "ECCM Improvement Factors (EIF)," *Electronic Warfare*, vol. 6, no. 3, May–June, 1974, pp. 41–45.
9. P. J. Klass, "GAO Study Expands AWACS Data," *Aviation Week and Space Technology*, September 15, 1975, pp. 47–51.
10. P. R. Dax, "Noise Jamming of Long Range Search Radars," *Microwaves*, vol. 14, no. 9, September 1975, pp. 52–60.
11. M. A. Johnson and D. C. Stoner, "ECCM From the Radar Designer's Viewpoint," *Microwave Journal*, vol. 21, no. 3, March 1978, pp. 59–63.

12. J. V. Difranco and C. Kaiteris, "Radar Performance Review in Clear and Jamming Environments," *IEEE Transactions on Aerospace and Electronic Systems*, vol. AES-17, no. 5, September 1981, pp. 701–710.

13. S. L. Johnston, "Hostile ECCM/ESM—Potential Achilles Heel for U.S. EW," *Journal of the Electronic Defense*, vol. 7, no. 6, June 1984, pp. 41–48.

14. E. K. Reedy, "Millimeter Radar, Fundamentals and Applications," Military Electronics and Countermeasures, vol. 6, no. 8, August 1980, pp 62–65.

15. S. L. Johnston, "MM-Wave Radar Challenges and Benefits EW Applications," Microwave Systems News and Communications Technology, vol. 16, no. 6, June 1986, pp 95–110.

16. S. L. Johnston, "MM-Wave Radar: The New ECM/ECCM Frontier," Proceedings of Military Microwaves Conference (MM-84), London, October 1984, pp 424–435.

17. S. L. Johnston, "Guided Missile System ECM/ECCM," *Microwave Journal*, vol. 21, no. 9, September 1978, pp. 20–24.

INDEX

"A" scope display, 233, 234, 448
Accuracy
    angle measurement, 387, 584
    frequency measurement, 601
    range measurement, 561
Active arrays, 172
AFC (*see* Automatic frequency control)
AGC (*see* Automatic gain control)
Agility, frequency, 442
Air-to-ground applications, 399, 502, 560
Alpha-beta (α-β)
    filter, 584
    tracker, 609
Altimeter, 542, 551
Altitude
    effect on sidelobe clutter, 420
    return, 426
Ambiguities
    consideration in choice of PRF, 419, 519,
        524
    Doppler, 420, 425, 433, 494
    range, 420, 430, 494
    resolution, Doppler, 430, 431
Ambiguity function, 425, 434, 438
Amplifier
    crossed field, 110, 116, 117, 118, 119, 120
    extended interaction, 127, 132
    IF (Intermediate Frequency), 208, 209,
        218, 219, 262, 403, 404, 406
    IMPATT, 110
    klystron, 110, 121, 122, 123
    low noise preamplifier, 220, 221, 222
    traveling wave tube, 110, 123, 124, 125,
        126, 128, 129
Amplitude-comparison monopulse, 569
Amplitude scintillation, 586
Analog
    filters, 445, 457
    signal processing, 513

AN/FPS-85, 475
AN/FPS-115, 475
Angle
    gate, 582
    measurement, 387
    noise, 586
    resolution, 504
    scintillation, 586
    tracking, 567
    tracking error, 584
    tracking error measurement, 386
    tracking filter, 584
AN nomenclature system, 23, 24
Antenna, 3, 9, 148
    aperture, 9
    array, 172, 504 (*see also* Array)
    beam, 152
    blockage, 159
    effective area, 9, 11
    electronic counter-countermeasures
        (ECCM), 686, 691
    gain, 9, 150
    radiation explained, 148
    radiation pattern, 148
    search scan, 433
    sidelobes, 152, 688
    size, limitations on, 151
    tolerance effects, 162, 180
    tracking control, 567
    tracking feeds, 568, 571
Antiradiation missile (ARM), 681, 692
Aperture
    definition of, 151
    effective area of (A_e), 151, 522
    efficiency (η), 149
Area of radar cell,
    beam limited, 286
    pulse limited, 286
ARM (*see* Antiradiation missile)